Sustainability Matters

Environmental Management in Asia

Sustainability Matters

Environmental Management in Asia

editors

Lye Lin-Heng

George Ofori

Malone-Lee Lai Choo

Victor R Savage

Ting Yen-Peng

National University of Singapore

World Scientific

NEW JERSEY • LONDON • SINGAPORE • BEIJING • SHANGHAI • HONG KONG • TAIPEI • CHENNAI

Published by

World Scientific Publishing Co. Pte. Ltd.

5 Toh Tuck Link, Singapore 596224

USA office: 27 Warren Street, Suite 401-402, Hackensack, NJ 07601

UK office: 57 Shelton Street, Covent Garden, London WC2H 9HE

British Library Cataloguing-in-Publication Data
A catalogue record for this book is available from the British Library.

SUSTAINABILITY MATTERS
Environmental Management in Asia

ISBN-13 978-981-4322-90-4
ISBN-10 981-4322-90-3

Typeset by Stallion Press
Email: enquiries@stallionpress.com

Printed in Singapore by World Scientific Printers.

PREFACE AND ACKNOWLEDGEMENTS

The Master of Science (Environmental Management) [Msc (Env Mgt)] Programme was launched in July 2001, by the National University of Singapore (NUS). Often referred to as the "MEM" programme, it is a multi-disciplinary graduate programme to provide education in environmental management for senior and mid-level managers in corporations, institutions, and government and non-government organisations. It aims to equip graduates with the necessary inter-disciplinary and multi-disciplinary knowledge to properly manage the environment and to deal with the challenges of an environmentally conscious society and international market. It enables graduates to assume responsible and influential roles in the public and private sectors and to make environmentally sound decisions that support sustainable development and livelihoods.

The MEM Programme is jointly mounted by seven Faculties and Schools in the NUS, namely, the Faculties of Arts and Social Sciences, Engineering, Law, and Science; and the NUS Business School, the Yong Loo Lin School of Medicine, and the School of Design and Environment (SDE) which hosts the programme. It is administered by a Director, from the staff of the host Faculty/School, the SDE, under the leadership of a Programme Management Committee which comprises a representative from each Faculty/School. The Programme also draws on the talents of the teaching staff of the Lee Kuan Yew School of Public Policy which was established in 2004 with the mission of educating and training the next generation of Asian policymakers and leaders. The MEM Programme has formal links with the Yale School of Forestry and Environmental Studies (FES) and Duke University's Nicholas School of the Environment and Earth Sciences, providing for collaboration in teaching and research as well as academic staff (faculty) and student exchanges. The program draws its inspiration from the Yale Master of Environmental Management (MEM) programme. Much as we would have liked to call our programme the MEM in a similar vein, we decided to name it more conventionally as an MSc programme, this being more readily acceptable by industries. And so the programme remains formally today as the MSc in Environmental Management, although we refer to it as the MEM programme.

It is a requirement of the MEM Programme that each student must pursue research in a chosen field and, under guided supervision, complete either a dissertation of 20,000 words or a study report of 10,000 words. This book is the first volume of a collection of some of the best research papers done by students in the programme for the initial years from 2001/2002 to 2006/2007. They have been edited and abridged for publication. In each chapter, the name of the student appears first, and that of the supervisor/s follows thereafter. Future volumes will soon follow. The programme wishes to acknowledge those students whose works do not appear in this volume, emphasizing that every student in this programme has contributed to its success.

Apart from the students, many people have also contributed to the success of this programme. We wish to thank them most sincerely. They are:-

1. Former NUS President Prof Shi Choon Fong, former Provost Prof Chong Chi Tat of NUS and the NUS Board of Graduate Studies for approving this unique programme in 2001.
2. The Deans of the seven Faculties and Schools in NUS who have joined us in this programme. They have supported us by mounting relevant courses in their faculties/schools and recognising the contributions of their staff in teaching, supervision of research and in examining our students.
3. Each and every member of the teaching and administrative staff.
4. The representatives of each Faculty, who serve on the Programme Management Committee. Their enthusiasm and passion for the environment and for this programme has helped in immeasurable ways to ensure its sustainability and its constant improvement.
5. The Deans of the School of Design and Environment (Prof Cheong Hin Fatt [2000–2007] and Prof Heng Chye Kiang [2008 to date]) for hosting the programme and giving it administrative and financial support.
6. The Directors of the programme, Dr Malone-Lee Lai Choo (2001-February 2009), Dr Ong Boon Lay (March 2009-31 December 2009) and Prof George Ofori (1 January 2010), who actually administer this programme.
7. The members of our Advisory Committee, particularly its Chair, Prof Tommy Koh.
8. Our partners in this venture, the Yale School of Forestry and Environmental Studies. Dean Prof Gus Speth signed an MOU with the programme at its inception in 2001, and was Lee Kuan Yew Distinguished Visitor to NUS in 2005. Each year, Assoc Prof Marian

Chertow, Director of the Industrial Management Programme teaches our students in the course "Business and the Environment", and recently, students were also taught by Dr Weslynne Ashton, Director of the Yale programme on Industrial Ecology in Developing Countries.

Last but not least, the programme wishes to gratefully acknowledge the support of Shell Companies in Singapore. Since the programme's early years, Shell has contributed financially, enabling us to fund bursaries for needy students and prizes for the best students. Shell also bears the costs of this publication. We wish to thank its Chairman Mr Lee Tzu-yang for his continued encouragement and support. Mr Lee has been a member of the MEM Advisory Committee since June 2002. Thanks are also due to Shell's Mr Warren Fernandez, Regional Director (Asia Pacific), who is also on our Advisory Committee; Ms Mavis Kuek, General Manager, External Affairs & Communications, and Ms Caroline Loke, Head, Social Investment, External Affairs. They have taken a keen interest in our programme, attending our seminars and other events, sharing their expertise and facilitating internships for our students.

As has been mentioned, this is the first volume of student papers and future volumes will follow. It is hoped that these papers will be a contribution towards enhancing our knowledge and understanding of the many challenges that the environment can present, and the solutions that can be found by enlightened and responsible environmental managers.

Assoc Prof Lye Lin Heng
Chair, M.Sc. (Env. Mgt.) Programme Management Committee
School of Design and Environment, &
Deputy Director, Asia-Pacific Centre for Environmental Law,
Faculty of Law
National University of Singapore

MESSAGE FROM SHELL COMPANIES IN SINGAPORE

Shell is honoured to have been able to contribute to development of the NUS Master of Science (Environmental Management) "MEM" Programme from its inception in 2001.

The MEM Programme arose out of a pressing need for an integrated and multi-disciplinary curriculum on environmental management. I am glad the MEM Programme is even more relevant today than when it was introduced nine years ago. The impact of this course has spread beyond Singapore with many alumni from the region. They return home with a better understanding of the inter-disciplinary approaches towards building environmental policies, and will have gained strong friendships and good personal contacts to support inter-dependence and connectedness between countries.

As countries in our region continue to grow, more and more of our people will come to live in cities and demand a better quality of life, thereby increasing energy demand and the impact on the environment. Government policies and blueprints and sustainability to meet these demands will be important, but governments alone cannot provide all the answers. The awareness, understanding and onus for action lie with every individual and organization in our societies to take responsibility for a more sustainable future.

In Shell, we believe in working with others to find shared solutions for a more sustainable world. On one level we look at how best to meet the growing demand for energy while minimizing CO2 emissions in our own operations. From our upstream activities, we seek to develop and bring to markets cleaner and more carbon efficient energy sources such as natural gas. We develop cleaner fuels for millions of people, and we work with different customers like motorists on the road to help save energy consumption. There is much we learn and apply across our businesses from the different disciplines represented in the MEM Programme.

I applaud the MEM faculty for this initiative to produce a publication of the best research works over the years. I hope it will encourage many to adopt a spirit of discovery and respect for our environment. I am sure you will find in these papers a common thread — that we each bring our unique talents but have a shared responsibility for managing and protecting our environment.

Lee Tzu Yang
Chairman, Shell Companies in Singapore

MESSAGE FROM CHAIR MEM ADVISORY COMMITTEE

This collection of essays, based on some of the best research theses of the NUS Masters of Science (Environmental Management) "MEM" students, is a worthy tribute to the success of the Programme. The 24 chapters in this book, based on the synopses of the students' theses, provide important statements in three areas.

First, the multidisciplinary nature of environmental challenges and their solutions is strongly affirmed. The NUS Environmental Management Programme has, since its inception in 2001, maintained its multidisciplinary focus (based on the academic inputs of the Faculties/Schools of Arts and Social Sciences, Business, Engineering, Law, Medicine, Design and Environment, and Science) and the chapters in this book highlight this unique academic endeavour. Few environmental programmes in the world have been able to harness faculty members from so many disciplines to collaborate in the teaching of a graduate programme, and to sustain it over the years. In the face of our growing environmental crisis, the emphasis on interdisciplinary teaching and research is all the more pertinent and necessary in trying to understand the complex environmental interrelationships in society. While some students had initially expressed difficulties in understanding these complex interrelationships, however, as alumni, they are grateful to the Programme for this wide exposure which has enabled them to function more effectively as environmental managers and leaders.

Second, these essays focus on the many sub-regions, countries, cities, provinces and societies of Asia, and reflect on the environmental issues of the different national developmental agendas. Unlike other academic programmes which tend to be more conceptual and theoretical, the content of these papers demonstrate the practical orientation of the MEM Programme. Our students, resident or working in Asian environments, bring to the Programme the real world environmental realities of their own working experiences. The mix of students, who have prior working experiences from the public, private and non-government sectors, have enlivened classroom discourses and produced much mutual learning. These essays are testimony to the pragmatic and realistic nature of the MEM Programme.

Third, as the environment and climate change challenges become more important and urgent, I hope the Programme will continue to remain, not only as a beacon of environmental research and teaching, but also as a foci for training new generations of environmentally conscious and competent leaders in administration, politics, business, industry and services. I hope that the graduates from this Programme will make a difference and lead their companies, organizations and governments towards a more environmentally sustainable world.

Prof Tommy Koh
Ambassador-At-Large
Chair, Advisory Committee
MSc (Env Mgt) Programme
National University of Singapore

MESSAGE FROM DEAN
SCHOOL OF DESIGN AND ENVIRONMENT

The School of Design and Environment (SDE) was chosen to host the university-level M.Sc.(Environmental Management) programme when it was being considered for approval. Two reasons were given for this choice. First, the management of the environment has been the main mandate of the SDE after it was restructured from the former Faculty of Architecture, Building and Real Estate, and given its current name. Second, the school was the only one which was to offer two core modules on the programme. My predecessor as Dean of the SDE, Professor Cheong Hin Fatt, guided the programme through its initial years, and our faculty member, Dr Malone-Lee Lai Choo, was the first programme Director. Dr Malone-Lee stepped down in March 2009, and was succeeded by Dr Ong Boon Lay who left NUS in December 2009. Professor George Ofori is now Director of the programme as from 1 January 2010. The MEM programme is administered like our other taught master's programmes in the school, and we have found the synergies as well as differences between this special programme and the others quite instructive.

We at the SDE have been very pleased to see the programme grow from strength to strength, to become what it is today. We are gratified that it has been able to attract students from around the world. We believe the MEM is unique. The enthusiasm and genuine interest of the students in the subject of environmental management, the commitment of, and camaraderie among, the teachers on the programme, and the dedication of the alumni have been very special to see. Also special is the hard work which the programme director and the Programme Management Committee (PMC) puts into the MEM. Members of the PMC are dedicated staff members from the seven faculties/schools that are involved in this collaboration. They have enthusiastically and with great dedication, nurtured this unique programme. The Committee is chaired by Assoc Prof Lye Lin Heng of the Law Faculty, who has been the programme's chair since its inception; with Prof Ofori as Deputy Chair (till 31 December 2009). Indeed, the five editors of this publication are all members or former members of this Committee. Assoc Prof Ting Yen Peng was a member of the PMC until 2006, representing the Faculty of Engineering. Assoc Prof Victor

Savage is from the Department of Geography, representing the Faculty of Arts and Social Sciences; and is Deputy Chair of the PMC.

Another striking feature of the programme is the relationships which have been developed with collaborating institutions overseas, especially the Yale School of Forestry and Environmental Studies; and Duke University's Nicholas School of the Environment. We have had great advice and assistance from members of the Advisory Committee and in particular, its chairman, Professor Tommy Koh. We are indebted to Professor Koh for the help he has given us in our efforts to raise funds for the benefit of the students on the programme. It is evident that there is an MEM community which is held together by a very strong bond among its members. We at SDE are proud that we have provided the home and the administrative support which has enabled this to be realized. I would also like to recognize the contributions made by the Deans of the other schools and faculties at the NUS who have given us their support, and encouraged their staff members to contribute in teaching, supervision of research and examinations.

The chapters of this book provide a good illustration of the nature, features and strengths of the MEM programme. The works cover a broad area of environmental management. Each research study had a strong practical element backed by a sound review of the theory. Organizations and practitioners will find much in this book that is important to their work. The bond which unites the members of the MEM community is evident in the various examples of collaboration which led to the book. First, the relationship between the students and the supervisors enabled the research studies to be completed. The collaboration between these same persons, but this time, as alumni and former supervisors, resulted in the production of the chapters. Third, the editors of the book worked together to produce this commemorative piece. Finally, the generous sponsorship of the Shell Group of companies, with which the MEM programme has been associated almost since it was started, made the publication of the book possible.

This book marks the end of a phase in the development of the MEM programme. We at SDE would be pleased to contribute to efforts by the programme' management and advisory committees to take it to a higher level during the next stage of its life.

Professor Heng Chye Kiang
Dean, School of Design and Environment
National University of Singapore

CONTRIBUTORS

Ms M.S.T. Fardausi AKHTER holds a BSc (Class 1) Hons, Engineer Centre and School of Military Engineering, Qadirabad Cantonment, Natore, Bangladesh. She graduated with the MSc (Env Mgt) from NUS in 2005 and also holds an MSc in Soil Science from the University of Manitoba, Canada. She is now a PhD candidate in Biology at the University of Western Ontario, Canada. She is working closely with metals in the Human Environment Strategic Network (MITHE-SN, Canadian network of toxicology centres) to investigate the physiological mechanisms involved in cadmium uptake, localization and distribution in plants. Her main research interest involves trace element uptake in plants and mycorrhizal fungi. She has published several research articles and presented her work at international conferences. She was Associate Program Officer, Young Power in Social Action, Bangladesh

Ms ANBARASI Boopal holds a MSc in Life Sciences in Biotechnology 1st Class, from Bharathidasan University, Tamil Nadu, India. With a keen background in biology, having worked as Junior Research Fellow in the Salim Ali Centre for Ornithology, and work experience in wildlife protection in India, she joined the MEM programme, graduating in 2006. Her passion in animal welfare and environmental protection made her volunteer with a local animal protection charity called ACRES (Animal Concerns Research and Education Society). As a full-time staff at ACRES since 2007, she is currently the Director of the ACRES Wildlife Rescue Centre and runs the ACRES Wildlife Crime Unit, working towards tackling wildlife trade issues in Singapore and the region.

Mr R.S. ARUN Kumar holds a Bachelor of Engineering, Kuvempu University, Karnataka, India and graduated with the MSc (Env Mgt) in 2006. He then became Executive Director in the World Toilet Organization (WTO), Singapore and successfully completed a 2-year sustainable sanitation project funded by Tidal Waves Asia Fund in the tsunami affected towns in Indonesia. Before working with WTO, he worked as Team Leader, Enviro Group in Technology Informatics Design Endeavour (TIDE), an NGO based in Bangalore for 4 years and was involved in renewable energy and water management projects for rural development. Arun relocated to India in September 2008 from Singapore. Since October 2008, Arun is also the CEO of Ecosan Services Foundation (ESF), Pune, India, a non-profit Foundation working in the field of ecological sanitation. He is Founder CEO of Energy and Environment Resource Group Inc. (eERG), a social enterprise based in Bangalore, INDIA. With the ideology of providing technically sound and meaningful solutions to issues faced by the common man, eERG is presently focusing on water management.

Mr Laxmi Dutt BHATTA has more than fourteen years in the field of sustainable natural resource management. He was Conservation Officer at the Mountain Institute, Nepal, and now serves as Livelihood and Biodiversity Advisor, SNV (Netherlands Development Organization). His research interest is the governance of biodiversity linking to sustainable community livelihoods in developing countries. Presently, he is working with SNV in Nepal where he provides strategic and programmatic advisory support to strengthen the capacity of timber value chains and payment for environmental services. In addition, Laxmi is involved with managing, facilitating and brokering knowledge on forests, biodiversity and ecological services in Nepal. He has presented in many international seminars and conferences. He is co-author of a book on Payment of Environmental Services in Nepal (in Nepali). His new work, *Unlocking the Eucalyptus for Poor — a Value Chain Analysis* is in press. Laxmi graduated with the MSc (Env Mgt) in 2004, and holds a BSc in Forest Science from Tribhuvan University, Nepal.

Mr Christian Rae P. BUSTAMANTE holds a BSc (Environmental Science) from the Ateneo de Manila University, Philippines. He graduated with the NUS MSc (Env Mgt) in 2006. In 2002, he started his career as an environmental professional in the power generation industry both in Philippines and Singapore. He was Environmental Technician/ Specialist, Mirant Sual Corporation. In 2008, he moved to Unilever Asia as its Environmental Coordinator. His work involves integrating environmental sustainability in Unilever's supply chain across Southeast Asia and Australasia. These initiatives range from manufacturing ecoefficiency, packaging sustainability, and energy efficiency. Christian hails from the Philippines but is currently based in Singapore with his wife.

Ms CHAN Yen Sen graduated with a Bachelor of Engineering in Environmental Engineering (2nd class Hons (Upper Division)), National

University of Singapore, and the MSc (Env Mgt) in 2005. She has over 5 years of consultancy and audit experience in the environmental, health and safety field with government and private organizations. She was Environmental Health Officer, National Environment Agency, Singapore. She was also a consultant with NOVO Environmental Technological Services. She has always maintained a passion for current environmental affairs and is always keen to promote and share eco-friendly practices and health and safety tips with colleagues, friends and family members.

Ms DANG Thuy Trang holds a BA (Class 1 Hons), University of Social Sciences and Humanities, Vietnam. She graduated with the MSc (Env Mgt) in 2004 and completed the Master's Programme in Urban Planning and Policy Design at the Politecnico di Milano, Italy in 2008. She is now Mekong River Eco-region Coordinator, WWF Greater Mekong programme, and leads the regional planning and management of WWF Greater Mekong Programme's effort to conserve the Mekong River running through China, Myanmar, Laos, Thailand, Cambodia and Vietnam. Prior to taking this position, Trang was working mainly in the development sector with the European Commission, the Asian Development Bank and the Norwegian Agency for Development Cooperation. She was also Interpreter and Translator, Department of Science, Technology and Environment Project of Hazardous Waste Master Plan for the Southern Focus Economic Zone in Vietnam.

Mrs DAO Mai Anh obtained the NUS MSc (Env Mgt) in 2002. She is currently a Ph.D scholar at the National University of Singapore. Her main research interests are in organizational behaviour and corporate environmentalism. She holds a Bachelor of Arts (English Studies), Hons. Class 1, School of Foreign Language, Vietnam National University, Hanoi, Vietnam. She was a Lecturer with the School of Foreign Languages in Vietnam.

Mr Timothy HROMATKA holds a BA in Biology and International Studies, University of Oregon. He was a member of the initial intake of the National University of Singapore's MSc (Env Mgt) Programme, studying on a part time basis and graduating in 2004. He has grown up with forests, timber, and woodworking and, having manufactured and exported furniture from Indonesia and Malaysia, has been keenly interested in tropical forest degradation and conversion, and alternatives including sustainable forestry and plantations. After the Asian Tsunami, he was involved in the Aceh reconstruction using sustainable lumber for low cost housing initiatives which has led to green building projects in Singapore and the Middle East. He is President, Globex Green Company, Minnesota.

Ms Silvia IRAWAN is currently a PhD student at Crawford School of Economics and Government, Australian National University (ANU). Her research is focusing on decentralized forest management and the scheme of Reducing Emissions from Deforestation and Forest Degradation (REDD) in Indonesia. She holds a Bachelor of Animal Husbandry (Livestock Breeding) from the Udayana University, Bali. She was Technical Consultant, Directorate of Environmental Affairs, Ministry of National Development Planning, Indonesia. Following her graduation from the MEM program in 2004, she has worked as an individual consultant with a number of international organizations such as the World Bank, United Nations Development Programme (UNDP) and WWF in Indonesia. She also worked closely with several ministries in Indonesia including the Ministry of Development Planning, the Ministry of Forestry and the Ministry of Environment.

Ms Rawadee JARUNGRATTANAPONG is currently a PhD student at Ritsumeikan Asia Pacific University, Japan. She holds a BA (Hons) in Statistics from Srinakarinwirot University and an MA in Economics from the National Institute of Development Administration, Thailand. She graduated with the MSc (Env Mgt) from the National University of Singapore in 2003. She was a researcher for Natural Resource and Environmental Program at Thailand Development Research Institute (TDRI) since 1996 before she left for her PhD studies. Her specialization is Environmental Economics and her current research focuses on climate change adaptation and payment for ecosystem services.

Ms Mantazh KHANNA is currently completing her Masters in Business Administration at the Harvard Business School with a focus on clean energy investing and sustainability in corporations. With prior work experiences at Investment Banking at Credit Suisse, New York, and Energy Investing at Acumen Fund in India, she is particularly interested in the intersection of corporate valuation and environmental risk. She obtained her MSc (Env Mgt) from NUS in 2005 and joined SembCorp Industries, Singapore, where she managed the company's multimillion dollar Carbon portfolio. Prior to that, she worked at the Ministry of Environment & Water Resources, Singapore, as a Policy Planner. She continues to find joy in learning how to link science, economics and law to create innovative, equitable and cost-effective solutions to our most urgent environmental problems. She also has a BSc.(Environmental Policy & Behaviour), from the University of Michigan.

Mr Pak-Juan KOE has worked with various organisations to establish regional operations, manage new businesses and meld different perspectives into decision-making. Besides being the management system controller with Shell Eastern Petroleum, he leads improvement and integration efforts in the management of health, safety, security, the environment and social performance. Pak-Juan serves on the Australian School of Business Alumni Advisory Board and volunteers as a guide at the National Museum of Singapore and Singapore Biennale. He has visited over 70 countries. He has a BEng (Electrical Engineering), MEngSc, and MBA from the University of New South Wales; and graduated with the NUS MSc (Env Mgt) in 2006.

Ms Gina KUM holds a BA from the National University of Singapore, and graduated with the MSc (Env Mgt) in 2003. Gina's love for the sea is a major reason to why she has been in the shipping and marine industry for over a decade. After graduating in 1999, she was a Senior Executive with the Chien Chun Fishery Co. Ltd, a Taiwanese long-liner company that captures the controversial bluefin tuna. She attended several international meetings of the International Commission for the Conservation of Atlantic Tuna. It reinforced her belief that sustainable solutions are necessary. Currently, she is Marketing Manager, ICSB Singapore, and passionate to reach out to the shipping and marine industry with various green and sustainable solutions. She is happily married with two beautiful children, Denzyl and Taryn and would like them to be able to continue to enjoy the beautiful underwater world.

Mr NG Chun, Justin holds a BSc (Real Estate), 2nd Class Upper Hons, National University of Singapore, and graduated with the MSc (Env Mgt) in 2005. As a young officer with the National Parks Board (NParks), Justin was involved in planning for Singapore's greenery and natural environment. He enrolled in the MEM programme to gain a better understanding of how global environmental issues and challenges have shaped governments' policies. He credits the programme for providing unique opportunities for passionate students from the region to come together in pursuit of environmental interests. He was a former Planning Manager, National Parks Board and is currently Senior Assistant Director, Resource Division, Ministry of Trade & Industry. He is still keenly involved in the advancement of environmental sustainability in Singapore.

Ms NG Kay Leng is a chemist with PowerSeraya Ltd for 5 years and is currently heading its Multi-Utilities Services (MUS) Department which handles all water treatment and environmental issues. PowerSeraya Ltd is a utilities company in Singapore supplying electricity, steam and water to industries, and MUS manages and operates the seawater desalination plant, water treatment plants and a laboratory that carries out fuel, water and environmental testing. It also handles all environmental issues such as the maintenance of ISO 14001 management system, ensuring the company meets all environmental requirements, monitoring of plant emissions, air quality, groundwater quality, wastewater discharge, and waste management. Prior to joining PowerSeraya, she was a consultant with IRM Consultants, Singapore, specializing in environmental law and ISO 14001 management system. She holds a Bachelor of Engineering from NUS and graduated with the MSc (Env Mgt) in 2003.

Mr Ashok Prasad OJHA a Nepalese citizen, holds a BSc (Hons) Forestry (First Division), from the Institute of Forestry and Environmental Sciences, University of Chittagong, Bangladesh. He graduated with the MSc (Env Mgt) in 2005. During the course of study and work in Nepal, he has published over a dozen articles in national and international journals and newspapers in the field of biodiversity conservation, participatory planning, and community development. Professionally, Mr. Ojha has more than 8 year's progressive work experiences in the field of biodiversity conservation, alternative livelihood and community development in Nepal. He is Natural Resources Conservation Officer, King Mahendra Trust for Nature Conservation Project in Nepal. Currently, he has been involved in the Western Terai Landscape Complex Project as a Manager. The project is the joint undertaking of the Government of Nepal, GEF, UNDP, SNV, WWF, Bioversity International, NARC and LIBIRD.

Ms Amanda ONG Hwee Fang holds a BSc (Hons) (Real Estate) from NUS and graduated with the MSc (Env Mgt) in 2004. She will complete her Masters in Public Administration at the Lee Kuan Yew School of Public Policy in July 2010. Amanda had pursued a professional consultancy career in energy and international development, including feasibility studies and planning of specialized economic zones (SEZs), the setting-up of regulatory agencies and institutional framework for strategic environmental assessment. Her research paper, "A Schematic Plan for A Green Energy Policy in Singapore" was inspired by her work in energy procurement consultancy in the early 2000s when Singapore was liberalizing its electricity market. Recognising the important link between sound policy frameworks and development success, Amanda is now with Singapore's Ministry of Trade and Industry, as Senior Assistant Director under the Trade Division (South East Asia).

Mr Manoj Kumar PUTHIA holds a Bachelor of Veterinary Sciences and Animal Husbandry, First Class, from the G.B. Pant University Agriculture & Technology Pentagon, India. He graduated with the MSc (Env Mgt) in 2003. He was a former Veterinary Officer, Animal Husbandry Department, Rajasthan, India. He is now a PhD student, Laboratory of Molecular and Cellular Parasitology, Department of Microbiology, Faculty of Medicine, NUS.

Mr Gunawan TANUWIDJAJA holds a Bachelor of Architecture (ST) from Bandung Institute of Technology (Institut Teknologi Bandung). He graduated with the NUS MSc (Env Mgt) in 2006. He has practiced as a professional urban planner, architect and researcher in Indonesia and Singapore since 2001. He is currently lecturing in the Department of Architecture, Petra Christian University and managing Green Impact Indonesia, an Integrated Urban, Drainage and Environmental Planning & Design Studio. His research interests are in Sustainable Urban Development, Eco-Cities, Integrated Ecological Planning and Sustainable Architecture.

Ms TEO Wen Jing, Jasmine Ann obtained the NUS MSc (Env Mgt) in 2006, and was awarded the Shell Prize for Best Dissertation for her thesis, "The Challenge of Environmental Sustainability in Wafer Fabrication Plants: A Framework for Implementation". Jasmine received her Bachelor of Science in Building from NUS in 2002, and also published a paper on "Implementing Total Quality Management in Construction Firms" in the Journal of Management in Engineering. Since 2002, she has been involved in both environmental engineering and corporate social responsibility work in the telecommunications, real estate and semiconductor industries. She was Environmental Engineer & EHS Coordinator, STMicroelectronics Pte Ltd; Corporate Social Responsibility Manager for CapitaLand Ltd, and is presently Group Corporate Social Responsibility Manager for Singapore Telecommunications Limited.

Ms TRAN Thi Minh Hang holds a BSc (Class II) Hons from the National University of Singapore, and graduated with the MSc (Env Mgt) in 2005. Thereafter, she returned to Vietnam and took a position at Hanoi School of Public Health, teaching Environmental Health. Currently, she is pursuing a Ph.D at the Department of Biological Sciences, Faculty of Science, National University of Singapore. Her research interests are coastal management and marine conservation. She wishes, upon completing the programme, to be equipped with the passion, knowledge and skills to be able to take part in her beloved country's efforts in managing her coastal areas sustainably.

Dr Scott VALENTINE is currently Assistant Director of the Masters of Public Policy International Program at the University of Tokyo's Graduate School of Public Policy. He has over 20 years of diverse international business experience, specializing in business development and organizational development. Past positions include Executive Vice-President of Vektor Multimedia in charge of Japanese operations, Managing Director of Asia Pacific Development International, Dean of the British Education College in Tokyo (affiliated with the UK Northern Consortium), and Director of the British Council Training Centre in Taiwan. He has a BA from Lakehead University, Ontario; MA in Advanced Japanese Studies, University of Sheffield; MBA (International Management), Asia Pacific International Institute, Thailand; Doctor of Business Administration, University of Southern California and graduated with the NUS MSc (Env Mgt) in 2006.

SUPERVISORS

CHANG Youngho is Assistant Professor of Economics at the Division of Economics and the S. Rajaratnam School of International Studies (RSIS), Nanyang Technological University (NTU), Singapore. He is also an Adjunct Senior Fellow at the Energy Studies Institute (ESI), National University of Singapore. He was a degree fellow at the East-West Center, Hawaii and received his BASc (in Landscape Architecture) from the Seoul National University, MA (in Economics) from the Yonsei University and Ph.D. (in Economics) from the University of Hawaii at Manoa, U.S.A. Dr Chang specializes in the economics of climate change, the economics of renewable resources, energy and security, oil and economy, and electricity market deregulation. His current research interests are oil price fluctuations and macroeconomic performance, the economics of energy security, the transition of resource use in an economy, the economics of sustainability, energy use and climate change, and the effectiveness of a new market structure in a deregulated electricity market. He has published research papers in various academic journals and has carried out consultation projects for the public and private sector.

Audrey CHIA is Associate Professor of Management and Organisation at the NUS Business School. Her research and teaching interests are multi-disciplinary. For the UCLA-NUS Executive MBA and the NUS MBA, she offers courses in Leadership, Change Management and International Management Strategies. She has published in various journals such as the Academy of Management Executive, Science Communication, Jurimetrics and the Journal of Business Ethics. She has also co-authored two books, *The Three Paradoxes: Working Women in Singapore and Culture and Management*: A Casebook. Her research has been presented at international conferences on Management, Applied Psychology and Socio-Economics. Her current interests are in workplace diversity, and the repair of reputation and relationships.

CHOU Loke-Ming is Professor in the Department of Biological Sciences, National University of Singapore. His research interests include coral reef ecology and restoration, and integrated coastal management. He studied coral reefs in Okinawa and the ASEAN region and has been a member of the Science & Technical Advisory Committee of the Global Coral Reef Monitoring Network (International Coral reef Initiative) since its formation. Recently, he was invited to be a member of the Joint Group of Experts on the Scientific Aspects of Marine Environmental Protection. He has provided consultancy services in the field of marine environment management to UNEP, FAO, World Bank, WorldFish Center and many national agencies. He has over 90 publications in international journals and over 40 book chapters focusing on the marine environment.

Joseph CHUN graduated from the Faculty of Law, NUS with an LLB (Hons.) in 1991, and from University College London with a PhD in environmental law in 2003. He was an Assistant Professor with the Department of Real Estate, NUS, from 2003 to 2007, teaching real estate law, environmental law, and environmental management, with particular research interests in the environmental impact of real estate law, and the openness and transparency of the law and practice of public environmental decision-making. Joseph has returned to private legal practice as a corporate real estate partner in Shook Lin & Bok LLP since 2007.

KOH Kheng Lian is Emeritus Professor at the Faculty of Law, NUS, and a founding member and the Director of its Asia–Pacific Centre for Environmental Law (APCEL), which was established in 1996 in partnership with the World Conservation Union's Commission on Environmental Law (IUCN-CEL), and in collaboration with the United Nations Environment Programme (UNEP). She was the IUCN-CEL Regional Vice-Chair for South and East Asia, and a member of its Steering Committee from 1996–2004. She has presented numerous papers on environmental law at international and regional conferences and workshops. She was a visiting professor at the Universidad Metropolitana de Mexico, and the Richardson School of Law, Hawaii.

LYE Lin Heng (LLB (Sing), LLM (London), LLM (Harvard)) is an Associate Professor at the Faculty of Law, NUS. She chaired the Steering Committee for the MEM program and now chairs its Programme Management Committee. She is also Deputy Director of the Law Faculty's Asia-Pacific Centre for Environmental Law (APCEL). Her areas of teaching and research are in Environmental Law and Property Law. She is a member of the Board of Governors of the IUCN Academy of Environmental Law, and Co-Chair of its Teaching and Capacity Building Committee. Since 2004, she has been Visiting Associate Professor at the Yale School of Forestry and Environmental Studies, and also teaches at the University of Sydney's Law School. She was Honorary Legal Advisor to the Nature Society Singapore for many years and is currently a member of the Ministry of Environment & Water Resources' Committee on Clean Drinking Water Standards for Singapore, as well as a member of the Land Titles Strata Board.

Lai Choo MALONE-LEE (M.T.C.P. (Sydney), Ph.D. Tokyo Institute of Technology) is currently Director of the Centre for Sustainable Asian Cities at the School of Design and Environment, NUS. She was Director of the Environmental Management Program since from its inception in 2001 to 2008. Prior to joining NUS, Dr Malone-Lee held senior positions in various government planning authorities, working in the areas of planning policies, strategic planning and urban conservation. She has lectured, written and consulted extensively on urban and environmental planning subjects both in Singapore and internationally. Her research interest is in the area of urban policies and strategies for cities, with a focus on urban sustainability, city livability and resource conservation. She is currently co-leading two major funded research projects on Benchmarking Cities' Sustainability and Evaluating Density Thresholds and Environmental Performance.

Peter K. L. NG is Professor in the Department of Biological Sciences, National University of Singapore. He graduated from NUS in 1983 and was with the Ministry of Education until 1990 when he joined the NUS. He is Director of the Raffles Museum of Biodiversity Research at the Faculty of Science, and Director of the university's Tropical Marine Science Institute. His research interest is mainly on the systematics of decapod crustaceans, but he has also published on fishes, rainforest and reef biodiversity and ecology, as well as conservation biology. He sits on the advisory and editorial boards of numerous regional and international biodiversity journals, and is also a member of the International Commission for Zoological Nomenclature which has oversight over animal names.

Jeffrey OBBARD is Associate Professor with the Division of Environmental Science & Engineering, NUS. He holds a joint appointment with the Science & Engineering Research Council (SERC), Agency for Science Technology & Research (A*STAR) as the Manager of its Bioenergy Programme. His doctorate investigated the impact of anthropogenic contaminants on the biogeochemical cycling of carbon, and he has worked extensively on the application of microbes and enzymes for the remediation of contaminated environmental media and, in more recent years, has become focused on microbial production of biofuels, as funded by A*STAR. He has conducted training courses in ASEAN countries on climate change science, mitigation and adaptation on behalf of the Ministry of Foreign Affairs, Singapore. He is currently taking a leave-of-absence to work with a Royal Dutch Shell Petroleum joint venture company, based in the USA, on the commercialisation of microalgae biofuels.

George OFORI (B.Sc.(Hons)(Kumasi), M.Sc. (London), Ph.D. (London), D.Sc. (London), is Professor at the Department of Building, School of Design and Environment, NUS. He is a Fellow of the Chartered Institute of Building, Royal Institution of Chartered Surveyors, and Society of Project Managers (Singapore), and Co-Director of the Centre for Project Management and Construction Law. His research is on construction industry development, international project management, green buildings and sustainability in construction, and leadership in construction. He worked as a senior quantity surveyor in Ghana before joining the National University of Singapore. He has been a consultant to the Australian Agency for International Development, Commonwealth Secretariat, International Labour Organisation and United Nations Human Settlements Programme, and several governments, on various aspects of construction industry development including sustainability in construction and small enterprise development.

Ruth O'RIORDAN ((B.A. (Mod.), (Zoology) 1986, Trinity College Dublin; Ph.D. (Zoology), 1992, University College Cork (UCC); B.A. (French and Italian), 2008, UCC, Ireland) is currently a Senior Lecturer in the Department of Zoology, Ecology and Plant Science, University College Cork. Between 2001-2004 she was an Assistant Professor in the Department of Biological Sciences, National University of Singapore. Her research interests include the ecology of temperate and tropical marine invertebrates; climate change – the use of intertidal species as indicators and the potential effect of climate change on shellfish species and their pathogens; the ecology of non-indigenous aquatic species and their potential impact on native fauna; the behaviour of vertebrates in the wild and in captivity. She has over 50 internationally refereed publications.

Maharaj Krishnan PANDIT was a member of the teaching staff at the NUS University Scholars Programme, where he taught a course on Biodiversity and Conservation Biology.

Victor R SAVAGE is an Associate Professor in the Department of Geography and teaches in the Southeast Asian Studies Programme, Faculty of Arts and Social Sciences, NUS. He is Deputy Chair of the MEM's Programme Management Committee; Co-Chair of the Faculty of Arts and Social Sciences environmental research cluster; Country Advisor for Southeast Asia at the International Relations Office (NUS) and Honorary Vice-President of the Commonwealth Geographical Bureau (2009–2012). His research interest is mainly on Singapore and the Southeast Asian region — historical and cultural landscapes, sustainable development, environmental education, sustainable urban development and cross-cultural issues. He has written many books and is Co-Editor of the *Singapore Journal of Tropical Geography* and an International Editorial Board member of the *Sustainability Science* (Japan) and *Geografiska Annaler* (Sweden).

Jesuthason (Dodo) THAMPAPILLAI holds a Personal Chair in Environmental Economics at Macquarie University, Australia and an Adjunct Professorship in the same field at the Swedish University of Agricultural Sciences at Uppsala. Presently he is an Associate Professor at the Lee Kuan Yew School of Public Policy, National University of Singapore. In 2005, he was included in a list of Eminent Environmental Economists by UNESCAP and has been a member of the UNEP Expert Group in Environmental Economics. He has over 100 publications including seven books and nine refereed monographs. Dodo's current research focus is on Macroeconomics and the Environment

TING Yen Peng (BSc Chem Eng, MSc UMIST, UK; PhD Monash, Australia) is an Associate Professor in the Department of Chemical and Biomolecular Engineering. Prior to his postgraduate studies, he worked as an environmental engineer in the petroleum industry. He teaches modules in environmental protection, industrial effluent control and treatment technologies. His research interests include wastewater treatment and solid waste management, with a focus on application of biological systems in pollution abatement and remediation. He recently completed a research project on plastic waste recovery which was funded by the National Environmental Agency (Singapore). He has also held a visiting position in Imperial College, London under a Commonwealth Fellowship.

M.S. UDDIN is Associate Professor at Department of Chemical and Biomolecular Engineering. He received his bachelor degree in Chemical Engineering from Bangladesh University of Engineering and Technology, Dhaka (BUET) in 1972 and joined the department as a lecturer. He obtained Master and PhD degrees in Chemical Engineering from the University of Manchester Institute of Science and Technology (UMIST) under Commonwealth Scholarship plan in UK. He then taught five years in BUET before joining the National University of Singapore in 1985. His research interests are liquid membrane process, surface functionalized nano-magnetic particles for separation and purification of bio-molecules and heavy metals, micro-encapsulation of phase change materials for thermal energy storage, drying of food materials and bio-accumulation of heavy metals.

CONTENTS

Biodiversity and Planning

Marine Environment

Environment and Economic Development

Energy Sustainability

Urban Pollution and Waste Management

INTRODUCTION

SUSTAINABILITY MATTERS: ENVIRONMENTAL MANAGEMENT IN ASIA

VICTOR R SAVAGE, LYE LIN HENG and GEORGE OFORI

I. Introduction

Ten years ago it would have been hard to imagine that environmental issues would be a subject of discussion in major global leadership meetings, but in the last two years they have been tabled at the G20 2009 meeting in Pittsburg and the 2009 APEC meeting in Singapore. In addition, the United Nations has held climate change conventions in Bali (2007) and Copenhagen (December 2009). Even in bilateral meetings between major developed countries, environmental and climate change issues have been part of the agenda. Clearly the deteriorating global environment arising mainly from climate change issues has precipitated global concern and action. Each month, new scientific material and visible natural catastrophes seem to underscore the growing impact of climate change and its consequences on the global economy, livelihoods, food security and cities. We have arrived at a critical crossroad on the environment and at all levels (global, regional national, local) there is now an urgent need for public awareness, government forward planning, cultural adaptation, mitigation efforts and implemention of national environmental management systems. Yet unfortunately the global community still remains divided in terms of the scientific veracity of climate change and its impacts. In the United States, a country with much access to global information, recent survey results show a reduction in believers in climate change on two fronts. In 2008, 71 percent of Americans believed there was solid evidence that the world is warming, but by 2009 only 57 percent believed this. In 2008, 47 percent of Americans believed global warming was caused by human activities, in 2009 only 36 believed this (Begley, 2009: 38–39). Among the reasons for this American perception, ideological persuasion, economic problems, and conflicting scientific errors have undermined the public knowledge of climate change in the US. In Europe, a survey by the German Marshall Fund published in 2009 found similar dwindling public perceptions on climate change. One year ago, 3 in 4 Europeans (75 percent) felt climate change was a "very serious" problem, today only 67 percent feel this way (EU Centre in Singapore, 2009:4).

Yet despite the skeptics in some quarters, the impact of climate change is already unfolding in many countries and cities. But the global and national communities have been slow in recognizing the connection between environmental impacts and growing natural hazards. In different parts of the global ecosystem, the frequency of cyclones, typhoons, floods, drought, heat waves and rising sea water seem to be causing havoc in many countries around the world. For many countries in Asia, the impact has been devastating: in 2004 violent storms in the Philippines killed 669 people and left another 695 people missing; Mumbai (India) received in July 2005 in 24 hours, 944 mm (37.2 inches) of rainfall that caused massive floods in the city; in Northern India the ground water has fallen by 20 percent between 2002 and 2008 which threatens agricultural output and millions of livelihoods; typhoon Nargis hit Myanmar in 2008 and killed 140,000 people and devasted the lives of another 2.4 million people; China suffered its worst drought in 50 years in 2008–09 which created water shortages for 4.37 million people; the drought in the Middle East and Central Asia has led to a 22 percent drop in wheat production in 2009; typhoon Ketsana in September 2009 killed over 330 people in the Philippines, Cambodia, Laos and Vietnam; and in November 2009, floods from monsoon rains wracked havoc for thousands of inhabitants in the Malaysian states of Terengganu and Kelantan and the Indonesian province of Aceh.

II. Asia: The Environmental Challenge and Prospect

If we are to believe Kishore Mahbubani's (2008) thesis that the 21st century will become the new Asian Hemisphere in global affairs, then there is reason to believe that the node of human civilizational impetus and economic dynamics is moving back to Asia after going one cycle from Asia to Europe and then to North America. One indicator of Asia's rising economic prowess is her increasing percentage in the world's GNP — from 26 percent in 1990 to 38 percent in 2007. Obviously the notion of Asia as some organic constituent is more perception than reality. Asia was defined, formulated and commodified though western perceptions. As Edward Said (1979) argues, the West found its own identity in its sustained cultural development of *Orientalism.* In reality the term Asia only functions as a geographical demarcation. Europe is a more cultural holistic entity than Asia. Yet this large region is now the theatre of global economic impetus, political muscle-flexing and cultural effervescence. The key regions in 'Asia Rising' are East Asia (China, Japan, South Korea), India and the Middle East with the Southeast Asian region riding on their coat-tails. Yet while there is cautious optimism about the 21st century belonging to Asia, the region is still embedded with major challenges economically, politically and environmentally.

In the environmental sphere, Asia is a cause, victim and benefactor of environmental problems and climate change issues and outcomes. In rising to the challenges of development, and securing better qualities of living and standards of life, Asian governments have not been equal to the task in delivering the best environmental goods for their communities and citizens. Among the top 10 poorest countries in the world, five come from Asia (China, India, Pakistan, Bangladesh and Indonesia) (Kotler and Lee, 2009:8–9). The challenges to sustaining environmental quality in countries and cities are products of traditional and new factors. Firstly, the demographic factor is a primary variable in the human-nature equation which will continue to scourge Asian countries for many decades ahead. Large populations still plague the environmental and developmental deliverables for China (1.3 billion), India (1.1 billion), Indonesia (240 million), Pakistan (172 million), Bangladesh (147 million), and the Philippines (90 million). In other Asian states, rapidly growing populations (Lao PDR, Cambodia, Timor Leste) and aging populations (Japan, Singapore, South Korea, Thailand) are causes of concern for their governments and these have different environmental implications and impacts. Small countries like Brunei, Singapore, and Kuwait which are wealthy oil producers or refiners have large per capita carbon impacts, much higher than large countries like India, Indonesia or Thailand. The neo-Malthusianists of the developed world are less sympathetic and patient to the way populations have exploded in many developing countries in Asia and Africa. Their response to this predicament is that it reflects the demographic choices of governments and communities within these large populous countries and hence they have to bear the consequences of their demographic challenges. The most well known advocate of this position is Garrett Hardin (1980) in his life-boat scenario, promethean ethics and 'triage' solutions.

Secondly, the shift from rural to urban living has created major environmental stress in many Asian cities. In 2007, the global population reached 50 percent in urban living. In Asia while many countries have not crossed the 50 percent threshold in urban population, the phenomenal growth of capital cities has made urban living a major environmental nightmare for urban governments. The most challenging scenario is the development of mega-cities (10 million and above) which dominate many Asian countries: Bangkok, Calcutta, Dhaka, Mumbai, Jakarta, Manila, Osaka, Shanghai, and Tokyo. Apart from the Japanese mega-cities, none of the other mega-cities in Asia has followed the correct eco-logic; all have been unable to implement environmentally sustainable urban systems; and do not have the right urban governance to manage their brown issues. The urban environment, with its artificially controlled cultural landscapes, needs more attention. Much discourse has been in sustainable cities, the ecological footprint and now eco-cities. But environmental attention to urban arenas must be balanced. Attention to eco-cities cannot be based solely on intra-urban environmental issues because the survival and sustenance of urban populations depend more on extra-urban ecosystem relationships for water, food and natural resources (Savage, 2006). The cities of Asia will suffocate if hinterlands and the ecological footprints are not given sufficient environmental attention.

Thirdly, capitalism and globalization has widened the disparities of wealth of peoples between and within countries. Despite the fact that capitalism is a failed system for sustainability in the developed world, it is the model for development for the rest of the world. In the developed world, capitalism will increase the debt of the 10 richest countries from 78 percent of GDP in 2007 to 114 percent by 2014 (*The Economist*, 2009:11). The rich thrive on tapping resources from a global environment while the poor have to eke out a living from recycling and reusing goods. Ironically, it is the poor that are the best exemplars of the three 'R's in conservation: reuse, recycle and reduce. Given these disparities of wealth between and within countries, it seems difficult to enact policies to reduce carbon emission and pollution because both the rich and poor of cities and countries are contributors to environmental problems: the *rich* leave a massive ecological footprint around the world by their environmentally unfriendly conspicuous consumption; while the *poor* are denied usage of clean water, modern sanitation and proper refuse disposal systems which undermine the cities' general public health. Hence the continuing disparities of wealth and status in many cities in Asia will remain a challenge for the development of environmentally sustainable cities. Widening disparities of wealth amongst populations, underscoring capitalistic and globalizing processes are in the long run a destabilizing factor economically, socially and environmentally.

Fourthly, despite the rising standards of living in general, poverty still exists in many Asian states. Of the top 10 poorest (below US$1 a day) countries in the world, five are in Asia (India; China; Pakistan, Bangladesh; Indonesia) (Kotler & Lee, 2009:8). In 1990, 29 percent of the world's population was classified as living in extreme poverty (less than US$1 a day) and the UN's Millennium Development Goal (MDG) was to bring this down to 10 percent by 2015. By 2004, the figure seemed promising as extreme poverty had dropped to 18 percent (The World Bank, 2008:31) but since the 2008–09 global financial crisis, there is now pessimism about the MDG meeting its 2015 target. After all, support from the developed countries is likely to diminish over the next decade given that the 2008–09 global financial crisis has cost economic losses of an astounding US$26 trillion according to the Bank of England (Laszlo, 2009:17). This is a tragic outcome for Asian states especially India and China aspiring to lead the Asian renaissance. Hence when developed Western countries pressurize China to upgrade its currency, they fail to recognize that the country has the second highest percentage of people living in extreme poverty in the world. As Philip Kotler and Nancy Lee (2009:xix) argue in their book, of all the problems facing mankind, poverty is "among the most persistent and shameful". So long as poverty remains a major issue in developing Asian countries, the conflict and contestations of resources and environmental goods will never be equal and will never be managed in an environmentally sustainable manner. As Joan Martinez-Alier (2005:1) in her thought provoking book, *The Environmentalism of the Poor* argues, the environmental movement has grown as a reaction to economic growth and not as against economic growth. And Nicholas Stern (2009:8) notes climate change and poverty are "inextricably linked". Hence in developed and developing countries, environmentalism has different meanings, embraced by both the wealthy and the poverty-stricken. The result is three environmental movements: the cult of the wilderness; the gospel of eco-efficiency; and the environmentalism of the poor.

Fifthly, most Asian countries are in the throes of developmentalism (economic development, modernization) and statism (national identity, nationalism) and are thus seeking trajectories like the developed countries in the west, even though such economic and political models have failed and are environmentally flawed. Hence while globalization is said to make borders porous, in reality new Asian states are cementing their national credos, developing nationalism and are hyper-sensitive to border infringements and contestations. As Paul Kennedy (1993: 134) argues, while the autonomy and functions of the state have been eroded by transnational trends, there is no adequate substitute to replace the state in responding to globalization. Globalisation is thus a convenient term for selective usage whenever it suits individual governments. Indeed if states accepted the most globalised system, the global ecosystem, we could have a more concerted and effective basis of international cooperation.

The global economic system of capitalism is certainly a major driver of development in Asia but the environmental consequences are likely to be felt later. Development thus needs to ensure ecosystem vulnerabilities, Green risks and environmental sustainability considerations. As Nicholas Stern (2009:62) contends, development is a way to "strengthen a society's ability to adapt" and hence development without adaptation is "misguided". China's appetite for natural resources around the world to feed her massive industrial output and consumer demands is a case in point. Unfortunately while the west dominated this unequal system for decades, without much sanction and criticism, the rise of Asia's economic dynamos seem to be the target of increasing criticism, concern and containment policies. Suddenly there seems to be a frightening realization that the global powerhouses are changing, corporate buyouts and natural resource consumption is in new Asian hands, and that Gaia will not be able to support an American quality of life and standard of living without dire environmental consequences globally. Developing countries like China, India and the Southeast Asian states are all caught up in the *nouveau riche* frenzy for materialism and the desire for a better quality of living. Tata's introduction of the 'People's Car', the Nano at a rock bottom price of US$2,500 might be a commercial coup for Tata but an environmental disaster for India.

The periodic warnings of the Worldwide Fund for Nature (WWF) that the world is already consuming 20 percent above its capacity and by 2050 will be 100 percent above what Gaia can support seems to be publicly unheeded. What are the boundaries of humanity's load on our planet seems to be now a concern with some scientists trying to quantify the earth's limits (Rockstrom, 2009). On the other hand, each country seems to accept a finite world in which every national stakeholder wants to maximize its share of the constant global pie. The end reality is what Hardin (1968) predicted four decades ago, the tragedy of the global commons. Unfortunately this scenario is delayed because the rich and wealthy in developed and developing countries have the political muscle and economic power to control and buy-in scarce natural resources. The global poor as Jeffrey Sachs (2005) notes still remain an embarrassing large number (one-sixth of global population live in extreme poverty) despite 50 years of United Nations, World Bank and IMF poverty alleviation schemes and programmes.

The choices countries have before them in tackling environmental and climate change issues are not easy nor are they zero-sum options. In climate change, the most pressing environmental challenge for the 21st century (Rees, 2003; Martin, 2007), the general mantra for solutions lies in two areas: adaptation and mitigation. Given that few Asian countries have the economic clout, scientific capacity, and research and development abilities, the options for mitigation seems remote. Yet Asian states are large emitters of carbon dioxide. Globally, China is the largest emitter of greenhouse gases due largely to her rapid economic development, her massive industrialization and high reliance on fossil fuels for energy. The Chinese government will need to reduce coal-based energy supplies but the options for alternative energy are not without problems. Her large hydropower project, the Three Gorges Dam has been the butt of severe environmental and social criticisms. The dam has led to the displacement of 1.3 million people, and the flooding of 13 cities, 140 towns, 1,350 villages, while the environmental fallout is yet to be fully calculated and experienced. Furthermore China's proposed eight dams in the upper reaches of the Mekong River (Yunnan) has been seen as an ecological disaster for countries in the lower Mekong River (Lao PDR, Thailand, Cambodia and Viet Nam) and a likely cause of future political tensions between her smaller Southeast Asian neighbours.

The best prescription from James Lovelock (2007) for an interim halting of carbon emissions is the use of nuclear energy; which seems a controversial energy solution for the author of the Gaia perspective. But for the global community these have had mixed reviews. China is following the French model and racing towards building numerous nuclear energy plants across the country without much global attention. And India is equally interested in boosting its nuclear energy output with hopefully the help of the United States. Yet for other Asian states like Iran and North Korea, any attempt at nuclear energy developments is greeted with suspicion and fear. Nuclear power is a neutral issue, but countries have the choice for peaceful or belligerent usage. In Southeast Asia, Thailand, the Philippines and Indonesia seem bent on following a nuclear energy path but this raises other concerns because of the tectonic instability in the region.

Indonesia is another major Asian contributor of greenhouse gases but for different reasons from China. Indonesia's greenhouse gas emissions stem from natural and landuse changes. Globally, Indonesia has the highest emissions of carbon dioxide arising from degrading peat – some 500 million metric tonnes released in 2008 excluding losses caused by fire. The country has also the third largest peat carbon stocks in the world after Canada and Russia. Much of Indonesia's carbon releases also stem from landuse changes due to forests fires, the spread of oil palm plantations, rural development and reforestation of commercial timber plantations. If Indonesia is going to reduce its carbon emissions it needs to rework its development strategies. At the moment the government hopes to be a major global biofuel supplier and its competition with Malaysia in this area seems to be misplaced and environmentally irresponsible.

To date one can list hundreds of adaptive suggestions by communities, cities and countries with regard to climate change. But all these micro and meso scale ad-hoc, piece-meal adaptive mechanisms seems unlikely to change the course of increasing carbon emissions towards 550 ppm, reducing the build-up of greenhouse gases, and slowing down global warming and its likely impacts on ecosystems and human populations around the world. For the smaller countries, local adaptive measures might provide temporary solutions and solace, but climate change is an environmental issue of global proportions that cannot be solved at the local levels. There are two fundamental guidelines in dealing with adaptive measures for environmental and climate change issues: i) adaptation must take place at the local and regional levels (Stern, 2009:60); ii) adaptation must be grounded in 'risk management' involving diversification, flexibility and education (Stern, 2009:61–62). Given its strong agriculture base, Asian states will need to adopt both 'hard' (drought resistant crops; seawalls; irrigation technologies) and 'soft' (crop rotation patterns) adaptive technologies.

Hence the December 2009 United Nations Framework Convention for Climate Change (UNFCCC) meeting had much global anticipation of delivering global solutions and a comprehensive action plan based on its 192 government members, but in reality the meeting betrayed mixed responses. There was the ugly confrontation between the US and China, the deadlock between the developing and developing countries over responsibilities, and on the other hand the European Union states' offer of monetary assistance to developing countries for their adaptive strategies. Clearly the Copenhagen meeting demonstrated deep divisions amongst states and global agreement on solutions seem much more difficult. In short, the global community demonstrated a lack of responsibility, poor leadership and selfish national motivations. Both India and China had different perspectives to the Copenhagen meeting. Both countries are contributors and likely victims of climate change hence both India and China had vested interests in the Copenhagen meeting. The 2009 Copenhagen meeting as expected led to a clash of interests between the developed and developing states. Already the G77 (the Group of 77) in Bangkok made it clear they want to maintain the Kyoto Protocol which is binding for developed but not developing countries. The European Community and the US hoped to change that perspective in Copenhagen (Tollesfson, 2000:1034), but as it turned out it was not an easy or acceptable proposition. India's position is for "common but differentiated responsibility" (Pachauri, 2009:1054) while China expected the developed countries to demonstrate leadership in Copenhagen especially in providing technical and financial assistance to developing countries (Pan, 2009:1055). Yet both the Asian powerhouses set aside their diplomatic differences and signed a five-year agreement in October 2009 to increase cooperation in tackling climate issues. As Raul Estrada-Oyuela (2009:1056) argues, any agreement at Copenhagen would require a strong negotiator with exceptional skill, knowledge and diplomacy. The Copenhagen dialogue has left the global community still divided and Asian states are likely to feel its impacts and implications in the coming decades. The bottom line at Copenhagen is that the Summit fell short of its goal to cut carbon emissions sharply enough to hold off a 2-degree-Celsius rise in global temperatures. Asian states and cities must now put greater emphasis in activating their adaptive systems in response to the increase in carbon emissions, rise in global warming, and varied environmental impacts it is likely to unfold on their communities.

III. Commentary on the Collection of Essays

This collection of 24 essays on Asia's environmental issues is not a *prescription* for environmental problems *per se* but reflects rather a broad spectrum of interdisciplinary interventions, processes and viewpoints on environmental management in the Asian region. Given that environmental problems are multi-faceted and multi-dimensional challenges, there can be no single 'one size fits all' explanation or solution. The environment, after all, covers all aspects of human endeavours, living environments, leisure places and work spaces. While studies in this book might use single variables and perspectives to analyse issues, these are done for the ease of analysis rather than a reflection of the complex reality people and communities confront in environments.

In this study, the idea of environmental management is a complex issue which requires definitions of three elements: what is meant by environment, what is the nature and process involved in management and who comprises and decides management. The question of environment is a general term which covers a range of associated terms and labels: ecology, ecosystem, nature, Green, natural resources, physical landscape, flora, fauna, environmental sustainability, urban environments and cosmos. In this book, we use the term "environment" in the way Ian Simmons (2008:3) defines it as "those elements of nature which are in an ecological relation with humans, that is, where there exists a possible transfer of energy and materials between them". The use of the term 'management' tends to underscore the work places of business where organization and employees are involved. We manage corporations, shopping malls, hospitals, and restaurants. In nation-states and countries the word "management" is replaced by "governments and governance". Hence depending on the scale of environment management in this book, the papers cover the corporate levels, specific bioregions and ecosystems, cities and national territories. In short, these articles address environmental issues at both the private and public sectors. At one end of the scale '*management*' might be the appropriate term, but at national levels *governance* might be better suited. The people responsible for environmental management cover a broad spectrum from political leaders, corporate chieftains, national administrators, urban planners, to NGO personnel, pressure groups and the concerned public volunteer. Since the environment concerns everyone, the responsibility of the environment has to be shared. Every citizen is a stakeholder of the environmental health and sustainability within communities, neighbourhoods, cities, countries, regions and globally. We all have multiple citizenships at various scales and hence must exercise responsibility in what we do.

While few articles make cross comparisons between cities or states, one can make independent cross-comparisons because the varied articles cover a range of scales from regions (Southeast Asia; East Asia) to countries (Vietnam, Thailand, India, Nepal) and cities (Hong Kong, Singapore, Ho Chi Minh City) in Asia. The articles also cover a plethora of environmental management themes dealing with legal environmental issues, conservation methods, policy matters, Green planning, environmental measurements, enforcement matters, environmental education and environmental monitoring. There is also a wide range of ecosystems and Brown issues to meet the environmental interests of varied specialists, corporate honchos and Green activists dealing with forestry, water issues, varied types of wastes (solid, hazardous, electronic), fisheries and aquaculture, agriculture, and energy in urban, rural, marine and pristine areas. Certainly no environmental management book would be complete without discussions on adaptive and mitigation strategies. And in this collection of essays one can find all sorts of environmental strategies from fiscal management to legal frameworks, from specific breeding programmes for endangered species to forest plantations.

While this collection of environmental essays are abridged and edited versions of the Masters' theses and Study Reports of the Masters in Environmental Management (MEM) students they cannot be viewed as purely academic conceptual statements situated in a vacuum and without experiential realities. Indeed, we would argue that these essays reflect important environmental statements in three ways.

Firstly, many of the students are locals from the countries and cities in Asia that they are writing about and hence have often a lifelong experience with their subject matter. Their writings demonstrate a candid sincerity of what is happening on the ground and often the topics picked underscore a certain familiarity that students have had over time. In short, students' theses are rarely random academic exercises but reflect the individual interest or familiarity each student has with an environmental theme of his or her country or environmental theme.

Secondly, many of the Masters students have working experience in the corporate, government and non-government organizational (NGO) sectors and hence are not discussing issues in a vacuum. The working experience of students more often reflects a working engagement with environmental issues and hence many are cognizant about the administrative processes, economic limitations and political underpinnings of managing environmental problems. This working knowledge with the environment provides a realistic and applied dimension to the analysis. In short, these papers are grounded in grassroots realism even though they are framed within academic-derived concepts and theories.

Thirdly, given that environmental issues are complex, holistic problems covering a wide array of issues, disciplines, and perspectives, these papers cover varied interdisciplinary as well as disciplinary perspectives and analysis. In these papers one will see evidence of the disciplinary dialogue and philosophy of the MEM programme which educates students to adopt interdisciplinary and multidisciplinary perspectives on environmental issues. Unlike the 'culture wars', the environmental themes in these contributions demonstrate that the scientific, social science, legal and policy perspectives need to be understood and analysed within a holistic and integrated manner before realistic and pragmatic prescriptions, solutions and mitigation issues can be implemented. One of the tragedies of environmental debates is the lack of dialogue between the hard and soft sciences. In these papers there is glimmering hope of unlocking disciplinary territories between the sciences and non-sciences and a genuine and realistic dialogue of interdisciplinary perspectives.

IV. Reflections

Given the importance of environment and climate change issues to newly developing states in Asia, this book underscores the complex environmental themes which resonate in the academic literature as well as manifest themselves in grassroots realities. Environment is as much an integral component in the developmental processes of many Asian developing communities, cities and countries as it is an important variable which could undermine development programmes and trajectories.

The urgency of environmental attention caused by the speed of climate change issues unfolding globally warrants our immediate attention to how societies and governments need to manage their environments. Unfortunately Asia has its own spread of successful and failed states. Each failed state adds to the litany of accumulating environmental woes at the regional and global scales. Unfortunately unlike political issues, environmental factors do not observe political boundaries. The ecosystem issues transcend economic, political and cultural borders. Each variable of the environment cannot be studied in isolation, nor can they be addressed without a broader understanding of the complex forces in which they find expression. This collection of papers does not offer magic formulas for environmental solutions and mitigation, but hopefully, this collection of case studies can underline important correlations, present new data, identify areas for environmental monitoring, alert promising areas of research, define environmental problems, situate environment within development programmes of countries and companies, explore new areas for environmental engagement and innnovation, and shed new light on promising areas for adverting environmental catastrophes.

References

Begley, Sharon (2009) "The evolution of an eco-prophet" *Newsweek*, vol CLIV(19):34–39.

Estrada-Oyuela, Raul (2009) "Copenhagen needs a strong lead negotiator" *Nature*, Vol. 461: 1056–1057.

EU Centre in Singapore (2009) *Road to Copenhagen 2009* Singapore: EU Centre in Singapore.

Hardin, Garrett (1980) *Promethean Ethics: Living with Death, Competition, and Triage* Seattle: University of Washington Press.

Hardin, Garrett (1968) 'The tragedy of the commons' *Science*, 162:1243–1248.

Kennedy, Paul (1993) *Preparing for the Twenty-First Century*, New York: Random House.

Laszlo, Ervin (2009) *WorldShift 2012*, Rochester, Vermont: Inner Traditions.

Lovelock, James (2007) *The revenge of Gaia*, London: Penguin Books.

Mahbubani, Kishore (2008) *The New Asian Hemisphere*, New York: Public Affairs.

Martin, James (2007) *The Meaning of the 21st Century: A vital Blueprint for Ensuring our Future*, New York: Riverhead Books.

Martinez-Alier, Joan (2005) *The environmentalism of the Poor: a study of ecological conflicts and valuation* New Delhi: Oxford University Press.

Pachuari, Rajendra K. (2009) 'India pushes for common responsibility" *Nature*, Vol. 461, p. 1054.

Pan, Jiahua (2009) "China expects leadership from rich countries" *Nature*, Vol. 461, p. 1055.

Rees, Martin (2003) *Our Final Century: Will the Human Race Survive the 21st Century?* London: Heineman.

Rockstrom, Johan (2009) 'A safe operating space for humanity" *Nature*, Vol. 461:472–475.

Sachs, Jeffrey (2005) *The End of Poverty*, London: Penguin Books.

Said, Edward (1979) *Orientalism*, New York: Vintage Books.

Savage, Victor R. (2006) 'Ecology matters: sustainable development in Southeast Asia' *Sustainability Science*, Vol. 1(1):37–63.

Simmons, I.G. (2008) *Global Environmental History 10,000 BC to AD 2000* Edinburgh: Edinburgh University Press.

Stern, Nicholas (2009) *A Blueprint for a safer planet*, London: The Bodley Head.

The Economist (2009) 'The Biggest Bill in History' *The Economist*, 391(8635):11.

The World Bank (2008) *Global Monitoring Report 2008: MDG and the Environment*, Washinton DC: The World Bank.

Tollefson, Jeff (2009) "Time running out for climate talks" *Nature*, Vol. 461: 1034–1035.

1

A STRATEGIC ENVIRONMENTAL MANAGEMENT FRAMEWORK: EVALUATING THE PROFITABILITY OF BEING GREEN

SCOTT VICTOR VALENTINE and VICTOR R SAVAGE

The extent to which corporate environmental governance is correlated with financial performance remains a topic for debate; however, consensus of opinion seems to be that there are some environmental management initiatives that add value to a firm's bottom-line and other environmental management initiatives that adversely impact a firm's bottom-line. This paper puts forth a new Strategic Environmental Management Framework (SEM Framework), which ties together much of the existing theory on corporate environmental management. The SEM Framework provides analysts and corporate strategists with a conceptual model that demonstrates how various forces influence a firm's environmental governance commitment. Factors which influence a firm's commitment to environmental governance are extracted from the SEM Framework and employed to guide assessment of publicly available corporate environmental disclosures for 88 firms listed on 5 major stock market indices in Singapore and Canada. The environmental governance commitments of these firms were then compared to aggregated financial performance data in order to provide the comparative data for evaluating the environmental governance — financial performance relationship. The empirical research fails to confirm a significant correlation between environmental governance and financial performance for firms within any of the industry indices studied. Acknowledging the positive contributions that certain environmental initiatives can have on firm profitability, this paper concludes that a weak correlation between environmental governance and financial performance could exist; however, the correlation is not strong enough to out-weigh the other elements that influence a firm's profitability. The implication of these findings for policy makers is that mechanisms other than free-market incentives are necessary if policy is to guide corporations toward adopting more sustainable environmental governance practices.

1. Introduction

The main intent of the empirical research presented in this paper is to contribute to efforts to clarify the extent to which improved corporate environmental governance supports and enhances a firm's financial performance. Specifically, the research attempts to empirically measure the strength of the correlation between corporate environmental governance and reported financial performance.

Existing research concerning the corporate environmental governance and financial performance relationship has skirted the main issue by focusing on overly-narrow proxies for evaluating the corporate environmental governance financial performance relationship. The current body of research in this field supports claims that *some* elements of improved corporate environmental governance support improved financial performance; however, a holistic link between corporate environmental governance and financial performance has yet to be supported. Studies have been carried out which show that;

- There appears to be a link between reducing pollutant emissions and financial performance (King and Lenox, 2001[b]; Stanwick and Stanwick, 1998).
- There appears to be a link between lean production (less waste) and financial performance (King and Lenox, 2001[a]; Porter and Van der Linde, 1995).
- There appears to be support for the premise that some firms adopt the ISO14000 environmental management system to improve financial performance (Berthelot and Coulemount, 2004).
- There appears to be a link between superior environmental governance as perceived by independent evaluation bodies and market valuation (Connelly and Limpaphayom, 2004; Kiernan, 2001).
- There appears to be a link between significant environmental events and market valuation (Klassen and McLaughlin, 1996).
- Many CEO's perceive a link between financial performance and environmentally-friendly technological investments and innovations their firms have made (Karagozoglu and Lindell, 2002).

Despite such promising indications that aspects of corporate environmental governance are positively related to financial performance, there is a notable absence of research to prove that a significant correlation exists between *overall* corporate environmental governance and financial performance. It is posited that the reason for the dearth of such studies stems from the complexities involved in measuring overall corporate environmental governance. Quantifying a firm's corporate environmental governance commitment requires identification and measurement of environmental initiatives related to inputs (i.e. material usage), processes (i.e. energy use, production efficiencies) and outputs (i.e. waste and emissions). This implies that a high degree of insider knowledge is required to understand a firm's corporate environmental governance commitment. In the absence of sufficient insider knowledge, assessing corporate environmental governance poses a significant challenge.

In order to evaluate the corporate environmental governance — financial performance relationship from a macro perspective, a promising approach would be to develop a method to permit the analysis of corporate environmental governance from external sources of information. Consequently, the main conceptual goal of this paper is to develop a framework that will guide analysts in understanding the forces which influence corporate environmental governance. By understanding what causes firms to embrace corporate environmental governance initiatives, an interpretivist model can then be created to help analysts better evaluate information provided in environmental disclosures to assess corporate environmental governance.

This paper contributes to the body of knowledge in corporate environmental management from three perspectives. First, the development of the conceptual SEM Framework represents a first attempt at collating existing research to develop a model for understanding the exogenous and endogenous strategic forces which influence a firm's environmental governance policies. Second, the paper also demonstrates how the SEM Framework can be utilised to guide evaluation of corporate environmental strategy. This is demonstrated by using the SEM Framework to guide development of an evaluative tool for assessing the properties of a firm's environmental disclosures. Third, the empirical results presented in the latter part of this paper contribute to a growing body of empirical data on the corporate environmental governance — financial performance relationship. The empirical research which evaluates environmental disclosures and financial data from 88 firms in Singapore and Canada fails to confirm evidence of a statistically significant linkage in the corporate environmental governance — financial performance relationship.

2. A New Conceptual Framework

The Strategic Environment Management Framework was constructed from previous environmental management and environmental reporting research.

Thus, it represents a compendium of existing knowledge. As no such framework currently exists, it is believed that the creation of the SEM Framework represents a significant contribution to corporate environmental governance research from both academic and applied perspectives.

It is believed and hoped that the academic value of the SEM Framework stems from the graphic clarity with which the strategic forces which influence corporate environmental governance have been collated. The framework allows academic scholars to begin to consider the inter-relationships between these forces. As the empirical section of this paper will demonstrate, by seeing these forces from a holistic perspective, a better understanding of a firm's environmental governance strategy can be ascertained.

From an applied perspective, the SEM Framework represents a first attempt to graphically organise the existing research on elements of corporate environmental management strategy into one cohesive model. Thus, the SEM Framework can serve as a guide map for practitioners who wish to clarify their firms environmental management strategies.

i) *Developing the SEM framework*

The fundamental premise upon which the SEM Framework has been conceptualised stems from the widely recognized tenet that forces influencing the development of corporate strategy are both exogenous (not within the control of the corporate strategist) and endogenous (within the control of the corporate strategist) (Grant, 2005). As will be seen as the model is clarified in this section, forces that influence the development of corporate environmental management strategies are similarly manifest in both the exogenous and endogenous realms.

Accordingly, in conceptualising the development of a framework to aggregate these forces into one model, it was decided that the forces would be best grouped according to the extent of control that strategist had over the forces- working from exogenous to endogenous realms. The framework is presented in Fig. 1.1 with the realms summarised below:

- The first realm of influential forces are those broad forces that occur in the macro-economic environment. Unsurprisingly, these forces are also forces which the corporate strategist has the least control over.
- The second realm of forces are the forces that arise through the efforts of secondary stakeholders (community members, interest groups etc.). The

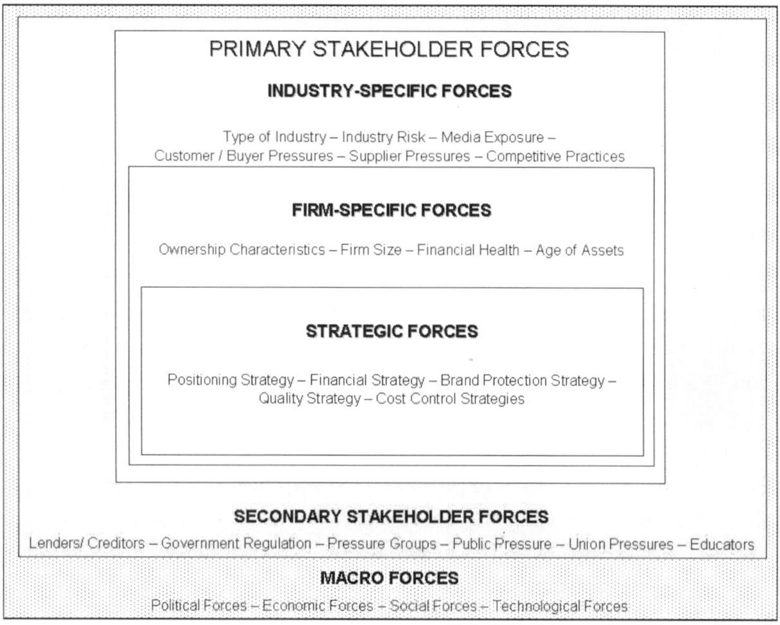

Fig. 1.1. The strategic environmental management framework.

strategist may be able but manipulate these forces; but by and large, these forces are also exogenous to firm control.

- The third realm of forces are industry-specific forces. Since firms have the ability to adjust strategy to enter and exit market niches, there is a degree of strategic control implied when strategists consider how to respond to these forces.
- The fourth realm of forces are firm-specific forces. These forces are predominantly structural in nature and as such can be manipulated by the firm. However, the structural foundation of these forces implies that manipulation of these forces does not come without significant strategic repositioning and organisational re-engineering.
- The final realm of forces are those which are endogenous and strategic in nature. In short, this final strategic realm is the realm within which most corporate environmental management currently takes place because these elements are strategically controllable.

In subsequent paragraphs these five realms will be clarified with examples from prominent extant research which exists in each realm.

Furthermore, an intuitive review of the impact that the forces from each realm should have on environmental reporting will be provided in order to guide conceptual development of an analytical tool that will be used later in the paper for conducting the empirical research.

ii) *Realm 1- macro forces*

Macro forces are the broadest forces that influence how a firm approaches environmental management. In strategic management theory, these forces are referred to as PEST forces: Political, Economic, Social and Technological (Grant, 2005). The tenet is that the PEST forces in each country influence the extent to which firms within industries approach environmental governance (Kolk, 2005).

Macro forces act as "expectation dampeners". The macro forces which influence an industry frame the degree of environmental governance that stakeholders expect of a firm. A comparison between macro forces influencing corporate environmental governance in China and Canada is insightful as an example. In China, civic reaction to poor corporate environmental governance is less intense for a number of macro reasons (Iu and Batten, 2001). Economically, the country is in the throes of an economic boom and; therefore, sustaining economic growth is driving public policy. Although there are recent indications of change,[1] economic growth currently takes precedence over environmental protection. Moreover, as the Tiananmen Square protests of 1989 exemplify, the Chinese government does not respond well to unauthorised mass public protest. On the other hand, in Canada, civic reaction to poor corporate environmental governance is relatively more robust because the environmental watchdog community is well-established and has better access to communication networks to voice dissent. Moreover, Canadian citizens are comparatively more affluent so they exhibit a higher propensity to accept higher taxes to facilitate improved environmental policies (Costanza *et al.*, 1997). In short, it can be concluded that environmental governance expectations are currently higher in Canada than in China because of the macro forces which act on society (Kolk, Walhain and Wateringen, 2001). In summary, macro forces influence the environmental governance contributions expected of corporations and

[1] For example, China now boasts the second largest number of ISO14000 certified firms in the world.

therefore, frame the lower limits of environmental disclosure done in a country (Kolk, 2005).

iii) *Realm 2- secondary stakeholder forces*

A significant amount of research has emerged over the past decade in support of the contention that firms must balance strategic objectives to ensure that a broad spectrum of stakeholders' expectations are adequately satisfied (Neely, Adams and Kennerley, 2002; Kaplan and Norton, 1996). Secondary stakeholders may have less direct contact with a given firm; but nevertheless, they possess formidable power in influencing the fortunes of the firm. Six secondary stakeholder groups which have been identified in previous research as having a particularly strong influence on how a firm approaches environmental governance include:

- Creditors (See for example: Deegan, 2002; Wilmshurst and Frost, 2000).
- Government Regulators (See for example: Yakhou and Dorweiler, 2004; Patten, 1992).
- Interest Groups (See for example: Adams, 2004; Wilmshurst and Frost, 2000).
- General public (See for example: Van Tulder, 2005; Cormier and Magnan, 2003).
- Educators (See for example: Russell, 2006; Cormier, Gordon and Magnan, 2004).
- Unions (See for example: Van Tulder, 2005).

Secondary stakeholders represent significant threats to firms in regard to environmental governance issues (Khanna, 2005). Lenders and borrowers who are displeased with a firm's environmental stewardship can punish the firm by denying the firm access to funding (Wilmshurst and Frost, 2000). Government regulators who are displeased can impose stricter regulations, levy financial penalties, and even force businesses to close (Beets and Souther, 1999). Union displeasure can result in plant closures. Most significantly, public pressure and pressure from watchdog groups can severely impact the market viability of a firm's product offerings (Van Tulder, 2005; Carter, 2001). Accordingly, given the severity of these threats, firms are motivated to embrace stakeholder theory and ensure that secondary stakeholder expectations are, at a minimum, satisfied.

In environmental reporting, signals that a firm is attempting to appease secondary stakeholders are indicated through qualitative exhortations which seek to positively laud isolated but comparatively minor environmental achievements. This type of reporting is the realm within which legitimization attempts would most likely occur. It is also the most unreliable area in which to judge actual corporate performance because firms that subjectively claim a commitment to superior environmental management may or may not be committed to such goals behind corporate doors.

iv) *Realm 3- industry-specific forces*

In 1980, Michael Porter, an authority on corporate strategy introduced his 5 Forces framework which demonstrated how forces in a firm's industry influence the attractiveness of the industry and define the parameters within which firms in the industry must operate if they are to succeed in the long run (Porter, 1980). Since industry forces influence overall strategy, it should come as no surprise that industry-specific factors related to environmental governance also impact a firm's environmental strategy. Research has identified six groupings of industry-specific forces that impact the extent to which firms embrace environmental governance initiatives. Previous research has identified the following elements of industry structure as influencing corporate-level environmental governance:

- Types of Industry (See for example: Shirley, 2005; Campbell, 2003).
- Risks associated with specific industry tasks (See for example: Cormier, Gordon and Magnan, 2004; Cerin, 2002).
- Media exposure (See for example: Adams, 2004; Cormier *et al.*, 2004).
- Customer (buyer) pressure (See for example: Kolk, Walhain and Wateringen, 2001; Wilmshurst and Frost, 2000).
- Supplier (vendor) incentives (See for example: Wilmshurst and Frost, 2000).
- Industry business practice (See for example: Kolk, 2005; Deegan, 2002).

Overall, industry-specific forces have the capacity to either raise the bar in terms of corporate environmental governance in an industry or dampen progress. At the end of the day, a firm's continued viability is directly related to its ability to compete successfully with rival firms and to profit in doing so (Porter, 1980). Consequently, when it comes to industry best practice, there is a tendency for firms to adopt a herding mentality. For this reason, comparison of corporate environmental practices in firms from different industries should be undertaken with the knowledge that an industry-type will impact the extent of commitment to corporate environmental governance.

In environmental reporting, the presence of industry-specific forces will often induce similar reporting patterns for firms from similar industries. In fact, it would not be unreasonable to assume that environmental report writers at one firm use the disclosures of competing rivals as benchmarks to guide the presentation of content in their own firm's environmental disclosures. Therefore, when analyzing environmental disclosures, analysts should be aware of the possibility that disclosures which seem on the surface to be of superior quality may, in fact, simply be a product of common practice within the industry.

v) *Realm 4- firm-specific forces*

Firm-specific forces can be defined as influences on corporate environmental governance that are related to the unique structure of a given firm. In particular, the type of ownership and characteristics associated with the firm's asset base influence the extent to which firms can financially support environmental governance initiatives beyond those initiatives that actually produce positive returns to the firm. For example, firms that are facing cash flow constraints and which possess an ageing asset base are in a less viable position to upgrade to new technologies that would reduce pollution and waste levels. There are five main firm-specific structural characteristics that have been identified in the literature as influencing how a firm approaches corporate governance.

- Ownership characteristics (See for example: Cormier and Magnan, 2003; Cormier and Gordon, 2001).
- Firm size (See for example: Cormier and Gordon, 2001; Kolk, Walhain and Wateringen, 2001).
- Financial health (See for example: Cormier, Gordon and Magnan, 2004; US EPA, 2000).
- Age of assets (See for example: Cormier, Gordon and Magnan, 2004).
- Environmental reputation (See for example: Cerin, 2002; Patten, 1992).

Two observations concerning the influence that firm-specific forces have on corporate environmental governance are worth highlighting.

First, similar to the impact that macro forces have on corporate environmental governance, firm-specific forces act as *constraints* to corporate environmental governance. This characteristic is important to understand because it implies that all firms do not share the same capacity to commit to superior environmental governance practices. Larger firms with a more concentrated ownership base, newer assets and healthy levels of cash flow should be financially able to adopt more proactive environmental stewardship strategies.

Second, firm-specific forces represent the only cluster of forces that can be assessed through quantitative analysis of a firm's financial statements. The breadth of the ownership base, the relative size of the firm, a firm's financial health and the age of assets can all be determined through quantitative financial statement analysis.

Thus, knowledge of a firm's structure can play an influential role in interpreting a firm's commitment to corporate environmental governance. Quantitative analysis of financial reports can identify constraints if any, that a firm is facing which would limit its ability to adopt more proactive environmental management strategies. Overall, analyzing a firm's capital structure (ownership, size, past profitability, cash flow, age of assets etc.) can tell an analyst what upper limits to environmental governance can be expected to emerge in environmental disclosure.

vi) *Realm 5- strategic forces*

Strategic forces represent the activities that firms undertake in the environmental governance sphere which directly impact a firm's market valuation, revenue prospects or cost performance in either the short or long term. These forces stem from the principle of "economic rationality"- firms will adopt environmental management practices that contribute positively to a firm's financial health (Deegan, 2002; p. 290).

The ability to strategically manipulate these forces underlie the foundation of Porter's Hypothesis (Porter and Van der Linde, 1995). Previous research have identified the following areas where issues related to environmental governance can be strategically managed:

- Green positioning strategies (See for example: Kiernan, 2001; Hawken, 1992).
- Financial strategies (See for example: Khanna, 2005; Beets and Souther, 1999).

- Brand Protection strategies (See for example: Ruf *et al.*, 2001; Kolk, Walhain and Wateringen, 2001).
- Quality strategy (See for example: Stanwick and Stanwick, 2001; Berry and Rondinelli, 1998).
- Cost-control strategies (See for example: Adams, 2004; Porter and Van der Linde, 1995).

In environmental reporting, strategic environmental initiatives are often highlighted in order to convey the message to shareholders and stakeholders alike that the firm is achieving balance between environmental responsibility and profitability. For example, in Xerox's 2005 Environment Health and Safety Report, quantitative data is provided which illustrates progress over the previous year in reducing the firm's ecological footprint. However, in the same document, Xerox also draws attention to the fact that its improved environmental governance efforts also resulted in significant cost savings to the firm.

vii) *Pulling the elements together*

Systems which are comprised of numerous elements that are inter-related and that themselves are constantly evolving are known as complex adaptive systems (Brown and Eisenhardt, 1997). As the name implies, the main distinguishing characteristics of a complex adaptive system (CAS) are the complexity of interrelationships that impact the evolution of the system and the dynamic nature of the system's components which results in goal posts that are constantly moving (Stacey, 1995). As should be evident at this point in the paper, the forces discussed in this paper which influence a firm's environmental governance strategy and environmental disclosures should be considered collectively as elements within a CAS. Thus, the SEM Framework which was presented in Fig. 1.1 represents an attempt to visualize this complex CAS in a clear conceptual framework.

Extrapolating from the observations made earlier regarding how each of the five groupings of forces identified in the SEM framework influence environmental disclosure, a conceptual model can be derived to demonstrate how the strength of a given firm's commitment to environmental governance strategy relates to the type of environmental disclosure it makes.

Figure 1.2 illustrates that semantic patterns in environmental disclosures can be used to approximate a firm's commitment to environmental

Fig. 1.2. Strategic commitment to environmental governance and the relationship to environmental disclosure.

governance. Specifically, working left to right along the spectrum:

- The presence of *quantitative* environmental performance data in environmental disclosures indicates that not only has a firm thought deeply about environmental governance policy, but it has also established systems and structures to monitor commitment. Such firms can be considered as highly committed to corporate environmental governance strategy.
- Less committed, but still strategically superior firms can be identified by a commitment to maintain internationally certified environmental management systems (EMS) such as ISO14000 or EMAS. These standards commit the organization to a process of formally adopting the processes and structures necessary to receive external certification. Although these standards do not commit firms to specific environmental performance standards, these independently-verified environmental management systems are representative of a high level of formalized commitment to improving corporate environmental governance (King and Lenox, 2005).
- References in environmental disclosures to the adoption of in-house EMS are indicative of firms that are showing signs of adopting more sophisticated strategic environmental governance policies. The disparate nature of in-house EMS imply that these firms may not be as strongly committed to improved environmental governance as firms

with ISO14000 or EMAS certification; however, in-house EMS regardless of their scope, take time and effort to develop. Accordingly, such firms should be considered to be above average in terms of commitment to environmental governance.

- Similarly, firms which produce stand-alone environmental reports should be considered to be above average in terms of commitment to environmental governance practices. Although for some firms, the production of stand-alone environmental reports may be partly a marketing exercise (Cerin, 2002); creating a stand-alone document that will be vetted by external stakeholders requires significant consideration of environmental issues.
- The level of commitment to environmental governance strategy for firms that limit public disclosure of environmental governance to environmental policy statements published on corporate web-sites or in corporate public disclosures is difficult to assess. Such policy statements may in fact guide concrete environmental initiatives but policy statements may also simply be attempts to legitimize a firm's environmental governance to external stakeholders. Accordingly, in the absence of specific examples of environmental initiatives to support the policy claims, these types of statements should be considered as extemporaneous legitimization efforts because stakeholder theory and legitimacy theory imply that firms which can publicly laud environmental governance achievements will do so.
- Finally, in some Annual Reports, firms that do not have a strong commitment to environmental governance may present vague, qualitative exhortations which declare a concern for the environment and the corporate intent to be responsible corporate citizens. Again, in the absence of specific examples of environmental initiatives to support these declared intentions, such statements should be considered as legitimization efforts which lack substance.

With a clearer conceptual understanding of how the five realms of Strategic environmental management forces influence the nature of corporate environmental disclosure, we can now turn to the empirical research.

3. Research Objective and Methodology

The main premise of the empirical study presented here is to test the strength of the corporate environmental governance — financial performance relationship. Formally stated, the central hypothesis is affirmatively stated as follows:

A significant positive correlation exists between a firm's commitment to corporate environmental governance and its financial performance.

Prior to selecting subjects for inclusion into the research project, it is necessary to establish two baseline standards to operationalise the variables. First, a standard for measuring a firm's commitment to corporate environmental governance through the interpretation of environmental disclosures must be determined. Second, a standard or set of standards for measuring a firm's financial performance must be established. Standardizing measures in these two areas will enable comparison of firms to establish relative performance.

i) *Baseline standards for measuring environmental performance*

As per the conceptual foundation outlined in Fig. 1.2, strategic commitment to environmental governance can be measured on a scale from deep to shallow commitment and this commitment can be assessed by analyzing the contents of environmental disclosures. The presence of quantitative environmental benchmarks or other environmental performance data reflects a firm with a comparatively deep commitment to environmental governance. At the other extreme, the presence of unqualified environmental exhortations in environmental disclosures is indicative that a firm's commitment to environmental governance is more style than substance- it is trying to legitimize a largely ambivalent commitment to environmental governance. This conceptual understanding permits the creation of standardized measures.

Table 1.1 demonstrates that conceptual knowledge of how environmental governance strategy and environmental disclosure interact permit the development of operational measures that can be applied to environmental disclosure analysis in order to interpret a firm's commitment to environmental governance. The five levels of commitment identified in Table 1.1 represent the standards by which the firms in this research study will be measured to evaluate commitment to corporate environmental governance. The deepest form of commitment to environmental governance is represented by Level 1- Leaders. Similarly, the weakest form of commitment is represented by Level 5- Avoiders.

Table 1.1. Standards for measuring environmental governance.

Level 1- Leaders
Are defined as...
- *Firms committed to setting quantitative benchmarks in environmental initiatives.*

Which appears in environmental reporting as...
- *Quantitative progress indicators related to environmental initiatives.*

Level 2- Contenders
Are defined as...
- *Firms which have adopted externally accredited environmental management systems.*

Which appears in environmental reporting as...
- *Specific reference made to ISO14000 or EMAS accreditation.*

Level 3- Talkers
Are defined as...
- *Firms that are currently experimenting with the effectiveness of environmental management initiatives.*

Which appears in environmental reporting as...
- *Qualitative stand-alone environmental reports.*

Level 4- Pretenders
Are defined as...
- *Firms that understand the threat posed by poor environment governance. These firms therefore, endeavour to demonstrate that they meet all binding regulations.*

Which appears in environmental reporting as...
- *Qualitative environmental disclosure over 50 words amidst Annual Reports or on web-sites.*

Level 5- Avoiders
Are defined as...
- *Firms that obey environmental regulations because they must. However, they do not see a need to strategically address any environmental issues.*

Which appears in environmental reporting as...
- *Empty space. Firms classified as Avoiders do not disclose any information on environmental issues.*

ii) *Baseline standards for measuring financial performance*

There has been considerable debate over which financial metrics should be used for judging the competitive performance of a firm (Ruf *et al.*, 2001). For the purposes of this study, four metrics have been established as collective indicators of financial health: 1) % change in sales volume, 2) Profit margin, 3) Return on equity and 4) Return of assets.

The justification for using these four metrics as financial performance measures is largely provided by Ruf *et al.* (2001). Growth in sales was selected because of the significant role it plays both in strategy formulation and in investment valuation models. Moreover, sales growth is not as prone to the criticisms often applied to return on investment measures that such measures can be manipulated by accountants (Ruf *et al.*, 2001). Profit margin represents the most commonly referenced measure of a firm's overall operating effectiveness. Return on equity was selected because it is the primary measure of evaluating profitability from an investor's perspective (McLaney and Atrill, 2002) and return of assets was selected because it provides a measure of profitability that aggregates both equity and debt sources of capital. It is conceivable for all four measures to vary from one another.[2] Therefore, in order to leverage the benefits of each of the four measures while concurrently avoiding the disadvantages associated with the specialised perspectives on financial performance that each measure represents, it was deemed that an equally-weighted aggregate of all four measures would be used to evaluate financial performance.

It was decided to use corporate annual reports as the basis from which to gather audited data related to financial performance. On the other hand, data related to public disclosure of environmental governance initiatives were gathered from three sources: (1) corporate annual reports, (2) stand-alone environmental reports if available on a given firm's web-site and (3) information on environmental activities presented on the web-sites of the subject firms. Campbell (2003) confirms that for the sake of assessing public disclosure strategy, annual reports and other documents available on corporate internet sites can provide sufficient information for assessing a firm's social reporting practices.

The target firms selected for inclusion in the study came from two geographically-defined sources. Group 1 firms were all from Singapore. Firms from one country were analyzed together in order to eliminate the disparate influences on corporate environmental governance activities caused by macro forces and secondary forces as described in the ELF conceptual framework. Singapore was chosen as the target country for Group 1 firms because its political environment "constrains citizen demands for information and minimizes regulatory pressure on companies" (Perry and Sheng, 1999: 310). In other words, secondary stakeholder influence is comparatively light in Singapore. Accordingly, it was anticipated that firms from Singapore will have a lower propensity to produce environmental disclosures that attempt to legitimize environmental governance performance.

---

[2]In fact the data from the empirical study bears witness to the validity of this claim. Many firms in this study reported disparate performance in the four areas of financial measurement.

Group 2 firms were all from Canada. Once again, the rationale for grouping firms from one country together was to reduce environmental reporting distortions caused by macro and secondary forces. Canada and Singapore have similar levels of affluence in terms of per capita GDP and both countries are Commonwealth nations so political structures are similar. Therefore, the allure of choosing Canada as a foil to Singapore is that the similarities permit comparison of whether or not the comparably higher presence of environmental watchdog groups in Canada (Perry and Sheng, 1999) influence environmental governance.

In total there were 88 firms chosen for the study: 60 firms from Singapore and 28 firms from Canada.

In order to isolate industry-specific forces which influence environmental governance (as outlined in the ELF framework), it was decided to select target firms from common industry groupings. Two environmentally-sensitive industrial groupings were selected because previous studies indicate that environmental governance and environmental disclosure will be most robust in environmentally-sensitive industries (Shirley, 2005; Campbell, 2003; Deegan and Gordon, 1996). The two broad industrial areas selected were (1) construction/industrial production and (2) transport, shipping and telecommunications.

In order to try and minimize distortions caused by disparate sized firms (a concern identified in the SEM framework), all the firms chosen for the study were leaders in their respective fields. In order to determine which firms were recognized as industry leaders, all the firms selected for the study were part of stock market industry indexes.

In summary, the firms chosen for the empirical study included:

- All the firms comprising the Singapore Stock Exchange Construction Index (20 firms in total).
- All the firms comprising the Singapore Stock Exchange Transport, Shipping and Telecommunications Index (29 firms in total).
- All the firms comprising the Toronto Stock Exchange Industrials Index (22 firms in total).
- All the firms comprising the Toronto Stock Exchange Telecommunications Index (6 firms in total).

In total, this represents 77 of the 88 firms included in the study. The other 11 firms were all firms comprising the Singapore Stock Exchanges Hotels Index (11 companies). They were included to provide a reference benchmark for service industry firms. It was felt that earmarking a major service industry as a reference benchmark would provide an opportunity to compare correlation results in order to determine whether or not environmental-sensitivity in a given industry truly does influence environmental disclosure.

As a first step, after reviewing all the publicly available information from 2002 relating to environmental activities,[3] each firm was classified into one of the five environmental governance levels outlined earlier:

- L1 = Leaders (quantitative data on corporate environmental governance).
- L2 = Contenders (adoption of ISO14000 or EMAS).
- L3 = Talkers (publishing stand-alone environmental reports).
- L4 = Pretenders (producing qualitative environmental discourse of over 50 words).
- L5 = Avoiders (making no public mention of environmental activities).

For each firm, the highest level of environmental governance commitment identified through environmental reporting analysis was regarded as the defining level for that firm. For example, if a firm published a stand-alone environmental report (L3), was ISO14000 accredited (L2) and included quantitative data on corporate environmental governance in its stand-alone environmental report (L1), it was listed as an L1 firm.

Next, total revenue and operating profit data were extracted from audited Income Statements for years 2002–2005 in order for sales growth[4] and profit margin[5] statistics to be determined. Furthermore, total assets and total equity data for 2003–2005 were extracted from the audited balance sheets of each firm in order to calculate return on equity[6] and return on asset[7] data.

For each year, averages for all the firms within a given index were calculated for sales growth, profit margin, return on equity, and return

[3]2002 public disclosures were used for this process in order to link environmental reporting to future financial performance. Financial data from 2002–2005 were used for this purpose.
[4]Sales Growth = (Current year revenues — Previous year revenues)/Previous year revenues.
[5]Profit Margin = Operating Profit/Total Revenue.
[6]Return on Equity = Operating Profit/Total Shareholder Equity (including reserves).
[7]Return on Assets = Operating Profit/Total Assets.

on assets. A smoothing methodology was used to calculate these averages. In the smoothed average approach, 10% of the firms exhibiting the most extreme performance variances (either above or below industry average) were removed from the pool before the industry average was calculated. This was done separately for the Singapore Stock Exchange Construction Index firms, the Singapore Stock Exchange Transport, Shipping and Communications Index firms, the Singapore Stock Exchange Hotel Index firms, the Toronto Stock Exchange Industrials Index firms, and the Toronto Stock Exchange Telecommunications Index firms.

Once smoothed averages for the firms in each of the five index categories were calculated, the performance of each firm was then compared to the smoothed index average in the respective indices. This facilitated the identification of firms that outperformed the average for each financial performance metric used.

The number of times each respective firm outperformed the industry average (within the 12 performance metrics) along with the firm's environmental management commitment rating (L1 to L5) were then plotted through correlation analysis in order to determine if there was correlation between environmental commitment and financial performance either in the base year or in two subsequent years. Trend analysis that included identification of the coefficient of determination (r^2) was applied to each of the five indices in order to assess the strength of correlation between corporate environmental governance and financial performance.[8]

4. Results of the Environmental Reporting Comparison

A) *Singapore*

As Fig. 1.3 outlines, only 17% of the Singapore firms included in the study disclosed information on environmental activities in a public manner and only 14% could be considered substantive reporters (L1 and L2) of environmental initiatives. This parallels findings in a 2002 survey of environmental reporting conducted by ACCA-Singapore which found that only 10% of the Singaporean firms surveyed produced substantive environmental disclosures (ACCA, 2005a).

[8]In this instance, r-squared represents the fraction of variability in financial performance that is attributable to a firms level of environmental governance (L1 to L5).

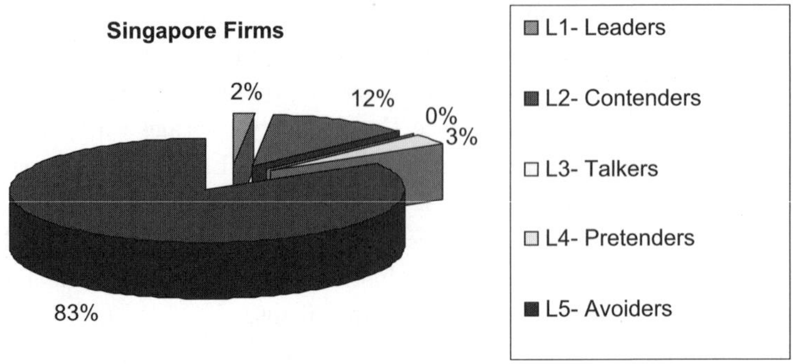

Fig. 1.3. Percentage of singaporean firms falling within each L-classification.

Some specific points relevant to the findings from Singaporean (Group 1) firms include:

- There was only one "L1-Leader" (Singapore Airlines) which produced an environmental report with quantitative data.
- Six firms were classified as "L2-Contenders" on the basis of ISO14000 certification.
- There were no "L3-Talkers", because the only firm that produced a stand-alone report (Singapore Airlines) was also ISO14000 certified (classifying it as a L2 firm) and included quantitative data in its environmental disclosure (classifying it as an L1 firm). As stated earlier, the end rating assigned to each firm was based on the highest attainment achieved on the L1-L5 scale.
- Only two firms (CWT and SMRT Corporation) were classified as "L4-Pretenders". These two companies made qualitative mention of their intention to operate in an environmentally responsible manner. However, the disclosures failed to identify any significant initiatives designed to support such a quest.
- None of 11 Singapore hotels included in the survey provided corporate environmental disclosure of any type. They were all "L5-Avoiders". It is likely that this reflects the underlying mind-set of many service industry firms that their ecological impact is negligible and that, therefore, there is no need to report on environmental governance (Cerin, 2002).

Overall, the results of the research support previous findings that environmental initiatives in Singapore are, by and large, limited to

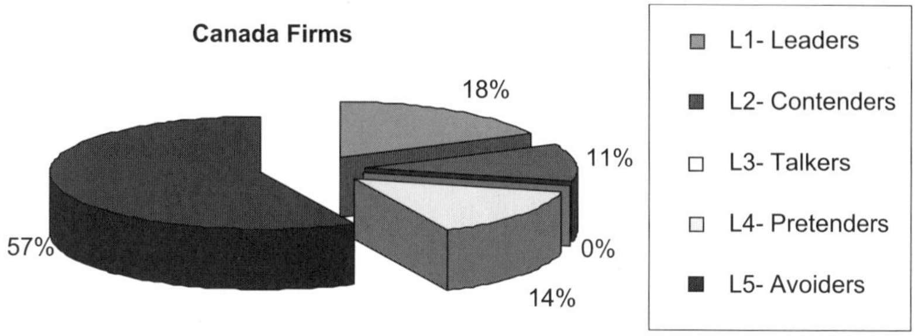

Fig. 1.4. Percentage of canadian firms falling within each L-classification.

compliance initiatives (Perry and Sheng, 1999); and consequently, corporate reporting in also limited (ACCA, 2005).

B) *Canada*

As Fig. 1.4 indicates, 43% of the Canadian firms included in the research study provided public disclosures concerning environmental activities. Although this appears to be considerably superior to the percentage of firms in the Singapore group that produce environmental disclosures (17%), if the L4-Pretenders (firms which produce disclosures that are not supported by substantive examples) are factored out, the ratio of firms that provide substantive reporting (L1-Leaders and L2-Contenders only) narrows to 29% for the Canadian firms and 14% for the Singaporean firms.

Some specific points relevant to the Canadian findings include:

- Of the five "L1-Leaders" which produced an environmental report with quantitative data, only two were ISO14000 certified. In other words, the other three generated quantitative reports based on in-house environmental management systems. This supports previous research which states that in-house Environmental Management Systems (EMS) are as prevalent as ISO14000-based EMS in North America (Berthelot and Coulmont, 2004).
- The three firms that were classified as "L2-Contenders" on the basis of ISO14000 certification, included Bombardier (the airplane manufacturer), Finning (the sales agent for Caterpillar tractors) and Quebecor (the global printing company). These three firms have likely sought ISO14000 certification because they operate in international markets where ISO14000 certification is a competitive strength.

- As with the Singapore cluster of firms, there were no "L3-Talkers" in the Canadian cluster because the three firms that produced stand-alone environmental reports (CN Rail, Bell Canada-BCE and Manitoba Telecom) also included quantitative data in their environmental disclosures (classifying them as L1 firms).
- There were four firms classified as "L4- Pretenders". These companies publicised policy statements that they intended to operate in an environmentally responsible manner. However, the disclosures failed to support these exhortations with examples of supporting initiatives. It is possible that there were twice as many Pretenders in Canada as there were in the Singapore study because leaders of Canadian firm's deem it necessary to legitimize their environmental governance efforts due to pressure from numerous environmental watchdog organizations in Canada.

Contextual inferences

The data presented above, cannot be used as a proxy to represent the overall level of reporting undertaken in Canada or Singapore. This is because the scope of the firms included in the research sample was not representative of the entire scope of business in each respective country.[9] As was pointed out during the development of the SEM Framework, factors such as firm size, financial health and industry type play roles in influencing the extent to which firms commit to environmental governance. In the case of the firms included in this study, the firms were all market leaders in their respective industries. As a result, they were all comparatively larger than most rival firms and in comparably better financial health because the equity shares of the firms were included on major indices which attract funds more easily. Furthermore, these firms were from the same industries; therefore, extending the results to firms from other industries would likely result in attribution errors (Shirley, 2005).

The true value of the data presented above is to provide a contextual background to the reporting done by firms in Singapore and Canada. Given that substantive environment reporting (L1-Leaders and L2-Contenders) was significantly higher concerning the Canadian firms (29%) than the

[9]Note- It was not the intention of this paper to undertake a study of broader scope. The main thrust of this paper is to test the corporate environmental governance — financial performance link; therefore, limiting the scope of the study was necessary in order to eliminate industry distortions.

Singaporean firms (14%), there are reasons to conclude that Canadian firms have a greater propensity to disclose environmental governance issues when compared with Singaporean firms. What the exact ratio is between firms releasing environmental reports in Singapore and firms releasing environmental reports in Canada would require a study of much broader scale and scope. For the purposes of this study it is merely of importance to understand that more Canadian firms report on environmental issues than compared to Singaporean firms.

The reason that this knowledge is of contextual importance to the study relates to the findings regarding the correlation between environmental governance and financial performance. If a significant positive correlation is identified, the observation that more Canadian firm's produce environmental reports implies that Canadian businesses might better understand the link between environmental governance and financial performance. In other words, more education would be necessary in Singapore to enlighten Singaporean businesses about the link between environmental governance and financial performance.

Evaluating the environmental governance-financial performance correlation.

The Y-Axis represents the percentage of financial indicators outperforming the industry average (aggregate total of the index firms). The 12 financial indicators include:

- 2003 Return of Equity, 2004 Return of Equity, 2005 Return of Equity.
- 2003 Return of Assets, 2004 Return of Assets, 2005 Return of Assets.
- 2003 Revenue Growth, 2004 Revenue Growth, 2005 Revenue Growth.
- 2003 Profit Margin, 2004 Profit Margin, 2005 Profit Margin.

For example, if a firm's performance in 2003 exceeded the industry average in Return on Equity and Profit Margin, its performance in 2004 exceeded the industry average in Return on Assets, and in 2005 it did not exceed the industry average on any of the 4 financial standards, then its total would be 3 indicators exceeded out of 12. Accordingly the Y-Axis measure would be 25%.[10] Three years of financial data following the environmental reports of 2002 were selected to ensure that the impacts

---

[10]It should be noted that for a handful of firms, data was not available for certain years. In these cases, the denominator for calculating the Y-Axis percentage was adjusted accordingly.

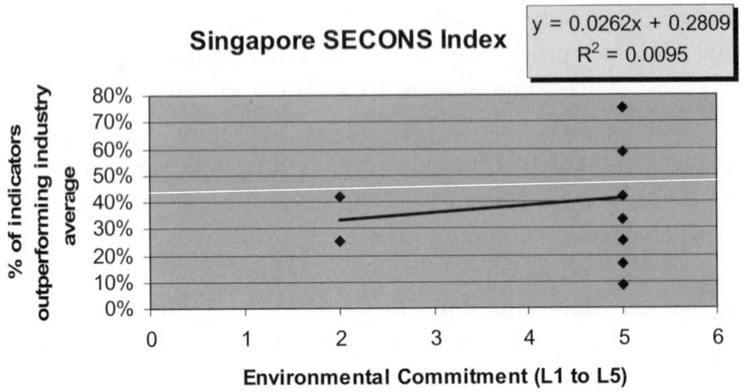

Fig. 1.5. The environmental commitment-financial performance link for Singapore exchange construction (SECONS) Firms.

of the environmental initiaitives were incorporated into the financial performance of the firms.

Singapore exchange construction index

Upon first glance, the upward sloping trend line in Figure x could be interpreted to mean that superior corporate environmental governance gives rise to inferior financial performance for firms in the SECONS index. However, with an R^2 value of 0.0095, it can be concluded that no significant correlation is evident between corporate environmental governance and financial performance for the firms studied in the Singapore exchange construction index.

Both of the ISO14000 accredited firms (L2) underperformed in over half of the financial indicators when compared to index financial performance averages.

Singapore exchange transport, shipping and communications index

As a visual fit purview of Fig. 1.6 indicates, there is a significant correlation between environmental governance commitment and financial performance is not evident for the 29 firms included in Singapore's Transportation, Shipping and Communications Index. Mathematically, an R^2 value of 0.0015 confirms this assertion.

Of particular interest, the firm with the best overall commitment to environmental management initiatives (Singapore Airlines) performed below the financial performance index averages 80% of the time.

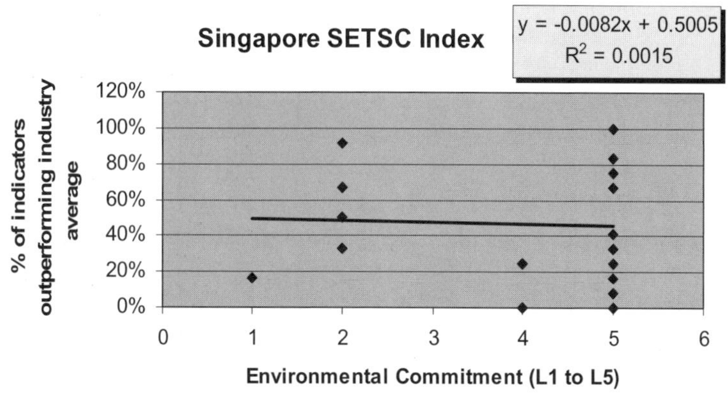

Fig. 1.6. The environmental commitment-financial performance link for Singapore exchange transport, shipping and communication (SETSC) Firms.

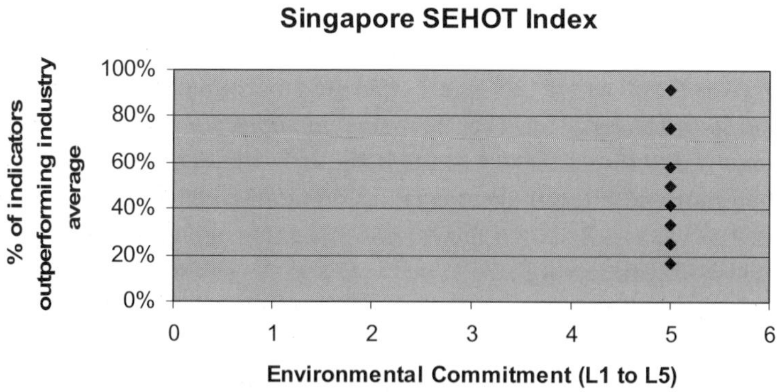

Fig. 1.7. The environmental commitment-financial performance link for Singapore exchange hotels index (SEHOT) firms.

Singapore exchange hotels index

Given that none the 11 firms included in Singapore's Hotels Index published environmental disclosures, creating a trend line was not possible. Figure 1.7 shows the dispersion of corporate data. With no disparate corporate environmental governance standards for comparison, a correlation between environmental commitment and financial performance could not be assessed.

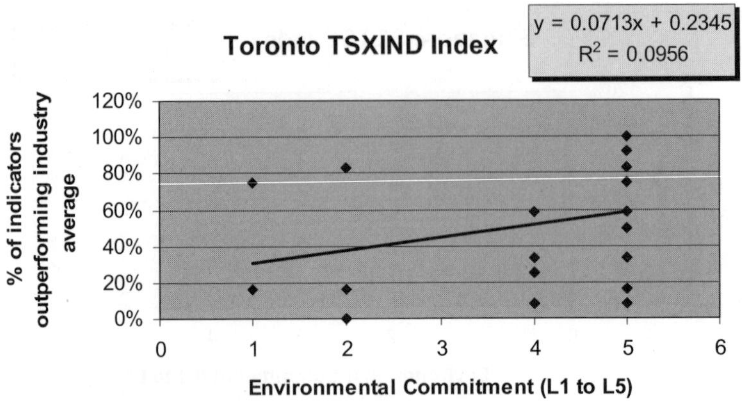

Fig. 1.8. The environmental commitment-financial performance link for toronto stock exchange industrial (TSXIND) firms.

Toronto exchange industrials index

The trend line in Fig. 1.8 also implies at first glance that an inverse correlation exists between corporate environmental governance and financial performance for the 22 firms included in the Toronto Stock Exchange Industrials Index. Although the R^2 value for the TSXIND firms of 0.0956 represents the highest correlation of all the five index correlation studies, the value is still too low to conclude that a significant correlation exists between the variables.

It is worth highlighting that the firms in Fig. 1.8 which produced quantitative corporate environmental governance measures (L1) and ISO14000 accredited firms (L2) achieved widely disparate financial performance results. Two firms in these two groupings (L1 and L2) reported financial performance which exceeded the index average in over 75% of the financial performance markers; however, 3 firms reported financial performance which fell below the index average in over 80% of the financial performance markers.

Toronto exchange telecommunications index

Unsurprisingly given the results from the other indices, evaluating the 6 firms listed in the Toronto Stock Exchange Telecommunications Index also produced evidence of no significant correlation between environmental commitment and financial performance (Fig. 1.8). Mathematically, the R^2 value of 0.002 supports this conclusion. Although the sample size was too

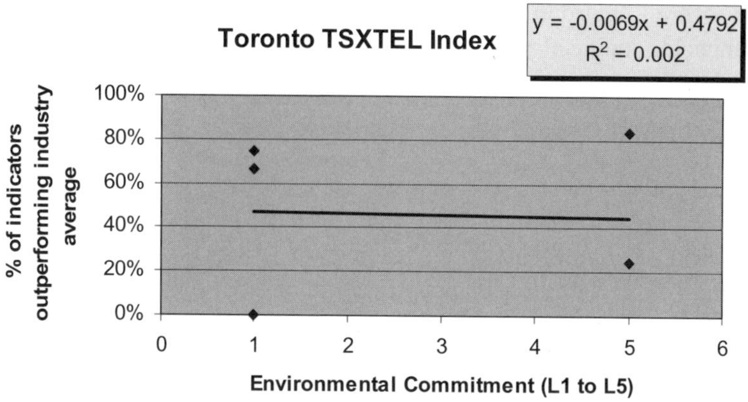

Fig. 1.9. The environmental commitment-financial performance link for toronto stock exchange telecommunications (TSXTEL) firms.

small for statistical reliability, the data strongly implies that other forces appear to influence financial performance to a far greater extent than corporate environmental governance does.

Based on the results of the research, it is valid to conclude that for the firms included in the indices researched in this report, no significant positive correlation can be proven between corporate environmental governance and financial performance over the range of years covered in this research study (2003–2005). Therefore, results from the research do not support the central hypothesis.

It should be noted that failure to confirm the central hypothesis does not necessarily mean that environmental commitment has *no* impact on financial performance. It does appear from the research study that there are other more significant influences on financial performance that render overall environmental governance commitment negligible in terms of influencing financial performance.

This interpretative conclusion is supported by the observation that the research findings also indicate that no significant *negative* correlation exists between corporate environmental governance and financial performance. If the abundance of anecdotal research which identifies financial benefits to some environmental initiatives is to be believed (Reinhardt, 1999; Porter and Van der Linde, 1995), and there is no significant positive or negative correlation found in this study between corporate environmental governance and financial performance, then it is justifiable to conclude that there are other more significant influences on financial performance that render overall environmental governance commitment negligible in terms of influencing financial performance.

5. Conclusion

From an environmental conservation perspective, it would have been heart-warming to conclude from the research undertaken in this study that a firm's commitment to environmental governance is positively and significantly correlated to financial performance. Clearly, if such a positive correlation did exist, a rational profit-maximizing firm would be negligent in its duties if it did not adopt enhanced environmental governance strategies. Proving that such a correlation exists could catalyze a new era of corporate environmental benevolence. However, the research presented in this paper indicates that such an absolute correlation likely does not exist. In all probability, any positive influence of environmental governance on financial performance is overpowered by other variables that influence financial performance.

This conclusion is in itself a useful contribution to the study of environmental governance policy. If it is a goal of environmental governance policy to promote improved corporate environmental governance practices, then policy makers must understand the realities which impinge upon this outcome.

With no significant correlation between environmental commitment and financial performance, a profit-maximizing firm can at best be expected to generally adopt *satisficing* environmental governance strategies (Doyle, 2003). Profit maximizing firms will seek to: (1) implement environmental initiatives that are profitable, (2) comply with regulations governing unprofitable areas of environmental governance and (3) satisfice stakeholders groups to defend against unwanted stakeholder protest.

If policy makers are to successfully develop policies to motivate improved corporate environmental governance, they need to make the effort to understand the realities underpinning the challenges that corporate leaders face. Policy makers need to understand that the vast majority of corporate leaders do not get up in the morning with the intent of carrying out policies to erode the planet's environmental capital. The vast majority of corporate leaders get up each morning with the knowledge that they must satisfy the profit expectations of their owners and the broad needs of various internal and external stakeholders. In other words, each day, management performs a precarious balancing act. True there are some corporate leaders that neglect environmental issues because they do not see environmental problems as part of the corporate remit. However, the majority of business people likely seek balance if only to reduce the risk of civil reprisal against poor environmental governance.

As the discussion leading to the development of the SEM Framework demonstrated, there are a number of justifiable business reasons to improve environmental governance. There are also a number of endogenous and exogenous forces that catalyze better corporate environmental governance. Until change comes to the economic, free market, capitalistic structure that guides corporate activities in most countries around the world, corporate leaders will continue to pursue profit-maximization within the limits prescribed by legislation, regulation, and stakeholder pressure. Within such bounded rationality, sustainable environmental management practices are unlikely to be attainable. Accordingly, rather than trying to change mindsets by appealing to the social consciences of corporate leaders, policy makers should be trying to change the structural flaws that give rise to this dilemma. In the short run, this may imply more regulatory control over environmental governance.

References

ACCA. 2005a, *Report Summary: The State of Corporate Environmental and Social Reporting in Singapore 2005*, Association of Chartered Certified Accountants, Singapore.

Adams, C. 2004, The Ethical, Social and Environmental Reporting-Performance Portrayal Gap, *Accounting, Auditing and Accountability Journal*, Vol. 17, No. 5, Pg. 731–757.

Beets, S.D., Souther, C. 1999, Corporate Environmental Reports: The Need for Standards and an Environmental Assurance Service, *Accounting Horizons*, June 1999, Vol. 13-2, Pg. 129–145.

Berry, M., and Rondinelli, D. 1998, Proactive Corporate Environmental Management: A New Industrial Revolution, *The Academy of Management Executive*, May 1998, Vol. 12-2, Pg. 38–50.

Berthelot, S. and Coulmont, M. 2004, ISO14000- A Profitable Investment? *CMA Management*, Nov. 2004, Vol. 78-7, Pg. 36–39.

Brown, S. and Eisenhardt, K. 1997, The Art of Continuous Change: Linking Complexity Theory and Time-Paced Evolution in Relentlessly Shifting Organizations, *Administrative Science Quarterly*, March 1997, Vol. 42-1, Pg. 1–34.

Campbell, D. 2003, Intra- and Intersectoral Effects in Environmental Disclosures: Evidence for Legitimacy Theory? *Business Strategy and the Environment*, Vol. 12, No. 6, Nov-Dec 2003, Pg. 357–371.

Carter, N. 2001, *The Politics of the Environment: Ideas, Activism, Policy*, Cambridge University Press, UK

Cerin, P. 2002, Communication in Corporate Environmental Reports, *Corporate Social Responsibility and Environmental Management*, Vol. 9, No. 1, Mar 2002, Pg. 46–66.

Connelly, J.T., Limpaphayom, P. 2004, Environmental Reporting and Firm Performance: Evidence from Thailand, *The Journal of Corporate Citizenship*, Vol. 13, Spring 2004, Pg. 137–149.

Cormier, D., Gordon, I., and Magnan, M. 2004, Corporate Environmental Disclosure: Contrasting Management's Perceptions with Reality, *Journal of Business Ethics*, Vol. 49, No. 2, Jan 2004, Pg. 143–165.

Cormier, D. and Magnan, M. 2003, Does Disclosure Matter? *CA Magazine*, Vol. 136, No. 4, May 2003, Pg. 43–45.

Cormier, D. and Gordon, I. 2001, An Examination of Social and Environmental Reporting Strategies, *Accounting, Auditing & Accountability Journal,* Vol. 14-5, Pg. 587–616.

Costanza, R., Cumberland, J., Daly, H., Goodland, R. and Norgaard, R. 1997, *An Introduction to Ecological Economics*, CRC Press, USA.

Deegan, C. 2002, The Legitimizing Effect of Social and Environmental Disclosures- A Theoretical Foundation, *Accounting, Auditing and Accountability Journal*, Vol. 15, No. 3, Pg. 282–311.

Deegan, C. and Gordon, B. 1996, A Study of the Environmental Disclosure Policies of Australian Corporations, *Accounting and Business Research*, Vol. 26-3, 187–199.

Doyle, P. 2003, *Marketing Management and Strategy*, 3rd Edition, Prentice-Hall, UK.

Grant, R. 2005, *Contemporary Strategy Analysis*, 5th edition, Blackwell Publishing, UK.

Hawken, P. 1992, Ecology of Commerce, *Executive Excellence*, October 1992, Vol. 9-10, Pg. 3–5.

Iu, J. and Batten, J. 2001, The Implementation of OECD Corporate Governance Principles in Post-Crisis Asia, *Journal of Corporate Citizenship*, Vol. 4, Pg. 47–62.

Karagozoglu, N. and Lindell, M. 2000, Environmental Management: Testing the Win-Win Model, *Journal of Environmental Planning and Management*, Nov 2000; Vol. 43-6, Pg. 817–830.

Kaplan, R. and Norton, D. 1996, *The Balanced Scorecard: Translating Strategy into Action*, Harvard Business School Press, US.

Khanna, M. 2005, *Measuring Corporate environmental governance: A Delphi Study on Ranking Corporate environmental governance of Companies in Singapore*, MSc Environmental Management Dissertation, Graduate School of Design and the Environment, National University of Singapore.

Kiernan, M. 2001, Eco-Value, Sustainability and Shareholder Value: Driving Corporate environmental governance to the Bottom Line, *Environmental Quality Management*, Summer 2001, Vol. 10-4, Pg. 1–12.

King, A. and Lenox, M. 2005, ISO14001 Helps Buyers Push Suppliers to Clean Up Their Acts, *Duke University News and Communications*, 16 December 2005.

King, A. and Lenox, M. 2001a, Lean and Green? An Empirical Examination of the Relationship Between Lean Production and Corporate environmental governance, *Production and Operations Management*, Fall 2001, Vol. 10, No. 3, Pg. 244–256.

King, A. and Lenox, M. 2001b, Does it Really Pay to be Green?: An Empirical Study of Firm Environmental and Financial Performance, *Journal of Industrial Ecology*, Vol. 5-1, Pg. 105–116.

Klassen, R. and McLaughlin, P. 1996, The Impact Of Environmental Management On Firm Performance, *Management Science*, Aug 1996, Vol. 42-8, Pg. 1199–1215.

Kolk, A. 2005, Environmental Reporting by Multinationals from the Triad: Convergence or Divergence? *Management International Review*, Vol. 45, Spring 2005, Pg. 145–166.

Kolk, A., Walhain, S. and Van de Wateringen, S. 2001, Environmental Reporting by the Fortune Global 250: Exploring the Influence of Nationality and Sector, *Business Strategy and the Environment*, Jan-Feb, 2001, Vol. 10-1, Pg. 15–28.

McLaney, E. and Atrill, P. 2002, *Accounting: An Introduction*, 2nd Edition, Prentice-Hall, UK.

Neely, A., Adams, C. and Kennerley, M. 2002, *The Performance Prism*, Prentice-Hall Financial Times, UK.

Patten, D.M. 1992, Intra-Industry Environmental Disclosures in Response to the Alaskan Oil Spill: A Note on Legitimacy Theory, *Accounting, Organizations and Society*, Vol. 15-5, Pg. 471–475.

Perry, M. and Sheng T.T. 1999, An Overview of Trends Related to Environmental Reporting in Singapore, *Environmental Management and* Health, Vol. 10-5, Pg. 310–320.

Porter, M and Van der Linde, C. 1995, Green and Competitive: Ending the Stalemate, as reprinted in Porter, M. 1996, *On Competition*, Harvard Business School Press, USA.

Porter, M. 1980, *Competitive Strategy: Techniques for Analyzing Industries and Competitors*, Free Press, USA.

Reinhardt, F. 1999, Bringing the Environment Down to Earth, *Harvard Business Review*, Jul-Aug. 1999, Pg. 149–157.

Ruf, B., Muralidhar, K., Brown, R., Janney, J. and Paul, K. 2001, An Empirical Investigation of the Relationship Between Change in Corporate Social Performance and Financial Performance: A Stakeholder Theory Perspective, *Journal of Business Ethics*, Jul 2001, Vol. 32, No. 2, Pg. 143–156.

Russell, J. 2006, Breaking into the Mainstream: The Rise of Ethics Education, *Ethical Corporation*, May 2006, European Academy of Business in Society, Pg. 4–6

Shirley, D. 2005, Acting Responsibly, *European Chemical News*, Sept 26-Oct 2, 2005, No. 2164, Pg. 24.

Stacey, R. 1995, The Sceince of Complexity: An Alternative Perspective for Strategic Change, *Strategic Management Journal*, Sept 1995, Vol. 16-6, Pg. 477–495.

Stanwick, P. and Stanwick, S. 1998, The Relationship Between Corporate Social Performance and Organizational Size, Financial Performance, And Corporate environmental governance: An Empirical Examination, *Journal of Business Ethics*, Jan 1998, Vol.17-2, Pg. 195–205.

US Environmental Protection Agency (US EPA), 2000, *Green Dividends?: The Relationship Between Firms' Corporate environmental governance and Financial Performance*, May 2000, EPA-100-R-00-021.

Van Tudler, R. 2005, *International Business-Society Management: Linking Corporate Responsibility and Globalization*, Routledge Publishing, UK.

Wilmshurst, T. and Frost, G. 2000, Corporate Environmental Reporting: A Test of Legitimacy Theory, *Accounting, Auditing & Accountability Journal*, Vol. 13-1, Pg, 10–19.

Yakhou, M., Dorweiler, V. 2004, Environmental Accounting: An Essential Component of Business Strategy, *Business Strategy and the Environment*, Vol. 13, Mar-Apr 2004, Pg. 65–77.

Xerox Corporation, *Environment, Health and Safety: Progress Report 2005.*

2

DEVELOPING A COMMON FRAMEWORK FOR TRACKING CORPORATE ENVIRONMENTAL PERFORMANCE AND ESTABLISHING AN ENVIRONMENTAL CODE OF CONDUCT FOR THE HOTEL INDUSTRY IN SINGAPORE

PAK-JUAN KOE and AUDREY CHIA

Hotels form a major component of the tourism industry. From its operational activities that require lighting, heating, cooling, cleaning and washing, to the provision of auxiliary services such as restaurants, entertainment and retailing, the hotel industry bears characteristics that can impact any social or ecological order through its demands on energy, water, food supply and waste disposal, and its influence over land use decisions and real estate prices. If not properly managed, hotels can adversely affect the environment, society and cultural heritage of the places they operate in.

A variety of large and small hotels are located in Singapore. In recent decades, Singapore-based companies have also ventured overseas, to develop or operate hotels and resorts. There are now a number of Singapore-headquartered or Singapore-managed hotel groups with operations in Singapore and/or overseas.

The context of Singapore as the corporate headquarters or operational headquarters for these hotel companies provides the basis of this study. The objective of the study is to develop a new framework for tracking corporate environmental performance and to establish an industry-based self-regulatory mechanism for Singapore hotels and resorts.

This study contains descriptions of the proposed framework and of the key supporting approaches that are vital to the implementation of the framework at the corporate level. It also explores key factors concerning the use of an environmental code of conduct as a way to extend environmental best practices from the corporate level to the industry level.

Introduction

In the hotel industry, globalisation is popularly regarded as a process which establishes a hotel company's presence on a global or international basis.[1]

[1]Pine, R. and F.M. Go (1996), "Globalisation in the Hotel Industry", in Kotas, R., R. Teare, J. Logie, C. Jayawardena and J. Bowen (eds.) *The International Hospitality Business*, p. 97.

The concept of a global hotel corporation, that sees the need to penetrate foreign markets and to be competitive across borders, is increasingly being questioned,[2] not least because it has an effect on local business, and also on local socio-cultural and environmental conditions of the foreign markets.

Globalisation has a standardising effect. In the 1990s, a central issue for many hotel corporations was how to organise, integrate and manage their activities to respond to the simultaneous need for a sense of global strategic intent and a focus on local conditions of the host country.[3] The industry today is experiencing a new phenomenon riding on the back of broader changes being experienced in the tourism sector. As nations democratised, borders opened, living standards improved (income gaps notwithstanding) and transportation choices widened, more sectors of society have gained the freedom to move across the country and across borders.

Global travel and tourism spending is expected to grow to US$6.5 trillion in value in 2006 - up 4.6% in real terms from an estimated expenditure exceeding US$6 trillion in 2005. The forecast for annualised growth between 2007 and 2016 is 4.2% per annum. In 2006, the hotel industry is expected to account for 3.6% of global gross domestic product (GDP) and to produce 2.5 million new jobs comprising a total 76.7 million jobs – the latter representing 2.8% of total world employment. When taking into account both the direct and the indirect contributions to the world economy, such as growth of tourism-related businesses like cleaning companies and caterers, the industry is expected to total 10.3% of the world's GDP in 2006 and create almost 10 million new jobs globally, for a total of 234.3 million jobs or 8.7% of total employment, making it one of the world's largest industries, if not the largest.[4] For some countries, tourism has become the foundation of the economy.

The focus on the tremendous potential of the hotel sector has both unmasked and masked inter-related trends and dimensions of this global phenomenon. From an environmental perspective, the need for actors in

---

[2]ibid.

[3]ibid.

[4]World Travel & Tourism Council (WTTC) (2006), *World Travel & Tourism Climbing to New Heights: The 2006 Travel & Tourism Economic Research*, p. 2, 6–21.

World Travel & Tourism Council (WTTC) (2006), "*Global Travel and Tourism Exceeded US$6 Trillion in 2005; Strong Performance Expected in 2006*", Press Release 6 March 2006, London: WTTC, http://www.wttc.org/frameset5.htm, accessed 23 July 2006.

the hotel sector to move from being environmentally conscious and passive to behaving in an environmentally responsible way is vital.

It has been acknowledged that the effective integration of environmental management practices into an organisation's operations can lead to a variety of benefits, such as cost reductions, improved resource productivity, competitive advantage from service or product differentiation, better servicing of a niche customer segment, access to financing, improved staff morale, and increased customer loyalty. Much has been written about the competitive advantage and benefits of applying environmental strategies in business, albeit generally with a product bias or in a manufacturing context. With the few written for services, most have been anecdotal or case studies rather than empirical evidence.[5]

The hotel, as a service organisation, faces a number of "unique challenges that result from the distinctive characteristics of services vis-à-vis goods".[6] The major distinctive characteristics that can impact on environmental efforts are:

- that the hotel is a high-contact (high touch) operation with the ubiquitous presence of the customer as a 'co-producer' in the service system,
- the resulting simultaneity of service production and consumption.[7]

From an operations flow perspective, the accommodation service is the hotel's core system comprised of four sub-systems, namely reservations, reception, overnight stay (housekeeping) and payment (billing system). Depending on the size of the hotel and the customer segment served, there may or may not be a number of ancillary systems such as laundry, restaurants, bars, businesses and leisure services. The core system is essentially a customer-processing operation.

On the other hand, the food-service operation is simultaneously a 'materials processing operation' and 'customer processing operation'. The ten sub-systems of food-service include storage, preparation, cooking, holding, transport, regeneration, service, dining, clearing and dishwashing.[8]

---

[5]Kassinis G.I. and A.C. Soteriou (2005), "Greening Service Organisations: Environmental Management Practices and Performance", in Sharma, S. and J.A. Aragón-Correa (eds.), *Corporate Environmental Strategy and Competitive Advantage*, pp. 115–117.
[6]ibid, p. 117.
[7]ibid.
[8]Ball, S., P. Jones, D. Kirk and A. Lockwood (2003), *Hospitality Operations: A Systems Approach*, pp. 30–31.

The challenge and the opportunity in such a 'high touch' system is to meet the customer demand and involvement in environmental management without compromising the quality of service the hotel provides. Service simultaneity, i.e. the simultaneous production and consumption of services, also presents another challenge. In a manufacturing environment, the most significant environmental management activities are conducted at the back-end system (i.e. production) where fulfilment of environmental management objectives depends on cooperation from employees.

Due to the hotel's 'high touch' characteristic, for any environmental management programme to be meaningful would naturally require a much higher in-system involvement from the customer in a hotel than would be the case in manufacturing. That is, the hotel's corporate environmental management activities must overlap with and extend deeper to the front-end experience where the customer and service (the hotel product) interface. An environmentally conscious customer may not only apply pressure on management to change company policy but must also be involved as a 'co-producer' in the firm's environmental practice, such as energy or water savings practices.[9]

Of course, certain environmental activities in the hotel are hidden from the customer as they take place in the back-end. This is the case in, say, food-service where waste recycling or waste disposal tends to occur out of customer view.

There are also challenges in the introduction, implementation and availability of environmental technologies or approaches in a hotel, compared with manufacturing plants. A manufacturing facility can choose between end-of-pipe pollution control techniques, such as use of filters, and pollution prevention solutions, such as source reduction, reusing and recycling programmes. At a hotel, the pollution control approaches are available to operations at the back-end. The front-end is limited to pollution prevention approaches only. Because of the presence of the customer in the system, the implementation of pollution prevention at a hotel can be challenging in its own right. No doubt, the involvement of the customer is a far more vital, if not more integral, component in the environmental management equation in a hotel than in manufacturing.

The importance of the customer was supported by the findings in an empirical study by Kassinis and Soteriou (2005).[10] The research which

---

[9]Kassinis G.I. and A.C. Soteriou (2005), op cit, p. 118.
[10]ibid, pp. 131–133.

involved hotels in at least 11 European countries found that the use of environmental management practices was positively related to market performance, but only through the mediating effect of customer satisfaction and customer loyalty, i.e. specifically through the positive relationship between corporate environmental practices and customer satisfaction, between customer satisfaction and customer loyalty, and finally between customer loyalty and market performance. However, the direct link between environmental management practice and market performance could not be proven in the study. The importance of customer satisfaction and customer loyalty as the mediating effect is noteworthy in the context of rising consumer awareness of environmental responsibility in the European market.

In a separate study, Crespí-Cladera and Orfila-Sintes (2005)[11] found that the propensity to invest in technological innovations for the management of environmental quality was higher in hotels that belonged to a chain, hotels which were owner-managed, larger hotels, and in those that had an intended purpose to serve the targeted customer segment. Affiliated hotels tended to have an image or reputation that was more positively related to environmental management and also to its access to management know-how. The owner-managers made technical improvements because they had more certainty of costs while the larger hotels did so to cater to customer preferences. By orientating itself toward a market segment that valued qualitative aspects such as environmental protection, the propensity of a hotel to spend on environmental improvement was commensurate with the targeted customer's greater willingness to pay.

The Environmental Dimensions of the Hotel Industry

Although perceived as less 'polluting' than most industries, the hotel industry, by the very nature of the business, nevertheless has a potential impact on the environment, notably through its consumption of energy and water, waste generation and land use. This section highlights some environmental issues that relate to the hotel sector. Some illustrative solutions or examples of how some hotels handle the issues are also discussed.

[11] Crespí-Cladera, R. and F. Orfila-Sintes (2005), "Environmental Innovation in the Hotel Industry of the Balearic Islands", in Sharma, S. and J.A. Aragón-Correa (eds.), *Corporate Environmental Strategy and Competitive Advantage*, pp. 139–156.

Water is perhaps a hotel's most important resource. Guests typically consume more water than local residents. More than 50% of water is used in guestrooms and kitchens while the other major users are laundries and public toilets. A hotel can consume 60 m^3 - 220 m^3 of water per guest room per year depending on the facilities provided and whether water conservation practices such as flow restrictors and on-site wastewater treatment are in place.[12] Excessive water usage can stress out local water resources and supply. Water can account for a considerable portion of a hotel's purchasing costs.

Water in most hotels is supplied by utility companies although some businesses especially in rural areas may have facilities on-site to purify water drawn directly from surface waters or aquifers, or to desalinate seawater as in the case of some resorts in the Maldives.

Bathrooms, kitchen, laundry, toilets and HVAC (heating, ventilating and air-conditioning) need both cold and hot water. Cold water is needed for drinking and fire-fighting. Hot water is supplied at different temperatures for different purposes, such as 50°C for guest rooms, 40–80°C for laundry and 60°C for kitchen. Larger hotels will have separate boilers and storage tanks to heat and hold water supplies at different temperatures. An alternative is to use location-specific water heaters to increase temperatures of water as required. With much energy being used to heat and store hot water, reducing hot water consumption will also help to reduce energy costs.

Hotels produce significant amounts of wastewater, including graywater which mainly comes from washing machines, showers, sinks, baths and roof runoff, and blackwater which comes from dishwashing and toilets. Wastewater should be treated before being discharged into the environment. Yet, large volumes of wastewater are discharged without treatment because of the high cost of treatment or because there is no space for a biological treatment facility. Treated graywater can be re-used — for washing floors, flushing toilets and irrigating gardens and lawns.

Often, energy accounts for the largest share of operating costs in a hotel after payroll. It is used for lighting, ventilation, heating and cooling, and powering appliances. Most hotels draw their electric supply from the national grid, whose power in most cases is produced by the burning of

[12] World Travel & Tourism Council, International Federation of Tour Operators, International Hotel & Restaurant Association and International Council of Cruise Lines (2002), *Industry as a Partner for Sustainable Development - Tourism*, p. 36.

fossil fuels such as fuel oil, coal and natural gas, thereby contributing to local air pollution and global climate change. Besides electricity, hotels may also use other types of energy sources for cooking and to heat up water and space, such as fuel oil and natural gas.

The intensity of energy usage can vary according to building design and configuration, climate conditions, heating and cooling requirements, levels of occupancy and activity, energy sources used and the energy-efficiency of appliances. Air-conditioning is used to lower temperatures and humidity in hot climates, while in cooler climes, heating systems are required to raise temperature and control humidity. The potential savings linked to energy efficiency improvements can be significant. Regular repair and good housekeeping measures can reduce energy costs by 10% and retrofitting and refurbishment options may bring savings of at least 30% in some countries.[13]

To conserve energy, degree-day thresholds can also be used to guide property managers on cooling and heating required. Degree-days are a representation of outside air-temperature data which are used for calculations relating to the effect of outside air-temperature on building energy consumption. Comparing outside temperatures with the degree-day thresholds will indicate the levels to which buildings have to be cooled or heated.[14] Timers or controllers can be set according to these thresholds and level of activity inside the building.

Hotel Nikko Hongkong estimated that about a third of its guests forgot to turn off the room master switch which controlled all electrical outlets when leaving the room. The hotel replaced the button switch with a card-controlled master switch which would automatically switch off all electrical units when rooms were vacant. Hotel Nikko estimated that the key card system brought a saving of US$0.30 (1995) per day per room, with the cost of the new switch recovered within 70 days. Daily thermometer readings were also taken to ensure that the indoor temperatures were maintained at 20°C in summer and 21–22°C in winter. Water temperature was reduced from 60°C to 55°C - hot enough for personal use and to prevent legionella growth. The hot water boilers were switched off between 1.00am and 5.00am. These measures combined brought the hotel 6% reduction in

[13]EUHOFA International, International Hotel & Restaurant Association (IH&RA) and United Nations Environment Programme (UNEP) (2001), *Sowing the Seeds of Change: An Environmental Teaching Pack for the Hospitality Industry*, p. 78.
[14]ibid, p. 77.

electricity costs and more than 9% reduction in fuel costs in 1995, the year the measures were implemented.[15]

Hotels generate large quantities of solid waste including bottles, cans, food scraps, oils and fat, packaging, garden waste, paper, old furniture and equipment and potentially hazardous wastes such as batteries and solvents. Waste disposal costs money and because of this, it would make business sense for hotels to minimise waste by minimising the materials used in the first place, recycling and reusing waste materials wherever possible, and by safely disposing of residual wastes.

As a major consumer of goods and services, hotels can play a role in steering change in the supply chain by adopting a waste management strategy around 'avoid, reduce, reuse, recycle and recover'. Through their collective efforts, hotels can contribute towards increasing recycling volumes, driving down recycling costs and slowing the need for landfill and incineration. It is believed that corporate environmental stewardship can help to motivate employees and reduce employee turnover as well as encourage customer loyalty.

Hotels purchase large quantities of products - including food and beverages, cleaning and laundry supplies, toiletries, office supplies, equipment and bedding. The purchases can have negative environmental impacts through their manufacture, distribution, use and disposal. A purchasing decision may also impact on the amount of waste produced by a hotel and the potential of polluting air, water and soil through the use or disposal of these products.

A standard cake of hotel soap weighs 15g. An overnight guest uses about 3g and the rest is discarded. To reduce waste, Scandic Hotels replaced individual bars and bottles with dispensers made out of recycled plastic (polyethylene terephthalate). The dispensers contained soap and shampoos that used natural products and that were biodegradable.[16]

The hotels can help promote environmental and social improvements in manufacturing by purchasing products made entirely or partly with recycled materials, materials with low toxicity e.g. non-solvent cleaners, products manufactured through cleaner production processes e.g.

---

[15]International Hotel & Restaurant Association (IH&RA) and United Nations Environment Programme Industry and Environment (UNEP) (1996), *Environmental Good Practice in Hotels: Case Studies from the International Hotel & Restaurant Association Environmental Award*, p. 11.
[16]ibid, p. 30.

unbleached paper, products that are more durable, products that require less energy during manufacture and use, reusable or recyclable products, products with reduced packaging, environmentally certified products and products manufactured locally and that do not require long-distance transportation. The use of environmentally-preferable products can help enhance the hotel's reputation with guests and others who are concerned about wasteful or harmful product use.

Like most other businesses, hotels contribute to emissions into the atmosphere through the use of fossil fuels, ozone-depleting substances and the transportation of supplies. The hotel industry is also a significant user of energy. As this involves the burning of fuels, the hotel industry is a contributor to emission of greenhouse gases, and therefore global warming. Direct emissions from hotels include combustion gases from fossil fuel and gas-operated boilers, stoves and generators, coolants from refrigeration and air-conditioning equipment, halons in fire-extinguishing equipment, vapours from dry cleaning solvents and vehicular emissions.

Comfort is a subjective experience and attempts have been made to measure it in terms of temperature, air quality, ventilation, humidity and noise level. Indoor environmental quality depends on the activities and emissions within the hotel building and the pollutants brought into it from the air outside. Common sources of indoor air quality contamination include combustion gases from stoves, boilers and other combustion equipment, tobacco smoke, photocopy emissions, microbial contaminants, pesticides, dust and particles, radon released from building materials, and VOC (volatile organic compounds) vapours from cleaning solvents, paints and varnishes.

Indoor environmental quality is of great importance and can impact the health, comfort and well-being of guests and employees. Poor indoor air quality is known to induce adverse health effects, such as headaches, dizziness and even respiratory irritations and allergic reactions. Long-term exposure can induce more serious illnesses among hotel employees.

Fresh air is needed to renew oxygen and to remove microbial contaminants, allergens, vapours and odours, excess heat and moisture. As adequate ventilation is essential for maintaining good indoor air quality, some hotels use mechanical ventilation systems. Larger hotels may use integrated heat, light and ventilation systems which are more efficient and more feasible for larger properties.

Prolonged exposure to undesirable or irritating sound can be disturbing, annoying and even detrimental to health. Continuous exposure to noise levels above 70dB can induce migraines, circulatory disturbances, high blood pressure, and ulcers, and if above 130dB, can even cause permanent damage to the ear.[17] High levels of noise in a hotel business can affect the value of the property and lead to the loss of business. Noise affects the health of guests, impairs employee productivity and can also be a source of conflict with neighbouring businesses and the community. Undesirable noise in a hotel can include noise from traffic, crowds, vehicles and equipment. Areas that generate the most noise in hotel businesses are mechanical rooms with fans, compressors, boilers and generators, kitchens, laundries, delivery areas, waste output areas, driveways, garages, car parks, bars and clubs, function rooms and lobby areas. The most noise-sensitive areas are guestrooms, meeting rooms, conference rooms and offices. New hotels should be designed to segregate noise-sensitive areas from noise-producing areas. Existing hotels should conduct regular noise audits and invest in noise control facilities where necessary.

Hotel developments can have significant impacts on biodiversity and local communities. Environmental damage from a hotel development may degrade or restrict access to resources such as forests, water or farmland which local people depend on. In some cases, people may be displaced by hotels or other guest infrastructure, or be deprived of access to what was previously a public beach.

Local people who initially welcomed a hotel may find that the development failed to provide the benefits they had anticipated. Although hotels generate jobs, they often do not go to the ones who are most affected by the development, particularly those who lacked skills or start-up capital.[18] Revenue from a hotel may, largely, not make it to the local community because of foreign ownership or the use of imported goods and materials. Insensitivity to local cultures, beliefs, traditions and practices on the part of hotel guests or staff may cause tension or anti-tourist sentiment among local communities.

On the other hand, hotel developments can hasten the development of much-needed water, power and transport infrastructure. Some developers have learnt from experience and are well aware that poor siting, design,

[17]EUHOFA International, International Hotel & Restaurant Association (IH&RA) and United Nations Environment Programme (UNEP) (2001), op cit, p. 86.

[18]Center for Environmental Leadership in Business (CELB) and Tour Operators' Initiative for Sustainable Tourism Development (TOI) (2003), *A Practical Guide to Good Practice: Managing Environmental and Social Issues in the Accommodations Sector*, p. 16.

engineering and construction of their hotels can run counter to their long-term interests by destroying natural protective features such as dunes and vegetation cover and causing erosion, landslides and flooding.

In many places, the local culture and nature are the basis of a holiday. Contributing to conservation in these places can help preserve tourism resources and destination quality. Some have also attempted to harmoniously integrate the architecture or design of their facilities with the natural or architectural characteristics of the environs. However, much has to be said about the blurring of what is Asian and what is local, and the incongruity of finding a Balinese-style resort or a Thai spa retreat in, say, the Indian Ocean.

Nonetheless, the construction or refurbishment of a hotel can be an opportunity to discover local materials, consult local people and apply local techniques to construct buildings that befit the local landscape and climate. This was the case with a small development near Zanzibar, Tanzania. The Chumbe Island Coral Park employed local skills and traditional practices to construct 'eco-bungalows' with materials that have near-zero impact on the environment. The use of 'eco-technologies' such as composting toilets to handle sewage, photovoltaic cells to power lamps and a roof design that harvests rain for freshwater further ensures that the on-going operations of the 7-bungalow hotel had little impact on the environment.[19]

It ought to be realised, however, that hotel "developers do not operate in a vacuum".[20] The control of environmental impacts also hinges upon the ability of a local government to implement and enforce environmentally sound development policies. This is particularly relevant as hotel businesses continue to foray and expand into rural or remote locales, and yet uninhabited islands.

Environmental Performance Management in Singapore's Hotel Industry

Singapore received 8,942,408 international visitors in 2005 — an increase of 7% over 2004 and the highest number of visitor arrivals ever achieved

[19]Chumbe Island Coral Park, http://www.chumbeisland.com/Accomodation/accomodation. html, accessed 25 July 2006.

[20]World Travel & Tourism Council, International Federation of Tour Operators, International Hotel & Restaurant Association and International Council of Cruise Lines (2002), op cit, p. 36.

in Singapore.[21] Reeling from the near-sequential shocks of SARS, tsunami, avian flu, and bombings that impacted the Asian region in prior years, industry watchers were quick to attribute the growth to resilience in the industry and a healthy economic outlook in the region and beyond. Indeed, the high number of arrivals could be attributed to several factors.

The entrance of low-cost carriers spurred regional travel with increased capacity and lower airfares (including retaliatory promotions from full-service airlines). Vacation and business travellers accounted for 34% and 15% of the arrivals respectively. Several major MICE (meetings, incentives, conventions and exhibitions) events including the 117th International Olympic Committee Session contributed 5% to the total. On average, a non-resident visitor (including day-trippers) stays in Singapore for 3.4 days.[22]

All indicators also pointed to a good year for the Singapore hotel sector with revenue reaching S$1.2 billion and ADR (average daily rate) at S$137 - an increase of 15.1% and 12.3% respectively over 2004. Occupancy averaged at 84%, a 3 percentage-point increase over 2004. In a typical Singapore hotel, room rentals normally account for 60–70% of the hotel's revenue with the balance revenue coming from F&B. In a MICE hotel (with convention facilities), F&B (food & beverage) can account for 45–50% of revenue and rooms 50–55%.[23]

There is a total of 220 hotels with a total of 36,765 rooms in Singapore as at end-2004, out of which 102 were gazetted hotels[24] accounting for 30,300 available rooms.[25] Approximately 38% and 43% of the gazetted hotels can be arbitrarily classified[26] as 5-star (if average room rate exceeds S$155) and 4-star (if the room rate is S$70–154).[27] Most of Singapore's 'international-standard' hotels are found in Singapore's Central Business

[21]Singapore Tourism Board (2005), *Tourism Focus 2005*, December 2005, p. 1.

[22]Millennium & Copthorne Hotels plc (2006), *Prospectus dated 10 July 2006 relating to the Offering of 425,000,000 Stapled Securities in CDL Hospitality Trusts comprising CDL Hospitality Real Estate Investment Trust and CDL Hospitality Business Trust*, p. IV–6.

[23]Industry interviews on 16 March 2006 and 28 March 2006.

[24]Being gazetted under Singapore Tourism Cess Collection Act requires the hotel to pay a 1% levy on all cessable items. This is usually borne by the customer.

[25]Millennium & Copthorne Hotels plc (2006), op cit, p. IV–18.

[26]There is no official star rating in Singapore adopted by Singapore Tourism Board (STB) or Singapore Hotel Association. Hotels may at times pitch themselves at some star rating as part of their marketing efforts. STB tends to classify hotels according to size, price tiers or customer segments such as family-friendly, boutique, MICE or hip hotel.

[27]Millennium & Copthorne Hotels plc (2006), op cit, p. IV–19.

District, Orchard Road, Marina Centre, Civic District, Bras Basah and Bugis precincts, are affiliated to an international chain or a home-grown hotel brand, and tend to target tourists, business travellers and MICE travellers. Boutique hotels and budget-priced hotels are dotted around Chinatown, Little India and Geylang. Resort-style hotels are found on Sentosa Island.

Singapore is targeting 17 million visitors by 2015 which the country will attract with a slew of strategies that include redevelopment, refurbishment, destination branding, investment promotion, marketing targeted at growth segments such as medical tourism and MICE, as well as infrastructure developments such as two Vegas-style integrated resorts that aim to take Singapore leisure, meetings and gaming to new heights. Considerable construction is imminent as Singapore doubles its rooms and exhibition space[28] in 10 years.

There is also outward movement. Since 2000, outbound departures by Singapore residents have exceeded 4 million annually.[29] Not necessarily in response to Singapore's outbound traffic, Singapore hotel companies have also expanded their operations internationally. Over the last decade or so, Singapore-based companies have ventured abroad, to develop or operate hotels and resorts, often in less developed countries.

Interestingly, a number of the Singapore-owned international hotel companies do not have hotels in Singapore. Even though they are headquartered in Singapore, their hotels are not subject to Singapore regulations as the hotels are not located in Singapore.

On the other hand, hotels in Singapore are subject to various laws and regulations, in particular the Hotels Act (Cap 127) under which hotels in Singapore are required to be licensed. There are also other compliance requirements under other laws such as the Building Control Act (Cap29), Fire Safety Act (Cap109A), Environmental Public Health Act (Cap 95), Public Entertainments and Meetings Act (Cap 257), Customs Act (Cap 70) and Singapore Tourism (Cess Collection) Act (Cap 305C) and standards such as "Singapore Standard CP:13 1999 Code of Practice for Mechanical Ventilation and Air-Conditioning in Buildings" and "CP 38:1999 Code of Practice for Artificial Lighting in Buildings". Oddly, even

---

[28]Business Times, June 10-11, 2006, "Singapore's MICE Space Set to Double by 2009".
[29]Singapore Tourism Board (2004), *Annual Report on Tourism Statistics 2003*, Singapore: Singapore Tourism Board, p. 31. In 2003, a total of 4,221,464 departures took place by air, land and sea.

given the amount of hotel construction or refurbishment activity anticipated for the next decade, there still exists no regulation in Singapore that would require hotel developers to conduct environmental impact studies before construction. There has not been any programme, guideline or policy targeted specifically at promoting environmental management or performance measurement, and hence performance improvement, in the operation of hotels in Singapore.[30]

The National Environment Agency's primary concern in the hotel sector is oriented towards ensuring that food hygiene and waste disposal standards are maintained in the hotels, although Singapore hotels generally do not pose a problem in these areas. As a relatively non-pollutive industry, the hotel sector naturally does not warrant the type of priority accorded to pollution prevention in the manufacturing sector. Despite its high energy usage and significance to Singapore's economic future, neither the hotel industry nor the wider travel and tourism industry is represented in the National Climate Change Committee (NCCC) spearheaded by the Ministry of the Environment and Water Resources.[31] Neither is there any industry-driven initiative from Singapore Hotel Association (SHA) or industry-focused initiative from Singapore Tourism Board (STB) geared towards environmental performance.

Although there is no shortage of governmental agencies, non-governmental organisations, or industry associations in Singapore that are concerned with either the hotel industry or with the environment, there is not one that has taken the lead in advancing corporate environmental management or environmental performance measurement best practices in the hotel industry in Singapore.

There is, nonetheless, an attempt to identify good performers in energy efficiency. The Energy Sustainability Unit of the National University of Singapore has been working on benchmarking energy efficiency in office buildings and has extended the study to include hotel buildings. The benchmarking system is still under development.[32]

---

[30]Verifications made through interviews, face-to-face discussions, telephone interviews and written correspondences between January 2006 and July 2006. Verifications were also supported by review of information found of websites or in publications associated with these organizations.

[31]National Environment Agency (2006), http://app.nea.gov.sg/cms/htdocs/category_sub.asp?cid=71, accessed 28 July 2006.

[32]However, preliminary findings (March 2006) pointed to electricity consumption varying from 200 to 500 kWh/m^2/year in Singapore hotels.

A New Framework

Given this context, a hotel company in Singapore has considerable freedom in coming up with its own process for managing environmental activities and measuring corporate environmental performance. Although companies should be granted flexibility in approach, there is also a need to balance that freedom with a set of generally acceptable principles.

As exemplified by the move toward EMS Standards (Environmental Management System) and GRI Reporting Guidelines (Global Reporting Initiative), there are benefits in adopting a harmonised framework. A harmonised approach serves to provide an implementation roadmap for companies that intend to take the path to better environmental responsibility. It also provides a basis for companies to compare the level of development and identify any potential gaps in the process that need attention or improvement.

A common framework is proposed for managing and tracking environmental performance in Singapore-headquartered or Singapore-based hotel groups. The new framework, called *SPITRAc*, treats environmental management and environmental performance measurement as an integrated concept. The thinking is simple: without environmental management, there is no need for performance measurement; without performance measurement, there is no effective management. Clearly, a shift is required to move us from mere *environmental management* to *environmental performance management*. This is the objective of *SPITRAc*.

Fact-Gathering

A series of interviews were conducted with a number of upscale/upmarket hotel companies in Singapore to identify the environmental dimensions and performance parameters that concern Singapore hotels and hotel companies, as well as to gauge the level of adoption of environmental management practices and performance measurement in the hotel industry in Singapore. The objective was to provide a basis for the harmonisation of performance criteria and environmental performance indicators and also to obtain useful information for the design of the SPITRAc framework.

Given the increasingly global character of the hotel sector and the global significance of sustainability, internationally oriented (i.e. which operate across borders) and high-end hotels and resorts (i.e. luxury or deluxe categories) were selected for the study. They comprised the following types of hotels and hotel companies:

- Singapore-based hotel company with operations overseas but not in Singapore.
- Singapore-based hotel company with operations overseas and in Singapore.
- Singapore subsidiary of international hotel chain headquartered outside Asia.
- Singapore subsidiary of international hotel chain headquartered elsewhere in Asia.

The interviews were conducted using face-to-face discussions, telephone interviews or email exchange, or a combination of the three. The interviewees consisted of personnel from top management, marketing communications, business development and engineering departments. If the hotel company were headquartered in Singapore, the interview would involve someone from its corporate management team. Where the company was the Singapore hotel of an international hotel chain, the interviewee would be a senior manager of a particular functional department. This was in order to obtain, as close as possible, a corporate-level management perspective.

Almost all of the hotels which participated in the study have been highly placed or top-ranked in international tourism surveys or publications and have, between them, garnered a vast variety of awards and accolades of national and international standing.

The respondents were also asked if they preferred to be acknowledged for their participation. Most did not express a preference for being acknowledged in this report. Other hotels were also approached for the study but had declined participation. The research was conducted during between February 2006 and July 2006, inclusive. Eight hotel companies participated.

Results

Corporate commitment to environmental protection and environmental management

Most respondents (6 out of 8) revealed that their respective hotels had either an environmental policy or a mission statement that communicated the company's commitment to environmental protection or environmental management.

This is somewhat surprising as, in some cases, a review of some of these respondents' websites or company publications would not reveal any overt expression of the company's corporate environmental policy or environmental commitment - as least not as holistically oriented or to the extent of detail as the overarching statement on sustainable development by a hotel company like Accor or an elaboration of its policy guidelines and initiatives by a hotel like Hotel Nikko Hong Kong. If there was any, it would be an invitation to the reader to be immersed in the hotel's natural surroundings or a mention of the company's charity events, philanthropic gestures, or fundraising towards the community or nature conservation. Two of the respondents' websites do provide relatively more details - one on on-going programmes, albeit concentrated on conservation, and the other on the extent of adoption of certified environmental management practices among the company's worldwide operations.

Nonetheless, the fact that most of the respondents have indicated a commitment to environmental management is a piece of positive news, be it the existence of an internal communiqué that does convey a message of environmental commitment, or an evolved phenomenon where the mere inclusion of the word "nature", "environment" or "natural surrounding" in a company's positioning or marketing copy is sufficient to imply a rallying call for respect and doing more for the environment.

Presence of corporate environmental management plan

Only 2 respondents thought there were no corporate environmental plans in place. They were the same two respondents who did not think that an environmental policy or statement existed in their hotels.

The purpose of a corporate management plan is to reinforce the management commitment and to set out clear objectives, targets and plan of action. There could be varying interpretations as to what an environmental management plan was. Three respondents provided some elaboration. One of them referred to the objectives of Green Globe, which is a certification and improvement process that covers 9 environmental impact areas.

Use of environmental management system

Even with a policy and plan in place, the respondents do not use a certified environmental management system, e.g. ISO 14001 or EMAS to assist with environmental management. In fact, a check with the list of ISO

14001-certified companies[33] has revealed that no hotel or hotel company in Singapore has obtained the ISO 14001 certification.

One explained that it was adopting the Green Globe methodology while another pointed out that the ISO 14001 was the company's preferred standard for environmental management system although the Singapore hotel has yet to be certified. Another hotel uses a total quality management approach that included environmental aspects.

Potential impact areas for performance measurement

All the respondents indicated that environmental management was carried out with activities that fall into the various dimensions suggested in the interview questionnaire. Energy conservation, water conservation and using biodegradable cleaning materials are almost standard practices. Each dimension of management is practised by at least half of the respondents. Noise is normally not a major problem to some of these respondents especially in small developments. As the situation differs from one hotel to the other, the decision on which dimensions to manage is usually left to the hotel-level management.

Environmental performance indicators

The use of environmental performance indicators or quantitative measures is not commonly practised. Only one company alluded to the adoption of a standard set of environmental performance indicators. This was in consequence to its recent plan to have some of its hotels certified according to Green Globe requirements. Another hotel, which is well-regarded for its service quality, uses a variety of indicators that also include the customers in overall performance measurement by way of guest satisfaction surveys. Another hotel compares energy performance according to technical standards. It also carries out energy efficiency audits and is benchmarking the energy efficiency of the company's properties in Singapore, firstly with each other.

Raising awareness and support from immediate stakeholders

The preferred way to motivate staff to think of the environment in a proactive fashion is to have direct line meetings on a regular

[33]NEA (2006), http://app.nea.gov.sg/cms/htdocs/article.asp?pid=1134#b1, accessed 28 July 2006.

basis involving management and employees. This could be supported by training and a policy of empowerment to allow employees to take initiatives and make proactive contributions towards environmental performance. The organisation and participation in charity events by the company is also a good way to motivate staff involvement and alignment to environmental performance, and create awareness of the hotel's environmental performance in the community. There is also a need to demonstrate to management and staff the benefits of environmental management, such as cost savings. This can be done through the direct line meetings or by directly passing environmental costs back to the department. Guest awareness could be increased through specific programs or brand communications that encompass disclosing the company's environmental initiatives.

In a certain hotel, the employee's contribution toward environmental performance was linked to the individual's work performance review. For example, if one was found to leave a computer turned on at the end of the workday, a negative score could be recorded in his work report. The employee would not be penalised but consistent underperformance could likely affect his or her share of the wage incentive bonus at the end of the year. In the same hotel, employees were discouraged from using the electric lift if they were to move no more than 3 floors, unless they were accompanying guests. The wide deployment of resource-saving measures within the hotel, such as dimming devices and time switches at lift landings, would no doubt help to reinforce the environmental and cost-saving philosophy among the staff.

Benchmarking

There are reservations among the respondents to participate in environmental performance benchmarking and be benchmarked against competitors. The hesitation is mainly due to uncertainty over what the benchmarking entails in relation to its benefits. There are also reservations about how a company's information will be shared. In some cases, a market leader's best practices are regarded as the company's competitive advantage and its environmental leadership provides the basis for differentiation from other companies. These market leaders will have a motivation to safeguard their 'trade secrets'.

It ought to be remembered that the group of respondents is comprised of companies which have ranked well in tourism surveys. Hence, the incremental benefit of the environmental performance benchmarking has to be many times more than the effort the benchmarking entails.

Major challenges in implementation

The interview question about implementation challenges was framed around the issue of environmental management rather than specifically on environmental performance because of the need to understand the more fundamental challenges. The concept of environmental management system is new to many and the concept of environmental performance measurement system is even newer. It is assumed that the issues surrounding the roll-out of environmental management systems are similar or have a direct impact on the adoption of environmental performance measurement.

From the interviews, diversity appears to be the major challenge in rolling out any program in environmental management at a hotel company. Diversity of operations and the operating environment requires a flexible solution that can cater to all situations. There is also a diversity of education levels and a diversity of environmental awareness within the company that need to be addressed, so that everyone can work in tandem. There is also a diversity, and even divergence, of opinion over the benefit of environmental management between operating staff and non-operating staff, between hotel manager and hotel owner-developer, and between the environmental cause and the business need of meeting and exceeding guests' service expectations.

Implications and Limitations of the Research Findings

As this was not a survey where statistically appropriate random samples had been selected, the responses gathered in the study are by no means representative of the hotel industry as a whole or even of any particular segment within. Its purpose was to determine if there were any parameters that were already utilised in environmental performance measurement. The premise was that if there were already a widely used set, then their adoption of the performance indicators would better facilitate the better acceptance of a common framework for tracking corporate environmental performance. The types of performance indicators used would have also provided a clue to the environmental dimensions that are of relevance to the hotel industry in Singapore. This information could have been useful for short-listing and prioritising the dimensions for measurement in such a framework.

Unfortunately, the interviews revealed low utilisation of performance indicators among the participating hotels and limited clues about performance measurement. Nonetheless, if at all indicative of the situation, the interviews revealed some interesting points that can be useful in the development of a common framework for environmental performance measurement. They include:

- the potential aversion to industry benchmarking.
- the need for flexibility to cater to diverse situations and environments.
- the significance of direct line communication and staff-public interaction in motivating staff towards good environmental behaviour.
- the low adoption of a structured approach in environmental management (e.g. EMS) or environmental performance measurement.
- the presence of some level of belief in the need for, or some degree of management commitment to, environmental management and environmental good behaviour.

There could be an environmental management plan or policy in place, but the plan or policy had not been implemented — at least not fully and not systematically implemented, monitored or measured.

Some of these observations have formed the basis of the new framework proposed for corporate environmental performance management in a Singapore hotel company. The new framework is called *SPITRAc* referring to the seven sub-processes of the framework, namely 'State', 'Plan', 'Implement', 'Track', 'Recommend', 'Announce' and 'Compare'.

1. State

From the interview results, it appears that some kind of intent to perform environmental action is already present in some hotel companies, at least among the upscale hotels interviewed. 'State' is a top-management sub-process that begins with an official declaration of a corporate environmental policy both within the company and to external parties. The internal declaration serves to align management and staff towards planning, acting and working towards a common environmental purpose and better overall environmental management by the company. The policy must also be appropriate to the overall corporate strategy and direction of the hotel company and can be understood by all employees in the company. It is in the interest of the corporate office of the company to make available translated versions of its policy, in local languages, to its hotel operations

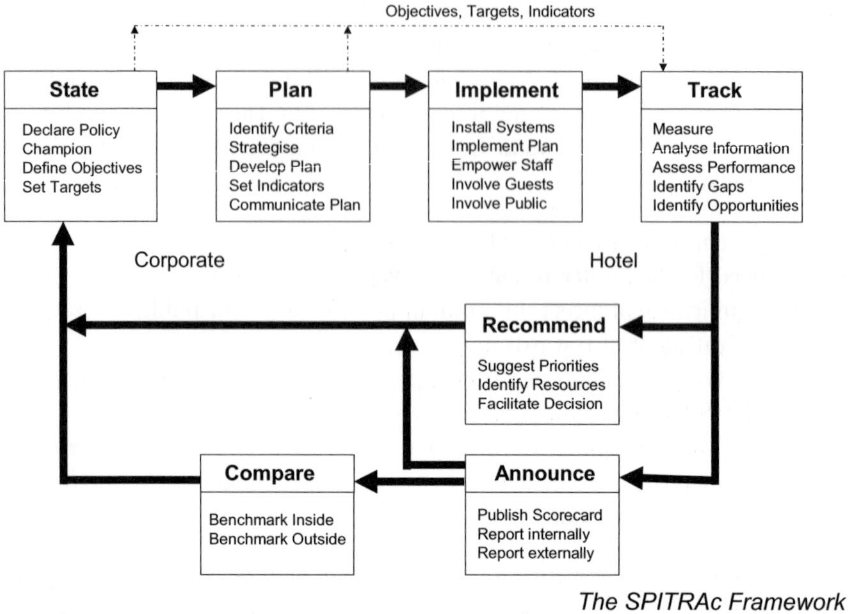

Fig. 2.1. A new framework for corporate environmental performance management.

outside Singapore. An environmental champion is also appointed in this sub-process. To signal corporate-level commitment, the Champion should be someone from top management or a newly appointed position within the corporate executive team.

2. Plan

The 'Plan' sub-process is where the corporate environmental policy, objectives and targets get developed into a plan of action. Based on an assessment of the current performance and targets, the environmental performance criteria are determined and, if necessary, harmonised. The criteria need to be established at both the corporate and hotel level. The Champion will then identify and develop environmental strategies that must be deployed to meet the performance criteria.

The strategies may include a strategy to be the first-mover in adopting a new environmental technology that is being developed, a strategy to change to a corporate culture that is oriented toward environmental improvement by using management training and staff coaching, the repositioning of the company's brand to a particular customer segment that is more aligned to the environmental objectives, making changes to the organisational structure, incorporating environmental activity into staff performance assessment and strategy for proactive material substitution.

3. Implement

The action plan developed in the earlier sub-process is put in place in this 'Implement' sub-process. At the hotel, a plan may be 'operationalised' into different actions, e.g. the plan calling for a reduction of piped water consumption by 20% in 5 months may involve multiple actions such as installing volume reducers in toilet cisterns, harvesting rainwater, treating and reusing graywater, repairing water leaks, running washing machines only when fully loaded and placing tent cards in bathrooms inviting guests to save water.

To assist the implementation, the company may want to develop and implement an environmental management system that is appropriate to the size and scope of the company's operations. The system may be implemented in some hotels first and progressively implemented throughout the network. Such a system can help a hotel to evaluate, manage and control the negative impacts by providing a methodology to integrate environmental management into the hotel's operations in a systematic manner.

4. Track

The 'Track' sub-process involves the monitoring and measuring of data that can be used to calculate key environmental performance indicators (KEPIs). Assessing the performance will entail comparing the KEPIs to the performance criteria as determined in the 'Plan' sub-process, as well as identifying performance gaps and opportunities where performance could be improved.

5. Recommend

This 'sub-process' is the principal objective of performance tracking. Performance tracking is useless if it is not used to facilitate management decision-making and to assist with driving future action. The SPITRAc Framework requires that the hotel management which tracks and analyses KEPIs to be empowered to make recommendations for improvement, to recommend prioritised activities and to identify resources that are required for performance improvement.

6. Announce

Environmental performance reporting and disclosure of quantitative data also has a role to play in encouraging environmental performance. To that end, SPITRAc treats performance reporting as an integral part of the framework. Environmental performance can be reported using a number of methods such as a newsletter, press release, a section in the financial report, or a stand-alone corporate environmental report. External reporting will help in building dialogue or communication with legislators, local communities and non-governmental organisations and also serves as an indicator of the company's environmental commitment. For long-term credibility, the company may also want to work towards adopting the Global Reporting Initiative (GRI) Framework and GRI's "In Accordance" scheme. The report must include quantitative data to ensure greater transparency.[34]

7. Compare

This is the sub-process where the company's results are compared with benchmarks. This is an optional phase, given the current aversion to benchmarking against competitors in the same industry. Just like airline companies which have an aversion (and an unwritten understanding) against being pitted against each other on safety issues, hotel companies in Singapore may also have a similar aversion against being pitted against each other on environmental performance. The 'Compare' sub-process in the SPITRAc Framework does not entail ranking of hotels in Singapore in terms of environmental performance. However, for a multi-site hotel company, the SPITRAc Framework encourages internal benchmarking to be performed so that the company's best practices could be shared.

[34]Accor (2006), *Accor Hotels Environment Charter Practical Guide*, pp. 2–62; EUHOFA International, International Hotel & Restaurant Association (IH&RA) and United Nations Environment Programme (UNEP) (2001), *Sowing the Seeds of Change*: An *Environmental Teaching Pack for the Hospitality Industry*, pp. 59–207, Global Reporting Initiative (GRI) (2002), *Sustainability Reporting Guidelines*, pp. 1–84; International Hotel Association (IHA), International Hotels Environment Initiative (IHEI) and United Nations Environment Programme Industry and Environment (UNEP) (1995), *Environmental Action Pack for Hotels: Practical Steps to Benefit Your Business and the Environment*, pp. 9–64; www.greenglobe21.com.

Developing Indicators for Performance Tracking

The interview findings showed that although there was some level of observance of environmental management, the use of quantitative measures or performance indicators was limited among the participating hotels. This limited the objective of the interviews to derive a basis for the harmonisation of environmental performance indicators. An alternative approach was attempted, by comparing the key criteria used by other organisations or hotel chains overseas, in order to draw up a set of indicators that could be relevant to hotels in Singapore.

From Table 2.1, it is clear that there is some variation in the way organisations categorise their environmental performance indicators and environmental management action items. There are however some themes which have appeared consistently among all the categorisations, namely water management, energy efficiency and waste management. One would assume, therefore, that these three are the most major, or most common, environmental impact areas in the hotel industry. These categorisations could be considered, among others, to select key environmental indicators that may be used with SPITRAc. Purchasing (or suppliers), wastewater management and emission control are also featured rather consistently.

Establishing an Environmental Code of Conduct for the Hotel Industry in Singapore

The inherent characteristic of an industry-specific code of conduct also fits into the underlying philosophy of the SPITRAc Framework. The framework is intended as a guide to help Singapore-based and Singapore-managed hotel companies to develop a system to manage the environmental performance of their local or international hotels. The concept of a voluntary, self-regulating initiative is relevant in an industry that transcends borders and laws.

The framework by itself is insufficient to ensure improved performance by the whole industry. The SPITRAc Framework is only one component within a wider context. There are other essential elements. For it to be adopted as an industry-wide framework, it would require both a mindset shift in the industry as well as a push from within. Establishing an environmental code of conduct for the hotel industry may help to propagate holistic environmental approaches. A code of conduct is meant to be simple and clear, but establishing the code may not be.

Table 2.1. Categorisations of environmental dimensions or action areas.

KEY IMPACT AREAS **Applicable to the travel and tourism industry** Greenhouse gas emissions Energy efficiency, conservation and management Management of freshwater resources Ecosystem conservation and management Management of social and cultural issues Land use planning and management Air quality protection and noise control Waste water management Waste minimisation, reuse and recycling *Green Globe programme (www.GreenGlobe21.com)*	**KEY ACTION AREAS** **Applicable to the hospitality industry** Reducing water use Reducing energy use Reducing waste output Purchasing environmentally-preferable products Lowering emissions Improving indoor air quality Reusing waste water Reducing noise Internal communication, delegation and training Communicating environment performance to guests Monitoring and documenting environment performance *EUHOFA International, IH&RA and UNEP (2001)*
ENVIRONMENTAL ASPECTS FOR SUSTAINABILITY REPORTING **Applicable to all industries** Materials Energy Water Biodiversity Emissions, effluents and waste Suppliers Products and Services Compliance Transport Overall *Global Reporting Initiative (2002)*	**KEY ACTION AREAS** **Applicable to the hotel industry** Energy Solid waste Water Effluents and emissions Contractors and suppliers Business issues *IHA, IHEI and UNEP (1995)*
KEY ACTION AREAS **Applicable to hotels in Accor Group** Informing and raising awareness Energy Water Wastewater Waste Ozone layer Biodiversity Green purchases *Accor Group (2006)*	

Table 2.2. Initial list of environmental dimensions (used when conducting the interviews).

Power conservation
Fuel conservation
Water conservation
Wastewater treatment
Waste minimization
Environmentally responsible purchasing
Emissions and effluent prevention or control
Indoor environment/air management
Noise management
Nature conservation
Raise employee awareness
Raise guest awareness
Other environmental dimension

It is not the goal of this study to draft a code of conduct for the Singapore hotel industry. The aim here is to examine the various issues that need to be considered in establishing such a code for the industry.

On the basis that environmental performance is the objective and SPITRAc is the framework, then the content of the code will include a principle that requires the hotel to declare its senior management commitment, to define and communicate its environmental policy, define its performance management objectives, to adopt systematic approaches and technologies in environmental management and performance management, to make a commitment to train and empower staff for implementing performance improvements, to engage stakeholders (including staff, guests and community), and to make a commitment to regularly track and evaluate its performance.

The code may also include a principle to remind the hotel to go beyond being compliant to regulations and generally accepted standards, and to exceed them. It may also extol the benefits of reporting with words that are supported by quantitative measures, and to encourage the hotel to exercise openness in disclosure as well as to being open to public opinion and feedback. The code may also elaborate on the performance criteria and key environmental action areas, and the need for the hotel to explore new areas of improvement. The code may also allude to the benefits of comparing one's performance with its peers within and outside the company's network and encourage the hotel to do so.

Theoretically, there is no shortage of organisations in Singapore which can take up the cause of encouraging corporate performance management.

However, to see the whole process from inception at a hotel to implementing it in the industry and through implementation across industries and borders would probably require a prerequisite amalgamation of the disperse resources that already exist into a consolidated force that has the clout and ability to proactively take a leadership position in driving environmental leadership behaviour in the Singapore hotel industry.

The code is not a prerequisite to improved performance, particularly for leading companies which are ahead of the norm that the code of conduct is encouraging. Signing onto a code may demonstrate the company's awareness of the code but it does not mean that the company has actively implemented the principles of the code. It is the company's statement of environmental commitment or policy, implementation of environmental systems or announcements of environmental achievements that will indicate if the principles of the code have been implemented. This is where *SPITRAc* comes in.

References

Accor (2006), *Accor Hotels Environment Charter Practical Guide*, Paris: Accor Group Sustainable Development Department, 2–62.

Ball, S., P. Jones, D. Kirk and A. Lockwood (2003), *Hospitality Operations: A Systems Approach*, London: Continuum, 30–32, 85–110.

Business Times, June 10-11, 2006, "Singapore's MICE Space Set to Double by 2009".

Case, P. (1999), *Environmental Risk Management and Corporate Lending: A Global Perspective*, Cambridge: Woodhead Publishing, 9–13.

Center for Environmental Leadership in Business (CELB) and Tour Operators' Initiative for Sustainable Tourism Development (TOI) (2003), *A Practical Guide to Good Practice: Managing Environmental and Social Issues in the Accommodations Sector*, Washington, DC: CELB and Paris: TOI, 2–20.

Crespí-Cladera, R. and F. Orfila-Sintes (2005), "Environmental Innovation in the Hotel Industry of the Balearic Islands", in Sharma, S. and J.A. Aragón-Correa (eds.), *Corporate Environmental Strategy and Competitive Advantage*, Cheltenham: Edward Elgar Publishing, 139–156.

EUHOFA International, International Hotel & Restaurant Association (IH&RA) and United Nations Environment Programme (UNEP) (2001), *Sowing the Seeds of Change:* An *Environmental Teaching Pack for the Hospitality Industry*, Brugge: EUHOFA, Paris: IH&RA and UNEP, 59–207.

Global Reporting Initiative (GRI) (2002), *Sustainability Reporting Guidelines,* Amsterdam: GRI, 1–84.

International Hotel Association (IHA), International Hotels Environment Initiative (IHEI) and United Nations Environment Programme Industry and Environment (UNEP) (1995), *Environmental Action Pack for Hotels: Practical Steps to Benefit Your Business and the Environment*, Paris: IHA and UNEP, London: IHEI, 9–64.

International Hotel & Restaurant Association (IH&RA) and United Nations Environment Programme Industry and Environment (UNEP) (1996), *Environmental Good Practice in Hotels: Case Studies from the International Hotel & Restaurant Association Environmental Award*, Paris: IH&RA and UNEP, 5–51.

Kassinis G.I. and A.C. Soteriou (2005), "Greening Service Organisations: Environmental Management Practices and Performance", in Sharma, S. and J.A. Aragón-Correa (eds.), *Corporate Environmental Strategy and Competitive Advantage*, Cheltenham: Edward Elgar Publishing, 115–138.

Millennium & Copthorne Hotels plc (2006), *Prospectus dated 10 July 2006 relating to the Offering of 425,000,000 Stapled Securities in CDL Hospitality Trusts comprising CDL Hospitality Real Estate Investment Trust and CDL Hospitality Business Trust*, Singapore: Millennium & Copthorne Hotels plc, 1–31, 111–135, IV6–19.

National Environment Agency (2006), *List of ISO 14001 Certified Companies*, Singapore: NEA, http://app.nea.gov.sg/cms/htdocs/article.asp?pid=1134#b1, accessed 28 July 2006.

Pine, R. and F.M. Go (1996), "Globalisation in the Hotel Industry", in Kotas, R., R. Teare, J. Logie, C. Jayawardena and J. Bowen (eds.) *The International Hospitality Business*, New York, NY: Cassell, 96–103.

Singapore Tourism Board (2004), *Annual Report on Tourism Statistics 2003*, Singapore: Singapore Tourism Board, 1–60.

Singapore Tourism Board (2005), *Tourism Focus 2005*, December 2005, Singapore: Singapore Tourism Board, 2–33.

World Travel & Tourism Council (WTTC), International Federation of Tour Operators (IFTO), International Hotel & Restaurant Association (IH&RA) and International Council of Cruise Lines (ICCL) (2002), *Industry as a Partner for Sustainable Development — Tourism*, London: WTTC, IFTO, IH&RA and ICCL, 11–42, 59–68.

3

MEASURING CORPORATE ENVIRONMENTAL PERFORMANCE: A DELPHI STUDY ON RANKING ENVIRONMENTAL PERFORMANCE OF COMPANIES IN SINGAPORE

MANTAZH KHANNA, LYE LIN HENG and AUDREY CHIA

Corporate environmental management in Singapore is not considered a new trend. As with many other developed nations, Singapore prides itself on its environmental achievements and its businesses have practiced sound environmental management for many years, much of it before most economies in the region. Many government and non-government environmental initiatives currently exist in Singapore and several companies have environmental management systems in place. However, unlike other environmentally-proactive countries, Singapore does not have an extensive benchmarking system that gathers, measures and ranks corporate environmental performance. This dissertation examines why it is so and whether there exists a demand for a corporate environmental performance ranking system or index in Singapore.

The thesis commences with an insight into environmental management and methodologies employed to measure corporate environmental performance. It also highlights viable ranking systems existing worldwide. To understand the Singapore context, a Delphi study comprising of seventeen experts is conducted through two rounds of questionnaires. Existing and future drivers of corporate environmentalism in Singapore are first explored. Subsequently, the feasibility of a corporate environmental performance ranking system, along with its design and execution considerations as well as advantages and disadvantages are studied in detail. The results indicate that a demand does exist and that most experts believe the benefits of such a system outweigh its costs in the long run. The thesis recommends a coordinating structure of possible involved bodies and highlights the importance of clearly defining the purpose and context of the system as the starting point. It concludes with the hope of instigating a future Corporate Environmental Performance Index which could be further expanded to include social and economic performance factors under an all-encompassing Corporate Social Responsibility Index.

Introduction

Businesses today are increasingly facing the challenge of behaving in a responsible manner. Corporate Social Responsibility (CSR) has been defined as "the integration of business operations and values whereby the interests of all stakeholders, including customers, employees, investors and the environment are reflected in the organization's policies and actions".[1] It has also been defined as "the obligation of the firm to use its resources in ways to benefit society, through committed participation as a member of society, taking into account the society at large, and improving welfare of society at large independently of direct gains of the company".[2] CSR is generally seen as the business contribution to 'sustainable development'[3] and is understood as focusing on how to achieve the integration of economic, environmental, and social imperatives. It also overlaps and often is synonymous with many features of other related concepts such as corporate sustainability, corporate accountability, corporate responsibility, corporate citizenship, corporate stewardship, etc. Although there are a number of definitions, each one takes into account the *social* and *environmental* impacts of corporate activity when making decisions. For the purpose of this study, the environmental impacts of business will take center stage.

The purpose of this study was to investigate and contribute a new body of knowledge in the realm of corporate environmental management in Singapore. It aims to serve as a stimulus for further research and development of a Corporate Environmental Performance (CEP) ranking system or index in Singapore in the future. The research questions are:

1. What are the current and future drivers of corporate environmental management in Singapore?
2. Is there a demand for the development of a corporate environmental performance ranking system in Singapore? Do experts and key industry players foresee such a system being adopted?
3. If so, what may be some advantages and disadvantages of such a system?
4. How should the system and its coordination be designed in order to make it acceptable to all?

This study argues that Singapore will need a corporate environmental performance rating system in the near future as the drivers for corporate sustainability increase locally and internationally. With environmental responsibility now becoming a key performance indicator underpinning

[1] Smith, K. "ISO considers Corporate Social Responsibility Standards". *The Journal for Quality and Participation* 25.3 (2002): 42.
[2] Kok, P.T., Wiele, R., McKenna and Brown, A. "A Corporate Social Responsibility Audit Within a Quality Management Framework". *Journal of Business Ethics* 31.4 (2001): 285–297.
[3] Defined by the Bruntland Commission as "development that meets the needs of the present without compromising the ability of future generations to meet their own needs".

management practices of the world's leading companies, it would be beneficial for Singapore to have an accurate system for tracking performance in this area. In addition to ranking, comparing and recognizing environmentally responsible companies, the index would also provide companies with a unique platform to communicate good environmental practices. The design and implementation of an uncomplicated, impartial, and credible ranking system will provide momentum and prominence to Singapore's economic and environmental progress. The study also hopes for the possible future expansion of such a system to include corporate social and economic performance under an all-encompassing "Corporate Social Responsibility Index".

As environmental sustainability takes greater precedence in the Asia-Pacific region, environmentally-proactive programs and initiatives are being increasingly introduced at all levels of the public and private sector. With Australia's recent introduction of the Corporate Social Responsibility Index, Japan's booming ISO14001 certification and Green Reporting movements, and Hong Kong's proliferating socially responsible investment funds, Singapore is surrounded by other developed economies in the region that are making great leaps in the realm of corporate sustainability. This study puts forth a much needed *Singapore* perspective to the state and drivers of CSR by presenting the collected opinions of local experts and key industry players. It also focuses specifically on the 'environmental' aspects of CSR so as to illustrate how companies in Singapore manage their environmental impacts. As a small-island nation balancing between its limited supply of land and natural resources and increasing population and industrialization, Singapore requires its businesses to maintain a leadership role in protecting the environment, by continually seeking opportunities to improve their environmental performance.

Methods

In order to study the Singapore perspective, the Delphi methodology was deemed appropriate as it provided a forum to gather expert opinion without the disadvantages of face-to-face discussions or random mass surveys. Each Delphi member of the panel was selected based on his/her expertise, experience and influence in Singapore's business-environment endeavors. The Delphi was conducted through two rounds of questionnaires and personal interviews. The first questionnaire (answered by 17 experts) served as a preliminary assessment of expert opinions on the need and design of a corporate environmental performance ranking system in Singapore. The second (answered by 14 of the 17 experts) provided members with a summary of the opinions from round one and allowed the experts to reassess their views. A final summary was then sent to all panel members.

The objectives of the two rounds were as follows:

Round 1

- To brief members about the study and its objectives;
- To collect preliminary information on whether CEP ranking is needed in Singapore;
- To learn details of how such ranking should be conducted; and
- To foresee the advantages and disadvantages of CEP ranking in Singapore.

Round 2

- To inform members of results of previous data analysis;
- To re-question whether CEP ranking is needed in Singapore; and
- To foresee the development of such a ranking instrument in Singapore.

Results

Round one involved 17 experts on the panel, of whom a total of 14 responded during round two (i.e. participation rate of 82.4%). This section strives to indicate the status of corporate environmental management in Singapore, and to present the views of the 17 Delphi experts on the development of a CEP ranking system. Each question included in the two rounds is analyzed consecutively, under separate thematic sections.

(1) *Present local drivers of corporate environmental management (CEM)*

Through the first question of Deplhi round one, experts were asked to rank 5 drivers of CEM in Singapore in order of importance (where 1 = most important) from a list of available options. The results showed the following:

- The most frequently selected drivers (in decreasing order) were Standards, Cost Savings, Competitive Advantage, Legislation and Branding (see Table 3.1 below). These drivers were selected by over half of the respondents.

Table 3.1. Most frequently selected present drivers of corporate environmental management in Singapore.

Driver	No. of Respondents	% of Respondents
Standards	14	82%
Cost Savings	12	71%
Competitive Advantage	11	65%
Legislation	10	59%
Branding	9	53%
Reputation & Trust	7	41%
Reduced Business Risk	4	24%
Other (e.g. Corporate Culture, Staff Driven)	4	24%
Green Consumerism	3	18%
Employee Motivation/Recruiting	2	12%
Community Pressure	2	12%
NGO/Media Pressure	0	0%

These results show that CEM in Singapore is strongly driven today by 'Opportunities' such as Cost Savings, Competitive Advantage and Branding, and by 'Threats' such as Legislation and Standards. NGO/Media Pressure was not viewed as a driver by any expert. Other drivers such as Corporate Culture and Staff Initiatives were also mentioned by 4 respondents.

- Furthermore, a per-rank analysis of each driver was also conducted. Based on the assigned rankings by the experts, individual drivers were assigned a score. An inverse scoring system with points was established, i.e. Rank 1 = 5 points, Rank 2 = 4 points, Rank 3 = 3 points, Rank 4 = 2 points, and Rank 5 = 1 point. A total score based on the frequency of responses multiplied by the ranking score was calculated. This score enabled accounting the weight of individual ranks for every driver. See Table 3.2 below for results.

  The total scores above suggest that opportunities drive CEM in Singapore more than threats do. However, when considering individual factors' scores, Standards and Legislation seem to be stronger drivers than any Opportunity. Nevertheless, it is important to note that most other Opportunities have a higher score than other Threats, hence it can be concluded that overall, opportunities seem to be the stronger drivers.
- Finally, a per-sector analysis was conducted to determine the level of agreeability between respondents representing the four sectors, i.e. Industry, Government, NGO and Consulting. 17 experts from senior management in Government (2 respondents or 12%), NGO

Table 3.2. Total scores for present drivers of CEM in Singapore.

	Score
Opportunities	
Cost Savings	40
Competitive Advantage	30
Branding	22
Other (Upper Mgmt/Corporate Culture/Staff Driven)	14
Reputation & Trust	12
Employee Motivation/Recruiting	5
Total	123
Threats	
Standards	51
Legislation	41
Reduced Business Risk	6
Green Consumerism	5
Community Pressure	5
NGO/Media Pressure	0
Total	108

(5 respondents or 29%), Consultancies (3 respondents or 18%) and Industry (7 respondents or 41%) were involved.

- Experts from all four sectors selected Cost Savings, Competitive Advantage, Legislation, Standards and Reputation & Trust as drivers for CEM in Singapore. All sectors, except Consulting, selected Branding as another driver. Only NGOs and Industry viewed Community Pressure and Green Consumerism as drivers, whereas only Industry and Consulting viewed Reduced Business Risk as a driver. Employee Motivation & Recruiting was ranked by Industry only.
- The per-sector results are consistent with the ranking scores as they too highlight the importance of Cost Savings, Competitive Advantage, Legislation and Standards as dominant drivers of CEM in Singapore. In addition Reputation & Trust was also deemed as an important driver.

(2) *Future drivers of corporate environmental management*

The second question invited experts to rank 5 future drivers of CEM in Singapore in order of importance (where 1 = most important) from a list of available options. The results and comparison to question 1 showed

Table 3.3. Most frequently selected future drivers of corporate environmental management in Singapore.

Driver	No. of Respondents	% of Respondents
New Legislation	12	71%
Competitive Advantage	11	65%
New Standards	10	56%
International Pressure	7	41%
Cost Savings	7	41%
Reputation & Trust	6	35%
Rankings/Indices	5	29%
Reduced Business Risk	4	24%
Branding	4	24%
Other	4	24%
Community Pressure	3	18%
Green Consumerism	3	18%
NGO/Media Pressure	2	12%
Employee Motivation/Recruiting	1	6%

the following:

- The most frequently selected drivers (in decreasing order) were New Legislation, Competitive Advantage and New Standards (see Table 3.3 below). These drivers were selected by over half of the respondents. However, experts also felt that Cost Savings and International Pressure may serve as compelling drivers. Other drivers such as Upper Management and Corporate Culture were also mentioned by 4 respondents.
- A comparison of present and future drivers by percentage of respondents (as shown in Table 3.4 below) indicates that new Legislation may become a stronger driver of CEM in Singapore in the future (an increase by 12%), while new Standards and Cost Savings may be less motivating (a decrease by 26% and 30% respectively). Furthermore, Competitive Advantage may remain just as strong a driver in the future as it is today. Other threats such as Community Pressure and NGO/Media Pressure may also press companies to manage their environmental impacts (an increase by 6% and 12% respectively). In addition, other drivers such as Branding, Reputation & Trust and Employee Motivation were predicted to decrease while drivers such as Reduced Business Risk and Green Consumerism were predicted to remain constant.
- Scores assigned by per-ranking weight, as demonstrated previously, show that in the future, Threats may be greater drivers of CEM in Singapore than Opportunities (see Table 3.5 below). Both New Legislation and

Table 3.4. Comparison of drivers of CEM in Singapore: Present & Future.

Driver	% of Respondents Q1 (Present)	% of Respondents Q2 (Future)	Change between Rounds
Standards	82%	56%	↓
Cost Savings	71%	41%	↓
Competitive Advantage	65%	65%	↔
Legislation	59%	71%	↑
Branding	53%	24%	↓
Reputation & Trust	41%	35%	↓
Reduced Business Risk	24%	24%	↔
Other	24%	24%	↔
Green Consumerism	18%	18%	↔
Employee Motivation/Recruiting	12%	6%	↓
Community Pressure	12%	18%	↑
NGO/Media Pressure	0%	12%	↑
Rankings/Indices	N.A.	29%	N.A.
International Pressure	N.A.	41%	N.A.

Table 3.5. Total scores for future drivers of CEM in Singapore.

	Score
Opportunities	
Competitive Advantage	29
Cost Savings	19
Branding	13
Other	13
Reputation & Trust	11
Rankings	9
Employee Motivation/Recruiting	2
Total	96
Threats	
New Legislation	59
New Standards	44
International Pressure	23
Community Pressure	12
Reduced Business Risk	8
NGO/Media Pressure	7
Green Consumerism	6
Total	159

New Standards may be the major driving forces, followed closely by Competitive Advantage and International Pressure.

These results were compared to those obtained from question one in to order detect any changes. A comparison of results, shown in Table 3.6

Table 3.6. Comparison of scores for drivers of CEM in Singapore: Present & Future.

	Score Q1	Score Q2	Change between Questions
Cost Savings	40	19	↓
Competitive Advantage	30	29	↓
Branding	22	13	↓
Other	14	13	↓
Reputation & Trust	12	11	↓
Employee Motivation/Recruiting	5	2	↓
Standards	51	44	↓
Legislation	41	59	↑
Reduced Business Risk	6	8	↑
Green Consumerism	5	6	↑
Community Pressure	5	12	↑
NGO/Media Pressure	0	7	↑
Rankings	n.a.	9	n.a.
International Pressure	n.a.	23	n.a.

below, demonstrates once again that Threats may be stronger drivers of CEM in the future than Opportunities. Threats such as new Legislation, Reduced Business Risk, Green Consumerism, Community Pressure and NGO/Media Pressure may increase. Rankings/Indices (e.g. Dow Jones Sustainability Indiex, Footsie4Good) and International Pressure may also serve as new drivers.

- Per-sector analysis shows that the drivers that were selected by all four sectors remain unchanged, i.e. New Legislation, New Standards, Cost Savings, Competitive Advantage and Reputation & Trust. All sectors except Government viewed International Pressure as a future driver of CEM in Singapore, whereas all sectors except Consulting viewed Branding as the same. Only Industry and Government selected Community Pressure and NGO/Media Pressure as future drivers, whereas only NGOs and Industry selected Green Consumerism and Rankings/Indices as the same. In addition, Reduced Business Risk was selected by Industry and Consultancies only in both questions. Employee Motivation & Recruiting was also selected by Industry only.
- In summary, the results of question one and two demonstrate that the strongest drivers of corporate environmental management today, i.e. legislation, standards, costs savings and competitive advantage, will remain strong in the future. Reputation and trust will also emerge as strong drivers in the future as society will continue to urge companies to balance business operations and profits with environmental responsibility.

(3) *Demand for CEP ranking system in Singapore*

Respondents were asked the following questions to determine the demand for a CEP ranking system in Singapore:

R1 Q3. Should environmental performance of companies in Singapore be ranked today?
R1 Q6. Will Singapore adopt a ranking system in the future?
R1 Q5 & R2 Q4. Will Singapore need a ranking system in the future?
The results of the first question, proposed in Delphi round one, are shown below.

76% of experts advocated (i.e. Agreed or Strongly Agreed) for environmental performance of companies in Singapore to be ranked, while only 24% were unsure. None disagreed/strongly disagreed.

The results indicate a strong support for such a system to be in place as most experts felt that transparency in corporate performance is needed and that only when performance is measured and ranked is it managed and valued respectively. It was also felt that such a ranking system will enable investors and stakeholders make better informed decisions and drive improvement through market pressures by pressurizing poor performers. Increased public awareness, the sharing of best practices within industry and possible global comparison were amongst other stated reasons. A Consultancy expert also felt that companies were already addressing environmental issues in their day-to-day practices without realizing that they could be recognized for their initiatives, thus a ranking system could provide them with an appropriate platform for such acknowledgment. However, an Industry expert furthered the aforementioned statement by stating that while companies in Singapore are becoming more aware of environmental management issues they may not be advanced enough to employ environmental management practices to warrant a CEP ranking.

Amongst the experts that opted for "Unsure", most expressed reservations about the complexity and equity of the design and execution of the system (discussed in the next section). Suggestions for an intra-sector analysis and a separate comparison for Small & Medium Enterprises (SMEs) and Multinational Companies (MNCs) were put forth. One Industry expert felt that as businesses in Singapore moved away from mainly manufacturing practices, they had relatively low environmental impacts. Hence there may not be much use of a ranking system today for such sectors.

When asked whether Singapore will adopt a CEP Ranking system, 59% of respondents Agreed or Strongly Agreed, while 41% remained Unsure. None Disagreed or Strongly Disagreed. These results demonstrate that most experts predict a CEP ranking system as a possibility in Singapore's future. All experts were in agreement that motivations for such a system will stem from Singapore striving to remain globally competitive and following international ranking trends (e.g. FTSE4Good, DJSI). Some also felt that Singapore could use such a ranking system — the first in Asia — to brand itself regionally and internationally. Others felt that it may only be a matter of time as the idea was already being discussed amongst expert circles and a small-scale preliminary system is currently being developed by the Singapore Environment Council to help the judgment process for selecting winners of the Singapore Environmental Achievement Award.

Experts that were Unsure expressed uncertainty over whether Singapore can overcome certain barriers such as corporate participation and fair design. The failure of Socially Responsible Investing was seen, by one Industry expert, as an example of what may happen to CEP ranking due to the lack of consumer awareness. Similarly, a Government expert felt that consumers will decide the demand for such a system and that will stimulate the adoption of such a ranking system. He went on to add that the system may experience a similar trend as the ISO14001, where corporate participation was slow initially but picked up exponentially after a few years.

While the first 2 questions showed some form of consensus in the responses, the third generated more debate. Hence, this question was repeated in the second Delphi round in order to gain some consensus. During round 2, the level of disagreement fell by 10% (or by 2 respondents) while uncertainty increased by 9% (or by 1 respondent). Furthermore, only Industry experts seemed uncertain in the second round, compared to NGO and Consultancy experts in the first round. Government no longer disagreed in the second round but instead strongly agreed.

Most experts expressed their uncertainty due to the unpredictable nature of what Singapore will 'need' in the future. Of those who disagreed, some felt that Singapore was too small a market for CEP ranking to ever be required by government or industry while others felt that the lack of consumer of awareness of the benefits of such ranking would serve as an impediment. However, a similar and major proportion of experts in both rounds Agreed/Strongly Agreed that Singapore will need a CEP ranking system. This was due to similar reasons stated in the previous question, i.e. international pressure, global competitiveness and the need for companies to performance beyond compliance standards. Hence, it can be concluded from responses to all three questions that there exists a significant demand for a CEP ranking system in Singapore, with uncertainty only over design and execution considerations.

(4) *Design & Execution considerations*

The following questions were proposed to collate opinions on issues pertaining to the design and execution of the CEP ranking system:

R1 Q4(a). What aspects of environmental management should be included?
R1 Q4(b). Which industries should be ranked?
R1 Q4(c). Who should conduct the ranking?
R1 Q4(d). Who should endorse/verify the results?
R1 Q4(e). Who should have access to the rankings?
R2 Q1. Who should fund the ranking system?
R1 Q2. Should participation in the ranking by voluntary or compulsory?
R1 Q4(f) & R2 Q3. How feasible is the development of such a CEP ranking system?

In response to the first question, 11 out of 17 (65%) experts believed that all aspects of environmental management should be included in the ranking's assessment criteria. These experts felt that the assessment would be incomplete if all aspects were not taken into account, therefore non-compliance events and environmental reporting should also be included. However, others felt that only Singapore's pressing environmental issues should be incorporated such as energy usage, water usage and green supply chain management, as including all aspects would be overwhelming and in some cases unnecessary. Many also suggested new aspects that could be included in the ranking, such as design for environment, eco-footprint of products and processes, and climate change data or carbon dioxide emissions. A Government expert suggested focusing on efficiency measures rather than absolute usage when developing the performance criteria. The inclusion of all sectors of the economy in the ranking system regardless of the extent of their environmental impact was deemed necessary by 9 experts. Most believed that if companies were compared to those within their own industry, the comparison would be fair and compelling to know. Others believed that only those industries with high environmental impact should be included in the ranking system, e.g. Oil, Gas & Chemicals and

Manufacturing & Packaging. These were followed shortly by Biomedical Sciences, Logistics & Transport, Electronics and Food. Suggestions to include Hotels and Tourism were also made by an Industry expert.

While experts were asked to select only one party they deemed fit to conduct the CEP ranking in Q4(c), many of them revealed two choices. While 9 experts (53%) believed that a NGO should conduct the rankings, including some NGO experts themselves, there was more agreement amongst all sectors that a combined steering/executive committee consisting of all parties, i.e. Research, Consulting, NGO and Government would be a more viable option. Most experts in this category felt that no one party had the sufficient resources (technical or financial) and public standing to carry out the rankings itself. Hence, a collaborative effort would be most fruitful and acceptable to all interested parties. Other experts urged for a research institution to conduct the rankings as they felt that it had the right level of expertise required to carry out complex and technical computations. Government was selected as the least suitable party to conduct the rankings as most experts felt it would suppress voluntary participation, prevent confidential disclosure and create avoidable paperwork. Government experts themselves expressed concerns over the lack of sufficient resources and corporate voluntary initiatives as reasons for not being a suitable ranking agency. Of those who did select Government, most feared that the ranking would not be taken seriously by interested parties unless the Government was involved in a significant way.

Eight experts (47%) supported a NGO to verify rankings, seven experts (41%) felt that a private consultancy would be equally acceptable. Experts from all sectors agreed to both these options as they felt that they had the credibility or technical resources to be able to conduct audits and verify the results as a third independent party. Some experts believed that Government endorsement of the rankings was pertinent to provide credence to the results. Experts that selected Other believed that either all parties of the combined steering committee should verify the results or the media should serve as an external verifying party. Conversely, a NGO expert believed there was no need for verification if the Government conducted the rankings itself.

Round two proposed a new question on the likely source of funding for the ranking system. Six experts out of 14 (i.e. 43%) proclaimed that Government should fund the rankings as an impartial and trustworthy source. Any other funding source was seen as introducing an element of bias to the credibility of the system by these experts. Another large group of experts (36%) believed that a sole funding party may be impractical, but a combination of parties could co-fund the ranking system instead. Combinations of (i) Government, Industry and NGO; (ii) Government and Industry; and (iii) NGO and Industry were suggested. A Consulting expert also suggested applying the Business in the Community model adopted in UK which is funded by Business members contributing membership fees to the organization. An NGO expert furthered the above suggestion by suggesting tax deduction incentives for those companies that fund the NGO conducting the rankings. Nevertheless, all experts felt that some form of Government support for such a system was crucial to its success.

All experts (100%) believed that all parties, including the public, should have access to the rankings results. Most felt that complete transparency was essential and the only way forward. Others believed that making the rankings public was the real purpose of having such a system in the first place. An overwhelming majority of experts (79%) stated that participation in the ranking should be voluntary. Most of these respondents felt that the system should start off as voluntary on the basis of good corporate citizenship and self-regulation. A Government expert reasoned that as most organizations in Singapore already met national standards, any further environmental initiatives were considered as being proactive. Therefore compulsory participation may seem unreasonable. Experts from all sectors suggested that voluntary participation should have certain incentives and benefits attached to it. For example, it may be integrated into tendering systems used by the government or used as a criterion for selecting industry partners for business ventures and joint-schemes. In the long run, participation may be made compulsory once a critical threshold of acceptance has been reached. It may also be linked to the company's stock listing process, requirements for business license renewal, or towards meeting future national environmental standards. However, the experts that opted for compulsory participation stated that voluntary participation would only attract the best performers and leave out the rest as they may shirk from participation in fear of being ranked low. This was seen as defeating the purpose of adopting a national CEP ranking system.

After taking into account all the above design and execution considerations, experts were asked to judge the feasibility of developing a CEP ranking system in Singapore. Due to the sizeable differences in views of the experts in round one, this question was repeated in round two. Although the differences still existed in round two, their proportions diminished. The number of experts agreeing that the ranking is definitely feasible increased from 12% to 36%. In addition, NGO and Consultancy experts joined Industry experts from round one in this view. While the proportion of experts that selected Probably Feasible dropped from 64% to 50%, the proportion of experts that selected Probably Unfeasible fell from 12% to just 7%. Similarly, the percentage of respondents that were unsure (i.e. selected 'May/May not be Feasible') dropped from 12% to just 7%. Hence, the second round brought further consensus amongst the experts with a majority optimistic about the feasibility of CEP ranking system in Singapore.

(5) *Potential problems*

Each expert was asked to list up to three significant barriers, negative repercussions or problems that they could foresee for a CEP ranking system. Their answers were collated and thematic coding was applied in order to generate a multi-level diagram summarizing the results. All responses fell into the following four categories:

(i) Lack of Resources
(ii) Lack of Support
(iii) System Design
(iv) Sustainability

See Fig. 3.1 for a diagrammatic summary of the above points.

(6) *Foreseeable benefits*

All experts listed up to three significant foreseeable benefits of the CEP ranking system. Once again, their answers were collated and thematic coding was conducted in order to generate a multi-level diagram summarizing the results. All analyzed responses fell into the following four categories:

(i) Improved Performance
(ii) Participation Rewards
(iii) Environmental Awareness
(iv) International Image

See Fig. 3.2 for a diagrammatic summary of the above points.

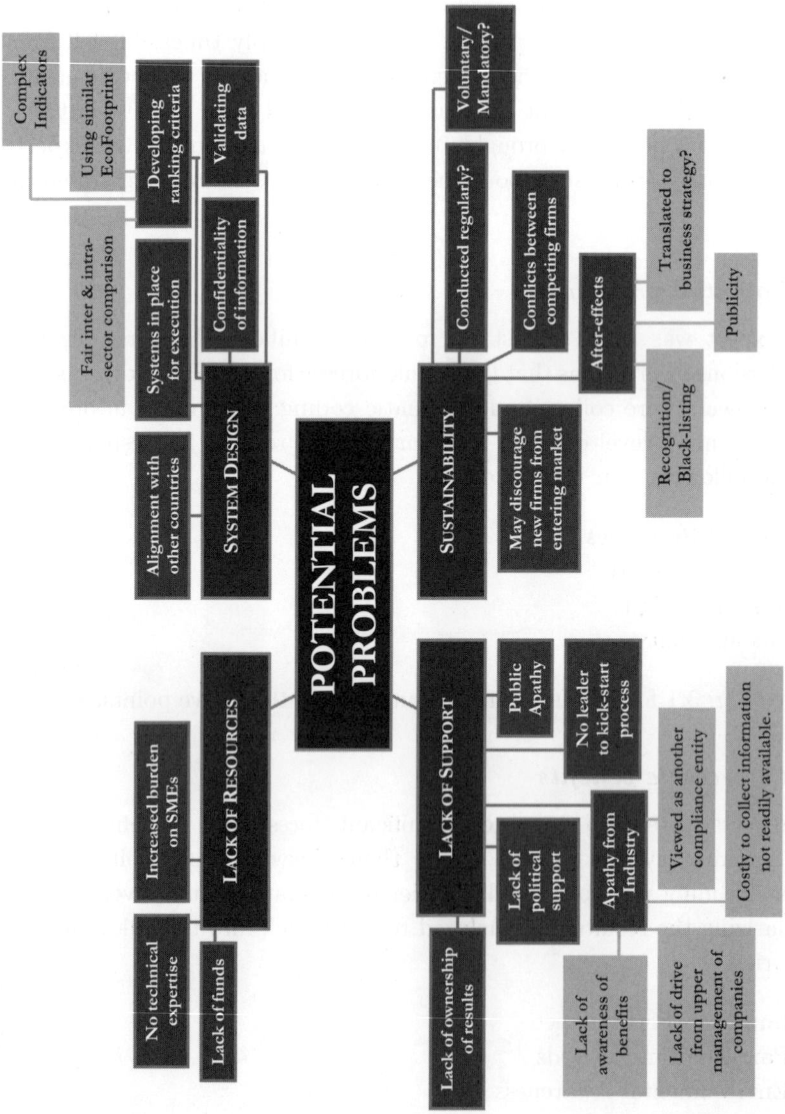

Fig. 3.1. Potential problems that might be encountered in forming a CEP ranking system in Singapore.

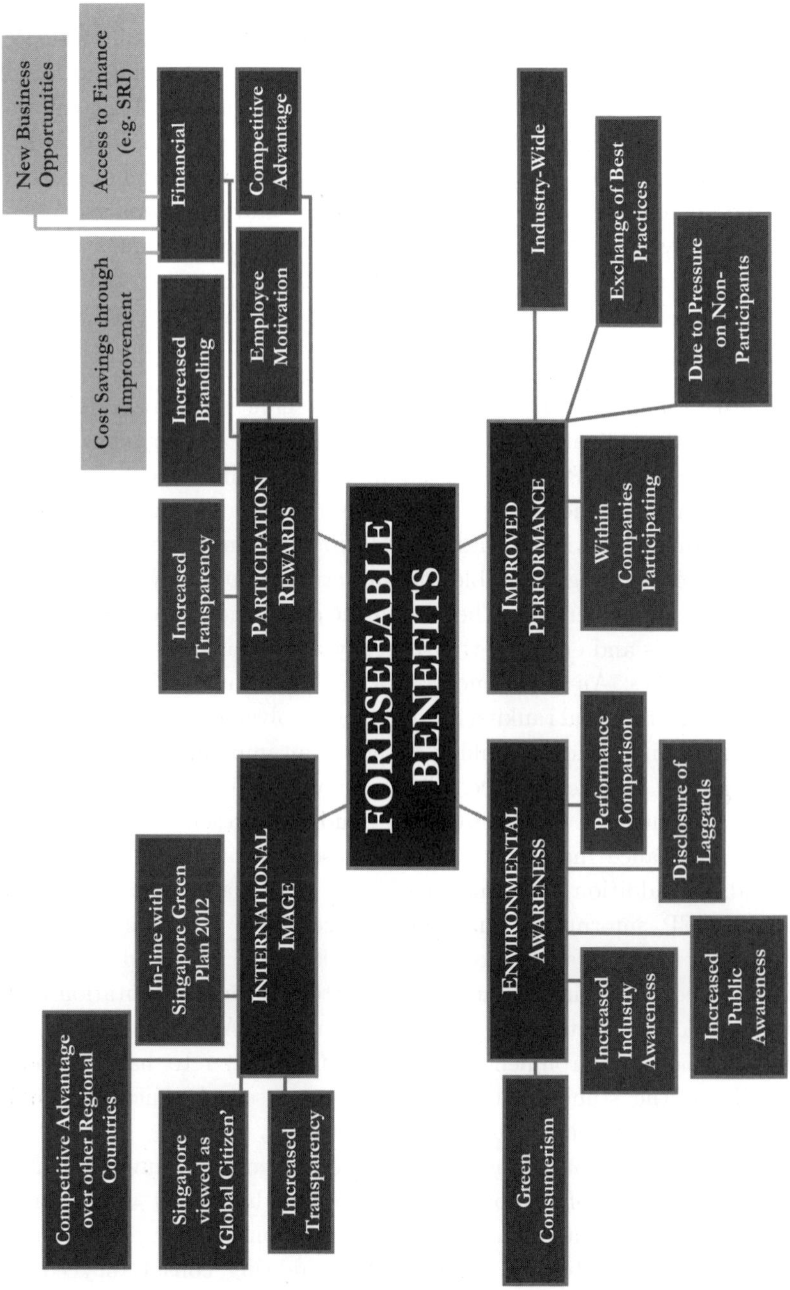

Fig. 3.2. Foreseeable benefits of a CEP ranking systems in Singapore.

Evaluation

This study has investigated local experts' views on the feasibility and design of a possible CEP ranking system in Singapore, keeping in mind the possible problems and benefits that may be encountered. Through the investigations aforementioned, it is evident that there exists a demand for CEP to be ranked in Singapore and that most reservations can be resolved after careful consideration. So how can Singapore start developing its own CEP ranking system? An answer as multi-facetted and complex as this requires further research, discussion and deliberation. However, the author attempts to make a few recommendations in this section that may serve as useful starting points.

i. *Executive Committee*: This committee could serve as the main decision-making and coordinating body for the development and implementation of an effective CEP ranking system. Senior management members from government, the rating NGO, industry, the auditing consultancy firm, and implementing bodies could form the committee and provide inputs on developing equitable ranking criteria, auditory compliance, and participatory benefits. The committee could also be responsible for sourcing funds and ensuring transparency in the ranking process.

ii. *Ranking Bodies*: An experienced NGO or environmental consultancy could pose as the ideal ranking agency. The involvement of a capable and trustworthy independent ranking agency is an imperative to maximize participation and acceptance. The agency would be responsible for the development of CEP criteria by studying present local business Key Performance Indicators and those used by ranking models used abroad. In addition it would reach out potential participants and collate CEP information in order to conduct the final analyses. Resources for the ranking organization could be obtained from the funds collected by the Executive Committee. With its reputation and recognition through the Singapore Environmental Achievement Award, the Singapore Environment Council (SEC) is seen to have already established the standing and trust of businesses. As such, it could possibly be an option for such a ranking agency.

iii. *Auditing & Consulting Body*: Verification of the collected environmental performance data of companies could be conducted by an external third party — i.e. a private environmental consultancy. This would provide credibility to the rankings. The auditing consultancy could provide its available technical resources and offer post-ranking services to companies that seek performance improvement, thereby expanding business development capabilities.

iv. *Endorsing Body*: Some form of support and/or endorsement of a relevant authority such as the Ministry of Environment & Water Resources or the National Environment Agency will provide additional credence to the system. While government endorsement will encourage or pressurize companies to participate, due to the sensitive nature of industry rankings, the authority may provide limited support in order to maintain an unbiased stand. Nevertheless, any form of government support is still imperative in Singapore's regulatory-driven business environment. Alternatively, an international body such as the Global Reporting Initiative could also endorse the rankings.

v. *Implementing Bodies*: In order to fully realize the benefits of corporate environmental ranking, the author proposes linking the results of the CEP ranking to indices in the Singapore stock market (i.e. the Straits Times Index), the DJSI or the FTSE4Good Indices. FTSE has already entered the local market with the recent announcement by the Singapore Exchange to team up with four other ASEAN bourses to create a regional index series using FTSE as the preferred Index.[4] Listed companies with top performance could be included in a new and exclusive CEP index similar to those adopted by the DJSI and FTSE4Good. Alternatively, companies on the Straits Times Index could be assigned a symbol or colour to indicate their environmental performance. For example, exemplary performers could be highlighted in green, while laggards shunned in brown. Furthermore, organizations interested in risk analysis such as Securities Investors Association Singapore (SIAS), credit rating agencies, banks and Socially Responsible Investment (SRI) investors could also be involved to utilize the results of the rankings in their assessments. Initiatives such as the Singapore Compact on CSR and the Center for CSR could also actively involve top performers and encourage the sharing of best practices through collaborations. These could serve as further impetus for the formation of an all-encompassing CSR Index in Singapore.

vi. *Promoting Bodies*: The index could be published and promoted in popular forms of media such as newspapers and television in order to make the results accessible to the public at large. The data could be used

[4] "Asean bourses to create index series", *The Straits Times* (22 July 2005) H33.

for further study by research institutions and environmental NGOs. It is hoped that these institutions will serve as future drivers to increase environmental awareness and capacity building in Singapore.

Conclusion

Corporate social and environmental responsibility is slowly taking center stage in Singapore. It is growing in many ways — from popular initiatives and awards such as the Singapore Compact on CSR and the Singapore Environmental Achievement Award, to the hidden yet integrated practices of companies (big and small) that employ environmental management systems and implement environmental policies. International pressure and standards are driving this message home — and companies in Singapore cannot afford to turn a deaf ear. They are driven on a regular basis to improve their environmental performance and find new ways to incur cost savings and competitive advantage. But despite all these scattered efforts, there is no single entity or instrument that brings together, measures and benchmarks environmental performance.

This study has shown that there exists a demand for such an instrument to be put in place as the benefits are immense and eventual. Delphi experts have rightly pointed out the design and execution barriers that may be encountered, but they have also suggested ways to overcome them. Most of them have highlighted that these benefits will outweigh any costs in the long run. What is needed is leadership in coordinating the necessary involved bodies and a clear definition of the purpose and context of a CEP ranking system/index in Singapore. Although this study is by no means a deciding factor, it is hoped that it will encourage possible involved parties to initiate further exploration into the topic. It is also hoped that Singapore business leaders will be encouraged to see the potential and impacts of a CEP ranking system and support any future endeavours that may come about.

4

THE CHALLENGE OF ENVIRONMENTAL SUSTAINABILITY IN WAFER FABRICATION PLANTS: A FRAMEWORK FOR IMPLEMENTATION

TEO WEN JING, JASMINE ANN and GEORGE OFORI

Wafer fabrication is a business which has the potential to adversely impact the environment through excessive consumption of natural resources, electricity, and materials such as chemicals and silicon. Large amounts of waste, toxic air, and water pollution could be generated if its operations are unsustainable and not properly managed. Thus there is a need to study if the wafer fabrication plants in Singapore are practicing environmental sustainability. An "Environmental Sustainability Framework" is proposed using environmental sustainability principles and tools, current best practices and global codes of conduct. Tools such as life cycle analysis, design-for-environment and environmental impact assessment, help companies to carry out their tasks more accurately. Combinations of codes of conduct and business best practices make the approaches of the companies more holistic. The framework is applied to the wafer fabrication industry in Singapore to test for evidence of environmental sustainability practices. Managers of six wafer fabrication companies were interviewed and two were selected for in-depth case studies; to analyze the differences between their environmental practices. The interviews and analysis of two of the wafer fabrication companies in detail, show that the presence of environmental sustainability in the daily business operations of wafer fabrication plants in Singapore is limited, and the areas for improvement are highlighted. The Environmental Sustainability Framework proposes strategies and practices that companies can adopt to achieve environmental sustainability.

Introduction

Global environmental issues are brought about by transnational companies such as the wafer fabrication plants and their large-scale business operations. Wafer fabrication is a business which has the potential to impact the environment adversely through excessive consumption of natural resources (water), electricity, and direct material inputs (silicon and chemicals), and the production of large amounts of waste, and toxic air and water if its operations are not sustainable and not properly managed.

Extensive research has been done on the need for corporations to be responsible towards the environment and on the possibility of sustainable development. However, there is a lack of research on the environmental strategies and management policies used by wafer fabrication plants to achieve environmental sustainability, with specific consideration of their manufacturing operations.

In addition, although much has been written about how companies can improve their environmental performance, there is a lack of an integrated and holistic approach on which sustainable businesses can be founded. Such an approach would encompass compliance with the requirements of international agreements and global codes of conduct. There is often a misconception that embarking on the sustainability pathway will lead to unnecessary costs and thus, companies are unwilling or unable to take up the challenge.

There are six wafer fabrication companies operating in Singapore, out of which five are subsidiaries or branches of trans-national companies. The sixth company is a locally owned enterprise. As large companies, the wafer fabrication manufacturers in Singapore are accountable to a multitude of stakeholders, placing them under public scrutiny.

Aim and objectives

This study aims to apply the concept of "Sustainable Development" using current corporate practice to formulate an "Environmental Sustainability Framework".

The objectives of this research are as follows:

(1) To analyse wafer fabrication business strategies and activities using sustainable development principles.
(2) To formulate an "Environmental Sustainability Framework" that can be implemented in the wafer fabrication industry.
(3) To examine and document the best practices of wafer fabrication plants in Singapore through Case Studies and Interviews.
(4) To apply the framework in wafer fabrication plants, and use it to set sustainability strategies.

Most of the wafer fabrication companies in Singapore have environmental practices but there is still the question of whether they encompass the completeness of environmental sustainability. Thus, the following hypothesis was formulated for the research:

> "Practicing Environmental Sustainability is evident in the business of Wafer Fabrication companies in Singapore."

Wafer Fabrication and its Environmental Impact

The semiconductor industry is an important pillar of Singapore's economy. It accounts for 13 percent of the total manufacturing output and employs close to 28,000 people. This is a significant contribution to the manufacturing sector. The semiconductor industry consists of activities such as research and design, supply chain management, front-end production, back-end testing and assembly. Singapore has successfully established and attracted 60 companies across the entire semiconductor value chain, including 14 operating wafer fabrication plants and 18 assembly and test facilities, and about 30 integrated circuits design operations (Singapore Economic Development Board, 2004). These span the entire spectrum of companies — foreign and local, large and small, mature and start-up.

The process of manufacturing semiconductors or integrated circuits (commonly called ICs, or chips) typically consists of more than a hundred steps, during which hundreds of copies of an integrated circuit are formed on a single wafer. Generally, the process involves the creation of eight to 20 patterned layers on and into the substrate, ultimately forming the complete integrated circuit. This layering process creates electrically active regions in and on the semiconductor wafer surface.

The semiconductor fabrication plant is a very complex facility built to support the making of the integrated circuits found in computers, cell phones, pagers, automobiles, and many other appliances. Each of the facility systems, i.e. gases, chemicals, environmental control, structure, exhaust waste, is vital to the final outcome of the manufacturing process. The manufacturing process for an integrated circuit can be generally divided into three phases:

- Materials — preparation of the silicon wafer.
- Wafer fabrication — processing the silicon wafer to make the integrated circuits.
- Assembly and testing — testing the final circuit and packaging the chip for installation in a product.

This study only focuses on the wafer fabrication portion of the process.

Semiconductor manufacturing begins with the production of a wafer — a thin, round slice of a semiconductor material, usually silicon. This is the starting material for manufacturing integrated circuits. The fabrication process takes place in a "cleanroom", and typically takes from 10 to 30 days to complete. All processes, except "assembly", are part of what the semiconductor industry calls "front-end manufacturing", done in wafer fabrication plants. Testing and assembly is carried out in "back-end manufacturing" facilities.

In this study, the environmental impacts of the wafer fabrication industry are classified into "inputs to production", "production process" and "outputs". These form the basis of an "Environmental Sustainability Framework" which is built. Resources and raw materials are needed for the wafer fabrication. Within each of the processes, potential environmental impacts exist and like all other production processes, there are outputs to the environment.

Potential ecological impacts that can be caused by wafer fabrication activities are categorised into eight parameters for the purpose of this study:

Inputs to production

(i) Consumption of natural resources

 a. Energy
 b. Water

(ii) Consumption of raw materials

 a. Chemicals and gases
 b. Silicon

(iii) Facilities

Production process

(iv) Etching and masking
(v) Diffusion and chemical vapour deposition (CVD)

Outputs from production

(vi) Pollution

 a. Air
 b. Water
 c. Land

(vii) Waste
  a. Toxic
  b. Non-toxic
(viii) Finished products.

Research Design

Research method

There are currently a total of 14 wafer fabrication plants belonging to six companies in Singapore. Because this industry is small, detailed interviews were conducted with relevant personnel in all the six companies. Case studies were conducted on a local wafer manufacturer and a Europe-based company. The objective of the case studies was to analyse the current practices of each company and to draw comparisons between the practices of these companies, as well as between those of each of them and the "Environmental Sustainability Framework". The purpose of selecting one local and one foreign wafer manufacturer was to compare differences in attitudes towards the concept of sustainable development.

Environmental sustainability framework

The proposed Environmental Sustainability Framework is shown in Fig. 4.1. The framework provides four main aspects of a wafer fabrication plant: (i) Organisational/Management; (ii) Product Development; (iii) Operational Management; and (iv) Communication and Marketing. For each aspect, principles and tools of environmental sustainability are proposed, taking the nature of the wafer fabrication business into consideration.

The elements of the framework provide: (i) a sound base for monitoring and measurement of environmental sustainability in wafer fabrication plants; (ii) a set of concepts and tools that would support the wafer fabrication industry's adoption of environmentally sustainable practices; and (iii) a set of processes that would help define goals and priorities.

Wafer fabrication companies in Singapore can apply the Environmental Sustainability Framework and the specific measures outlined to incorporate environmental sustainability into their business opeations.

Steps towards environmental sustainability

a) Step 1: Conduct a local "sustainability" assessment

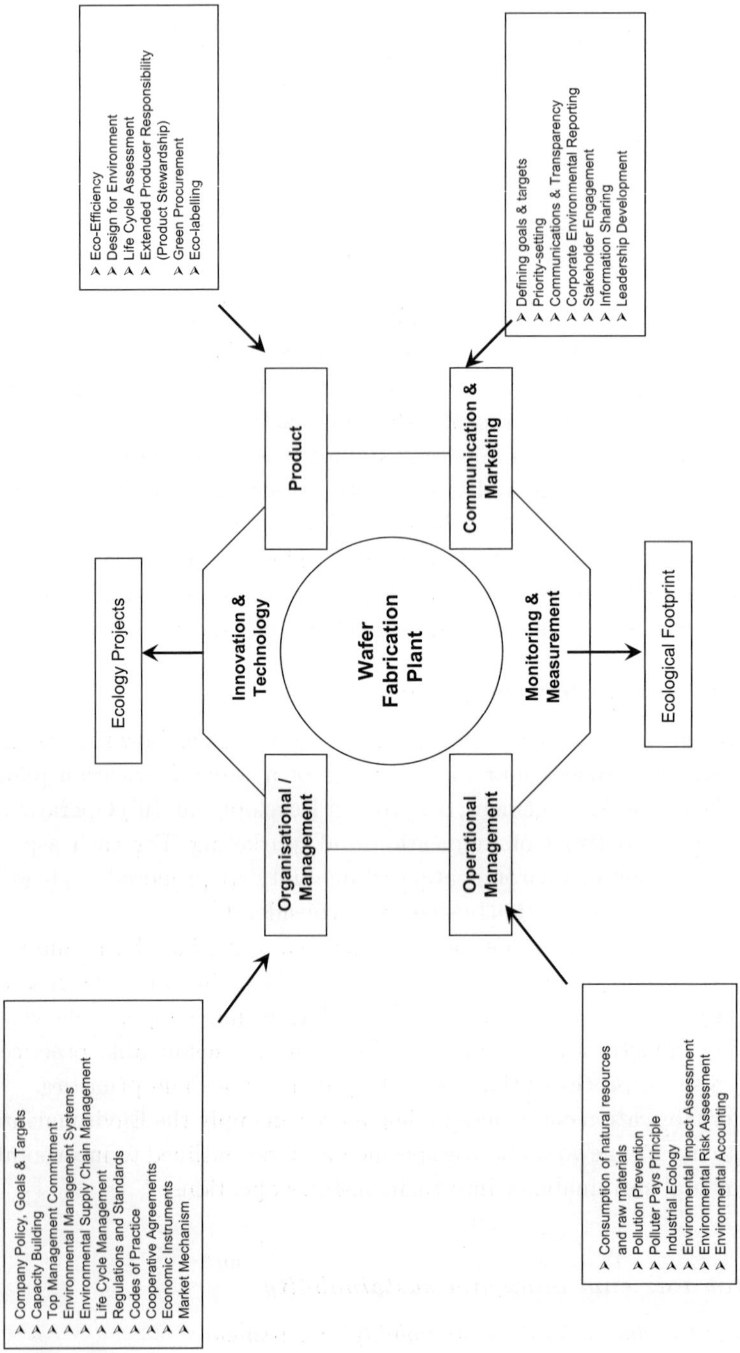

Fig. 4.1. Proposed Environmental Sustainability Framework.

Gather baseline information on such subjects as key environmental impacts and problems in the company; what the firm pays for energy and water; key economic, environmental and social issues. This information provides a baseline for measuring progress later, and can help identify the key goals of a sustainability campaign.

b) *Step 2: Get "stakeholder" concurrence on launching a sustainability program*
Using the assessment from Step 1, build support for a formal sustainability program, involving all people in the community, including customers; employees; neighborhood, environmental and business groups; the media; government agencies; and foundations.

c) *Step 3: Designate a sustainability champion*
The company should designate at least one individual to become the champion of and conscience for sustainable development.

d) *Step 4: Create a vision and mission statement*
Engage the entire company in a "visioning exercise", to define where the company would like to be 20 years from now. The vision should be specific and idealistic, but achievable. Best issued by the Chief Executive Officer (CEO), the vision(s) should be cascaded down to all employees and stressed repeatedly. Codes of conduct, such as the CERES, are good sources of ideas on how a company can formulate a vision. Alternative sources would be the mission statements and business practices adopted by other firms in the same sector.

e) *Step 5: Develop a roadmap for reaching the vision*
With the help of all stakeholders, identify what steps the company will need to take to achieve its vision. An environmental policy would be a good first step towards addressing the broader concerns of sustainable development, which would guide the planning process and put forward such values to the top management, employees and other stakeholders.

f) *Step 6: Develop sustainability indicators*
Based on the vision and roadmap, identify the "indicators" or yardsticks the company will use to measure progress towards attaining its objectives in these regards.

g) *Step 7: Incorporate sustainability into policy*
Conduct a thorough "audit" of policies to determine which advance sustainability and which stand in the way of progress. Remove policy barriers, and create policy incentives. Where business strategies may be in conflict with sustainable development objectives, strategies may have to be modified.

h) *Step 8: Carry out projects*
Start with "early success" projects to begin the implementation of the sustainability program, and involve the public in them.

i) *Step 9: Measure and report on progress and performance*
Using the sustainability indicators, evaluate the company's progress every two years or so, and make adjustments as necessary. Internal and external communications can be used by a company to report on its progress in achieving social, financial and environmental goals. By developing benchmarks to measure progress and reporting on successes and failures, the company can reinforce its commitment to stated objectives, and alert employees and other stakeholders to areas where more effort is needed.

j) *Step 10: Ensure external credibility*
An important part of achieving environmental sustainability is to ensure adequate communication to the stakeholders outside the boundaries of the company. Annual environmental or sustainable development reports are documents that highlight the progress made by the company towards improving environmental performance and societal well-being. When read by stakeholders (such as investors, customers, suppliers, regulator, and the public), these reports go a long way to demonstrate a company's commitment to environmental protection and societal responsibility. Businesses can also join trade associations or business networks such as the Singapore Green Business Alliance (SGBA) and the Singapore Manufacturing Federation.

Field Study

The field study had four objectives:

(i) To study the wafer fabrication business activities and strategies using environmental sustainability principles.
(ii) To examine and document the best practices of wafer fabrication plants in Singapore.
(iii) To analyse if the proposed Environmental Sustainability Framework can be applied in the wafer fabrication industry.

The Interviews

A list of interview questions was sent out to 5 wafer fabrication companies with operations in Singapore, followed by face-to-face interviews. Of the six wafer fabrication companies listed by the Economic Development Board, which produce 5-inch, 6-inch and 8-inch diameter wafers, only one company declined the request for a suitable employee to be interviewed on the basis of confidentiality. The pilot interview was carried out with an interviewee from one of the 6 companies. Table 4.1 shows an outline profile of the companies including the countries in which they are headquartered, the number of plants each has in Singapore and the size of the operations of the plants. It is not possible to provide further details of the companies and their plans owing to the sensitivity of the information.

The largest wafer fabrication company is a locally owned enterprise with five plants, followed closely by the European-based company with four plants. The rest of the companies, though small in size, still play a significant role in the semiconductor industry in Singapore.

The interviewees were at least the environmental manager or equivalent and this pre-qualification ensured that only persons who are familiar with the company's environmental commitments, programs and requirements were interviewed. Most of the companies had environmental programs and although their activities were not coined under sustainable development, some of the tools and methodologies mentioned in the framework were used.

Summary of interviews

A summary of the findings from the interviews is now presented. The key business drivers for the companies identified are mainly customer requirements and cost savings. The global companies face problems in implementing the initiatives started by their corporate headquarters. Apart from Company D, the business principles of the other companies are not determined on the basis of environmental sustainability issues but are, rather, based on stakeholders' requirements.

In the management systems of the firms, there is no specific indication of "sustainable development" or "corporate social responsibility". There is an inadequacy of the practice of "green procurement". Although lists of 'banned' chemicals are specified by the companies, suppliers and vendors are not obligated to practice environmental sustainability. One reason is that these companies are usually small in size and in their operations. The companies ensure that suppliers have a proper environmental management system but the concept and tools of environmental sustainability are not practiced. For example, suppliers and vendors are not assessed on their ability to design-for-the-environment regardless of their products.

The general conception of the local regulations in relation to environmental sustainability is inadequate. The respondents agreed that

Table 4.1. Details of companies interviewed.

Company Details	Company A	Company B	Company C	Company D	Company E	Company F
Country of origin	Singapore	Singapore	Singapore	Italy-France	Not available	Japan
Nature of Business Portfolio	Locally Owned	Joint Venture	Joint Venture	Subsidiary in Singapore	Subsidiary in Singapore	Subsidiary in Singapore
No. of wafer fab per company	5	1	2	4	1	1
Size of wafers produced	1 × 12-inch 4 × 8-inch	1 × 8-inch	2 × 6-inch	1 × 8-inch 2 × 6-inch 1 × 5-inch	Not available	1 × 8-inch
No. of Employees Locally	4,000	1,300	2,000	6,000	870	—
EMS Certification	ISO 14001	ISO 14001	None	ISO 14001 & EMAS	ISO 14001 (by end 2005)	Not available
Person Interviewed	Facilities Manager	Senior Manager (Risk Mgmt)	EHS Manager	EHS Manager	Senior Manager (ESH Dept)	*declined interview

Singapore should follow the example of European countries as their considerations for the environment are more extensive and stringent. They also feel that the reason for this is because Singapore is focused on attracting investments as a priority. There is also a general perception that economic instruments are not adequate in Singapore and penalties imposed for pollution are insignificant, especially to large companies. Apart from Companies B and D, the firms are not signatories to any global codes of conduct. Often, they await instruction from their parent company for direction.

Training courses for the employees often includes only the environmental aspects of the company's operations. No training is provided on the subject of environmental sustainability and its tools.

The companies do not practice "design-for-environment" because their facilities carry out the manufacturing, not the design of their products. The companies also do not practice life-cycle analysis which could be done for not just the products but also the processes within the company. This is one area in which the companies can improve. With the exception of Companies D and E, environmental accounting is not done as there is no requirement by their corporate headquarters or customers to do so. The companies are also unfamiliar with the method of environmental accounting. The companies give priority to pollution control issues over all others because of the legal implications. Other issues such as resource conservation, corporate responsibility and other broader aspects of sustainability take a backseat.

The level of transparency and the communication of the company's activities to the public is low in all companies except in Company D. Most of the companies publish environmental reports but these are not made available to the public. Also, a majority of these reports are corporate reports, which means that the information may not directly pertain to the operations of the local subsidiary in Singapore. All the companies experience a high degree of stakeholder scrutiny, and communication between the companies and these stakeholders (especially the residents living near the plants) is poor. Publishing and making the environmental reports public will be useful in this instance. Information sharing activities often take place between the National Environment Agency and each of the companies but there is a lack of industry sharing. Companies do not share environmentally-innovative technologies among themselves.

This summary shows some of the areas in which the companies can make improvements. The companies are taking adequate environmental pollution control measures, because of local regulations and also due to industry benchmarking. Environmental projects are also widely implemented in an effort to conserve water, electricity and chemicals and for cost efficiency. The companies also engage in environmental impact assessments for their internal processes and expansion projects. This is commendable as environmental impact assessments are not required to be done in Singapore. Targets set were usually met through the implementation of environmental projects which, naturally, reduced costs.

The companies should also be given credit for the effort to measure their environmental performance comprehensively, especially in terms of resource conservation.

Summary of case studies

In this section, the extent of environmental sustainability within two of the six wafer fabrication companies is reported. The case studies aim to examine the best practices of wafer fabrication plants and to study if the Environmental Sustainability Framework can be implemented in the industry. Companies A and D were selected for the case study and the purpose of choosing these organisations is to analyse the differences between the environmental sustainability practices of a Europe-based company and those of a local company. In addition, Company D was selected on the basis that, from the interviews, it was found that it had substantial environmental programs that could be studied.

Company A's objectives seem more like targets and only considers energy, water, materials and PFCs whereas Company D's objectives encompass economic, environmental and social aspects of sustainability. Its objectives address global issues and emphasize the use of design-for-environment and life-cycle analysis.

Top management commitment is an important component in a company's success towards achieving environmental sustainability. Company A, though a large firm, is run locally. Company D, on the other hand, is heavily influenced by its European headquarters. This results in a disparity between the two in terms of the level of environmental commitment. Regulations in Europe are stricter than in Singapore, and Company D is obliged to meet the most stringent regulations in the countries in which the company operates. Thus, its objectives and targets include a more comprehensive set of parameters that lead to environmental sustainability. Company A relies on ISO 14001 as a guidance document for its environmental management system whereas Company D's company-wide charter includes international standards and policies.

Company D is bound by its corporate requirements to meet "the most stringent regulations of any countries in which we operate". This means that the company sets targets that are often far better than what the local regulations require. This, again, is top management's commitment cascaded to all its subsidiaries. Company A is not a signatory to any global code of conduct. Company D is a signatory to the Global Compact and Kyoto Protocol. Its environmental programs are built around global standards. Thus, the company is always in touch with the global arena. All of Company D's subsidiaries are expected to follow the same requirements, regardless of their location.

The commitment of Company D to environmental conservation and preservation can also be seen from the extent of investment in environmental programs. Company D designs its products to be more energy efficient and conducts a life cycle assessment to ensure that all environmental impacts have been addressed. Training is also provided for employees to ensure that they are equipped with the appropriate knowledge to design-for-the-environment. Company D conducts environmental accounting on an annual basis and benchmarks all its plants worldwide for the best performance.

Company A's environmental report follows the ACCA Singapore reporting guidelines whereas Company D follows the format specified in EMAS. As it is listed on the Singapore Exchange, at the local level, Company A's communication with the stakeholders is more in depth than that of Company D. Company D's Environmental Performance Indicators (EPIs) are more comprehensive than Company A's. Their EPIs contain not only resource conservation but also extra parameters on air emission and wastewater discharge.

In summary, it can be seen that there is a large difference in the environmental commitments between Companies A and D. The main reason behind Company D's commitment towards the environment is due to the strong role of its top management in this task at both the corporate and local levels. The company has been successful in achieving its environmental goals and targets but one area in which it can improve is the area of life cycle analysis. So far, life cycle analysis is only done at the corporate level; the local subsidiaries are not required to do so.

Company A's environmental programs are also strong but are very much localised in terms of the operational aspects. Like most companies, Company A relies largely on cost-benefit analyses to determine the viability of environmental commitments. Being a relatively small enterprise with a limited budget for environmental programs, the company focuses more on its manufacturing productivity. It can be said that Company D's environmental sustainability practices are in place and the "Environmental Sustainability Framework" can be applied in this company. The framework can also be used in Company A to achieve environmental sustainability.

Discussion

The field study shows that the presence of environmental sustainability in the daily business operations of wafer fabrication plants in Singapore is limited. Although the majority of the companies interviewed have made environmental commitments through objectives, targets and projects, certain principles of environmental sustainability such as design-for-environment, life-cycle analysis and product stewardship are not evidently applied.

The only company that fulfilled most of the components of the Environmental Sustainability Framework was one that is headquartered in Europe. Its top management commitment towards environmental protection and conservation is very much present and cascaded down in the daily operations of its four wafer fabrication plants in Singapore. Thus, top management commitment to environmental sustainability as well as the availability of appropriate infrastructure is crucial to a company's success in achieving environmental sustainability. Principles and tools of environmental sustainability such as design-for-environment, product stewardship, life cycle analysis and environmental impact assessment methods can be used.

The hypothesis is not fully supported as environmental sustainability is not evident in the practices of the wafer fabrication companies in Singapore.

Figure 4.2 consolidates the research findings and suggests a route that wafer fabrication plants can follow in order to achieve environmental sustainability.

The problems that wafer fabrication plants may face during the implementation phase are, first, most of the companies have a lack of top management support for environmental sustainability commitments. Thus, although environmental sustainability has been proven in one company, the others find it difficult to implement due to the lack of commitment. Second, the companies interviewed are more inclined towards profit generation and improvement in manufacturing productivity rather than investing in environmental programs. Third, employees of the organisations, especially management staff, may be resistant to change, which will make awareness more difficult to attain. Finally, the wafer fabrication business ranges

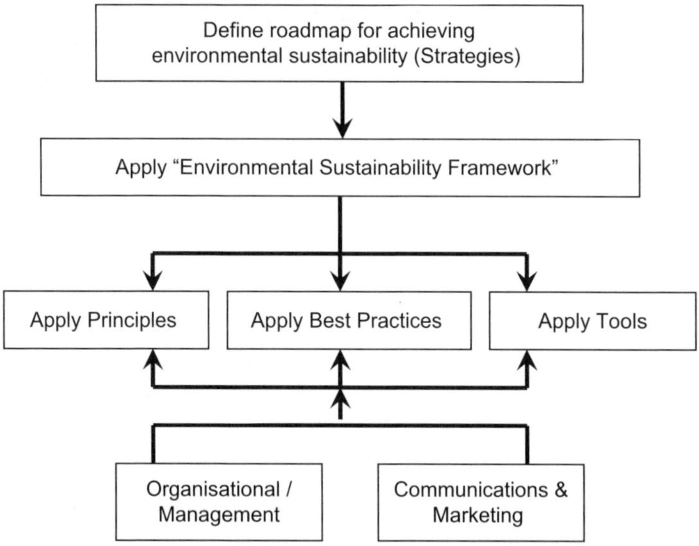

Fig. 4.2. Suggested route to achieve environmental sustainability.

from design to final product. Because of the size of its operations, a wafer fabrication plant may find it difficult to apply the principles of the framework for environmental sustainability offered in this study in its daily operations.

The study found that the wafer fabrication companies in Singapore are lacking in many regards. First, top management must be convinced that environmental sustainability is beneficial to the company, cost-wise, and with regard to recognition from stakeholders. Second, stakeholders are increasingly more aware of environmental issues. Thus, the driving force for environmental sustainability lies in the commitment of companies to meet stakeholders' requirements. Third, companies should apply green procurement by first imposing on the suppliers and contractors, environmental assessment procedures where they are assessed based on their environmental management systems. Training should be provided to educate the suppliers on the company's environmental requirements. Fourth, although a majority of the companies do not design the products they manufacture in Singapore, life cycle analysis can still be carried out on individual processes or facilities. Fifth, environmental accounting is an important tool which will aid each of the companies in its environmental inventory. Finally, the lack of transparency could be seen during the interviews as it was evident that the companies did not disclose environmental information to the public. Communication with the public can take place through dialogues where issues of concern can be raised. The companies' environmental reports can also be made available to the public so that they may understand the firms' operations better.

Conclusion

From the application of the "Environmental Sustainability Framework" to test the presence of environmental sustainability in the wafer fabrication plants in Singapore, it can be seen that the framework can be used most appropriately in a company that is committed to environmental sustainability and is willing to invest time, money and personnel to achieve it. There is limited literature on wafer fabrication activities.

In the study, the focus was placed only on the wafer fabrication industry; thus, the interviews were conducted for a small sample of five out of six companies. This is only part of the semiconductor industry; the suppliers and customers of wafer fabrication plants were not covered in the study. The companies were also reluctant to disclose monetary information such as costs and savings involved in achieving the environmental goals. Confidentiality of the wafer fabrication operations also limited the amount of information that could be gathered for the purpose of analysis.

The "Environmental Sustainability Framework" can be modified and applied to other industries such as the chemical, petroleum and semiconductor industries. Another area of study could be the semiconductor industry as a whole, encompassing its supply chain, customers as well as the back-end processes. Future research can include the study of industrial ecology in the wafer fabrication industry due to its high intensity in resource conservation and generation of large amounts of waste. Focus could also be placed on the design and the life cycle management of microchips.

Reference

Singapore Economic Development Board, 2004. Online. Available from: http://www.sedb.com. [30 November 2004].

5

IMPLEMENTATION OF ISO14001 ENVIRONMENTAL MANAGEMENT SYSTEM IN VIETNAM

DAO MAI ANH and GEORGE OFORI

This study examines the issues related to the implementation of ISO 14001 Environmental Management System (EMS) in Vietnam. Its objectives are to explore the current trend of ISO 14000 certification and implementation in Vietnam, its possible benefits and problems encountered by firms in implementing this standard. As Vietnam develops, attention is being paid to the implementation of ISO 14000 certification. In general, most of the benefits and problems in ISO implementation stated in the literature are applicable to Vietnam.

This research used a questionnaire-based survey to examine the validity of the factors identified in the literature. The method was face-to-face interviews with the managers of eight enterprises that have attained ISO 14000 certification in Vietnam. The hypotheses were tested using mean importance ratings and t-test of the means. Almost all the listed benefits were found to be applicable to the responding companies. Almost all the barriers to implementation that were identified in the literature review were found not to be applicable to the sample companies in Vietnam. Proposals are made on systematic ways of implementing the EMS to ensure maximum benefits to the enterprises.

Introduction

Since the Rio Earth Summit in 1992, the concept of sustainable development has been unfolding throughout the world. Businesses have recognized that there should be a common goal, not a conflict, between economic development and environmental protection, both now and for future generations. However organizations of all types have been struggling to find a set of practical tools to implement strategies supporting this concept. The ISO 14001 standard provides just such a set of tools.

The ISO 14000 EMS is a set of international standards for improving the environmental performance of organizations. It was published on October 1, 1996. Some countries and companies quickly embraced it, while others are still waiting to see if it becomes a requirement for doing international business and if there are benefits from adopting it (Haklik, 1997).

Over the last few years, in the country's transition to a market economy, the Vietnamese government and the entire society have been increasingly concerned about environmental issues. Many organizations and enterprises are aware that environmental issues are becoming urgent in all aspects of social life, and the government has been more concerned about environmental protection. The government has been taking various measures to protect the environment through an increasingly stricter legal system, particularly the enactment of the Law on Environmental Protection of 1993. Growing concerns about the environmental impacts of the operations of the enterprises push these firms to adopt pollution control measures. In this regard, EMS have emerged as an effective measure to protect the environment.

Although, to some extent, most companies have included environmental management in their operations, they have never incorporated it into their overall management framework. Environmental management is still not considered to be an issue which needs to be dealt with systematically. There is some degree of consensus that the environmental management systems and ISO 14001 can be useful in improving the environmental performance of companies and that they can assist in facilitating trade. However, it is recognized that there is little experience and empirical research on the implementation of ISO 14001. The level of awareness of ISO 14000 among businesses in Vietnam remains low.

International and regional experience of using the ISO 14000 EMS as a tool in environmental management policies encouraged this study of EMS and its application in Vietnam. Environmental management standards, particularly the ISO 14000 series, have been under consideration and expert study by different organizations in Vietnam. Such studies have covered aspects such as their possible effects on trade; their impact on foreign direct investment and the role of transnational corporations; issues relating to their implementation and certification; and the needs of industry, especially small and medium enterprises (SMEs).

Given such a scenario of EMS in Vietnam, there is a need to study the current trend of ISO 14000 use in Vietnam, its benefits as well as problems from its certification and implementation.

Research objectives

The objectives of this study are:

1. To explore the current trend of ISO 14000 certification in Vietnam
2. To assess the motivations behind ISO 14000 certification

3. To determine the benefits that ISO 14000 certification has brought to the companies
4. To identify the problems faced by companies in getting ISO 14000 certification and implementing ISO 14001 EMS
5. To assess the prospects for promoting ISO 14000 certification and implementation in Vietnam.

Findings from the study would help the authors to make recommendations regarding ISO 14000 certification and implementation in Vietnam. The study can also contribute to the knowledge of the ISO 14000 series standards, and thus, help businesses and organizations to obtain the greatest value from certification to, and implementation of, EMS.

Environmental Management in Vietnam

In 1989, the Government of Vietnam embarked on a reform named "doi moi" to guide the country from a centrally-planned toward a market economy. Vietnam has since opened up its economy and has made significant progress in the process of trade liberalization. While this has been successful in generating strong economic growth, it has created threats to the country's environment. In order to maximize the potential positive impacts while at the same time minimizing the negative effects of the process of modernization and industrialization in the country, it is necessary to take measures now to protect the environment. The ongoing reform process creates a good opportunity for the development and introduction of effective environmental management policies and instruments (Tran, 1996).

Together with the socio-economic achievements attained in Vietnam, the degradation of the nation's environment has become evident. Of the many environmental problems, there are six major ones including forest depletion, decline in biological diversity, soil degradation (marine and inland), water contamination, air pollution and the problems in solid and hazardous waste management (Dinh, 2001). Vietnam is also aware of the effects on the environment caused by industrialization. However, as the country is still poor and facing many immediate challenges, the Government of Vietnam has decided to gradually tackle these increasingly important environmental issues.

A number of measures have been proposed by the Government. According to Tran (1996), they can be grouped into *regulatory* (development of institutional and legislative frameworks, capacity-building, strengthening of international cooperation and signing of international treaties), *economic* (market-based instruments to be applied in parallel with "command-and-control" ones), and *communicative* (awareness raising, training and education, and information management).

In 1992, Vietnam participated in the United Nations Summit held in Rio de Janeiro, Brazil and committed itself to Agenda 21. In December 1993, the Law on Environmental Protection was adopted by the National Assembly and the Law was enacted in January 1994. In June 1998, the Politburo of the Communist Party of Vietnam promulgated the Directive 36-CT/TW on "Strengthening Environmental Protection in the Period of Industrialization and Modernization of the Country". In 2000, the "National Strategy for Environmental Protection for the 2001–2010 Period" was formulated by the Ministry of Science, Technology and Environment (MOSTE) and submitted to the Government for approval. These landmarks demonstrate that Vietnam is determined to develop its economy and protect the environment. The formulation and implementation of Vietnam's strategy toward sustainable development has been greatly and positively influenced by Agenda 21 in terms of viewpoint, methodology and experience.

Vietnam became the 65th member of the International Organization for Standardization (ISO) in 1977. Through its Directorate for Standards and Quality (STAMEQ), Vietnam is participating in the activities of ISO/Technical Committee 207 as an "Observer member" (before 1999) and a "Participating member" (since November 1999). Previously, Vietnam's participation in activities relating to ISO standards was limited since most of its standard systems were formulated in 1962 on the basis of either accepting or referring to the standards of the former Soviet Union (Nguyen, 1999). Up to 1993, Vietnam had no environmental standards and systems in the proper sense of the word. Among the 24 environmental protection standards, formulated separately and unsystematically to cope with immediate demands of state functional offices, none mentioned environmental management as a systematic approach.

As Vietnam is now engaged in a process of change towards a market economy and regional and international integration, the country's competent authorities need to review their perspectives on environmental protection in general, and environmental management in particular. It needs a new approach consistent with international regulations and Vietnam's particular situation.

Vietnam established a National Committee on Environmental Management Standards (TCVN/TC207), members of which were selected from institutions, industries, government agencies, trade organizations, and so on. The Vietnam Standards Centre, a subsidiary institution of STAMEQ, serves as Secretariat for both the National Committee and sub-committees. One of the major activities of the national committee and sub-committees has been to study and assess the relevant ISO documents, draft ISO standards, and formulate national views and comments on them. STAMEQ recommended and MOSTE adopted ISO 14000 standards for EMS and Environmental Auditing as national standards (TCVN) in 1997 and 1998. Other ISO 14000 standards were under consideration for adoption as national standards by the end of 2000. It is pertinent to note that the implementation of the ISO 14000 national standards by companies operating in Vietnam is not mandatory.

Field Study

This study uses a questionnaire-based survey to determine the current trend of ISO 14000 implementation in Vietnam, and the benefits obtained as well as problems faced by ISO 14000-certified companies. The objectives of the survey were to:

1. Determine the attitudes, values and beliefs of companies in Vietnam towards improved performance;
2. Assess the motivations behind ISO 14000 certification;
3. Assess the benefits that ISO 14000 certification has brought to the company;
4. Assess the problems faced by companies in getting ISO 14000 certification;
5. Assess the prospects for promoting ISO 14000 certification and implementation in Vietnam; and
6. For companies which are ISO 9000 certified, the survey aimed at determining advantages and disadvantages of the implementation of ISO 9000.

Survey and analysis methods

A survey questionnaire was chosen as the method of data collection. The most important advantage of a survey is that it can provide detailed descriptions of populations. A survey permits one to describe large and various populations accurately and economically through probability sampling (Singleton *et al.*, 1988).

The choice of methods for distributing the questionnaire is important to ensure the highest rate of return consistent with time and budget constraints. The target population of this research are enterprises that have attained ISO 14000 certification in Vietnam. There were 22 ISO 14000 — certified companies in Vietnam by November 2001 at the time the field study was commenced (VPC, 2001). All these companies were located in the industrial zones about 40 to 70 kilometers from the country's two main urban centres, Hanoi and HoChiMinh City, making it possible for them to be approached for the interviews. Since the target population was small and accessible, this research attempted to survey all 22 companies by face-to-face interviews with the managers of these companies.

The survey was conducted in three weeks in December 2001. These enterprises were approached through e-mail or telephone to request for meetings with managers of the environmental management departments. The survey was then carried out upon the interviewees' consent.

Each interview started by explaining its purpose and objectives. Respondents were assured that collected data would only be used for academic purposes and details of respondents and companies' profiles would be kept confidential. The respondents were told that the estimated time to complete the interview was 30 minutes, and that effort had been made for respondents to complete and return the survey.

When there are multiple attributes, such as in this research, there is a need to make a distinction between what are essential and what are desirable attributes. This is because different attributes are of different importance to decision makers. It is therefore necessary to find out the degree of importance of each attribute by assigning weights to them. The weight expresses the importance of each attribute relative to the others and indicates what the decision makers are most concerned about in a quantitative way (Edwards and Newman,1982).

Among the available methods, this researcher selected the Likert scale to elicit weights. This is a rating scale for decision makers to indicate their opinion on the level of importance of attributes. Likert scale was used in Ofori (1999). It was found to be the most suitable for this study because it is relatively straightforward to ask the respondents to indicate the level of importance of an attribute on a fixed scale. This research used a five-point Likert scale where 1 represents 'not important', 2 is for 'of little importance', 3 for 'moderate importance', 4 for 'high importance', and 5 for 'very important'. The Likert scores are then converted to a mean importance rating, then further converted to an attribute weight. The questionnaire is appended.

Section A *Company's Profile* involved questions about the respondent's company. There were eight questions concerning the company's background including its mission, year of formation, type of operation, turnover and personnel.

Section B *Company's Development Objectives* contained one question concerning the company's ranking of its development objectives.

Section C *Environmental Record* comprised four questions concerning the company's attainment of the Certificate for Meeting Environmental Requirements of Vietnam's MOSTE and other environmental records, if any.

Section D *ISO 9000 Certification* had six questions aimed at obtaining detailed information relating to ISO 9000 certification including the company's status, expectations, derived benefits and disadvantages in being certified to the standards.

Section E *ISO 14000 Certification* had six questions relating to ISO 14000 certification and implementation including the company's expectations, derived benefits, problems associated with certification to, and implementation of, the standard, and suggested measures to promote ISO 14000 implementation in Vietnam.

Data analysis

The mean importance rating of an attribute for a five-point Likert scale is calculated using Eq. (5.1) as shown below.

$$a_h = \frac{1(n_1) + 2(n_2) + 3(n_3) + 4(n_4) + 5(n_5)}{(n_1 + n_2 + n_3 + n_4 + n_5)} \tag{5.1}$$

where:

h is the attribute reference

a_h is the mean importance rating of attribute h

n_1, n_2, n_3, n_4, n_5 are the number of respondents who indicated on the five-point Likert scale, the level of importance as 1, 2, 3, 4 and 5 respectively for attribute h, where 1 represents 'not important', 2 is for 'of little importance', 3 for 'moderately important', 4 for 'highly important', and 5 for 'very important'.

The mean importance rating is standardized into a weight using Eq. (5.2) shown below.

$$\textit{Weight of an attribute} = \frac{\textit{Mean importance rating of an attribute}}{\textit{Sum of all mean importance ratings}} \tag{5.2}$$

The weight calculated in Eq. (5.2) is actually a relative index of the mean calculated in Eq. (5.1) (Holt, 1997). Holt (1997) showed mathematically that the relative index need not be calculated if the research output is merely to achieve ordinal sorting of the list of the attributes.

Having calculated the mean importance ratings from the information provided by the sample of ISO14000-certified companies, the next step is to assess how important the attributes are for the population of ISO14000-certified companies. Statistical tests of the mean (Newbold, 1991:364) were carried out. For each attribute, the null hypothesis that the attribute is unimportant and the alternative hypothesis that the attribute is important are set out below.

To test the null hypothesis $H_0 : \mu \leq \mu_0$

Against the alternative hypothesis $H_1 : \mu > \mu_0$, where μ is the population mean. The decision rule is to reject H_0 when the calculated t value is larger than $t_{(n-1,\alpha)}$ as shown in Eq. (5.3).

$$t = \frac{\bar{x} - \mu_0}{s_x/\sqrt{n}} > t_{(n-1,\alpha)} \tag{5.3}$$

where:

the random variable $t_{(n-1,\alpha)}$ follows a Student's t-distribution with $n-1$ degrees of freedom

$\bar{x}$ is the sample mean

n is the sample size, which is 22

μ_0 is the critical rating above which the attribute was considered important. In this study, μ_0 was fixed at 3 because by definition given in the rating scale, ratings above 3 (i.e. 4 and 5) represented 'important' and 'very important' attributes.

The significance level, α for this study was set at 0.10. This meant that there is a 90% certainty that the result was not due to chance and that the finding was significant at the 0.10 level. The probability of mistakenly rejecting the null hypothesis, or committing a Type I error is 10% and the probability of accepting the null hypothesis when it is true is 90%.

From the table of critical values of t-distribution, for degrees of freedom $= 21(22-1)$, and the level of significance for a one-tailed test at

0.10, the t value is 1.323. It meant that if the calculated t value is larger than 1.323, the null hypothesis is accepted. It is then concluded that the attribute is important.

Results of survey

During the interviews, the interviewer went through each question with the interviewee to ensure completeness, consistency and legibility of the response. Right after the interview, all responses were checked to ensure that they had been properly and substantially completed.

Out of 22 ISO 1400-certified companies approached, 8 companies were willing to participate in the study, showing a response rate of 36%. The questionnaires were completed by 7 Environmental Management Representative (EMR) (87.5%) and 1 Total Productivity Manager (12.5%).

As shown in Table 5.1, 100% foreign-owned companies comprised 25% of the total of respondents, while 75% were joint-ventures. In terms of

Table 5.1. Companies' profile.

	Company's name	Year of formation	Status of company	Area of specialization
1	Vietnam Stanley Electronic Co., LTD	Sep. 96	JV	Manufacturing auto/motor lamps/headlights, etc.
2	Unilever Bestfoods Co., Vietnam (Wall's)	1997	100% foreign-owned	Manufacturing ice-cream and tea products
3	Lever Vietnam JV Co.	Jan. 95	JV	Manufacturing cosmetics, detergents, shampoo, soap
4	Honda Vietnam	1996	JV	Manufacturing motorbikes
5	Fujitsu Computer Products of Vietnam, Inc.	1996	100% foreign-owned	Manufacturing computer products
6	Daeha Trade Centre — Hanoi Daewoo Hotel	1996	JV	Hospitality: Hotel, apartment and office for rents
7	Toyota Motor Vietnam Co., LTD	Oct. 1995	JV	Production and sales of automobiles and parts
8	Elida PS (refer to Wall's)		JV	Manufacturing cosmetics, detegents, shampoo, soap, toothpastes

the Area of Specialization, 7 out of 8 companies operated in the field of manufacturing (87.5%), while only 1 company specialized in hospitality business (12.5%). The manufacturing enterprises specialized in automobile and motorbike headlights and lamps manufacturing (12.5%); ice-cream and tea products (12.5%); computer products; and shampoos (25%). Table 5.1 presents the companies' profiles.

Company's development objectives

Respondents were asked to indicate their development priorities by ranking 6 stated objectives, 1 = 'most important' and 6 = 'least important'. The results of the ranking are presented in Table 5.2. They show that of greatest priority among companies' development objectives was to attain high quality, followed by minimizing the environmental impacts and meeting set project deadlines.

Company's environmental record

Only 3 out of 8 companies (37.5%) had the Certificate for Meeting Environmental Requirements of the MOSTE. For those companies, the main advantages of the Certificates were as shown in Table 5.3.

Table 5.2. Ranking of development objectives.

Development objectives	Ranking (%)					
	1	2	3	4	5	6
To minimize cost	0	0	0	12.5	**62.5**	25
To meet set project deadlines	0	25	**50**	12.5		12.5
To attain high productivity	0	0	0	12.5	25	**62.5**
To attain high quality	**62.5**	12.5	37.5	0	0	0
To avoid infringing any relevant regulations	12.5	12.5	12.5	**50**	12.5	0
To minimize environmental impacts of company's operations	25	**50**	12.5	12.5	0	0

Table 5.3. Advantages of the certification for meeting environmental requirements of MOSTE.

Advantages of the certificate	Number of responses	Percentage
Good relationship with the government agencies	3	37.5%
Protection of the environment	1	12.5%
More effective environmental management system	2	25%
Less environmental inspection	3	37.5%

Table 5.4. Respondents' environmental records.

Environmental records	Number of response	Percentage
Forestation	1	12.5%
Sponsoring green campaigns in HoChiMinh City	1	12.5%
Attending and organizing seminars	3	37.5%

Table 5.5. Reasons for seeking ISO 9000 — Mean importance ratings and t-test of the mean.

	Test value = 3		
Attribute	T	Mean	Standard deviation
To boost company's image	**1.746**	3.83	1.17
Because other companies had attained, or were seeking certification	−.277	2.83	1.47
To increase company's competitiveness	**4.000**	4.33	.82
Because it will soon be mandatory	−2.076	2.17	.98
To give clients greater quality assurance	**3.000**	4.50	1.22
To reduce company's operating costs	**1.581**	3.67	1.03
To improve company's operating procedures	**2.907**	4.17	.98

Three companies (37.5%) stated their other environmental records (refer to Table 5.4).

ISO 9000 certification

The majority of responding companies (75%) had obtained ISO 9000 series certification. There was one company which had not got ISO 9000 certified but had another quality certification instead (the Malcom Baldrige standards). One out of six ISO 9000-certified companies (16.7%) got the certification in 1996, one (16.7%) in 1999, and four (67%) in 2000. After processing the responses, mean importance ratings for all the attributes were calculated using Eq. (5.1); these are presented in Table 5.5.

The t values of the statistical tests of the mean ratings were also calculated. The same treatment was applied to all questions with hypothesis testing results (Questions 14, 15, 16, 21, and 22). The t values of the statistical tests of the mean ratings for reasons for seeking ISO 9000 are shown in Table 5.5.

From Table 5.5, the five attributes considered important ($t > 1.323$) were 'to increase the company's competitiveness', 'to give clients greater quality assurance', 'to improve company's operating procedures', 'to boost company's image', and 'to reduce company's operating costs' respectively. These results match Haklik's (1997) and Tibor and Feldman's (1997) findings. The difference was that attributes 'because other companies had attained, or were seeking certification' and 'because it will soon be mandatory' were considered by respondents in this study as not important. These differences may be explained by the conditions in Vietnam where ISO 9000 is still not popular or mandatory.

The t values of the statistical tests of the mean ratings for benefits derived from ISO 9000 certification are shown in Table 5.6. The results of Table 5.6 show that all attributes were considered important by respondents. The order of importance was: 1) Instilled pride in company's employees, 2) Reduced company's operating costs, 3) Increased company's competitiveness, 4) Enhanced the company's corporate image, 5) Improved company's operating procedures, 6) Helped to enhance company's productivity and 7) Improved communications among personnel and 8) Enabled company to reduce material wastage, respectively.

The t values of the statistical tests of the mean ratings for disadvantages from ISO 9000 certification are shown in Table 5.7 which indicates that there was no disadvantage from ISO 9000 certification encountered by respondents. Although there were actually some changes to the company's operation such as working procedures, operating costs, these changes were not considerable as perceived by respondents.

Since all responding companies were joint ventures or 100% foreign-owned, the process of obtaining certification was guided and assisted by their overseas parent companies which had obtained certification. Hence, those companies had few problems becoming certified.

Table 5.6. Benefits derived from ISO 9000 certification — Mean importance ratings and t-test of the mean.

	Test value = 3		
Attribute	t	Mean	Standard deviation
Increased company's competitiveness	6.325	4.33	.52
Enabled company to reduce material wastage	2.076	3.83	.98
Improved company's operating procedures	3.796	4.17	.75
Reduced company's operating costs	6.325	4.33	.52
Improved communications among personnel	2.739	4.00	.89
Helped to enhance company's productivity	3.000	4.50	1.22
Enhanced company's corporate image	3.873	4.00	.63
Instilled pride in company's employees	7.000	4.17	.41

Table 5.7. Disadvantages from ISO 9000 certification — Mean importance ratings and t-test of the mean.

	Test value = 3		
Attribute	t	Mean	Standard deviation
Reduced company's competitiveness	−11.000	1.17	.41
Increased amount of materials used by company to attain required quality levels	−4.000	1.67	.82
Lengthened operating procedures	−3.796	1.83	.75
Increased operating costs	−4.000	1.67	.82
Reduced company's productivity	−11.000	1.17	.41

Respondents were asked to suggest some measures to widen ISO 9000 Certification among firms in Vietnam. Their suggestions were:

- Financial incentives from government and sector;
- More pressure from customers, demands for higher quality products;
- Larger export market; and
- Awareness raising, education and consultant services.

ISO 14000 certification and implementation

Respondents were requested to indicate how they first got to know about ISO 14000 based on a list of 7 stated sources. Data on respondents' sources of information on ISO 14000 are presented in Table 5.8. This table shows that seminars/conferences were main sources of the respondents' knowledge of ISO 14000 in Vietnam. Internet and print media also played a considerable role in raising people's awareness of ISO 14000. One respondent first knew about the ISO 14000 series from the parent company; the Vietnam subsidiary's ISO 14000 certification program was directed by the parent company's environmental unit.

The t values of the statistical tests of the mean ratings of the companies' reasons for seeking ISO 14000 certification are shown in Table 5.9. The three most important reasons for seeking ISO 14000 as stated by respondents

Table 5.8. Source of information on ISO 14000.

Source	Percentage	Importance of source
Newspaper	33%	2
Journal	17%	4
Seminar/conference	67%	1
Television	0%	5
Radio	0%	6
Internet	33%	3

Table 5.9. Reasons for seeking ISO 14000 — Mean importance ratings and t-test of the mean.

Attribute	t (test value = 3)	t (test value = 4)	Mean	Standard Deviation
Increase company's competitiveness	**4.245**	1.158	4.38	.92
Enable company to reduce material wastage	**7.937**	**2.646**	4.50	.53
Improve company's procedures	1.323	−1.323	3.50	1.07
Reduce company's operating costs	**2.346**	.261	4.13	1.36
Help to enhance company's productivity	1.323	−1.323	3.50	1.07
Enable company to avoid infringing statutes and regulations	**10.693**	**4.583**	4.75	.46
Enhance company's public image	**3.989**	.798	4.25	.89
Be essential in company's overseas drive	**1.528**	−1.528	3.50	.93
Be insisted upon by stakeholders	−1.080	−3.240	2.50	1.31
Improve workers' health, safety and welfare	**4.245**	1.158	4.38	.92
Help company to contribute to efforts to protect the environment	**10.693**	**4.583**	4.75	.46

were: 1) Help company to contribute to efforts to protect the environment; 2) Enable company to avoid infringing statutes and regulations; and 3) Enable company to reduce material wastage. The attribute 'be insisted upon by stakeholders' was regarded as not important by respondents.

The t values of the statistical tests of the mean ratings for problems related to ISO 14000 certification and implementation are shown in Table 5.10. The results indicate that only two attributes 'There is little knowledge of the standards' and 'There is low awareness among staff members' were regarded by respondents as important problems related to ISO 14000 certification and implementation of the EMS.

It was noted by four respondents that Vietnam lacked the infrastructure to facilitate the implementation of ISO 14000 EMS. Almost all companies faced problems in sorting and disposing of solid waste. Only Honda Vietnam

Table 5.10. Problems in ISO 14000 certification and implementation.

	Test value = 3		
Attribute	*t*	Mean	Standard deviation
There is little knowledge of the standards	**1.594**	3.88	1.55
There is low awareness among staff members	**1.821**	3.75	1.16
There is shortage of personnel	−1.655	2.25	1.28
The standard involves high costs	−.664	2.63	1.60
There is employees' resistance	−8.881	1.38	.52
Low commitment of managerial staff	−10.693	1.25	.46
Lack of co-operation of or difficulties made by local government	−5.227	1.63	.74
There are difficulties in understanding the standards	−1.821	2.25	1.16
There are uncertainty costs (legal, e.g.)	.000	3.00	1.77
The clients do not recognize it	−7.937	1.50	.53
Lack of facilities to meet legal/other requirements	−3.055	2.00	.93
It is in conflict with the objectives of most of the company's clients	−10.693	1.25	.46
There are difficulties in integrating EMS and existing QMS	−3.813	1.88	.83
There is complexity in documentation procedures	.475	3.25	1.49

Table 5.11. Source of information on ISO 14000.

Measures	Percentage	Ranking
It becomes mandatory	40%	4
Training of auditors and certifiers	60%	3
Education of people's awareness	100%	1
Financial incentives for ISO certified organizations	80%	2
Other kinds of incentives rather than financial (e.g. less legal inspections)	20%	5

had a furnace for combusting solid waste generated by the firm. The legislation was also considered to be difficult to understand. There was no implementation guideline for solid waste management.

Respondents' suggestions on how ISO 14000 can be further promoted in Vietnam are presented in Table 5.11.

It is clear that all respondents considered enhancement of awareness a must in promoting ISO 14000. Other measures such as financial incentives for ISO 14000 certified organizations and training of auditors and certifiers were also deemed to be important. The two least important measures were mandatory certification (40%) and other kinds of incentives rather than financial ones (20%).

Discussion

With the total of 21 out of 22 the ISO 14000-certified companies in Vietnam being large foreign businesses, it is clear that companies that operate on a global basis are more concerned with environmental issues than Vietnamese ones. Sai Son Cement Company was the first state-owned company which obtained certification (in 2001). Joint ventures and companies owned outright by foreign investors indicated strong interest in conforming to ISO standards. For many others, especially domestic firms, the concept of ISO 14000 EMS is very new but there is a growing awareness that it will be an important tool for prevention of pollution by industry. These findings confirm the results of recent studies (i.e. Haklik, 1997; Tibor and Feldman, 1997; Radminton, 1997).

All respondents had international activities. Internal environmental concerns of other countries are being passed from international corporations down through their supplier networks in Vietnam.

Manufacturing enterprises appear to be more concerned about the environment than other sectors. Firms in that sector had taken greater interest in ISO 14000 certification than those in other business sectors, and its provisions had started to influence their attitudes, seven out of the eight companies operated in the field of manufacturing while only one company specialized in the hospitality business.

Quality is the most-valued development objective of the firms in the study. Environmental issues and time are much pursued by the businesses; minimizing environmental impacts and meeting set project deadlines take the second and third places in the ranking. This shows that market pressures relating to quality do have influence on businesses. The quality requirements of products need to be met while at the same time project deadlines are adhered to, and the environment is protected.

Three out of eight companies (37.5%) had the Certificate for Meeting Environmental Requirements of the MOSTE. Others also indicated their interest in having the Certificate. This is probably due to the perceived benefits of the Certificate such as good relations with governmental agencies, protection of the environment, more effective environmental management system, and less environmental inspection. By complying with the regulations, the companies will have a comfortable working environment in terms of their relationships with the local authorities.

The results of the survey show that ISO 9000 certification is beneficial to the businesses. No considerable disadvantages were indicated by respondents. This reflected in the high t-value of benefits from certification (from 2.076 to 7.000). Thus, it could be concluded that the respondents' views on ISO 9000 were consistent.

Seminars/conferences were the main sources of respondents' knowledge of ISO 14000 in Vietnam. Internet and print media also played a considerable role in raising people's awareness of ISO 14000.

Almost all the benefits of implementation of ISO 14000 that were identified in the literature were applicable to the sample companies. Respondents indicated that benefits from ISO 14000 certification and implementation are substantial. These were also the main reasons why the firms sought certification. In the study, 'help company to contribute to efforts to protect the environment'; 'enable company to avoid infringing statutes and regulations'; and 'enable company to reduce material wastage' were the most important reasons why the firms went for ISO 14000 certification and implementation. The cost-benefit concerns of firms when implementing ISO 14000 were dominant. For example, by reducing material wastage, costs are cut.

The results of the survey are consistent with the literature; the possibility of demonstrating conformity with legislation is considered to be among the most important motives for all businesses to achieve certification. Environmental management systems are believed to provide a means to meet mandatory standards effectively and efficiently.

Problems identified in the literature did not apply in Vietnam. Only two attributes: 'there is little knowledge of the standards' and 'there is low awareness among staff members' were regarded by respondents as important problems related to ISO 14000 certification and implementation of the EMS. Dissemination of the standards in Vietnam is quite limited. Although environmental awareness among managerial staff is high, the general workforce has very low awareness. Some respondents considered ISO 14001 to be just documenting the procedures they already had in place.

It should be noted that the survey sample of this research consisted of only eight multinational companies and no local companies. Thus, this study has not been able to capture the experiences of the Vietnamese SMEs, which may be significantly different from those of the large corporations studied.

Recommendations

Some actions to facilitate ISO 14000 certification and implementation which were proposed by the respondents include: (i) awareness building among the concerned stakeholders, especially employees; (ii) provision of infrastructure such as solid waste treatment facilities; (iii) training of auditors and consultants; and (iv) assistance in developing and implementing EMS; and (v) development of necessary regulations and legislation.

This research shows that there is still little knowledge of ISO 14000 series standards among the business community in Vietnam. In particular, environmental awareness of the employees is limited. As a result, the media should play a more active role in disseminating information on ISO 14000 EMS. Non-governmental organizations (NGO) and international organizations would also be an invaluable source of contribution to the knowledge of ISO 14000. Workshops and conferences should be organized to raise awareness and understanding of the relevant organizations and individuals, especially SMEs.

The responding businesses are evidently convinced by the benefits which may be derived by implementation of EMS and certification to ISO 14000. These benefits are mostly related to firms' environmental awareness, compliance with regulations and cost-benefit concerns. As these businesses are doing well in Vietnam, their performance can be used as evidence to help convince management and employees while selling the idea of implementation of EMS. Research on the benefits of ISO 14000 which gathers hard evidence should be encouraged in order to convince the businesses about the soundness of the ISO 14000 series standards.

As ISO 14000 certification has increasingly attracted the attention of the businesses, the government and international organizations should promote its implementation. There should be better coordination between STAMEQ as the national standards body, the relevant government institutions, trade unions, industrial sectors, transnational companies, NGOs and international organizations to promote the implementation of ISO 14000 in Vietnam. MOSTE, as the leading government environmental management organization, should propose the necessary policy and concrete measures to promote implementation of EMS in Vietnam.

Multi-national companies can play an important role in the implementation of ISO 14000 standards. Pressures from parent companies are put on developing country suppliers to attain certification. As a result, relevant authorities in Vietnam such as the Ministry of Trade may wish to consider the possibility of further cooperation between companies in developed countries and their suppliers in Vietnam in the area of EMS implementation.

The government should provide the necessary infrastructure, legislation and incentives for ISO 14001 EMS implementation. The Government should encourage SMEs to seek ISO 14001 certification through incentive programs. There is a consensus that for SMEs, lack of resources and limited access to technology are barriers encountered in the pursuit of ISO 14001 certification. Vietnamese SMEs may have similar resource constraints as those of other countries. Support from private and public sources can help to minimize the problems of the companies which are interested in pursuing certification but are constrained by resource limitations. The government agencies should also provide necessary technical assistance for solving legislative and other difficulties.

Conclusion

This research gave an overall view of environmental management and ISO 14001 EMS certification and implementation in Vietnam. It was able to explore the benefits, as well as problems encountered by firms in implementing ISO 14001 EMS certification. Based on the review of literature and collected data, this research was able to provide a number of recommendations for ISO 14001 implementation in Vietnam. However, it had some limitations. First, since ISO 14000 certification was very new in Vietnam, there was little information and research on this standard in Vietnam. Therefore, this research had to be based on information and from international sources which were not closely related to the conditions in Vietnam. Second, the survey population of this research consisted of eight multinational companies identified by convenience sampling. The data collected were therefore not complete. This study has not been able to capture the experiences of the Vietnamese SMEs, which may be significantly different from those of the large multinational corporations.

It is important to point out that each company is different, with different operations, unique specific products, as well as different customers, employees and business partners, so the nature and extent of barriers, benefits and critical success factors would probably be different for each company. The specific features of the country may also pose some problems to the firms in the implementation of ISO 14001 EMS. As the number of ISO 14001-certified companies is rapidly increasing in Vietnam, future studies may obtain a larger sample size and thus, produce more valid results. Such research works can also undertake comparative studies on companies with different backgrounds.

Appendix A

Survey questionnaire

Environmental management and ISO 14000 certification and implementation in Vietnam

A. COMPANY'S PROFILE

1. Company's name:
2. Company's year of formation:
3. Status of company (*please indicate with a 'tick' against the appropriate type*).
   a. State-owned (local)
   b. Private (local)
   c. 100% foreign-owned
   d. Joint-venture
   e. Others (*please specify*):
4. Company's area of specialization:
5. Company's initial investment:
6. Company's approximate annual turnover:
7. Company's personnel (*please provide numbers*):
   a. Professionals
   b. Technicians
   c. Administrative personnel
   d. Skilled workers
   e. Unskilled workers
   f. Others:
8. Designation of person answering this questionnaire:

B. COMPANY'S DEVELOPMENT OBJECTIVES

9. Please rank the following from 1 to 6, (*1 being 'most important' and 6 being 'least important'*) in terms of the company's priorities when it undertakes development projects:
   a. To minimize cost
   b. To meet set project deadlines

c. To attain high productivity
d. To avoid infringing any relevant regulations
e. To minimize environmental impacts of company's operations
f. Others (*please specify and indicate rank*):

C. ENVIRONMENTAL RECORD

10. Have your company got the Certificate for meeting national environmental requirements?

    a. Yes
    b. No

11. Organization's environmental record if any (i.e. awards, prizes, special organization policy regarding the environment, e.g. training, participation in events relating to the environment):

D. ISO 9000 CERTIFICATION

12. Does the company have the certification to ISO 9000 series of Quality Management Systems (QMS)?

    a. Yes
    b. No

13. If the answer to Question 12 is 'Yes', please indicate year in which company attained ISO 9000 QMS certification:
14. If the answer to Question 12 is 'Yes', please indicate the extent to which each of the following was one of the main reasons why the company sought certification by marking/circling appropriate number. (Note that 5 indicates 'most important' and 1 indicates 'not important')

a. To boost company's image	1 2 3 4 5
b. Because other companies had attained, or were seeking certification	1 2 3 4 5
c. To increase the company's competitiveness	1 2 3 4 5
d. Because it will soon be mandatory	1 2 3 4 5
e. To give clients greater quality assurance	1 2 3 4 5
f. To reduce company's operating costs	1 2 3 4 5
g. To improve company's operating procedures	1 2 3 4 5
h. Others (*please specify and mark appropriate number 1–5*)	

15. If the answer to Question 12 is 'Yes', please indicate the extent to which your company has derived any of the following benefits from ISO 9000 series certification. ISO 9000 has...

a. Increased company's competitiveness	1 2 3 4 5
b. Enabled company to reduce material wastage	1 2 3 4 5
c. Improved company's operating procedures	1 2 3 4 5
d. Reduced company's operating costs	1 2 3 4 5
e. Improved communications among personnel	1 2 3 4 5
f. Helped to enhance company's productivity	1 2 3 4 5
g. Enhanced the company's corporate image	1 2 3 4 5
h. Instilled pride in company's employees	1 2 3 4 5
i. Others (*please specify and indicate appropriate number 1–5*):	

16. If the answer the Question 12 is 'Yes', please indicate the extent to which your company has experienced any of the following disadvantages from ISO 9000 QMS certification. ISO 9000 certification has ...

a. Reduced company's competitiveness	1 2 3 4 5
b. Increased amount of materials used by	
c. company to attained required quality levels	1 2 3 4 5
d. Lengthened operating procedures	1 2 3 4 5
e. Increased operating costs	1 2 3 4 5
f. Reduced company's productivity	1 2 3 4 5
g. Others (*please specify and indicate appropriate number 1–5*):	

17. In your opinion, what would be a positive step in enhancing ISO 9000 certification in Vietnam?

E. ISO 14000 AND THE COMPANY

18. How did you first learnt about ISO 14000? (*please indicate with a 'tick' against the appropriate answer.*)

    a. Newspaper
    b. Magazine
    c. Journal
    d. Conference/Workshop
    e. Television
    f. Radio
    g. Internet
    h. Others (*please specify*):

19. Please specify company's necessary time period to meet requirements of ISO 14000
20. Please indicate approximate costs of certification and implementation (consultancy fees, annual costs of maintaining, technology, human resources, and etc.)
21. Please indicate the importance of each of the following factors to the company's decision to seek ISO 14000 certification. ISO 14000 helps...

a. Increase company's competitiveness	1 2 3 4 5
b. Enable company to reduce material wastage	1 2 3 4 5
c. Improve company's procedures	1 2 3 4 5
d. Reduce company's operating costs	1 2 3 4 5
e. Help to enhance company's productivity	1 2 3 4 5
f. Enable company to avoid infringing statutes and regulations	1 2 3 4 5
g. Enhace company's public image	1 2 3 4 5
h. Be essential in company's overseas drive	1 2 3 4 5
i. Be insisted upon by stakeholders	1 2 3 4 5
j. Improve workers' health, safety and welfare	1 2 3 4 5
k. Company to contribute to efforts to protect the environment	1 2 3 4 5
l. Others (*please specify and indicate appropriate number 1–5*):	

22. What are the problems which your company has encountered when seeking ISO 14000? (*Please answer by circling the appropriate number for each factor. Note that 5 indicates 'most important' and 1 indicate 'not important'*).

a. There is little knowledge of the standards	1 2 3 4 5
b. There is low awareness within staff members	1 2 3 4 5
c. There is shortage of personnel	1 2 3 4 5
d. The standard involves high costs	1 2 3 4 5
e. There is employees' resistance	1 2 3 4 5
f. Low commitment of managerial staff	1 2 3 4 5
g. Lack of co-operation of or difficulties made by local environmental government	1 2 3 4 5
h. There are difficulties in understanding the standards	1 2 3 4 5
i. There are uncertainty costs (legal, e.g.)	1 2 3 4 5
j. The clients do not recognize it	1 2 3 4 5

k. Lack of facilities to meet legal/other requirements 1 2 3 4 5
l. It is in conflict with the objectives of most of the company's clients 1 2 3 4 5
m. There are difficulties in intergrating EMS and existing QMS 1 2 3 4 5
o. There is complexity in documentation procedures 1 2 3 4 5
p. Others (*please specify and indicate appropriate number 1–5*):

23. What measures can be taken to promote ISO 14000 certification and implementation in Vietnam?

    a. It becomes mandatory
    b. Training of auditors and certifiers
    c. Education of people's awareness
    d. Financial incentives for ISO certified organizations
    e. Other kinds of incentives rather than financial (e.g. less legal inspections)
    f. Others (*please specify*)

Thank you very much!

Please indicate whether you would like to receive details of the findings

Yes
No

If 'Yes', please indicate:

Name:
Address:
Tel.:
Email:
Fax:

References

Dinh, V. Sam *et al.* (2001) Vietnamese National Industrial Overview. National Report Prepared For Rio+10 World Summit On Sustainable Development, United Nations Industrial Development Organization.

Edwards, W., & Newman, J. R. (1982) *Multiattribute Evaluation.* London: Sage Publication, Inc.

Haklik, J. E. (1997) "ISO 14000 Environmental Management: Benefiting Companies, Saving The Environment", Business Access Online. Http://Www.Clickit.Com/Newsweb/Nwsrch.Htm

Holt, G. D. (1997) Construction research questionnaires and attitude measurement: relative index or mean. Journal of Construction Procurement 3 (2), pp. 88–96.

Newbold, P. (1991) Statistics for Business and Economics, 3rd ed., Englewood Cliffs, New Jersey. Thaler, R., 1987.

Nguyen, V. Nam. (1999) Eco-Labeling, Ems Certification And Implications For International Trade Expansion Of Vietnam, In *Trade And Sustainable Development In Vietnam.* International Institute For Sustainable Development. Hanoi, Vietnam.

Ofori, George (1999) Implementation Of Environmental Management Systems (ISO 14000) In Construction Project Management Organization In Singapore. The National University Of Singapore, Singapore.

Singleton, Jr R. *et al.* (1988) Approaches to Social Research. Oxford University Press. New York.

Tibor, T. And Feldman, I. (1997) Implementing ISO 14000: A Practical, Comprehensive Guide To The ISO 14000 Environmental Management Standards. Mcgraw-Hill, New York.

Tran, T. T. Phuong (1996) Environmental Management And Policy Making In Vietnam. Australian National University, Canberra, Australian. Http://Coombs.Anu.Edu.Au/~Vern/Env_Dev/Papers/Pap03.Html

Vietnam Productivity Centre (VPC) (2001) http://www.vpc.vn/Accessed 11/2001.

6

STRATEGIC ENVIRONMENTAL ASSESSMENT AS A TOOL FOR SUSTAINABLE DEVELOPMENT IN HONG KONG

NG KAY LENG and JEFFREY OBBARD

Hong Kong is one of the few countries to implement Strategic Environmental Assessment (SEA) of major plans and proposals. SEA has provided decision makers with information on environmental impacts resulting from proposed plans or programs and has resulted in greater accountability and transparency in the decision-making process. SEA has also helped to increase environmental awareness of various governmental departments. SEA models for assessing developmental impacts on air, noise, and water quality are developed to aid in SEA application. Public participation takes place at various stages of the SEA process and has helped to increase the public environmental awareness and involvement. However, application of SEA in Hong Kong has notable limitations and difficulties. Generally, Hong Kong lacks a comprehensive strategic planning policy for the environment. Different departments often have conflicting interest when facing development. In Hong Kong, SEA is primarily used as a planning tool to evaluate the environmental implications of development plans or strategies but is seldom applied to major policies and programs. Hence, most SEA studies are development led and mainly consider new development areas instead of placing conservation value on important habitat and natural resources. Another limitation is that current land use plans and transportation strategies in force will largely determine the pattern of development over the next few years but are not apparently subject to SEA.

The lack of knowledge of Hong Kong's own environmental resource base has resulted in projects with adverse impacts on high ecological value areas being implemented without knowledge of the impacts. Furthermore, the intrinsic value of the natural environment and cultural heritage is subjective and is often underestimated or neglected. The indirect impacts caused by proposed developments may be unforeseeable during the time of assessment. Rapid population growth, rapid urbanisation, increasing unemployment rates coupled with a historic lack of planning controls, low public awareness, inherent conflicts among government departments, and vested interests in economic development have been major obstacles to achieving environmental and social sustainability.

There is a need for comprehensive overall strategic planning for the environment. Environmental objectives should be stated in all development proposals. Effective SEA can take place either through enforcement by a legal system or management of the process by a competent official body with sufficient power. There should be close interaction among relevant planning authorities, the environmental protection department and other relevant parties. Guidelines developed by the International Association for Impact Assessment (IAIA) should be closely followed to maintain a high standard of SEA processes. The public participation element of the process should also be strengthened and early public involvement encouraged. There should be a review of the need to conduct an SEA whenever there is a material change in development strategy. SEA application should move beyond a sectoral approach, to look at ways in which development decisions can not only prevent environmental damage but positively enhance natural resources.

Introduction

Strategic environmental assessment (SEA)

Environmental assessment (EA) has been widely applied at project-level since the early 1970s. However, project-level EA is often not sufficient as it reacts to development proposals rather than proactively anticipating them. EAs only take place after strategic decisions have been made and they often address only a limited range of alternatives and mitigation measures. Assessment of indirect, induced, and cumulative impacts are very often absent and there is a lack of detailed analysis of project alternatives as these are not considered at an earlier stage of planning. With the emphasis on sustainable development, there is growing recognition that environmental assessment should be utilised earlier in the project proposal stage. SEA is the environmental assessment of proposed and existing policy, plan and program (PPP) and their alternatives. SEA is able to ensure the environmental integrity of a country's PPP in order to achieve a more sustainable approach to development as it involves the evaluation of the likely significant environmental consequences at the early planning stage and hence increases the possibility of analysing and proposing alternative solutions and also choosing the most appropriate action based on environmental and socio-economic grounds, and to minimize any residual environmental impacts. SEA is able to incorporate sustainability criteria throughout the planning process, as it carries the principles of sustainability down from policies to individual projects (Arce and Gullón, 2000). Hence, SEA has gained wide acceptability as a supporting tool for decision making towards achieving sustainable development (Noble, 2002).

Hong Kong

Hong Kong is a developed economy with a high population density (6,375 persons per km^2, 2002) in a confined land area. Hong Kong has a population of 7 million and total land area of 1098 km^2. Despite Hong Kong's capitalist economy and overcrowded conditions, it contains a diversity of ecological habitats due to its geographical location, being at the boundary of the tropical and temperate climatic zones. Physical and climatic conditions combine to produce a variety of marine and terrestrial habitats of significant scientific and conservation interest. However, development projects are often in conflict with the natural environment especially in countries with limited land resources. The quality of both the natural and urban environments of Hong Kong has undergone enormous change in recent decades due to the rapid pace of urbanisation, economic growth, and development.

SEA is quite well established in Hong Kong and the application of SEA to plans and strategies in Hong Kong commenced in 1988. Up to now, the most comprehensive application of SEA was in the Territorial Development Strategy Review, completed in 1996, which made provisions to cater for an increase in the population of Hong Kong from 6.4 million to 8.1 million in 2011 (Au, 2002).

Research objectives

The objective of this study is to evaluate the effectiveness of using SEA as a tool for achieving sustainable development in Hong Kong. The methodology employed involves the evaluation of the SEA process in the Territorial Development Strategy Review (TDS Review) and the Third Comprehensive Transportation Study (CTS-3) based on the International Association for Impact Assessment (IAIA) performance criteria. The lack of proper SEA for the development of Disney Land Theme Park on Lantau Island and the associated problems that resulted are also discussed. The effectiveness of these SEA applications in helping Hong Kong to meet the sustainable development criteria developed by the Study on Sustainable Development for the 21st Century (SUSDEV 21) in Hong Kong are also assessed. Recommendations for improvements are given at the end of these evaluation processes.

Application of SEA in Hong Kong

Territorial development strategy review (TDS review)

The most comprehensive application of SEA to the long-term physical planning process in Hong Kong was in the TDS Review in 1996. TDS is the highest level of land-use planning in Hong Kong. Many of the principal components of Agenda 21 were considered in the TDS Review as shown in

Table 6.1 (Dubin, 2002). The framework for analysing sustainability issues in the TDS Review covered the four main themes described.

The original TDS was published in 1984 and a review of the TDS commenced in mid-1990 with the objective of providing a land use/transport/environment framework for the future development of Hong Kong to cater for an increase of population from the existing 6.4 million to 8.1 million in 2011. The review produced a long-term strategy for an assumed target year of 2011 and a medium-term strategy for the year 2005, for two broad scenarios, each using different assumptions of the population of the territory and range of economic influence within China. Several rounds of detailed multi-disciplinary evaluations (including environmental evaluations), studies and screening processes were undertaken for individual sectors such as industrial, residential, commercial, transport, recreational, land-use and port development.

One basic doctrine applied in the TDS Review is that environmental initiatives and objectives should be fully integrated into the overall planning process. The objectives of the SEA were to provide inputs for formulation and assessment of future developments within a broad plan, to identify any outstanding environmental issues requiring further investigations, to optimise the selection of development options in environmental terms and to ensure that the least environmentally damaging development options were chosen.

The SEA process sought to enhance and protect the quality of the environment with regard to air quality, water quality, noise, solid waste disposal, and potentially hazardous installations. It focused on cumulative environmental implications associated with the broad pattern and direction of growth and the selection of strategic growth areas. It involved carrying out a baseline study to establish the environmental profiles for the territory; identifying strategic environmental constraints; establishing environmental principles and criteria for generation and evaluation of development options; and assessing and evaluating the options generated.

There are three stages of option formulation and evaluation, from Initial Options, through Hybrid Options and finally to the Preferred Options and Refined Preferred Options. Each stage builds on the findings of the previous one to identify positive attributes and eliminate negative ones and to narrow down the range of choices. The two Preferred Options evolved from an iterative evaluation process using a series of performance criteria to screen out undesirable elements, to protect areas of specific conservation value and to focus on areas where development could be achieved in a controlled manner. Examples of the performance criteria used are presented in Table 6.2.

Table 6.1. TDS and sustainability.

Agenda 21 components	Application to TDS
Trade and environment mutually supportive	TDS integrates land use, transport and environmental components into overall strategies
Combat poverty to maintain sustainable livelihood	An objective of TDS is to create and maintain a high standard of living
Internalise environmental costs	The cost of sewage and waste disposal were incorporated into the TDS 'Polluter pays' principle is HK Governmental policy
Promote sustainable consumption patterns	TDS development strategies aim to provide a sustainable growth rate
Incorporate demographic trends in the analysis of environment and development issues	Consideration of demographic trends was explicit in the iterative TDS process Environmental impact predictions were used to produce improved strategies
Protect and promote human health	TDS aims to minimise air quality impacts and other health related environmental impacts, rationalise industrial development, reduce emissions, promote clean industry and improve recreational facilities
Protect the atmosphere	TDS aims to minimise air quality impacts, rationalise industrial development, reduce emissions and promote clean public transport and clean industry
Promote environmental education	Consultation exercise associated with the TDS generated considerable community interest
Implement land use planning for conservation strategy	TDS included existing conservation strategy and land use constraints. Protection was given to Country Parks, Agricultural lands and Sites of Special Scientific Interest
Protect marine resources	TDS screened out some development options which would adversely impact tidal flow regimes, and considered impacts on receiving water bodies as population increases
Protect fresh water resources	Implicit in TDS approach and in preservation of water catchment areas
Provide environmentally sound management of solid wastes	Explicit in TDS and in waste disposal plan and initiatives for waste reduction
Encourage participation by business and industrial sectors, trade unions, women and youth groups in achieving sustainable development	Not explicit in TDS but promoted through public consultation and debate on TDS recommendations
Strengthen role of non-governmental organisations (NGOs) in achieving sustainable development	Not explicit in TDS but implicit in public consultation process

Table 6.2. Examples of the performance criteria.

Environmental quality	Parameters measures
Air quality	• Kilometres of road within airshed • Projected numbers of vehicles on roads • Total NOx, SOx emissions/CO emission rates • Length of industrial/residential interface • Population within confined airshed
Noise	• Rail interface with residential areas • Road interface with residential areas • Population within airport NEF25 countour
Wastes	• Projected domestic/industrial wastes generated • Proximity to appropriate disposal
Water quality	• Areas of development without connection to trunk sewers or treatment facilities • Projected increase in BOD from all sources • Flows in excess of committed disposal capacity • Proximity of outfalls to confined water bodies/ sensitive receivers
Ecology	• Interface with ecologically significant sites • Extent of wetland depletion
Others	• Potentially hazardous installations • Compliance with planning guidelines • Sustainability

EA of the development strategies in terms of their effect on air and water quality, noise, solid wastes, ecology and potentially hazardous installations were conducted during the first round of evaluation of the Preferred Options. This identified components of the Preferred Options which required refinement and components which were synthesised into the medium and long term development strategies. The second round of evaluations of the Refined Preferred Options focused on the key issues which were identified in the first round of assessments relating to liquid and solid wastes arising and the implications for their disposal, air pollution from vehicular sources and cumulative effects of the industrial and transport strategies.

The SEA output indicates that on the basis of current policies and technology, potentially serious environmental impacts will result from the assumed levels of population growth set out in the long-term development scenarios. Even with all proposed mitigation measures in place to minimise the environmental impacts, serious human health-related residual impacts are predicted to arise from the Refined Preferred Options. The major impacts include deterioration of air and water quality; overloading of sewerage infrastructure; reduction in life spans of existing and proposed landfills; increased noise levels in both the urban areas and the New Territories; and both direct and indirect impacts on ecologically valuable habitats. Although it was proposed that some of the identified problems could be mitigated if adequate resources were applied; others were identified as requiring policy actions, no concrete mitigation measures were mentioned and such mitigation measures were left to be proposed at the project-level environmental impact assessment (EIA) stage. A set of guidelines were formulated for future development of sustainable transport, and industrial and commercial policy following the TDS Review.

A qualitative assessment was also carried out for the medium-term strategy, with a target year of 2006. It showed that some environmental problems identified could begin to manifest themselves before 2006. These include possible failure to meet air quality objectives, worsening of noise situation due to traffic flows, and water quality and sewage disposal problems in some areas.

The predicted environmental impacts from the TDS SEA analysis have cast doubt on the sustainability of the long-term TDS strategies. The results of the study highlighted the need for developing a territorial and strategic sustainability framework to form the basis for proposing further development without depleting the environmental qualities of Hong Kong.

Such impacts should not have surfaced if SEA had been applied right at the beginning of the planning process. All environmental impacts and policy alternatives should be taken into account at the start of the planning process when developing the initial options. In this case, SEA was conducted for the development options but not during the formulation of the options. As a result, EA showed that the natural capacity would be exceeded for the various development strategies. SEA application in this case did not aid the achievement of sustainable development.

The TDS Review report states that individual components have to be subjected to the EIA process and cumulative impacts will require further investigation through detailed studies as a basis for future policy decisions. This again proves that the application of SEA was not quite successful in the TDS Review. SEA is supposed to take into consideration cumulative impacts and residual effects of development proposals. By the time the individual project proposals reach the EIA stage, it will be too late to make any changes to environmentally damaging actions. Limited mitigation measures can be proposed at this stage, and such mitigation measures may not be feasible from the engineering or economic standpoints.

In certain cases, results of SEA were not seriously considered. For example, it was stated that the proposed long-term transport link across Deep Bay, the port back up land in San Tin/Lok Ma Chau and residential development in rural North West New Territories could affect the fragile ecosystem within this area both during and following construction. Despite this finding, developments are being undertaken in these areas; there seems to be no plan to reconsider these proposals.

Following the TDS Review, the government had proceeded with plans for extensive reclamation to cope with the demand for land and physical facilities. This includes reclaiming a substantial part of Victoria Harbour which will reduce its original size by 50 percent and turn the harbour into a narrow sea channel. An alternative plan to develop the New Territories was seen as requiring a longer lead-time prior to development due to the need for comprehensive feasibility studies and that assembly of land under fragmented ownership would be difficult, time consuming and costly. The harbour-based option was promoted as a more speedy option because harbour reclamation areas are either covered by detailed feasibility studies or are currently subject to such investigations. This proposal met with opposition from the public and professional groups including architects, planners, landscape specialists, engineers, academics, environmentalists, politicians, and even real estate developers. It would appear that the government has a vested interest in creating such prime site properties through reclamation as the government derives a major proportion of its revenue from land, including property rates and taxes, stamp duties, lease modifications, and premium on new land. The environmental cost of reclamation is not internalised. Of the three elements of sustainability, it would appear that the economic element has taken precedence over the others.

The plan to develop more container terminals to cope with the increase in the area of cargo sources and the volume of cargo flow passing through the Hong Kong port due to China's accession to World Trade Organisation (WTO) which would bring more trade and investment flows to China in the long term, despite the shortage of land, will bring with it worsening pollution as more container terminals will increase the number of diesel driven container trucks on the road. According to the Hong

Kong Government's estimates, the number of such trucks could increase a staggering 564 percent by the year 2011 (Loh, 1997). Diesel powered vehicles are the main air polluters in the urban areas in Hong Kong. The strategy of continuing to increase the number of container terminals should be reassessed to determine if the economic gain from container terminals is able to offset the economic loss due to deterioration of human health and the loss of the intrinsic value of the natural habitats that are to be removed to accommodate the terminals.

The implementation of environmental quality standards is not sufficient to prevent impacts on the environment. Continuous increase in population will continue to deteriorate environmental quality. There will be increased sewage discharges to the sea and the increase in the demand for land will cause further deterioration of water quality due to the dispersion of contaminants from dredging and reclamation activities. There should be a change in the policy with the targetof reducing population growth instead of implementing standards which are difficult or impossible to achieve.

Despite the various flaws, SEA application helped to identify potentially serious environmental consequences from the development scenarios, recommended provision of environmentally related infrastructure to meet the forecast medium-term development needs, and brought to the attention of the general public and decision-makers the potential adverse impacts on environmental quality and the urgent need for actions to deal with sustainability issues. Throughout the process, the findings of the SEA can also help to influence strategy formulation with a number of environmentally damaging options discarded or significantly modified. For example, some major potentially damaging development options in the East and South of Hong Kong were abandoned and such sites are preserved for nature conservation, amenity and recreation. Thus, Buffer Zones I and II in Deep Bay are protected from further development (HK Planning Department, 1995). The SEA also identified issues which need to be addressed from a regional perspective. Moreover, it resulted in the formulation of follow up actions to develop a sustainable development system for Hong Kong.

Although the TDS Review did not come up with a set of indicators and criteria to examine the extent to which sustainability has been achieved, it provided a platform for the development of the set of forward-looking indicators in the SUSDEV 21 study which was completed in 2001. Environmental sustainability issues associated with the long-term strategy were investigated in greater detail in the SUSDEV 21 study. It is hoped that the development of a strategic territorial sustainability framework and future developments will continue to be applied to long-term planning. The results of the TDS Review are currently being reviewed as part of the latest round of strategic planning under the 'Hong Kong 2030: Planning Vision and Strategy (HK2030)' study which provides a long-term land use-transport-environment planning framework for Hong Kong in the next 30 years.

However, there is a time lag between the completion of SUSDEV 21 and the implementation of the strategies developed by the TDS Review, and results from SUSDEV 21 may be too general to have immediate impact on policy. Meanwhile, if the Hong Kong government is to push ahead with the full implementation of the TDS Review, Hong Kong could end up with little left to develop in the future, with irreversible damage to the environment, reduced biological diversity, and as the overall result would be that the territory would be an unpleasant place for human habitation.

Third comprehensive transport study (CTS-3)

The objective of the CTS-3 study is to provide a framework for the Government to develop a balanced transport strategy, with due consideration of budgetary constraints and environmental issues and to facilitate the mobility of people and goods in Hong Kong by road, rail and ferry in an environmentally sustainable manner up to 2016. The primary product of the study is a transport plan for major infrastructure and policies for Hong Kong, quantified in terms of costs, preliminary alignments, benefits and timing, covering the planning horizon up to 2016. The study was commissioned by the Hong Kong Transport Department. Consultants from Wilbur Smith Associates Ltd and ERM Hong Kong Ltd were appointed to undertake the CTS-3. ERM undertook the SEA to address the environmental implications of different transport scenarios developed from the CTS-3.

The SEA reports the EA conducted on the Main Model Runs, and includes the Recommended Transport Strategy up to year 2016. The projects and policies in the CTS-3 are subjected to SEA for the first time. The comprehensive transport study addresses potential environmental impacts of different transport strategies to ensure that future CTS-3 transport developments are realized in an environmentally acceptable manner. It focuses on the air and noise qualities and the ecological aspects of the environment. The SEA also examines the environmental constraints on future strategic transport developments, and of air and noise pollution as well as general ecological profile to identify where strategic transport developments should be avoided. The Study evaluated the environmental benefits and limitations of various transport development options to help identify a transport framework which would contribute to more sustainable development in Hong Kong.

In order to reduce the need for travel, and hence the need for further infrastructure and adverse impact on the environment, an approach is required in which land-use, transport, and the environment are considered in an integrated manner in both strategic and sub-regional planning studies. The TDS Review was conducted in accordance with this theme, and provided key development assumptions to CTS-3.

EA was conducted to establish the baseline environmental conditions so that the potential deterioration of future transport scenarios could be identified. The SEA considered four population/traffic growth scenarios with a range of rail, road and traffic management options. Some 99 different transport scenarios were considered which employed air and noise indicators to evaluate the environmental 'performance'. Air quality assessment was conducted in terms of emission inventories and detailed modelling. The emission inventories compared transport scenarios on a district by district basis and identified districts where air quality could deteriorate. The detailed air quality modelling predicted the future air quality in terms of Air Quality Objectives. Noise assessment was conducted in terms of noise conditions relative to the baseline (1997).

The development of new transportation infrastructure has the potential to degrade natural habitats in Hong Kong, many of which support a diverse range of fauna and flora. Protecting such natural resources is a prime consideration in CTS-3. An overview of the ecological profile in Hong Kong focusing on identification of ecological conservation areas was undertaken at an early stage of CTS-3. Ecological appraisals were conducted when planning new infrastructure projects to identify potential ecological impacts at sensitive areas. The primary potential impacts related to the recommended strategic highways include the loss of habitat as a result of land-taking or slope-cutting processes. The potential loss of habitats has been identified for proposed infrastructure projects.

All projects proposed by CTS-3 which are classified as Designated Projects under the Environmental Impact Assessment Ordinance (EIAO) will undergo an EIA process during the project feasibility study stage. Alternative options and alignments would be studied and optimised to avoid and minimise environmental impacts before and during the formal EIA process under the EIAO or during the feasibility study of any specific transport infrastructure project to address project specific issues such as the ecological resource, direct and indirect air and noise mitigation measures required, visual and landscape impacts, impact on sites of cultural heritage, and other impacts arising from the proposed project. The environmental acceptability of the project should also be confirmed in the EIA. Preliminary environmental review (PER) during preliminary project feasibility study would also provide useful information to identify the best alignment before the commissioning of the EIA study.

An environmental monitoring and auditing framework was proposed to evaluate the effectiveness of the mitigation measures and assumptions used in the SEA. A precautionary principle approach was adopted for the implementation of the recommended transport strategy. Up-to-date information on vehicle numbers and classes, vehicle/kilometres travelled and related traffic emissions, ambient air quality, and noise levels at sensitive receivers could act as indicators to highlight the changing environmental conditions which could arise from the proposed strategy. Population, GDP growth, private and goods vehicle fleets, and cross-boundary traffic should also be monitored to determine if a review of the transport strategy and/or its environmental performance is required.

Based on the findings of the air quality and noise assessments conducted for the SEA, environmental conditions were expected to deteriorate due to increase in vehicle transportation in Hong Kong. This showed that the transportation strategy proposed in CTS-3 was not sustainable and the proposed mitigation measures may or may not be able to reduce or eliminate the environmental impacts caused by these development proposals. Alternatives provided by SEA can only apply downstream by recommending mitigation measures to reduce the impacts.

The new strategic highway projects recommended in CTS-3 have avoided some recognized conservation areas and areas of ecological interest identified as ecological constraints during the alignment development and evaluation stage, either through careful routing, or introduction of underground tunnels. It is estimated that 480 ha of cumulative terrestrial and marine habitat will be lost as a result of the development of the recommended strategic highway projects. This accounts for 5 percent of the total areas of such habitats in Hong Kong. However, it was stated that most of the habitats affected are mainly secondary in nature with low ecological significance, such as grassland and shrubland, and these impacts are considered to be generally low. Based on the broad new rail corridors, the overall potential ecological impact is expected to be low as most of the alignment will pass through either urbanised areas or secondary habitats, and tunnel options will be adopted to avoid conservation and other ecological sensitive areas. A more detailed rail alignments assessment was undertaken in the Second Railway Development Study (RDS-2).

In line with the study, feasibility study of an Electronic Road Pricing (ERP) scheme was commissioned in March 1997, with the objective of examining the practicability of implementing an ERP system in Hong Kong as a measure to control road usage during peak hours and on congested roads.

Following the CTS-3 study, the Intelligent Transport Systems (ITS) Strategy Review was conducted by the Transport Department of Hong Kong in 2001 to investigate how to reduce travel time, increase road capacity, reduce the accident rate and decrease fuel consumption.

Although ecological appraisal has been conducted, several proposed road alignments such as the Sai Kung and Ma On Shan Road Link, Eastern Highway, New Territories East-West Link and Route 81 do encroach upon the more important habitats such as natural woodlands and the associated ecological impacts are considered to be moderate to high. The Kam Tin Bypass, a multi-lane highway, bisects the land diagonally from north-east to south-west, leaving two smaller fragments of the School Marsh which forms part of the Kam Tin Wetland, an important bird habitat. The large roundabout at which the road terminates encroaches further upon the south-western corner of the marsh. The importance of the marsh was not identified at the strategic level (Kilburn, 2002). This problem resulted as SEA was applied only at the stage where development options were already formulated. It was proposed that these alignments should be refined to avoid or minimise the potential impacts to acceptable levels during the project EIA stage. However, mitigation measures at the project level may be too late to save the natural ecosystems at these sites.

The Deep Bay Link (DBL) and Shenzhen Western Corridor (SWC) had also resulted in encroachment upon ecologically sensitive areas. The site for the SWC was near the Mai Po Marsh Nature Reserve and the Chinese White Dolphins sanctuary and massive mitigation measures must be taken to reduce the impacts of these roads. Mitigation measures such as the implementation of a program to relocate exotic mangrove species and weeds, a plan to restore the function of Mai Po Gei Wais as bird feeding ground, and to recover the level of the bottoms of water channels connecting Gei Wai and Deep Bay in order to restore the function the of Gei Wai in Mai Po as the core feeding ground of the highly endangered Black-faced Spoonbill are costly. Moreover, the effectiveness of such mitigation measures is doubtful.

The DBL, which connects to the SWC at the coastline of Deep Bay, traverses the coastal areas where existing fishponds and wetlands form important foraging grounds for indigenous and migratory birds. As compensation, a wetland area of the same size will be created to remove the residual impact of the loss. Existing trees and bamboos habitat that served as nesting sites will be transplanted away from the limits of the works during suitable seasons to minimize the impact on the affected bird species. These mitigation measures may or may not be successful. Hence, there is a risk that the natural habitat of the birds will be permanently destroyed and this might threaten the survival of some of the species.

The development of the transport network in Long Valley has also affected the social and environmental sustainability of the area. Long Valley is the last remaining piece of land of any significant size representative of the natural habitat associated with the once extensive flood plains in the north-western part of the New Territories. It provides a diversity of fauna with a place to live and breed, and is a major source of biodiversity in the territory; more than 210 bird species have been recorded. Long Valley is also a site of great heritage value. Traditional wet agricultural practice is still alive, co-existing with the natural wildlife in harmony. The place provides a vivid and impressive picture of Hong Kong's agricultural past. It is a major resource for eco-tourism with a heritage sideline. Long Valley is also a site of great educational value for students, in terms of appreciation of nature and the traditional way of life in Hong Kong. The construction of highways and railway lines across the plain will result in habitat fragmentation and this may affect the birds and other fauna adversely and the open landscape which now forms the foundation of its aesthetic value will be changed. The environmental cost of this development has not been factored into the cost-benefit analysis for the project. This development clearly does not support the environmental and social objectives of sustainable development. Thus, SEA of the CTS-3 has failed to achieve the goal of sustainable development.

The total area of land and marine habitats that will be lost as a result of the proposed development strategy may be relatively small. However, some developments have actually resulted in habitat fragmentation and the impacts of such developments have affected a much larger area than predicted. For example, the original Greater Painted Snipe breeding site at a marsh near Kim Tin Road was totally fragmented and partly destroyed by the West Rail and the Kim Tin Bypass. This has forced the birds to move northwards to other less suitable sites at Sha Po wetland (Leung and Ng, 2002). Although the physical areas taken up by the rail and bypass may be small, the actual areas affected may be much larger.

In certain cases where ecologically sensitive areas are neglected at the SEA level, the project-level EIA helped to perform this function. For example, plans by the Highways Department to construct a road through the Tai Ho Valley (the most important catchment in Hong Kong for fish diversity) and some of the most sensitive areas of the Mui Wo Valley (including the marsh, which is a Site of Special Scientific Interest) were rejected by the Environmental Protection Department (EPD). The Highways Department is currently exploring other options including a possible widening of the Tung Chung Road. In another case, the EPD rejected the EIA of the Sheung Shui-Lok Ma Chau railway spur line which threatened to destroy the habitat of rare migratory birds (HK Bird Watching Society, 2001).

A massive investment of HK$200 billion (HK Transport Department, 1997) was allocated to the building of major highway and rail systems in the next decade but it is not clear if this will actually increase mobility or merely act as a holding strategy. The decline in traffic speeds over the last decade despite massive investment in infrastructure shows that this strategy is not effective. It was projected that transport demand will be growing at a rate that exceeds capacity. This investment may not improve the transport system; at best, it may only halt any further decline, and at worst, may even exacerbate the congestion problem. The continuous building of roads to accommodate the increase in number of vehicles is not environmentally and economically sustainable.

The Government is also planning to convert small diesel vehicles, including taxis and minibuses, to less polluting vehicles, i.e. gas-fuelled and electric vehicles. An example is the plan to provide HK$40,000 per newly purchased LPG taxi for taxi owners to make the switch as a measure to reduce air pollution as recommended by the CTS-3. However, at the same time, there are plans to build over 100 km of highways to attract more heavy-duty diesel vehicles. These two strategies seem to cancel out the effect of each other, resulting in a 'lose-lose' situation where economic cost is incurred without gaining any environmental or social benefits.

SEA application was development-led where development strategies were already formulated before being subjected to EA. The failure to apply SEA at the policy formulation stage of the study resulted in limited development alternatives being explored to avoid environmental degradation. This also resulted in inadequate assessment of cumulative impacts caused by habitat fragmentation and disturbance at the strategic level. Proposed development strategies seem to counteract the effects of each other as they do not attend to the same elements of sustainability. For example, as noted above, there is a contradiction between the strategy to build more highways to accommodate the increase in heavy-duty diesel vehicles in response to the economic gain of having more container terminals, and the strategy to reduce air pollution through the implementation of incentive schemes to promote the use of more LPG taxis and minibuses.

Disneyland theme park development on Lantau island

Under the North-East Lantau Port Outline Zoning Plan (OZP), North-East Lantau was zoned mainly for container terminals and port-related facilities as recommended by the Port and Airport Development Strategy Study in 1989. An SEA was conducted on the development of North Lantau in 1992. The original proposal then included the development of two new towns, with a total population of 250,000 over an area of 750 ha and the new Chek Lap Kok airport site of 1,000 ha. No provisions were made for the development of a large-scale international theme park. In the strategic planning framework for Lantau and the outlying islands as outlined in the TDS Review and the Recommended Development Strategy of the SWNT Development Strategy Review completed in mid 2001, the island would continue to perform a variety of key strategic economic, development and conservation functions.

However, due to the slow down in forecast port cargo throughput, the revised port development program and the tourism potential of North Lantau as identified by the TDS Review, the Northshore Lantau Development Feasibility Study was conducted to review the land use potential of the area and the change in planning emphasis was reflected in an amended draft North East Lantau Outline Zoning Plan which was gazetted under the Town Planning Ordinance in August 1999. The study states that North-East Lantau should be reoriented from a predominantly port setting to a major tourism area with an international theme park as a focus as well as a range of compatible and tourism-related uses (HK Planning Department 1999).

At about the same time, The Walt Disney Company was exploring the possibility of developing an international theme park in Hong Kong. In late 1998, The Walt Disney Company and the government identified Penny's Bay as the most appropriate location for a Disney theme park in Hong Kong and Disneyland was approved to be built in Hong Kong's Penny's Bay on North Lantau in 1999 (Hong Kong Tourism Commission, 2000). Disneyland is located on part of the proposed reclamation in the Penny's Bay area. To the north of the project site is Yam O Wan, while to the south across the bay is the sparsely inhabited Peng Chau Island and to the West is Discovery Bay, a residential community on Lantau Island. No SEA was applied to this massive development in which 280 ha of land within Penny's Bay and 10 ha of land at Yam O had to be reclaimed. The site was originally planned for container terminals and port related uses, including container back-up areas, a business park and other industrial uses under the Lantau Port and Western Harbour Development Studies conducted in 1993.

The proposed theme park was completed in 2005, and was expected to become a core tourist attraction in Hong Kong. The government announced that Disney would help Hong Kong to provide 8,000 jobs during the 5-year construction period, and a further 18,000 upon commencement of operations, gradually increasing to a total of 38,500. The proposed facilities include the Theme Park itself and other related facilities for entertainment, retail, dining outlets, resort hotels, a water recreation centre comprising a multi-functional lake and essential supporting services and utilities. New roads, a public transport interchange, a new railway (Penny's Bay Rail Link) and associated stations would provide direct transport links to the proposed park. Ferry access was also proposed, via two piers at the southern waterfront.

The entire development was divided into nine parts, each part made up a project, which by its characteristics, is an EIAO Schedule 2 Designated Project and hence, nine individual EIAs were required. The projects include the construction of the Theme Park itself, the Water Recreation Centre, two primary distributors and one district distributor, a 3.6 km long rail line, a public transport interchange, two public ferry piers and other general service infrastructure and associated works such as the storm drainage, sewerage facilities, irrigation, water supply and utility services; the reclamation of 280 ha of land within Penny's Bay and 10 ha of land at Yam O; and proposed slope formation and stabilisation, screening and landscaping works.

The EIAs of these projects were completed within three to four months and were submitted to the EPD in December 1999 and by April 2000, the EIAs were approved. The EIA process was shortened as some previous EIAs which were deemed appropriate for that function were used for baseline data purposes.

The nature and extent of the key environmental impacts and consequences of the construction and operation of the proposed project were examined. From the EIA reports, impacts on air quality, water quality, noise levels, waste management and both terrestrial and marine ecology were stated to be low and could be easily mitigated with the proposed solutions.

Although the government argued that the project will provide jobs, there are concerns that a large proportion of these jobs will go to imported labourers from the Asian subcontinent and mainland China. The government also predicted that between 2005 and 2045, Disney will generate HK$148 billion in profits, which will add around 0.4% to Hong Kong's GDP per annum. However, it was argued that the government has been too optimistic and the cost of HK$23 billion should be put to better use to improve Hong Kong's basic education and public housing (FOE, 1999). A better quality of life is essential in order to ensure social sustainability. This is a fundamental socio-economic argument, which is relevant to the debate on sustainable development in Hong Kong and the implementation of long-term planning objectives that strategic environmental assessment encompasses.

Social sustainability for the island has been neglected. Lantau Island is characterised by steep and scenic hills, little flat land, and a mostly undeveloped coastline. It much resembles the Hong Kong Island of 150 years ago, with the important difference that Lantau has far better forest coverage. Lantau Island has always been an important ecological, scenic, historical and recreational resource for the people of Hong Kong.

The cultural and heritage value of the underwater archeological sites at Penny's Bay was totally omitted from the assessment. Other culturally important sites around Lantau, such as Tai O, Yi O, Pui O, Tai Ho and Cheungsha will be deprived of resources to revitalize these dying communities and to preserve historical heritage. Many schools, youth centres and old peoples' homes need funds to improve their facilities and services. With such a volume of resources going to the theme park, these essential communities would be deprived of the funds they need to sustain a healthy society. The social sustainability element is not satisfied.

Other tourism-oriented development projects on the island have also met with opposition. For example, the government plans to build a HK$100 million cable car and theme village to attract visitors to Hong Kong's Po Lin Monastery which stands isolated on the hilltop amid the forests. Monks at the monastery warned that attempts to turn the site into a major tourist attraction would seriously diminish their religious role and duties and would offend the dignity of Buddhism (BBC News, 22 October 2002). Such a development has jeopardized the achievement of social sustainability.

Proposals for a detailed ecological impact assessment of the terrestrial and marine habitat with justification of a four-season survey; impacts on fishery due to loss of habitats, nursery and spawning grounds and food supply; and impacts on marine life, especially on the Chinese White Dolphin and Finless Porpoise due to loss in habitat and food supply were not well addressed. The World Wide Fund (WWF) Hong Kong reviewed the EIA reports, with particular focus on the ecological assessment, and concluded that the ecological assessment had underestimated the possible impacts of the construction works on the ecologically sensitive receivers and the proposed mitigation measures were inadequate. The effectiveness of the proposed mitigation measures had yet to be substantiated, and this finding by the WWF is contrary to the finding in the EIA report which states that there will be residential ecological impacts.

The independent review panel's report concluded that the Penny's Bay dredging work is directly related to the surge in suspended solids in Ma Wan, which is contradictory to the previous environmental monitoring and audit reports, and the conclusions of the government's investigations. Concerns about the impact of sediment plumes from reclamation activities on the Ma Wan Fish Culture Zone and possible sudden death of fishery stock were raised during the public consultation process. However, the government assured the public that fish kills are not expected even under the worst case scenario. The case of Ma Wan fish kills demonstrated that there is often a gap between the prediction and the reality when it comes to the environmental impacts of a project (FOE, 2002). A precautionary principle approach should be adopted in the decision-making process, especially in a case of great public health concern.

Based on government projections, approximately 28,000 people a day will travel to North Lantau to visit Disneyland which must place considerable strain on not only the infrastructure, but also the surrounding air, water and land. Waste generated by the visitors is but one source of this environmental strain. The natural carrying capacity of the island will be exceeded and hence the goal of environmental sustainability would not be achieved.

The building of a theme park does not involve simply the park itself. Infrastructure such as additional bus terminals, railway stations, new roads and railway and water treatment plants to support the theme park will exert pressure on the natural capacity of Lantau Island as well. Hence an SEA should have been conducted for the entire area to assess the cumulative, synergistic, indirect, long range, delayed and global impacts of the theme park and other planned and committed development projects in the same vicinity to determine if the carrying capacity of environmental stocks would be disrupted as it may affect the ability to meet the needs of present and future generations. If an SEA is applied at the beginning of a development proposal, it enables the EIA process to be streamlined by focusing it on the most significant issues.

Cumulative impact assessment of the transport load and associated air pollution from a number of large-scale projects planned in the North Eastern Part of Lantau Island such as the Lantau New Town Development, the Proposed Lantau North-South Link, the proposed reclamation work at Tai Ho Wan was not conducted. The cumulative impacts of the volatile organic compounds (VOCs) and heavy metals released from fireworks on the surroundings and the impacts of using chlorine as a disinfectant in the theme park which may generate some forms of organic chlorides that are toxic and carcinogenic were also not mentioned.

The cost and impacts of the storm water drainage system and its discharge to the surrounding natural habitat and the impacts of additional sewage from the operation of the theme park on the pollution loading of the North Lantau Water were also not addressed.

The agreement by the Hong Kong government with The Walt Disney Company to proceed with the theme park before a full EIA for the project had been completed represented a serious deviation from the principles of SEA, which emphasize the need for an upstream planning policy to prevent conflicts of interest between development and environment. This deviation highlights the vulnerability of strategic planning policy to alteration if development pressures require it. Whereas EIA is an integrated approach to sustainable development, whereby design changes can be made at an early stage in the planning process to avoid any unnecessary impacts, this situation was not observed in the case of the theme park proposal due to the lack of an SEA for the development.

The decommissioning of the Cheoy Lee Shipyard at Penny's Bay is another example of the issues that were not considered by the government in this project. In 2000, the Director of the EPD approved the original EIA report of the Disney Theme Park, which excluded a key aspect of the environmental impact assessment, i.e. the decommissioning of this shipyard. The risk estimate of clean-up costs of the shipyard was not included in the cost-benefit study presented to the Legislative Council. The shipyard was later found to contain 80,000 m^3 of contaminated soil, 30,000 m^3 of which were contaminated by dioxin. It was further evidence that the EIA report of the Disney Theme Park was simply a tool to legitimize the Administration's decision to gear up the project. The environmental sustainability element had been sacrificed for economic gain. The loopholes in the prevailing EIA process contributed to the Cheoy Lee Shipyard issue, in which the government might have to shoulder additional clean-up costs and the public could be exposed to unknown health and environmental risks. The coexistence of the Disney Theme Park and the dioxin treatment facilities close to Disney Theme Park and North Lantau Expressway creates tremendous potential risks to visitors and liabilities to the government. Had an SEA been conducted, the EIA for decommissioning of the shipyard would not have been left out and the impacts and the associated clean-up cost of the shipyard would have been foreseen at a much earlier stage.

The results of the TDS Review had actually identified Lantau Island for development for tourism. It would appear that the output of the SEA had not been seriously considered during the TDS Review. Such a large scale development on Lantau Island is not considered to be sustainable, socially and environmentally. The island is currently facing intense development pressure. Rapid population growth has forced the government to make plans for developing Lantau Island. For example, Tung Chung New Town was originally envisioned as a dormitory community for airport workers. However, it has now been called upon as a "strategic growth area", or population sink, for the predicted extra growth in the population of Hong Kong in the next several decades. This proposed added growth at Tung Chung takes little account of whether North Lantau is the appropriate place for further growth, still less of the impacts on the local environment. Similarly, the Chief Executive's October 1997 policy address identified strategic growth areas for housing on Lantau including South Lantau as well as Tung Chung and Tai Ho. More recent proposals for highways between Tai Ho and Mui Wo, and Tung Chung and Tai Ho, are also causes for concern.

All the above mentioned impacts associated with the building of the international theme park could involve substantial costs and the proposed mitigation measures in the EIA reports can be very costly to implement. Yet no thorough cost-benefit analysis of the project was conducted and the time frame for implementing these essential environmental measures is still vague. Due to the scale of this theme park and the long-term commitment in managing the park, it is important to assess the overall impacts on the sustainability of the project. Air, noise, waste, sewage, visual, ecological, fishery impacts are all essential elements that affect the sustainable development of Hong Kong. The government has not comprehensively evaluated the hidden costs that could be a major burden for the people of Hong Kong which will last for a hundred years. The economic gain may not be able to offset the economic loss due to these hidden costs.

Discussion

The enactment of the EIAO is a great improvement in protecting the environment and conserving the natural areas in Hong Kong. The EIAO has enabled the EPD to have greater power to minimise, if not prevent, the environmental impacts from development proposals, during the construction and operation phases. The EPD has more say over projects proposed by other government departments, and has the right to object to them if it finds that the impacts are not properly addressed. For example, the EPD rejected a proposed road by the Highways Department that runs through the Tai Ho Valley in the New Territory. The Environmental Monitoring and Audit element of EIA has been one of the strengths of the Hong Kong EIA initiatives and has partly resulted in the intensive use of intranet and internet for dissemination of data and helps to keep check on the impacts and the effectiveness of the proposed mitigation measures during construction and operation of the projects. The element of public opinion in EIAs is also commendable. Members of NGOs and the public have the power not to sanction projects with adverse environmental impacts and this gives a good insight of how the government views public opinions and integrates them into the decision-making process. During the EIA process, there is much pubic involvement including the public viewing of the documents, thereby making the planning process more transparent.

However, the EIAO alone is not sufficient to help Hong Kong to achieve sustainable development. It is essential to move environmental and social assessment upstream in the planning process to enable the likely significant environmental consequences of policies, plan and programs to be evaluated. This will facilitate the modification of development proposals in order to reduce adverse environmental consequences.

Hong Kong is one of the first few countries to implement SEA for major plans and proposals. Such a process was carried out even before the EIAO was enacted and this demonstrates the commitment of the government to environmental protection and to the achievement of the goal of sustainable development. SEA has been applied to several development proposals in Hong Kong, such as the TDS Review, the third comprehensive transportation studies, the second railway development study and the port and airport development strategy. SEA has provided decision-makers with an account of the implications of proposed policies, plans or programs and has resulted in greater accountability and transparency in the decision-making process. In the case of the TDS Review and CTS-3, the application of SEA helped to better integrate environmental concerns with land use and transport development requirements. For example, an objective was set in the CTS-3 study to enhance the mobility of goods and people in an environmentally and economically sustainable manner.

SEA models to assess air, noise and water quality have also been developed for the territory. For example, regional air and water quality modelling techniques for achieving the goal of sustainable development were developed and used in the Pearl River Delta Region (Lai, 2002). Noise indicators for two strategic level assessments in the CTS-3 and SUSDEV 21 studies have also been developed to provide useful and essential environmental information for decision-makers. The correct use of such SEA tools facilitates environmentally compatible development and should minimise adverse risks from cumulative impacts and ensure more compatible land use.

The public participation process in the SEA has helped to provide opportunities to examine whether the environmental degradation implicit in the present development strategies are acceptable to the community, and provide development alternatives, including the 'do-nothing' scenario. For example, in 1993, a report, "Territorial Development Strategy Review —

Development Options" was published for the purpose of seeking views from various NGOs and the general public. The useful comments received were compiled to summarise the principal areas of concern and to identify the adjustments to be made and new ideas to be taken forward in the next phase of the Review (HK Planning, Environment and Lands Branch, 1996). The application of SEA has also helped to increase the environmental awareness of the governmental departments and the public. Thus, the former should be able to formulate policies and plans at the strategic and regional levels. These are essential if conflicting demands of environmental needs and economic developments are to be reconciled in Hong Kong.

In spite of the many advantages of SEA in counteracting the limitations of project-level EIA, the application of SEA in Hong Kong has notable limitations. Generally, Hong Kong still lacks a comprehensive overall and strategic planning policy for the environment. There is a lack of coordination among relevant governmental departments. When faced with development issues, different departments have different stakes and are often pulled in different directions, leading to poor coordination and insufficient synchronization to overcome conflicting interests. The case of the Dolphin Sanctuary in the Sha Chau and Lung Kwu Chau Marine Park, where there are still many vessels passing through the area, and causing adverse impacts on the dolphins, is an example of a lack of coordination between the transport needs of the Marine Department and the conservation needs of the Agriculture and Fisheries Department.

In Hong Kong, SEA is primarily used as a planning tool to evaluate the environmental implications of development plans or strategies. However, SEA procedures are seldom applied to major policies and programs. Hence, most SEA studies are development-led and mainly look for new development areas or new transport corridors instead of placing conservation value on important habitats and natural resources. Many ecological assessments in SEA studies focus their efforts on finding spatial alternatives, for example, change of sites, change of alignments, or identification of mitigation measures. Ecological assessments rarely trigger exploration of other more radical alternatives, for example change of transport modes, change of engineering approach, or change of population targets. Given the development-led nature of most SEA in the territory, proponents are often unwilling to give way to the need to protect ecological resources, leading to a piecemeal and gradual denigration of the environmental stock of Hong Kong.

Another limitation of SEA in the prevailing context of Hong Kong lies in the fact that the current land use plans and transportation strategies in force in the territory will largely determine the pattern of development over the next few years. The applications of SEA in both the TDS Review and CTS-3 were constrained at the beginning of the planning process due to the vague details of alternative development proposals and therefore, decisions to update and improve on these strategies were made in an incremental manner. However, they were not clearly formulated early enough in the decision-making process. In order for SEA to be fully effective, it is essential that the development proposals which are offered fully take into account all impacts upon the environmental stock. This, in turn, is dependent on a systematic and comprehensive environmental monitoring strategy to determine environmental baseline conditions.

Another major problem facing the proper and effective application of SEA is the lack of knowledge of Hong Kong's own environmental resource base. Very often, projects with adverse impacts on areas of high ecological value were planned without adequate knowledge of the impacts. For example, the adverse impacts of the railway spur line through Long Valley and the extension of Tung Chung Road were only identified during the EIA stage, leading to expensive mitigation measures, and drastic modification of the design. If a comprehensive database for territory-wide evaluation had been available, it would have been most cost-effective to identify the ecological impacts of development projects on such valuable and sensitive sites at the planning stage.

One major obstacle to environmental and social sustainability in Hong Kong is the stress on the free market in the management of the economy and the preference for development. The Hong Kong government has set high goals for its commercial sector to be a regional leader and as one of the most developed economies in Asia. Historically, this has been built without consideration of the environment and, in some respects, this is still prevalent today. Examples of this are the development of the Disney Land theme park in the face of apparent environmental constraints, and of the Chek Lap Kok Airport in relation to the survival of the Chinese White Dolphins. The economic downturn in recent years and increasing unemployment prompted the government to once again sacrifice the fauna and flora in order to boost the tourism industry to create more jobs.

The intrinsic value of the natural environment and cultural heritage is subjective and there is no suitable or sure method to assess it. Hence, their values are often underestimated or even neglected. The indirect impacts caused by proposed developments may not be foreseeable during the time of assessment and hence the associated environmental cost may not be accounted for in the development proposal.

There is a need for comprehensive overall strategic planning for the environment so as to achieve the goal of sustainable development. The objective of taking the environment into account should be stated in all development projects to prevent the erosion of environmental quality as development takes place. For effective application of SEA to take place, either a legal system to enforce it, as in the case of the United States, or an official body with sufficient strength of purpose and professional competence, such as the EPD has to be responsible to determine the requirement for an SEA and oversee the quality of the SEA that is undertaken. There should be close interaction between the relevant planning authorities, the EPD and the parties responsible for conducting the SEA to ensure that all environmental issues are taken into account at all levels and stages of the SEA, and that the findings are translated into suitable development strategies and planning, especially at the start of the process. There should be stronger guidance on how to build up effective SEA processes and evaluate the effectiveness of existing SEA processes. The public participation element of the process should also be strengthened. Often, the government's consultation exercises have aimed to inform and notify rather than to invite and involve real participation by the public. The bureaucratic planning culture has a tendency to minimize public involvement for fear that it might impede the development process. As a result, the public becomes disinterested and remains uninformed (HK Friends of Earth, 1999). The guidelines developed by the IAIA should be closely followed to maintain a high standard of SEA processes in Hong Kong to reap maximum benefits from the application of SEA.

There should be a review of the need to conduct an SEA when there is a material change in development strategy to accommodate a specific development. For example in the case of Disney Land Theme Park development, an SEA was conducted in 1992 for the development of two new towns and the international airport. However, there was no revision of the SEA for the island when the addition of the theme park and the supporting infrastructure was proposed. Without proper environmental assessment, the natural carrying capacity of the island may, inadvertently, have been exceeded.

A comprehensive database for territory-wide evaluation of conservation value (ecological as well as cultural) should be developed to facilitate the assessment of subtle, complex, indirect or cumulative ecological and social impacts on important conservation sites in the planning stage of any development. Environmental baseline conditions should be established and the carrying capacity of the natural ecosystem should be quantified to help to achieve the goal of sustainable development where the most cost-effective solutions and alternatives for proposed development projects can be achieved.

The stage of recovery of habitat compensation and restoration projects should be assessed over a long period of time in the environmental monitoring and audit programs to compare the 'actual' versus 'predicted' impacts to assess the accuracy of predictions and to determine the feasibility and effectiveness of mitigation measures. Such monitoring will also help to identify ecological resources and functions in danger of non-sustainable use, providing triggers for action and a basis for initiatives to restore sustainability. However, it should be noted that these monitoring and audit programs are end-of-pipe measures and are not useful in cases where environmental stocks are already depleted or when the natural carrying capacity of the ecosystem has already been exceeded and further steps taken will not help to salvage the situation. Lessons learnt from monitoring and assessing cumulative, indirect, synergistic impacts in the monitoring and audit programs together with results from baseline environmental condition studies will provide useful environmental information for the environmental assessment of future development proposals. There should be a feedback loop, with the output of environmental monitoring and audit programs serving as inputs for future environmental assessments, to realise continual improvements.

The quantification of carrying capacity of the natural ecosystem is not easy as the ecosystem is extremely complex and unpredictable, with some being more resilient than others. Moreover, the types and potential uses of many natural resources are not yet well-known. Nevertheless, it is an important step as weighing the resilience of the ecosystem and resource stocks with population and consumption pressures provides an indication of sustainability. A valuation system which places a monetary value on the natural environment and cultural heritage can be developed to incorporate the benefits and costs of the environmental and social effects into the analysis of alternatives. In this way, the wider array of benefits and costs associated with a development proposal can be considered in order that the alternative that produces the largest net benefit to society could be chosen. An example of such an evaluation would be the assessment and comparison of the cost of using a brown field site and a green field site. Various methods such as the hedonic property values and contingent valuation methods (Tietenberg, 2000) can be used when conducting cost-benefit analyses. Guidelines for evaluating a range of environmental impacts that can be monetized have been developed by the World Bank (World Bank, 1998). The external cost of pollution to the society and the associated remediation and control techniques to be implemented in any development proposals should be internalised during the project proposal stage. For example, the cost of cleaning up the shipyard on Penny's Bay should have been internalised during the cost-benefit assessment of the project to reveal the true cost of building the Disneyland Theme Park. As it is not easy to assess the value and carrying capacity of the natural ecosystem and to predict the impacts of development projects, it is important to adopt a precautionary principle during the SEA process, especially when there are threats of serious or irreversible damage. Lack of full scientific certainty shall not be used as a reason for postponing cost-effective measures to prevent environmental degradation.

The application of SEA in Hong Kong should move beyond a sectoral approach, to look at ways in which development decisions can not only prevent environmental damage, but also, positively enhance natural resources. SEA should emphasize the need for a more integrated approach to potential land use strategies that looks at the land resource in terms of productivity, stability and sustainability and hence helps to determine the best options of land use, while accounting for the properties of the natural resource base. These can only be achieved if environmental factors are given adequate consideration at the earliest stages of the process of formulating development objectives.

The application of SEA may not be able to resolve all problems immediately, but the output of SEA is an important tool for defining outstanding issues requiring attention in future. However, whether the output of an SEA is implemented will depend on the commitment of the authorities to environmental sustainability and also the actual cost of implementing the proposed mitigation measures or adoption of the proposed alternatives, including the 'do-nothing' scenario. A balance should be struck, taking into account the interests of various stakeholders in the decision-making process.

Conclusion

The fact that the Hong Kong government is trying to inculcate sustainable development as a way of life across the entire population is encouraging. Waste minimization and resource conservation are examples of ways in which Hong Kong is moving toward sustainable development. Ideally, once a sustainable development approach is implemented, the importance of the environment, natural resources and cultural heritage will be on par with that of economic development, and the tendency to sacrifice the quality of the environment for development will decrease. However, the most important factor is implementation. The government has the power to weigh the relative importance of the three elements of sustainability and come out with a decision which may or may not be the best option for sustainable development.

Population growth and rapid urbanization, coupled with a historical lack of planning controls, low public awareness, inherent conflicts among government departments, and vested interests in economic development have put considerable pressure on the environment in Hong Kong. Developments during the next decade will be critical for Hong Kong's environment. If the territory is to maintain or achieve a higher quality of life, the government should no longer regard economic activity as independent of environmental protection and conservation. Influence and impacts from mainland China should also be carefully assessed and integrated into Hong Kong's development strategies. Currently, there are insufficient data on cross-boundary pollution and there is limited communication and data exchange between environmental officials in Hong Kong and their counterparts in China, and there should be improvements in these areas.

However, with Hong Kong's progress in environmental protection, for example, the enactment of the EIAO; application of SEA to major plans and programs; requirement for all government departments to provide environmental reports; development of sustainability indicators; establishment of the Environment and Food Bureau to remedy the fragmentation of food safety and waste collection and disposal among departments and municipal councils; long-term public education programs; and governmental provision of economic and other suitable incentives for environmental protection are signs that it is possible for Hong Kong to become a more sustainable city. Despite these promising trends, there is still a need for top-level commitment as piecemeal reforms are insufficient to rectify existing deficiencies in environmental protection and conservation should be given the priority they deserve.

SEA should be viewed as an integral step in the attainment of sustainability where it provides a macro-view of development based on proper assessment of environmental stocks and natural carrying capacity to support economic and social progress. The application of SEA should be in such a way as to allow the principle of sustainability to be carried down from policies to individual projects and beyond. SEA should set the theoretical requirements of sustainability in a practical context, and establish a framework within which project-level EIAs can be carried out.

Sustainable development involves environmental, social and economic sustainability. Hence, the values of these three elements should be balanced when considering any development policy, plan or program. SEA is only able to contribute in major ways to decisions concerning environmental and social sustainability, but not economic sustainability as this element is often irreconcilable with environmental and social sustainability. The use of SEA should be to provide full and quantitative information on the environmental and social dimensions of policy, programs and plans, during the PPP formulation and decision-making, feeding critical information into the sustainability debate. None of the three elements should be sacrificed for another or the others. Nevertheless, it is often difficult to put a value on the natural environment and cultural heritage and hence, in many cases, such values are underestimated and economic development has taken precedence over these elements of sustainability, and this is particularly true for Hong Kong. The government, which has the ultimate power to weigh the relative values of each of the sustainability elements may or may not arrive at sustainable development strategies despite having applied SEA to development proposals. SEA only provides for environmental and social sustainability issues, and there are many other factors such as peace and security, prevailing economic interests, political systems, institutional arrangements and cultural norms that should greatly influence the government's final decision.

References

Arce, Rosa and Gullón, Natalia. (2000) The application of Strategic Environmental Assessment to sustainability assessment of infrastructure development. Environmental Impact Assessment Review 20: 393–402.

Au, Elvis. (2002) EIA Newsletter 18 — Recent Developments in EIA and SEA in Hong Kong.

BBC News, 22 October 2002.

Dubin, B.I. (2002) Strategic Environmental Assessment for long-term sustainability planning. Conference on Reshaping Environmental Assessment Tools for Sustainability proceedings 1:189–194.

FOE 1999 — Friends of the Earth?

FOE 2002.

Hong Kong Bird Watching Society 2001.

Hong Kong Coalition of Service Indfustries 1997.

Hong Kong Planning, Environment and Lands Branch, Hong Kong Government (1996). A new, long-term development framework for Hong Kong, A response to change and challenges. A Consultative digest on the Review 1996 Territorial development Strategy.

Hong Kong Planning Department, The Government of Hong Kong. (1995) Territorial Development Strategy Review — Strategic Environmental Assessment of the Preferred Options.

Hong Kong Planning Department, The Government of the HKSAR. (1999) Northshore Lantau Development Feasibility Study.

Hong Kong Tourism Commission, The Government of Hong Kong (2000) Legislative Council Brief: Penny's Bay Reclamation Lantau Island.

Hong Kong Transport Department 1997.

Kilburn, M. (2002) The Kam Tim Wetland: A case study highlighting the need for cumulative impact assessment and compensation planning for ecologically sensitive areas. Conference on Reshaping Environmental Assessment Tools for Sustainability proceedings 1:217–220.

Lai, P.W. (2002) Strategic Assessment Tools for Sustainable Water Quality and Air Quality Management for the Pearl River Delta Region. Conference on Reshaping Environmental Assessment Tools for Sustainability proceedings 1:228–232.

Leung, C.N. and Ng, C.N. (2002) Cumulative ecological impacts of linear infrastructure projects: a case study of Kam Tin in Hong Kong. Conference on Reshaping Environmental Assessment Tools for Sustainability proceedings 1:275–283.

Loh, C.K.W. (1997) A vision for Hong Kong's future with regard to land use: sustainable urban living in the 21st century. Asian Journal of Environmental Management 5 (1).

Noble, Bram F. (2002) The Canadian experience with SEA and sustainability. Environmental Impact Assessment Review 22: 3–16.

World Bank Group, Environment Department. (1998) Environmental Assessment Sourcebook Update.

Tietenberg, T. (2000) Environmental and Natural Resource Economics. Fifth edition. Addison Wesley Longman, Inc.

7

POLICIES AND LEGAL FRAMEWORKS OF PROTECTED AREA MANAGEMENT IN NEPAL

LAXMI DUTT BHATTA, KOH KHENG LIAN and JOSEPH CHUN

The Protected Area (PA) is a common resource, and provides basic necessities for livelihoods. As a result, access to this resource is very much of interest to local communities. The success of PA management depends on the socio-economics and politics of the governing countries and communities. This study examines the policy implementation that encourages the conservation of biological diversity. It was carried out in the Langtang National Park and its surrounding buffer area in the central Himalayan region of Nepal.

It was found that local communities are heavily dependent on park resources for their livelihoods. Protected Area policies that provide access to grazing, and firewood collection from the park area are considered conducive to acceptance. However, the traditional rights over these resources are still the issue of park-people conflict. Medicinal and aromatic plants (MAPs) are still "locked resources" for local communities. The present PA policy fails to incorporate the issue of wildlife crop damage and livestock depredation. The rights and responsibilities of buffer zone (BZ) institutions are poorly delineated. There is a need to encourage local participation in buffer zone management activities.

It is necessary to review the existing root causes of park-people conflict. The roles, responsibilities and authorities of BZ institutions need to be clearly delineated. It is suggested that the buffer zone management council (BZMC) be strengthened as an apex decision making body, whose role should not be limited to the BZ boundary. Furthermore, the policy should address the issue of wildlife damage and mechanisms to compensate such loss would be an asset in PA governance. Local communities should be given management rights, at least to forest areas near the settlements. A long term and prescribed management plan for MAPs is necessary. The equitable benefit-sharing mechanisms help to overcome the major setbacks in sustainability of PA, a common property resource.

I. Introduction

Nepal has an enormous wealth of biological and cultural diversity. The country has been successful in establishing an excellent network of protected areas in order to safeguard its natural heritage and biological resources, covering about 18% of the total area of the county (DNPWC, 2001). The National Park and Wildlife Conservation Act 1973 (NPWCA), with its 4th amendment in 1994, provided the legal framework to institutionalize the entire protected areas (PAs) network of the country. Similarly, regulations and guidelines enforced under the NPWCA 1973 provided specific implementation frameworks to enact the programs and activities related to wildlife and biodiversity conservation. The recently endorsed National Biodiversity Strategy 2002 (NBS) and National Wetlands Policy 2003 (NWP) also focused on the local participation for long term social benefits.

The issue of park and people conflict, mainly on issues of customary rights, wildlife damage, and access to resources, has long been discussed. People living adjacent to the protected areas are vulnerable to wildlife damage. The traditional park management policy denies local people access to park resources (Budathoki, 2003), while local people have always been pressing on park authorities to allow them to collect basic forest resources and graze their livestock inside the park area (Silwal, 2003). The tension of this park-people relationship has been a threat to both the protection of park resources and local subsistence. The existing buffer zone management policy addressed some key issues such as "ploughing back" 30 to 50% of the total park revenue for the overall development of buffer zone, which helps to mitigate the park-people conflicts (Sharma, 1998). However, this provision is mostly helpful in areas where revenue is higher (for example in Royal Chitwan National Park). In most mountain protected areas, where revenue allocation for buffer zone management is very low in terms of financial support, this provision does not help significantly to mitigate the conflict situation. In these cases, without the meaningful involvement and support of local communities, threats to sustainable biodiversity conservation would exist in different forms and with differing severity.

The present strategy of empowering local governments to play a central role in managing their resources is considered an important step towards the decentralization and devolution process. The Local Self Governance Act (LSGA) 1998 emphasizes and empowers local governments to manage natural resources and guide them to integrate forest, biodiversity, soil conservation, land use and environmental planning as an integral part of the local area development. This national strategy to empower local governments has been reflected in many policy documents including the government's five year plan and National Sustainable Development Agenda (NSDA). However, the provision in LSGA relating to natural resources contradicts the existing NPWCA and related regulations. For example, Section 218 of LSGA stipulates that the natural heritage is the property of a Village Development Committee (VDC). Natural heritage generally includes forests, lakes, ponds, and rivers. This is against the definition of protected area as stipulated in NPWCA 1973, which defines the protected area to include forests, lakes, ponds, stones and river inside the protected area boundary.

This study was carried out to determine what are the policies and legal frameworks that encourage local people to protect their resources (investment) and expected return (community infrastructure, development). It examines the policies and legislative measures that encourage local communities to conserve natural resources and assesses the effectiveness of local institutions in conserving their biological resources in Langtang national park, in the Central Himalayan region of Nepal.

Langtang national park

The Langtang National Park (LNP) was established in 1976 as Nepal's first among mountain national parks, representing the Central Himalayan ecosystem. With an area of 1710 sq. km., it lies between 85°15′ east to 28°20′ north and extends from 32 km north of Kathmandu to the Nepal-Tibetan boarder. The area was selected for this study from among the six national parks and four declared buffer zones in mountain areas.

The park area is predominantly occupied by permanent snow peaks, ice, and rocks. Only a small portion of the area is cultivated by the communities residing within the park. Large numbers of livestock heavily depend upon grassland, which covers 4.9% of the total area of park. About 30% of the total area is covered under different types of forest vegetation, providing homes for different species of wildlife.

Buffer zone area coverage

His Majesty's Government (HMG) declared the buffer zone of the Langtang National Park (LNP) in 1998, covering an area of 420 km^2. It extends across 3 districts of the central development region (LNP, 2001). The Buffer Zone (BZ) area covers the peripheral park impacted area and 45 villages in Rasuwa district located inside the park boundary. These 45 villages were previously managed as the settlement park enclaves. These villages are the most vulnerable to wildlife damage and affected by controlled park policies.

Study sites

This research was carried out in three Village Development Committees (VDCs) of Rasuwa district. The study sites fall under the jurisdiction of the LNP. The study sites are part of the great Langtang valley, bordering the Tibet Autonomous Region and China in the north. The area is famous for its natural and cultural heritage. Two out of three VDCs are accessible by car. Public bus provides transportation regularly from Kathmandu. Basic services and amenities like public schools, health posts, telephone services, basic veterinary care, are available. Tourist lodges, hotels and camping grounds are also available along the Syaphru-Langtang and Dhunche-Gosaikunda tourist trekking route.

(1) Dhunche VDC

Dhunche is the district headquarters of Rasuwa district. The entire area of the VDC falls within the park boundary, except the settlements and private property, which are under the buffer zone of LNP. About 50% of the total households in the VDC have access to electricity (DDC, 2002). A feasibility study on a medium size hydropower plant from Trishuli River is completed, which will provide additional economic opportunities. NARA, a foreign mineral water factory has been in operation. *Nau Kunda* (lake) and *Wangul Danda* (ridge) are the major tourist places. Five district based Non Governmental Organization (NGOs), and two International Non Governmental Organizations (INGOs) have been working for the development process in the VDC.

(2) Syaphru VDC

The entire area of the Syaphru VDC falls within the park boundary, except the settlements and private property, and ward number 9, which are gazetted as the buffer zone of LNP. Forty percent of the total households in the VDC have access to electricity (DDC, 2003). The Chilime hydropower project has been operating in the area, and an EIA of hydropower from the Langtang River has been completed. However, the implementation of this project has not been confirmed yet, as the Department of National Parks and Wildlife Conservation (DNPWC) has not granted permission, citing as cause that the project will significantly impact the park resources in the area (personal communication from the Chief District Officer, Rasuwa, 2004).

(3) Bridim VDC

Bridim VDC is famous for its richness in Tibetan culture. The LNP covers the entire area of the VDC, except major settlements and private property. Tourism for Poverty Alleviation Project (TRPAP) under the support from UNDP has implemented a village tourism pilot project, where tourists and outsiders are expected to stay with local people rather than in tourist hotels.

II. Governance for Protected Areas

This research is primarily based on the governance principles for protected areas in the 21st century as discussed in the Fifth World Parks Congress, held in Durban, South Africa (Graham *et al.*, 2003). These principles are applied as an analytical tool in Langtang National Park and the surrounding buffer area. However, specific criteria for each principle are applied in local conditions and situations.

Before the field work, the VDCs very close to park areas and completely reliant on the parks' resources were considered the most appropriate for the sampling design. The justification behind this approach was based on the philosophy of "nearer the resources, greater the ownership". These VDCs were selected with a view that tourism in the area provides economic opportunities to the local people. Furthermore, Tourism for Poverty Alleviation Project (TRPAP) in collaboration with LNP has been working for the last 5 years in all VDCs, thereby providing a greater opportunity in resource mobilization and livelihoods. The Sample size for each VDC was set at 10% of total population. Households within the VDC were randomly selected to proportionally represent all groups, castes and occupations. The available adults in the households were interviewed regardless of their gender.

Data collection was carried out using: direct observation, meeting with key informants and group discussions, household surveys and analysis of secondary data.

1. Data analysis

Qualitative data on respondent's response such as perception, participation, decision making process, were interpreted quantitatively. The Likert's Scale in the Index of Relative Ranking (IRR) was used to calculate such data, and categorized in different strata and order before applying the IRR, which is the relationship of:

$$IRR = \frac{R_1S_1 + R_2S_2 + \cdots + R_nS_n}{nr} \quad \text{(Miller, 1986)}$$

Where,
R_1 = Rank of 1st order, S_1 = Score of 1st order
R_n = Rank of last order S_n = Score of last order
n = no. of observations r = no. of orders

A livestock unit (LSU) is measured as described by Neupane, 1995: Buffalo = 1, Cattle = 0.8, Goat = 0.08

2. Results

The key program areas that the LNP organized for local communities are training, workshops, and exposure visits. With the aim of providing a capacity development program, the LNP formed different functional groups such as tourism and hotel management, herders and grazing area management committees. However, resource scarcity in the LNP restricted the implementation of such programs. During the fiscal year 2003/2004, three major conservation-related training and awareness programs were organized. This included herders' awareness, and nursery management training in Dhunche VDC, and nursery management in Syaphru.

The study shows that the majority of the training is related to hotel and tourism management. About 90% of the total sampled population which is involved in tourism or hotel business received at least some training. The Mountain Institute (TMI) stated that more than 90% of all hotel and lodge owners in Langtang valley received training on cooking, sanitation, and lodge management (Thapa *et al.*, 2003). This training was mostly organized in collaboration with different NGOs, particularly TMI. Similarly, some NGOs are involved in capacity development and awareness programs, independently of the national park framework.

The impact of these programs is encouraging. It not only encourages communities to mobilize their resources, but also makes them aware of their responsibilities. However, as most capacity building programs are focused on social mobilization, tourism, and hotel management, LNP now needs to focus on sustainable management of natural resources, including forest management, and non-timber forest product harvesting. Similarly, indigenous knowledge on the use and harvesting of medicinal plants should be the focus of future training programs.

3. Participation and decision making

The present 10th Five-year Plan emphasized greater participation and decision roles for local communities in forestry sector management. It stated that "full governance shall be given to the system in which there will be direct participation of consumers and concerned people" (National Planning Commission, 2003). The Sustainable Development Agenda for Nepal (SDAN) also agrees that the success of biodiversity conservation programs depends upon effective broadening of community based conservation programs (HMG, 2003). SDAN also stressed the need for application of such an approach to other protected areas which have been managed under the 'wilderness' approach, where the army has been deployed for protection of park resources.

More than 90% of the households are affiliated to a group. More than 38% of the respondents always attend group meetings whereas only 2.4% have never attended the meeting. Similarly, 48% of the total sample size are active during meetings and decision making processes. Males dominate the decision making process. Similarly members in the age group of 18 to 35 years have the most influence on decision making processes. Respondents who have attained education levels above the School Leaving Certificate (SLC) also have a higher influence in decision-making.

Case 1: Hotel management inside the park area

Fifty four lodges and tourist hotels have been managed inside the park area under the lease agreement between the LNP and the hotel owner. The LNP provides space for the construction and operation of such facilities. The owner needs to pay royalty for the space leased from the park. A ministerial decision fixed the royalty rate seven years ago on the basis of different categories of hotels and the facilities they provide (Decision of Forest Minister, 2054/8/29 BS). Hotels were divided into 4 grades. The lease agreement between the LNP and owner is the key document governing the operation of services. This agreement provides "veto" power to the park warden and states that the warden shall forbid the operation of lodges and hotels if the terms and conditions are not fully implemented. But in practice, the warden, and his designated staff rarely visit the field and may not be fully informed of field situations. In this regard, a participatory monitoring mechanism would be useful. The representatives of local groups and VDC, if allowed to be involved in the monitoring mechanism, would be more effective and fair. Otherwise, there may be misuse of such veto power.

Case 2: Stone and sand collection from park area

Local communities need to acquire a permit from the LNP to collect stones, gravel and sand. They need to submit recommendations from the relevant VDC and affiliated group(s) along with the application to obtain such a collection permit. Without the recommendation of VDC and groups, the LNP may not issue such permit. The warden also has the authority to refuse the issue of a permit if he feels that such work negatively impacts park resources. An approval is also required from the LNP even to construct the house in privately owned land, which also needs to be approved by the relevant VDC.

Five years ago, when a 20 MW Chilime hydropower construction was started in the area, it needed large amounts of sand, stones, and gravel. The power house, penstock pipe, and base camp of the project, which is now in operation, lay inside the buffer zone area of the LNP. Official data showed that a collection permit for 2000 cubic metres of stone, and 800 cubic metres of sand was issued to a privately owned company (LNP, 2001). Similarly, a large quantity of firewood and timber was supplied from the park area, with permission from the LNP. Local communities blamed the park authorities for not consulting with local groups or VDC prior to issue of the permit. They also mentioned that park authorities require many paper proofs from the poor farmer who wants to build a house, but never felt the responsibility to acknowledge the view of that poor farmer. During the period of construction, a case was filed against local villagers from Sypharu, charging collection and selling of stones to Chilime Hydropower Company.

The participation of local communities greatly impacts on resource conservation. If this is not acknowledged, the issues of ownership and property rights become debatable, leading to resource deterioration rather than conservation.

4. Coordination and networking

The National Biodiversity Strategy pointed out the lack of adequate cross-sectoral coordination of programs within the government and aims to develop better coordination with relevant organizations and government agencies (HMG/NBS, 2002). The buffer zone policy tries to address the issue of coordination while implementing programs and projects in the area.

Despite the policies, there is a wide gap between program planning, and implementation. This gap and contradiction was expressed during the writer's meeting with the Chief District Officer (CDO) of Rasuwa, who is also responsible for overall administration. He criticized the park policy, stating that "the present national park policies do not encourage major development programs". The CDO pointed out the need for greater coordination from park authorities, saying that "The park either needs to re-allocate settlements outside the park area or coordinate for development activities". The opinion of an official at the DDC, Rasuwa, who was also the officiating local development officer (LDO) at the time of discussion, was similar to that of the CDO. He argued that the "LNP is a subsidiary unit of the DDC to carry out biodiversity conservation in the district" and agreed that there was lack of coordination with park management.

The tendency of forming separate groups rather than working with existing ones has been the major problem for local communities. NGOs and donor funded projects have their own mechanisms to mobilize and support the groups for programs and services. As a result, many different community based groups have been formed.

Case 3: Community-based group formation

Before the declaration of buffer zones, District Soil Conservation Office (DSCO) Rasuwa, formed 52 community based groups (known as Conservation and Development Groups, CDGs) in Lahrepauwa, Dhibung and Bhorle VDCs under grant assistance from Danish International Development Agency (DANIDA) watershed management project. These groups were formed in accordance with watershed management guidelines and were registered in the District Administration Office (DAO) under the Organizations Registration Act. The groups held the status of NGO under this Act. The Soil and Watershed Management Act does not have any provision to register groups, unlike in Buffer Zone Management Regulation, 2052. Before the termination of the watershed project, these groups had approximately Nepali Rupees (NRs) 6 million in group funds (DSCO, 2003). Then, the area was declared as a buffer zone of the LNP. Buffer zone institutions were to be formed under the new management. CDGs, formed by DSCO could be considered buffer zone institutions and registered under buffer zone regulations to save resources and time. In fact, DSCO had already spent a huge amount of resources to establish and strengthen CDGs. There are however three major issues concerning the classification of CDGs as buffer zone institutions. Firstly, they were already registered under a different law, which has different provisions from that relating to buffer zones. Secondly, the buffer zone policy requires only 3 executives at the group level, whereas an NGO needs more than 7, and thirdly, local

CDGs were not interested to be considered as buffer zone institutions, as they enjoy more power and access as a NGO than BZ user groups. In the circumstances, the LNP went on to form separate groups while these CDGs continue to operate as NGOs in the area (personal communication, Kamal Gautam, District Soil Conservation Officer, 2004).

5. Benefit sharing and equity

Nepal's tenth Five Year Development Plan (2003–2008) is directed towards the sustainable use of forest and biodiversity resources whilst providing access and benefits to local communities. One important policy change in the present plan is that it focuses on the process for the registration of local and indigenous knowledge on natural and biodiversity resources. Such registration became necessary when Nepal became a member of the World Trade Organization (WTO). The parks and buffer zone management policy also provides opportunities for benefit sharing with local beneficiaries. One important policy adopted is the allocation of 50% of the total park revenue for the development of buffer zone areas. However, there is much government intervention in managing protected areas, and the government still holds the reins of the decision making power, and control over resources. The government interventions are costly and conflicting, as they treat all equally whether they contribute to any development activity or not (Janssen *et al.*, 2001).

The traditional practice on decision making and sharing of benefits in most of the mountain communities in Nepal is still based on trust amongst the local elites (*lamas, priests, mukhiya*), and the LNP is no exception. As PA management authorities have the tendency to work with these local elites and formal groups, the marginalized and poor sections of the community do not have equal access to benefits. As Fisher (2002) pointed out, the community based conservation approach provides more space to include the marginalized and poor farmers. The issue of equity and benefit sharing is debatable in the present "command and control" PA management system.

The LNP disburses 50% of the total park revenue through BZ committees on the basis of numbers of user groups. The amount is disbursed equally to all 315 user groups through the respective user committees. The disbursed fund is used to complete pre-planned programs (operational plans), approved by the LNP. This disbursement mechanism by some means is based on the population size or household coverage. There are more user groups in highly populated VDCs. For example, Dhunche VDC (named as Naukunda User Committee (UC) under BZ institutional framework) has 10 user groups, where as Langtang VDC (also named as Langtang UC) has only five. But this mechanism fails to address the impact of wildlife and park dependency. It would be better to divide the wildlife impact zones from high to low. If funds are disbursed addressing the issue of wildlife depredation and vulnerability, together with the household coverage, it seems more equitable than the existing mechanism. VDCs covered by this research are more vulnerable to wildlife damage and totally depend upon the park resources, with very low population density. Based on the present funds disbursement mechanism, this area receives less funds than those that have low vulnerability and alternative resources (such as buffer zone forests).

III. Protected Areas and Rural Livelihoods

Rural livelihoods in Nepal primarily depend upon forests resources and subsistence agriculture farming. Mountain communities in particular primarily rely on livestock farming, non-timber forest products (NTFPs), and medicinal and aromatic plants (MAPs). An estimation in Humla district showed that 38% of the total income of this community is derived from NTFPs (MFSC, 2000). However, the livelihood of local people residing in and around the mountain PAs is different than that in other settlements. Wildlife damage and PA policies that provide very limited and restricted rights on access to resources, are among the various factors. As a result of this policy, local communities around the PAs are deprived of the kind of income that the people of Humla are earning from forest resources. However, livelihood does not only include the income of individuals or households but also the combination of institutions, social interaction, relations, and mechanism of access to resources (Ellis, 2000). All these factors, which influence the livelihoods strategy, have been discussed earlier. The success of the LNP management to achieve underlined objective is primarily based on its relation with local communities. For them, livelihoods options should be well addressed in policies and their implementation.

1. Use and access to park forest resources

There is hardly any community managed forest, except three patches in ward 9 of Syaphru VDC. Local communities rely almost solely on the park to supply their daily need for timber, firewood, and forage. Communities are allowed to collect dead and fallen trees from the park areas within the prescribed time and area. The cutting of green trees is not allowed in the park. However, local people choose to cut the nearby forests, and focus on the species good for firewood (i.e. less smoke, longer lasting).

The increasing number of tourists, though providing a significant income source for local livelihoods, has also increased the demand for firewood and timber for hotels. An estimate showed that hotels in the Langtang area consume about 136.5 metric tonnes of firewood annually (TTS, 2003). Similarly, two cheese factories in the area also have a high demand for firewood. In recent years, the cheese factory in Chandanbari consumes about 12,500 cubic feet of firewood annually, which is nearly double the years from 1997 and 1998 (personal communication from Padam Biswakarma, Manager of cheese factory, 2004). The Park staff and the Royal Nepalese Army designated to protect the park itself, also use a significant amount of firewood.

Timber is being used mostly for the construction of houses, hotels and offices. The pole-sized trees are primarily used in house beams, *goths* (cattle sheds), and for raising *lingo* (prayer flags). The LNP monitors and designates the tree to be cut for this purpose. Fremier (1998) pointed out that the present royalty system significantly decreased the timber extraction. He also mentioned that "the then timber use is not threatening". However, there is still a lack of effective monitoring, especially in the present context of insecurity when movement of the LNP staff is minimal. The researcher observed the transporting of pole-sized trees from nearby park forests in the Chandanbari area.

Livestock grazing is noticed almost everywhere in the park area, except in Dhunche, the district headquarters (HQ). Because of the lower number of stall-fed livestock, livestock farming is largely based on the pasture and forested areas in the park. The livestock rearing practice needs to set up temporary *goths* along the grazing route. These *goth* areas are high-use zones, where herders use poles for construction, firewood for heating and cooking, and graze their livestock for two months. Customary practices in livestock farming and pasture management are separately discussed later.

These resource uses are evident on the forest covers, and edges of larger settlements and some localized places. In larger settlements like Dhunche, and Syaphru, the health of the forests can be easily traced, where there is no or very little forest cover. Similarly in localized places like Chandanbari (where a cheese factory is in operation), and places where herders keep their goats for livestock grazing, the forest understory has been destroyed.

There is the need to encourage alternatives to forest resources. Alternative energy may be the only viable option for meeting the requirements of increasing demand for firewood in the area. Local communities have already initiated some alternative energy programs in the area. Most of the hotels in the Langtang trekking route are equipped with back boiling and heating systems. Similarly, local users have mostly replaced wooden *lingo* (wooden prayer flag) with iron poles (Case 4). Local users from the Dhunche area have started plantations in degraded lands and nursery management, which will help to meet the demand for firewood in the long term, reducing the impact on park resources.

This discussion leads to two important issues. One, despite the "command and control", there is an increasing supply of resources from the park area and secondly, local communities can fare better by imposing social sanctions. One example in this regard is in the Gorkha district, where forests around the area are managed well by local people despite the increasing population and livestock pressure (Fox, in HMGN, 2000). The case of Bagan Chagan in Kavre district, where local communities manage the oak forest by restricting the lopping of trees (Miller, 1999) also provides evidence of such management. Thus, the PA authority now needs to move towards a new strategy of directly involving local communities to manage the nearby park areas.

Case 4: Use of iron poles for lingo (prayer flag)

Traditionally, Tamang and other Buddhist followers need to raise the *dharche* (local name for prayer flag) at the top of their roof or in front of their house. They believe that with such a flag they would not need to face any troubles. Mostly, they use a wooden pole, *darsing* (also known as *lingo*) to raise the flag. Culturally, they are compelled to change and replace the *dharche and darsing*, stacked on the roof, every year during the period of Tibetan New Year, the *Lhosar*. The flags stacked in front of each house are normally replaced during special ceremonies like weddings, or funerals. Every year, each household cuts a pole-sized tree from nearby forests for the purpose of *darsing* (or lingo), which must be straight, free from any defect and knots. In practice, they normally cut two to three pole-sized trees, and select the best to raise the *dharche* during *Lhosar*. Collectively for this purpose, local communities cut thousands of pole-sized trees annually, which severely impact the forest cover and growth. In the year 2003, user groups decided to use iron poles instead of wooden poles to raise the *dharche*. This year, most households in the area have already replaced their wooden poles with iron ones, with support from buffer zone development funds. Local communities also contribute about 25% of the total cost. This simple but long-term decision, not only helps to protect the forest area in the LNP, but also proves that local knowledge and practice is more fruitful in PA management, if incorporated well.

Firewood consumption

Firewood is the main sources of energy for cooking, heating,, and making local *raksi* (local wine) (an estimate shows that about 17% of the demand for firewood is due to *raksi* making). The consumption of fuelwood within a household is mostly based on family size, frequency of *raksi* making, and duration of heating mostly during winter. Almost all firewood is collected from national parks. Only about 5% of the total supply is from private lands, and terraces.

Firewood consumption by households since 1998/1999 has decreased slightly. The average consumption of firewood in the past year was 336.8 *bharies* (one *bhari* is equal to approximately 20 kgs), which is 9.8% less than that in 1998/1999. On average, about 18.5 kgs of firewood is consumed per day per household. This does not include supplies to cheese factories, army and other staff working in the area. Hoteliers and lodge owners use the most firewood ranging from two to four *bhari* per day. The decrease in the number of tourists in recent years due to political instability and installation of fuel saving stoves are the major causes for the decrease in consumption.

Timber use

Unlike the consumption of firewood, the consumption of timber during the period has increased significantly. Despite the control and royalty mechanism, the extraction of timber has increased because of the construction of new buildings in the area. However, this information does not include the timber consumed by the development projects and government office building construction at the district HQ. The average consumption of timber during the past year is about 55.78 cubic feet, which is about 34% more than that for year 1998/1999. The official data on the issue of timber collection permits is not available from the LNP. Most of the timber is extracted from the park area, with other sources from outside national parks and three buffer zone community forests in Syaphru VDC. Some hoteliers and lodge owners in Dhunche, and Syaphru Besi use the timber from Kathmandu and Trishuli.

Forage and fodder consumption

Forage is mainly collected for stall-fed cattle such as bulls, cows and water buffalo. Some farmers also use grass in house roofing, and shelters for cattle (cowsheds). The number of stall-fed cattle in the area is relatively less than the *chauri* (cross-breed cow-yak) herds, and sheep, which are not stall-fed. With the decrease in livestock population, the total forage used in the past year is significantly reduced by 15%. In year 1998/99, the average forage consumption was 213 *bhari*. This amount was reduced to 179 in 2004. The livestock unit owned by local communities was also significantly reduced. The average Livestock Unit (LSU) owned by a household in 1998/1999 was 4.64, and this was reduced to 3.58 in year 2004.

Medicinal and aromatic plants

The traditional practice is to collect medicinal and aromatic plants (MAPs). During the Tibetan New Year, (*Lhosar*), they go to the nearby market to sell these valuable plants. With the income from the sale of MAPs, they buy essential commodities to celebrate the New Year. This has not been practiced for a long time, as present PA policy does not allow collection and harvesting of wild MAPs from national park areas. Many studies show the high potential for both sustainable wild harvesting and commercial farming in private lands, which would be the major source of income for local farmers. However, the present policy does not encourage local farmers to even cultivate their private lands. They need to acquire export permits (*chhod purji*) before selling these products in the market, even if they are extracted from private lands. A meeting between Park authorities and the local community was held 2 years before, to issue such permits. These have not yet been approved by the relevant authorities (personal communication from Nawang Tamang, former member at the DDC, Rasuwa, 2004).

The present study shows that local communities collect a tiny amount of MAPs from the nearby park area. In 1998/99, they collected about 17 kgs. of different species, reduced to 11 kgs. in the past year. However, in informal discussions with local traders, an MAP businessman pointed out that a large quantity of MAPs had been sold in the Dhunche or Kathmandu market. Local communities informed the researcher that there is an increasing trend in illegal collection of MAPs in the last three years from the park area. Local traders and collectors acquire exit permits (*chhod purji*) from the District Forest Office (DFO).

Grazing and livestock rearing practices

Local families from the Langtang area graze their yak, *chauri* and sheep in park pasture lands for 5 to 6 months in a year. Historically, most of the pasture lands were managed under local ownership. At present all pasture lands are under the LNP management, which also provides access to graze livestock. The LNP declares some areas as "the area of significance" and imposed prescribed and control grazing (such as Red Panda conservation area in Cholangpati and Gosaikunda Cultural site around Gosaikunda Lake). However, the practice of prescribed and controlled grazing has not been implemented effectively.

The traditional practices of fire burning to promote palatable grass species, and construction of *goths* in every season mostly impact the forest area in the park. White clover, an exotic grass species, introduced in the area to improve the pasture quality has now become a problem for livestock and local grass diversity. Local herders, during their meeting with researcher, mentioned that if this grass is grazed during its flowering season, livestock (mostly *chauri*) become sick and sometimes die. Presently, herders do not graze their livestock in the area during flowering season. The death of *chauri* and yak because of the white clover is documented (LNP, 2001). This may be because of the high protein value (protein content is even high during flowering season). The loss of diversity of local grass species is also noticed in these pasture lands. Local herders mentioned that there is no local species in Tekche pasture land because of the prevalence of white clover.

Sixty six herds, having an average number of 18 to 22 Yak and *Chuari* in each herd; and 30 herds of sheep, having an average number of 150 to 200 have been using the same pasture lands (Personal Communication from Padam Biswakarma, Manager for Cheese factory, 2004, and informal meeting with herders). They collect and sell milk to the local cheese factory in Chandanbari. An informal herders group was formed in the area to coordinate with the park authority in the registration of livestock and pasture lands. The then Park Warden himself initiated the formation of such a committee and registration process. However the park has neither formally registered the herders committee nor the livestock owned by them yet. A clear and long term management strategy is required to manage livestock grazing in the area. This strategy should focus on the registration of livestock and local herders committees, provide veterinary and other technical support, zoning for rotational grazing, and a mechanism to coordinate and network with other stakeholders such as cheese factories, and local veterinary offices.

Case 5: Traditional ownership on Kharka (Pasture lands) of Gosaikunda area

Traditionally, the local residents of Barkhu Village have management ownership over pasture lands around Gosaikunda Lake. Before this area was declared a national park, these pasture lands were managed under the control and supervision of local elites.

Historically, Rana Ruler Chandra Samser issued a royal seal to the *Mukhiya* (village headman) of Thimi, Bhaktapur to manage the *kharkas* around Gosaikunda (including the Lake) and to establish a *guthi* (trust) from which *kut* (tax) is collected. Because of the great distance between Bhaktpur and Rasuwa, the *Mukhiya* authorized the *Lama* (priest) of Yarsa village of Rasuwa district to look after the management of these *kharkas.* The *Lama* of Yarsa was the brother of the *Lama* of Barkhu village. Finally, the *Lama* from Barkhu village received the management authority from his brother. Since this time, the area was managed under trust until the government declared the area a national park. *Chiura* (bitten rice) was paid as tax (*kut*) in lieu of the use of *kharkas.*

Use of alternative energy and fuelwood saving stoves

Promotion of alternative energy and firewood saving devices in order to reduce the impact on forest cover has been the key strategy in the LNP area. The LNP prohibits the use of firewood energy for trekkers and tourists. Similarly, the commercial collection of firewood is also controlled. As a result, hoteliers and lodge owners have to pay a prescribed royalty for the collection of firewood.

However, the majority (about 87%) of the respondent households rely on firewood. Some hotels and lodge owners in Dhunche, Thulo Syaphru and Syaphrubesi use petroleum gas. Similarly, 90% of the respondents involved in hotels and tourism have installed firewood saving heating stoves. Eight percent of the households have solar electricity, used to heat water. Five percent of households have a biogas plant. Only a few among the selected households use kerosene for cooking. Six percent of the total respondents use solar heating, petroleum gas, and kerosene.

Rural infrastructure development

Infrastructure such as road access, electricity, drinking water supply, and irrigation are important for rural livelihoods to make their life efficient and more productive. The present policy encourages public-private partnerships and local participation. Development projects are mostly implemented under two different mechanisms. Large projects such as hydropower and road networks are mostly undertaken centrally, whereas micro projects are implemented locally.

An environmental impact assessment (EIA) is mandatory for the large projects like hydropower and road networks. The Environment Protection

Act 1997 provides for a public hearing of the EIA results before the implementation of these projects. However, the government can issue an EIA order even for small projects and services in a national park or buffer zone area, if it feels this is necessary (see Forest Secretary-level decision on operation of helicopter services in LNP, 2052/11/20 B.S.). A discussion is underway to require EIA for hotels and lodges inside the LNP (personal communication, Tika Ram Paudel, Officiating Warden, LNP 2004). Approval from the DNPWC is necessary to implement any kind of such large projects in the area of the LNP and its surrounding buffer zone.

The issue of EIA and the impact of development projects in the LNP and its surrounding area is sometimes debatable. The Chief District Officer (CDO) of Rasuwa stressed the need for a community oriented, rather than *jungle* oriented, policy. The cases of the mineral water factory and proposed hydropower project are two examples of such debate. The DNPWC rejected a proposal of Kantipur Hydropower, a privately owned company to construct a 2 km long tunnel and a dam in Syaphru VDC to generate 20 MW energy. An EIA was conducted by the company and a public hearing held on the possible impacts and mitigation measures. Local communities and the DDC endorsed the proposal, noting that there are no valuable timber and endangered species in the proposed area and effective measures would help to largely mitigate the impact. Karnyal Tamang, former Chairperson of Syaphru VDC also mentioned to the researcher that there is not any significant flora and fauna in the proposed dam site. In fact, the area is bare land. Neither would the underground tunnel impact the national park area. The impact of Chilime hydropower is visible on the local economy. Many locals get employment from the project. Chilime hydropower is providing support for road maintenance and is also involved in soil conservation. The *nikunja* (national park) should encourage the project and get them (Kantipur company) involved in conservation activities too. This Act broadly benefits the interest of local and *nikunja* itself, which ultimately helps to improve the park- people relationship.

Different government and non-governmental agencies have been implementing community infrastructure projects in the area under their own supervision and guidelines. However, all need to endorse the program through the District Development Committee (DDC) council. The present policy deals with the execution of small scale projects through local user groups, but this has not been practiced on the ground. Still, a large number of projects are being executed through contract, and tender. The drinking water supply project in Bridim village is an example of construction through contract. During the researcher's field visit, he found that all the workers and technicians involved in construction of this project were from outside the village. This was in June, the monsoon season. In fact June is not a suitable time, as most locals were involved in agriculture work. If, at the start, local communities are not directly involved in the project implemented at their village, theoretically they will not also become involved in the maintenance of such project and they do not have the feeling of ownership over the resources, which then cannot be sustained.

The buffer zone support program under the LNP provides grant assistance to implement community infrastructure projects. BZ user groups have been directly involved in completing these projects and contribute about 25–30% of the total cost in terms of cash or labor. About 96% of the total respondents contributed to development projects. Seventy five percent of the respondents contributed labor only, whereas 42% contributed both cash and labor. Only 4, out of the 125 sample households responded that they are neither directly involved nor contributed any cash. This shows a great degree of participation of local beneficiaries in development efforts. However, the BZ fund allocated for infrastructure is sufficient only for very small scale projects.

Tourism and Hotel Management

Nepal's policy on PA management is also to promote eco-friendly tourism within these areas of wilderness, as these areas serve a major destination for tourists. Official data at the Ministry of Culture, Tourism and Civil Aviation (MoCTCA) shows that 36.1% of tourists visited different protected areas in Nepal in 2002 (HMG/MOPE, 2004). The present policy makes EIA a pre-requisite for the establishment of hotels and lodges inside a national park. However, the necessity of EIA for 54 hotels and lodges leased by the LNP to individuals in the Langtang area is debatable. Most of the hotels were established before the Environment Protection Act 1997 came into force. These leases need to be renewed every five years, and last year, the issue of EIA was raised when leases had to be renewed. At the same time, some hotels, established after the passing of this Act, were leased without prior EIAs. The lack of vertical coordination within government ministries is one of the reasons for poor implementation of this provision. The Ministry of Population and Environment (MOPE) is the executive agency for the implementation of the Environment Protection Act 1997, whereas the Ministry of Forests and Soil Conservation (MFSC) has jurisdiction over protected areas, and can issue licenses and permit for hotel facilities within the areas. The promotion of tourism in Nepal is ultimately the responsibility of the Nepal Tourism Board (NTB), an autonomous body under the Ministry of Culture, Tourism and Civil Aviation (MoCTCA). The close coordination among these agencies would lead to better implementation of policies at the local level.

The impact of tourism in protected areas has been widely discussed for a long time in Nepal. Article 33 of the BZ Management Regulation provides for the preparation of a detailed Tourism Development Plan. All the services carried out in protected areas should be managed under the Development Plan and the Code of Conduct, which need to be approved by MFSC prior to the implementation of such plan and codes. Some of the protected areas already have a detailed Tourism Plan. For example, the plan for the Bardia National Park was prepared with grant assistance from the UNDP-managed Participatory Conservation Program. However, the LNP does not have any detailed tourism management plan. The park authorities stated that the lack of resources and staff is the main reason behind it. The recent management strategy of hotels and lodges in the LNP is largely based on the outcomes of the recent study done in 2003. This study primarily focused on hotels inside the parkland.

There are two institutions formed to facilitate tourism development in the area. The Tourism Management Committee and Gosaikunda Management Committee, registered as functional groups under buffer zone provisions, are mostly voluntary local institutions. However, in recent times, the progress of these institutions is not encouraging. For the past two years, the Gosaikunda Management Committee has not organized a formal meeting and is virtually functionless. When this was raised with the former Chairperson of this Committee, he told the researcher: "there are not any concrete rights or responsibilities of this Committee. We do not have any resources to carry out programs and activities in the area. The LNP provides support to BZ groups, and we do not receive any resource from outside. If we do not have any resources and rights, what can you expect?"

This discussion leads to two important conclusions about local governance. Firstly, once power and responsibilities are clearly delegated, local institutions can play a greater role in resource mobilization. Secondly, a clear long-term vision of the program is necessary for management of resources and development. Therefore, a long-term tourism management plan with a clear role for all stakeholders involved in the LNP needs to provide for close coordination with local government (DDC). During the preparation of plans, participation of local communities and other stakeholders involved in this sector should be provided for.

Vulnerability

Local communities in and around the PAs are more vulnerable to human loss and injury, crop and animal depredation due to wildlife. This externality of wildlife damage, over which local communities have no or limited control, increases the threat to their subsistence livelihoods. There is no formal data or figures on crop depredation in the area. Loss of life has not occurred during last five years. However, 5 cases of human injury, all by caused by the Himalayan black bear, have been reported during the year 2001–2002. The presence of the snow leopard in higher pasture areas is another threat to livestock depredation. In year 2000/2001, fifteen livestock were predated by snow leopards (LNP, 2001).

There is no policy or legislative measure to compensate for human or crop damage by wildlife. This is the major reason for park and people conflicts and a common feeling of local people around PAs is that "If wildlife destroys crops and kill livestock, people will not be compensated against that loss, whereas if people harm wildlife, they will be prosecuted" (Budathoki, 2000). The issue of wildlife damage and human loss has been a major policy challenge for PA authorities. In recent years, some PA management have established endowment funds with support from the UNDP-funded Parks and People project, to compensate that loss. Similarly, the LNP has established a relief fund of NRs. 300,000 under the buffer zone support program. In the year 2001/2000, 50% of the total treatment costs for all 5 cases of human injury was compensated. (personal communication, Tika Ram Paudel, Officiating Warden, LNP, 2004). However, crop and livestock damage by wildlife has never been compensated for in the LNP.

The research area does not have sufficient agricultural field and internal food production is sufficient only for 4–5 months in a year. Wildlife crop depredation, therefore, has been a crucial issue for these communities, which ranges up to 80% of the total production. The land use pattern is also changed. Some agricultural lands, near the forest area and far from human settlement, have been kept fallow because of the damage. Crop damage is not limited to any particular crop and season. Wildlife damage occurs from the first day of seed sowing to the time of harvesting. They damage all major crops like maize, wheat, finger millet, and potato. Monkeys damage major fruits like apples in the area.

The wild boar is the main crop raiding animal. About 86% of the total household respondents mentioned that the wild boar always damages their crops. This is followed by the monkey, and the deer family. The frequency of appearance of the snow leopard and Himalayan black bear is limited in settlement areas, but they are the main predators of livestock. Other species that damage crops are the porcupines and various birds. However, the frequency of wildlife movement is not directly proportional to the extent of crop damage. During the discussion with local villagers, they mentioned that the mobility of wildlife has decreased the last 2 years. The movement of monkeys in some villages has been limited in recent years as compared to 1997/1998 (for example, in the Dhunche area, damage by monkeys was significant, but limited in year 2003/2004). Another example is porcupines, which causes damage mostly in the Bridim VDC.

The actual amount of crop damage is difficult to quantify as respondents do not have confidence in the quantification of damage. However, they estimate that the damage sometimes reaches up to 80%. The wild boar damage takes place mostly during the night, when preventive measures like shooting, and making noise may not be effective. Nurpu Lama, former VDC chairperson, informed the researcher that "wildlife crop damage equivalent to NRs.0.5 to 0.6 million takes place every year in the Bridim VDC". Except in Dhunche, and Syaphrubesi, which are relatively more populated and have access to electricity, and are therefore less vulnerable, all villages are equally likely to suffer damage.

The information collected on wildlife crop damage is relative to total production, because of the lack of confidence among respondents in quantifying the amount of crop damage. However, a few respondents provided estimates of the quantity. To make the data consistent, these are also converted to a percentage over production. The crop damage by wildlife during last six years has decreased by 9%. The average crop damage in 1998/1999 was 45.21% of the total production, and was reduced to 36.12% in year 2003/2004. This is because of the preventive measures taken by farmers, permission for wild boar hunting by the LNP in high crop damage areas, and changes in crop pattern in some places such as Bridim and Syphru, where farmers plant fruit crops instead of cereal.

Local communities have been adopting various strategies to prevent wildlife movement in their cropland. Most of the farmers, however, have only been able to rely on traditional methods, such as making noise, shooting, use of color plastics, shining plastics (used in audio cassette) around the croplands. About 34% of the households have constructed animal preventive infrastructures. Of these, 8 percent of farmers have stone walls around their cropland. Similarly, 26 *machans* (wildlife towers) have been constructed. A few people dig trenches around the cropland, but this is not effective. Traditional approaches in preventing wildlife crop damage have been found more effective (see Case 6).

Case 6: Local practice in preventing movement of monkey in croplands

In 2001, a meeting of local residents from Bridim village in North Rasuwa district discussed possible measures to prevent crop damage by monkeys. The monkey is the second most damaging animal after the wild boar. They also discussed the possibility of catching these crop damaging animals, and agreed that the monkey is easier to catch than the wild boar. Their knowledge on the behavior of the monkey, which mostly lives in groups, also helped to find the right strategy. One of the participants, Tashi Lama, proposed to catch the leader monkey from the group. (Local residents informed the researcher that a group of monkeys normally has one group leader). They caught and painted the whole body of the leader monkey like that of a leopard, then released it into the forest. Their assumption that other monkeys in the group would be afraid of the leopard turned out to be correct, and thereafter for a long time, the movement of monkeys in cropland was reduced greatly. This showed the need for incorporating local knowledge of animal behavior, which not only helps to conserve the indigenous knowledge, but also provides significant areas for scientific study.

The future of the Protected Areas (PAs) mostly depends upon the relation with local communities, residing nearby. The increase in wildlife caused vulnerability to local livelihoods, greatly influencing the relation. Therefore, the PA authority needs to seriously rethink the policy provision on compensation or relief support to communities and/or victims. Similarly, the scientific studies on animal behavior need to be focused, incorporating local knowledge and skills.

IV. Conclusion and Recommendations

This study has reviewed the implementation of Protected Areas (PA) policy at the local level in Langtang area. The researcher has found that the communities residing in this area have been widely dependant on natural resources for their subsistence livelihoods since time immemorial. The controlled PA policy implemented during the initial period of park history makes their lives more difficult. Furthermore, their traditional rights over forest resources have been controlled. On the one hand, these communities are deprived of benefits that they used to get from the forest area. On the other, wildlife harm their crop and livestock, and they are never compensated.

***Controlled Access to Forest Resources*:** Local communities have access to grazing inside the national park, though controlled in many aspects. This positive policy instrument provides opportunity in mitigating the park-people conflict in the LNP. Local herders enjoy grazing rights throughout the year, but their traditional ownership still needs to be addressed in policy instruments. The management of *kharka*, where there is a lack of regular monitoring by the park staff, if given to the local committee under the prescribed management plan, would produce better results. Medicinal and aromatic plants, on which the traditional health treatment system is based, have been the "locked resources" for local communities. These valuable species are also the major source of local income. Present PA policy does not encourage cultivation of medicinal plants

***Crop Depredation and Compensation*:** For long time, wild animals have posed a major threat to local livelihoods. Annually, wildlife damages large quantities of crops, and depredates livestock. Local communities are neither compensated for that loss nor provided any alternatives. More importantly, the Park authority neither develops appropriate control measures nor provides any technical or financial support in order to establish wildlife preventive measures.

***Lack of Long Term Tourism Plans*:** Tourism in the area has provided significant contributions to the local economy. Hotels inside the national park areas are consuming large amounts of firewood annually, which has a major impact on forest cover. The burning of junipers by hotels at high altitude has reduced the slow growth forest cover in the area. However, the increasing use of alternative energy shows some hope of reducing energy dependency in forest areas. A detailed plan covering every aspect of the tourism development in this area would be an asset for the future of the tourism industry. The Tourism Management Committee, which is walking lamely these days, needs external support in order to strengthen tourism activities in the area.

***Weak Institutional Arrangements*:** Traditionally, local people in the area set their institutional mechanisms to protect the forest resources and cultural heritage. Their tradition was at risk when government controlled their rights, imposing PA policies in the area. They were protecting these resources, because they received returns through this protection. The present buffer zone policy tries to set up another form of institutional mechanism. User groups formed under such mechanisms have limited rights over the management of buffer zone resources. However, these institutions do not have any direct role or rights over national park resources. Neither are they involved in major national park decision-making, nor do they have access to park resources. Because of the limited resources in buffer zones, these institutions are merely to use the buffer zone funds allocated for development of their villages.

***Inequity in Buffer Zone Fund Allocation*:** Buffer zone funds have very limited resources to implement community development projects which help make local communities more productive. The present park policy to distribute the revenue equally among all BZ user groups is unjustifiable, taking into consideration wildlife impact and vulnerability. User groups in the southern area of the LNP have much resources in comparison with groups in the northern area. At the same time, the northern area is a high wildlife impact zone, and communities are more vulnerable to wildlife damage. Communities in the south have access to community forests, whereas northern communities need to pay a government royalty to get timber and other forest resources. It would more equitable if the BZ fund was allocated to take into account of wildlife damage and the availability of resources.

***Lack of Coordination*:** Intra and inter-organizational coordination is a major problem in implementing biodiversity conservation and sustainable development programs. Despite the policy provisions to coordinate and disseminate information, there is a wide gap in reality. This may be because of the issue of authority and rights within different governmental organizations. NGOs, however, are more conscious of this issue. The DDC, being a local autonomous government needs to play a coordinating role in the district, which helps to minimize program duplication and save resources. A district-based coordinating committee would help to overcome such problems.

***Legislative Conflicts*:** The Local Self Governance Act (LSGA), exercised by local governments (DDCs and VDCs) and the National Park and Wildlife Conservation Act contradicts each other in many cases. The LSGA empowers the VDCs and DDCs to impose taxes on natural resources and other materials (such as gravel, stones, sand), whereas all these natural resources are considered part of national parks under NPWCA. Similarly, water resources are considered the property of national parks, but the local governments also have authority over them under the LSGA. The issue of the mineral water factory in Rasuwa has been widely debated. The challenge of government authorities is to coordinate well with each other and implement regulatory mechanisms for the welfare of local communities and environment.

***Superficial Plans and Group Constitutions*:** The LNP buffer zone management plan fails to address the key issues and programs to improve park-people relations. The social and economic character of local communities needs to be thoroughly incorporated. The grass-root causes of resources depletion should be addressed. Similarly, the constitutions of user groups fail to address the key issue of the particular community. In fact, these constitutions are copied from the Shey Phoksundo national park. The socio-economic issues and constraints in Shey Phoksundo and Langtang are, however, very different. For example, Shey Phoksundo is not linked with the road network. Unlike in LNP, the need for EIA is not a big issue in Shey Phoksundo. The district HQ of Rasuwa lies in the LNP area, where most of the government services are located where as SheyPhoksundo is far from the district HQ.

The effectiveness of international instruments is largely based on their implementation at local level. The LNP possesses some endangered and endemic species of international importance. These species need better protection. Awareness of the importance of such species play an important role in conserving these valuable species. The LNP needs to revise and incorporate some of these issues in its buffer zone plan and constitution of BZ user groups.

Recommendations

The present PA regime has been facing constraints in social mobilization, access to nature resources, and institutional arrangement. Furthermore, there are still some policy contradictions and inconsistency that need to be urgently reviewed. The following recommendations would help achieve better management of protected areas, especially in mountain areas.

1. Legislative framework

- A revision of BZ management regulation and directives is necessary to make them consistent. The roles, responsibilities and powers of BZMC should be clearly spelled out, making it a more autonomous and decision making body. The role of the park warden should be limited to that of facilitator only.

- Legislative measures should be developed to compensate wildlife damage to crop, human loss and livestock depredation. Based on the experience from other parks, the adoption of community insurance, through matching support from the park, could be a viable option for such loss and damage.
- Legislation should also provide management authority over certain areas of forests, near the park settlement, where there is no community or buffer zone forests. Communities from these areas should directly be involved in the use of dead and fallen trees inside the park area.

2. Institutional framework

- The BZ institutions should be involved in park area management. Some authority such as the issuing of permits to collect firewood, and the monitoring of timber harvesting, should be provided. A clear and transparent constitution needs to be developed.
- Strong coordination mechanisms are required between both district-based agencies and local communities.
- Tourism management committees and Gosaikunda management committees should be revitalized, with clear roles, responsibilities and powers. The technical back up support to these institutions would help to strengthen tourism development and resource conservation in the area.

3. Use and access to resources

- Regulatory and descriptive guidelines on medicinal and aromatic plants (MAPs) harvesting, and commercial farming in private lands need to be developed. Local communities should have access to MAPs on the basis of prescribed and long term MAP plan.
- Rotational and block grazing should be encouraged in high pasturelands. A proposed Herders Committee should be registered and institutionalized.

References

Budathoki, P. (2003) "Nepal's Conservation in Crisis: Empowering People to Secure Natural Resources" in *Journal of Forest and Livelihoods.* pp. 72–76. 2 (2), February, 2003. Forest Action, Kathmandu.

DDC, (2002) *Periodic District Development Plan, Rasuwa*, 2002/03-2007/08, Volume 1 District Profile Analysis. District Development Committee, Rasuwa, Nepal.

DDC, (2002) *Periodic District Development Plan*, Rasuwa, 2002/03-2007/08, Volume 2. District Development Committee. Rasuwa, Nepal.

DNPWC, (2001) *Annual Progress Report (2000–2001)*. Department of National Parks and Wildlife Conservation. Kathmandu, Nepal.

Ellis, F. (2000) *Rural Livelihoods and Diversity in Developing Countries*. Oxford University Press Inc., New York.

Fisher, R. (2002) "Common Property Regimes: Risks and Governance of Biodiversity". Paper presented at the High Summit-Asia, 6–10 May 2002, Katmandu, Nepal.

Fox, J. in HMG/MOPE, (2000) *State of the Environment in Nepal*. His Majesty's Government, Ministry of Population and Environment, Kathmandu, Nepal.

Fremier, Alexander K. (1999) Langtang National Park: Forest Inventory and Site Specific Management Prescriptions (Draft). American Peace Corps. Langtang National Park, Rasuwa, Nepal.

Graham, J., Amos B., and T. Plumptre (2003) *Governance Principles for Protected Areas in the 21st Century*. Institute on Governance, Canada.

HMG/DSCO, (2003) *Jilla bhu sanrakshyan parichaytmak pustika*. District Soil Conservation Office, Dhunche, Rasuwa, Nepal.

HMG/LNP, (2001) Buffer Zone Management Plan. Langtang National Park, Dhunche, Rasuwa, Nepal.

HMG/MFSC, (1997) *National Report on Implementation of Convention on Biological Diversity in Nepal*. His Majesty's Government, Ministry of Forests and Soil Conservation, Singh Durbar, Kathmandu.

HMG/MFSC, (2002) *Nepal Biodiversity Strategy*. pp. 37–43. His Majesty's Government, Ministry of Forests and Soil Conservation, Kathmandu, Nepal.

HMG/MoLSPA, (2000) *Environment Protection Act 1997 and Environment Protection Rules 1997*. English Translation, His Majesty's Government, Ministry of Law, Justice and Parliamentary Affairs, Law Book Management Board, Kathmandu.

HMG/MoLSPA, (2002) *Nepal Ain Sangraha, Khanda 2(ka)*. His Majesty's Government, Ministry of Law, Justice and Parliamentary Affairs, Kathmandu, Nepal.

HMG/MoLSPA, (2002) *Nepal Ain Sangraha, Khanda 7(ka)*. His Majesty's Government, Ministry of Law, Justice and Parliamentary Affairs, Kathmandu, Nepal.

HMG/MOPE, (2000) *State of the Environment of Nepal*. His Majesty's Government, Ministry of Population and Environment, Kathmandu, Nepal.

HMG/MOPE, (2002) *National Assessment Report 2002, Nepal, World Summit on Sustainable Development*. His Majesty's Government, Ministry of Population and Environment, Kathmandu, Nepal.

HMG/MOPE, (2004) Biodiversity and Ecotourism. in *State of Environment.* pp. 143–173. Ministry of Population and Environment, His Majesty's Government, Nepal.

HMG/MOPE, (2004) Protected Areas of Nepal and Ecotourism. in *State of Environment.* pp. 95–115. Ministry of Population and Environment. His Majesty's Government, Nepal.

HMG/NPC, (2003) *The Tenth Plan 2003–2008.* Chapter 10. National Planning Commission, His Majesty's Government, Kathmandu, Nepal.

HMG/NPC/MOPE, (2003) *Sustainable Development Agenda for Nepal.* His Majesty's Government of Nepal, Kathmandu.

Janssen, Marco A. and Elinor Ostrom, (2001) *Critical Factors that Foster Local Self Governance of Common-Pool Resources: The Role of Heterogeneity.* Paper presented at the Conference on Inequality, Collective Action, and Environmental Sustainability, Santa Fe Institute, New Mexico, September 21–23.

MFSC, (1997), Internal circular by the Ministry of Forests and Soil Conservation on decision of Forest Minister, 2054/8/29 BS, Ministry of forests and Soil Conservation, Nepal.

Miller, J. Daniel (1999) Rangelands Management Consultancy Report. Submitted to Nepal Australia Community Resource Management Project. October 1999.

MoLJPA, Forest Act 1993, Nepal Ain Sangraha, Khanda 2 ka.

Neupane, R. P. (1995) A Study on the Development of Agroforestry Systems in the Hills of Ramechhap Nepal: A Sustainable Resource Management Perspective. M. Sc. Dissertation. Asian Institute of Technology, Bangkok, Thailand.

NPWCA (1973), The National Park and Wildlife Conservation Act, Nepal Ain Sangraha Khanda 7(ka) (Collection of NEpal Acts, Vol 7 (a)

Personal communication, (2004), Nawang Tamang, Resident of Dhunche VDC and Former elected member, Rasuwa, Nepal.

Personal Communication, (2004), Padam Raj Biswakarma, Factory Manager, Langtang Cheese factory, Rasuwa, Nepal.

Personal Communication, (2004), Kamal Gautam, District Soil Conservation Officer, Rasuwa, Nepal.

Personal Communication, (2004), Tika Ram Poudel, Officiating Chief Warden, Langtang National Park, Rasuwa, Nepal.

Sharma, U.R. (1998) *Buffer zone Policy and Analysis of the Royal Chitwan National Park.* King Mahendra Trust for Nature Conservation, Nepal.

Silwal, T. (2003) Buffer Zone Management in the Royal Bardia National Park. n.p. Dissertation work for the partial fulfillment of the Master degree in Forestry, Institute of Forestry, Tribhuvan University, Nepal.

Thapa, D. B. and L. Bhatta (2003) Exploring TMI's Program in Langtang Valley, n.p. project internal report, The Mountain Institute, Kathmandu.

TTS Nepal, (2003) Status Report on Parklands Hotels in Langtang National Park. TTS Nepal, Kathmandu.

Abbreviations

ACA	Annapurna Conservation Area
BPP	Biodiversity Profile Project
BZ	Buffer Zone
BZCF	Buffer Zone Community Forest
BZMC	Buffer Zone Management Committee
CA	Conservation Area
CBD	Convention on Biological Diversity
CBOs	Community-Based Organizations
CBS	Central Bureau of Statistics
CDO	Chief District Officer
CDC	Conservation and Development Committees
CITES	Convention on International Trade of Endangered Species of Wild Flora and Fauna
DDC	District Development Committee
DFID	Department for International Development, UK
DNPWC	Department of National Parks and Wildlife Conservation
DSCO	District Soil Conservation Office/Officer
EIA	Environment Impact Assessment
ha	Hectare
HHs	Households
HMG	His Majesty's Government of Nepal
HQ	Headquarters
IEE	Initial Environment Examination
IRR	Index of Relative Ranking
IUCN	The World Conservation Union
Kg.	Kilogram
KMTNC	King Mahendra Trust for Nature Conservation
LDO	Local Development Officer
LNP	Langtang National Park
LSGA	Local Self Governance Act
LSU	Livestock Unit
MAP	Medicinal and Aromatic Plants
MFSC	Ministry of Forests and Soil Conservation
MoCTCA	Ministry of Culture Tourism and Civil Aviation
MOPE	Ministry of Population and Environment
msl	Mean Sea Level
NBS	National Biodiversity Strategy

NCS	National Conservation Strategy
NPC	National Planning Commission
NPWCA	National Parks and Wildlife Conservation Act 1973
NRs	Nepalese Rupees (currency)
NTFP	Non Timber Forest Products
NWP	National Wetland Policy
PAs	Protected Areas
RCNP	Royal Chitwan National Park
SDAN	Sustainable Development Agenda for Nepal
SLC	School Leaving Certificate (High School Graduate)
UNDP	United Nations Development Program
VDC	Village Development Committee
WCMC	World Conservation Monitoring Center

Glossary

Bhari (bundle)	Unit of measurement, mostly used in fuelwood, forage. Normally, one *bhari* of fuelwood is approximately 20 kilogram, but differs by place to place.
Brahmin	Member of the highest priest caste under Aryan caste system.
BS	Bikram Sambat. Nepali calendar is based on B.S era. New Year of the Bikram Sambat falls around mid April and is 57 years ahead of AD.
Buffer Zone	An area set aside to adjacent of protected area in order to provide additional protection to the PAs, while providing access to neighboring communities.
Chauri	Cross breed of Yak and Cow.
Chhod purji	Exit permit. A legal document provided to export the forest products outside the area.
Danda	Ridge
DDC	District Development Committee. A local government at the district level, elected by the VDC representatives. Nepal has 75 districts. Each district is governed by DDC, consisting of a President, Vice President, and Members.
Gomba	Buddhist Monastery
Goths	Cattle shed. Herders construct temporary *goths* in high pasture areas for livestock rearing.

Guthi	A type of public trust. Traditionally, many places in Nepal have been managed under *guthi.* Government owned *guthi sasthan (corporation)* is responsible to look after all *guthi* areas.
Kharka	Pastureland, where herders keep their livestock for grazing.
Kut	A type of tax system during ancient Nepal.
Lama	Priest in Buddhist religion. Religious leader of a village or society.
Lhosar	Tibetan New Year.
Lingo	Stick, also refers to prayers flag, mostly used for religious purpose in Buddhist culture.
Machan	Tower. More specifically, wildlife viewing tower mostly constructed to monitor the wildlife movement in crop land. A kind of wildlife preventive infrastructure.
Mukhiya	Village Headman. During Rana regime, *mukhiya* was responsible for collection of tax within the authorized area.
Muluki Ain	The National Code of Nepal, on which most legislative measures area based on.
Nikunja	National Park.
Panchyat	VDC were known as Panchayat during Panchayat ruling system (before 1990)
Raksi	Local wine, mostly prepared from finger millet, through distillation process.
Ropani	A unit of measurement. One hectare of land consists of 19.79 *ropani.*
SLC	School Leaving Certificate. Graduates after 10 years of formal education
Tamang	An ethnic minority class of Tibetan origin, mostly engaged in livestock farming, and agriculture.
Ward	Smallest political unit. A VDC has nine wards, represented by Ward Chairperson who is directly elected by citizens.

8

CAPTIVE BREEDING AS A TOOL FOR CONSERVATION OF ENDANGERED SPECIES: THE LION-TAILED MACAQUE (*MACACA SILENUS*) — A COMPARATIVE CASE STUDY OF CAPTIVE BREEDING PROGRAMS WORLDWIDE

ANBARASI BOOPAL and MAHARAJ KRISHAN PANDIT

Lion-tailed Macaques or LTMs (*Macaca silenus*) are an endangered primate species, endemic to the Western Ghats of southern India. Due to their declining population in the wild and habitat degradation, captive breeding has been considered as a potential tool to re-establish populations in the wild. LTMs are captive bred in zoos principally in North America, Europe and India, and kept as exhibits in various other zoos. The captive breeding programs in North America and Europe were successful when compared to the captive breeding program in India.

In this report, features of the successful captive breeding programs were compared with the captive breeding program in India. The possible reasons for their failure in India were elucidated in the report from various angles such as the number of participating institutions and the viability of populations. Following discussions on the above areas, recommendations have been made towards successful captive breeding by cooperation of all these institutions and incorporating scientific management of these captive bred populations. This report comments on the endemic status of the Lion-tailed Macaques in India which restricts reintroduction programs only to Western Ghats of South India. The biology of these species has been taken into account using the research studies done on Lion-tailed Macaques to describe possible solutions. For example, the phenomenon of male migration has been discussed to mimic the same in captivity for successful breeding and also to maintain the natural gene flow.

Managerial recommendations have also been included in terms of studbook maintenance which is important for consistent monitoring of captive populations. Other viable methods to preserve their genetic pool have been proposed in this report, like the assisted reproductive techniques such as cryo preservation of sperm, artificial insemination and *in vitro* fertilization.

Thus, this report functions as a comparative analysis of the successful and failed captive breeding programs and proposes recommendations on the basis of review of available data on the same. The vital importance of preserving the habitat is also discussed in various sections.

Introduction

Biodiversity conservation is a multi-dimensional and challenging task for every nation in the world. Species conservation is an important and integral part of environmental management. With increasing rates of species extinction and threats to natural habitats, innovation in management of wildlife is an imperative. Future rates of species extinctions vary from less than 5% of all animal species to more than 25% with trends in forest clearance rates till 2025. (Olney *et al.*, 1994). The causes for the projected and documented extinctions have been analyzed. Human activity has contributed to 45% of those extinctions. Animal introduction and direct habitat destruction are the two major causes (Magin *et al.*, 1994).

Efforts to save species from extinction can be classified as *in situ* and *ex situ* methods. *In situ* (meaning "on-site") methods of conservation can be defined as methods to protect and improve the natural habitats. Protected areas, wildlife reserves and sanctuaries are examples of *in situ* methods of conservation. *Ex situ* methods of conservation are off-site methods of species preservation. Zoological and botanical gardens, germ plasm banks and gene banks are examples of *ex-situ* conservation methods. This report concentrates on the conservation of fauna, in particular mammals, as this is more relevant to the study of the Lion-tailed Macaque, which is a primate. The objective of captive breeding and management is to maintain a long-term viable population of a species that is endangered in the wild, and is meant to supplement *in situ* conservation.

The zoos and aquaria also can and do serve as potential 'arks' where successfully bred progenies can be reintroduced into their natural habitats for population restoration. At the same time, there are serious concerns and downsides to captive breeding, gene banks and reintroductions. These *ex situ* techniques when not properly integrated with *in situ* methods, can hardly help preservation of species in the wild. Nonetheless, if the proper expertise is not available, the whole exercise can be unfruitful. In the context of a developing country, the lack of expenditure and expertise can become prime reasons for the failure of captive breeding and reintroduction projects.

The Central Zoo Authority (CZA), a statutory body under the Ministry of Environment and Forests handles the captive breeding programs in India. In addition to the breeding programs, in 2000, the Laboratory for Conservation of Endangered Species (LaCONES) was also set up by CZA in collaboration with the Centre for Cellular and Molecular Biology (CCMB), Department of Biotechnology, Council for Scientific and Industrial Research (CSIR).

The legal setup for management of Indian zoos sprouts from the Indian Zoo Act, 1991. It is considered one of the most elaborate and organized legislation for zoos in South Asia. The Indian Zoo Act, its amendments and National Zoo Policy (2005) govern and provide guidelines for the conditions in which Conservation Breeding Programmes are carried out (www.envfor.nic.in).

Lion-tailed Macaque (*Macaca silenus)* is an endangered species, endemic to the Western Ghats of southern India, inhabiting the unique tropical wet evergreen forests. It is a monotypic species with the following taxonomic position, Order — Primates; Family — Cercopithecidae; Genus — Macaca; Species — *Macaca silenus.* Extensive studies on feeding, resting and reproductive behavior of LTMs have been done by research groups (Kumar, 1987), which will be discussed in the literature review.

LTMs are captive bred in three zoos in India viz. Arignar Anna Zoological Park (AAZP), Chennai, Trivandrum Zoo and Sri Chamarajendra Zoological Park, Mysore. They are also captive bred in North America and Europe. The Species Survival Program (SSP) of North American zoos and European Endangered Species Program (EESP) devised a separate breeding program for LTMs in 1989.

The endemic status of the LTMs introduces a lot of concern in the context of conservation. The Lion-tailed Macaques are captive bred in North America, Europe and India, but they are endemic to South India. While the captive breeding programs are successful in North America, they are a failure in India.

Thus, this study report aims to:

(1) compare the successful and unsuccessful captive breeding programs of Lion-tailed macaques at the different levels to elucidate the possible reasons for their failure in India,
(2) Discuss the issues relating to the captive breeding programs and reintroduction of an endemic species in a non-natural habitat, and
(3) Propose recommendations towards improvement of the current captive breeding program in India.

The Lion-tailed Macaque was chosen for this case study for the following reasons: (1) It is an south Indian endemic species that has been captive bred in North America and reintroduced there, a non-native habitat, which raises considerable concern on reintroduction issues and ethics; (2) The continuous failure of captive breeding programs in Indian zoos despite scientific advice, necessitates scientific rigor for the design of breeding programs, so a comparative analyses is necessary. Additionally, acquiring data for the rest of the species which have failed to breed in captive conditions was difficult; (3) Personally, I developed an interest in this spectacular species following their sightings in Silent Valley National Park, Western Ghats, India, and decided to examine the hurdles in conserving the Lion-tailed Macaques.

Methods

The captive breeding programs in North America, Europe and India are directly compared in a simple manner, by reviewing papers, comparing the size of the group in the program, the conditions maintained in the captive environment, facilities available and continuous monitoring done. Funding capacities, technical expertise of the team, educational background, and experience of the team members was considered during the development of the report. They were abandoned later due to the difficulty in procuring data from reliable sources. All the data for review and comparison are obtained from the scientific papers, CZA reports and International Zoo Yearbooks. However, at the end, based on the results from the comparison, the discussions are centered on relating the ecology of the species with the failure and success of breeding programs. In the case of successes, the long-term viability of the populations will be questioned on ecological and genetic grounds. The earlier introduction to the wild has not been considered seriously by research groups in the context of an endemic species. This issue will be addressed. The suitability and feasibility of such captive bred population for reintroduction to the wild is also discussed. Finally, based on the above discussions, possible recommendations will be summarized.

Lion-tailed Macaques have been found to be extremely difficult to breed in captivity due to their social structure. In this section, the initial position of their populations in North America, Europe and India will be discussed, followed by the comparison of captive breeding programs.

Phase I — Initiation of Lion-tailed Macaques (LTM) captive breeding

The first Lion-tailed Macaque caught from the wild in South India arrived in North America in 1899 (Lindburg, 2000). Information on the type of habitat from which it was caught was not available. Till 1965, this species has been kept only for display-based entertainment. Since 1990, retaining biologically viable captive populations has been incorporated into the objectives of captive breeding programs of Lion-tailed Macaques. By the end of 1965, there were 129 individuals imported from Western Ghats, India. Records on sex and age are unavailable for these individuals due to poor maintenance of studbooks and improper practices of husbandry. Thus for around 66 years, Lion-tailed Macaques in North America were not accorded the importance they deserved, as an endangered endemic species.. The zoos that possessed them in North America were located in San Diego, St. Louis, Chicago, New York and Milwaukee.

Following the work done by the Bombay Natural History Society on habitat fragmentation and the decline of Lion-tailed Macaques populations, a ban on unreported imports of animals was imposed. Lindburg (2000) describes the period after this (1983-present) as an era of scientific management. In 1982, the Species Survival Plan (SSP) for Lion-tailed Macaques was formed and the North American population of Lion-tailed Macaques was found to increase on its own. Scientific management and research in captivity helped to realize that these macaques are in an unfavourable position in terms of genetics and demographics. The very first step in captive breeding is to find out how viable they are in genetic terms. The population in North America was found to have 27.8% as genetic surplus (Lindburg, 2000). In 1982, there were 223 individuals distributed in 24 different locations, thus relocation was initiated to form a breeding group. The data on the groups formed in zoos were not available except for the San Diego Zoo. It consisted of 11 adult males, 12 adult females and 9 immature offspring, represented as 11.12.9 in Zoo studbooks.

In Europe, data on the Lion-tailed Macaques population is available only after the European Endangered Species Program (EESP) was launched in 1989. The number of participating zoos was 12 in 1989 which increased to 33 in 1999 and 41 in 2002. In 1989, there were 85–90 individuals in Europe (Kaumanns *et al.*, 2001), with 36 males and 49 females, represented as 36.49. The infant figure is not presented here due to the unavailability of that volume of International Zoo Yearbook (IZY) studbook. The numbers of captive individuals started at 730 in 1924 and only 237 individuals were alive in 2002 (Kaumanns *et al.*, 2006) after some imports. With this stock of 237 individuals, the captive breeding program was launched. The Coordinating Zoo for European Lion-tailed Macaques was the Cologne Zoo, Germany.

In India, a proposal to breed the Lion-tailed Macaques in captivity was submitted by the State Government of Tamil Nadu by June, 2002 to the Central Zoo Authority (CZA). Following that, three zoos *viz.*

Arignar Anna Zoological Park (AAZP), Chennai; Trivandrum Zoo and Sri Chamarajendra Zoological Park, Mysore, were selected for this breeding programme, whereas the rest of the zoos possess the individuals for purposes of display and exchange. The director of AAZP was the coordinator for this programme's execution. Two Indian primatologists Dr. Ajith Kumar and Dr. Mewa Singh were the scientific advisors for the programme. There were a total of 40 macaques during the initiation of the programme in these three zoos. The figures on sex and age groups were not available. The most recent data obtainable from CZA was from the 2003–2004 inventories of animals in Indian zoos (CZA 2004 Inventory). At the end of 2004, there were only 32 individuals in these three zoos.

Phase II — LTM captive breeding

The breeding of Lion-tailed Macaques in North American zoos was successful in the earlier stages in increasing the number of individuals. There were instances of keeping vasectomized males with females to avoid unwanted pregnancies (Lindburg, 2000). However losing a potential male from an endangered species is seriously questionable. The 11.12.9 (Males: Females: Unindentified) group formed by San Diego Center for Reproduction of Endangered Species was a mixture of imported individuals from Monaco. This new group was formed by introducing new males from the Monaco group because the first formed group since 1963 has not received any new males. This group has been vigorously monitored day and night for behavioral changes in females in response to new males in terms of calls, signs of menstruation and sexual skin through systematic observations. Analyses show that there has been erratic sexual swelling in females and no seasonality in births, which is in contrast to the occurrence of birth peaks in the wild from January to April (Singh *et al.*, 2006). From 223 total individuals, these were reduced during the course of the breeding program to 109 individuals in 22 institutions. In between, 39 individuals from SSP were exported to the European breeding programme. The problems encountered in these programs were high infant mortality rates, which were reduced through improved husbandry practices. Male-biased births and the failure of sexually mature females to reproduce were constant problems in these programmes. Introduction and release of the captive-bred Lion-tailed Macaques in a non-native habitat in St.Catherine's Island has been tried. Considering the endemic status, release in a completely different habitat is questionable though they were recaptured. No reports on this release and recapture exercise have been published to understand the observations made during this period of unknown length.

In contrast, Lion-tailed Macaque breeding in European zoos is an extensively documented and recorded programme. The number of individuals has increased from 90 in 1989 to 200 in 1998 (Kaumanns *et al.*, 2001). During this period 77 infants survived from 35 mothers. Results from the above study also show that 50% of the European Endangered Species Programme (EESP) output arrive from one-third of the female population. Similar to the results from North American Breeding programmes, the founder populations in EESP had more productivity in terms of the number of surviving infants per female compared to the F_1 and F_2 generations. The exact numbers are 5 surviving infants per reproducing female in founder generation to 2.5 surviving infants per reproducing female, which is a 50% reduction. However, it is not expected to reduce to 1.2 in the F_3 generation whose results have not yet obtained, but is expected to be close to 2.5 (Kaumanns *et al.*, 2001). It has been accepted by the relevant authorities that despite the increase in population, the reasons for the decline in productivity of females and the increase in infant mortality are not yet explained on a scientific basis. In addition to the breeding outputs, behavioral changes like self-directed hair pulling, eye-poking, saluting and similar stereotypic behaviors have been recorded in these captive social groups.

The Indian captive Lion-tailed Macaque population is of very small size despite its endemic status. It started with some 40 individuals but due to poor studbook maintenance, details on parentage are not clear. For example, there are 14 ancestors of unknown parentage. However, a genetic analysis has shown that there are 15 founders which should be increased to 30 founders (Gledhill, 1996) to increase the Founder Genome representation to 20 Founder Genome Equivalents. To achieve this, extensive genetic control must be exercised during the breeding programme. The present population in the Indian zoos is given in Table 8.1.

It can be noticed that there have been two deaths in the total of 40 individuals, with one female and one male. Details on the age of the dead individuals were not given in the inventory and the studbook was not available for reference. The annual report of CZA 2005 mentions 7 births from all the collections; however the reproductive potential of the population is estimated to be very low by research groups. The concluding remarks on the breeding program in India were that the groups have unexpectedly low levels of survival, which has been calculated by estimating

Table 8.1. The number of individuals sexed during 2003–04 in the zoos with LTM captive breeding program (CZA, 2004). (The numbers represent Male . Female . Unsexed with total in parentheses)

Name of the Zoo	Stock as on 01/04/2003	Births	Acquisition	Disposals	Deaths	Stock as on 31/03/2004
AAZP, Vandaloor	7.5.0 (12)	1.3.0	1.1.0	2.0.0	0.1.0	7.8.0 (15)
Trivandrum Zoo	9.6.0 (15)	—	—	1.3.0	1.0.0	7.3.0 (10)
Mysore Zoo	2.3.0 (5)	—	1.0.0	—	—	3.3.0 (6)

Table 8.2. Summary of number of individuals in the LTM captive breeding programs in three regions from 1982 to 2004.

Region/Country	No. of participating institutions	No. of individuals		Translocations/ Acquisitions
		During Launch	Present/Recent past	
North America	24	223 (1982)	109 (2002)	39 (−)
Europe	33	90 (1989)	200 (1998)	39 (+)
India	3	~40 (2002)	~35 (2004)	2 (+)

the probability of the groups surviving in the next two decades. However the Indian zoos aim at doubling their populations in about 7 to 10 years by translocations from other zoos.

The overall comparison of the captive breeding programs in these three regions has been tabulated. See Table 8.2. Although the data is available from different years, it is still possible to evaluate reasons for the successes and failures of the Captive Breeding program. It can be seen from Table 8.2 that there is an obvious difference between these regions in the number of participating institutions.

The number of individuals in captivity all over the world totals up to just more than 600 individuals, which also includes the individuals kept for display and not specifically for breeding purposes. This figure is shockingly 24% of the population in the wild which is close to only 2,500. The implications of this percentage will also be included in the discussion sections.

Discussion

On the basis of the above comparison and review of papers, the following discussion is centered on various factors like participation of zoos, scientific basis for breeding and long-term viability of captive populations and issues of reintroduction of endemic species in non-native habitats. These sections will be followed by the recommendations for the same.

(1) *Participation of zoos*

It is noted from the comparative analyses that the number of participating zoos in India is only 5% of the number of participating institutions worldwide. At the same time, the number of participating institutions in India is only 9% of the same in Europe and 12.5% of the same in North America. It is also noticed that the number of participating institutions in EEP had risen from 12 to 33 in 8 years (Fig. 8.1).

An international symposium for LTMs was conducted in 1982 in San Diego. The Coordinated Captive Breeding Program of India had an Information Sharing series in 1996 in Coimbatore, India, through the concerted efforts of the Central Zoo Authority, Forest Department of Tamil Nadu and Zoo Outreach Organization, a non-governmental organization, where experts from San Diego and Cologne Zoos met with experts of the Indian LTM breeding program for knowledge sharing.

However, it is observed that there is a clear demarcation between the academic research on Lion-tailed Macaque and practical exercises in the zoos. It was observed from the Indian Zoos' inventories that there were 15 unsexed Lion-tailed Macaques in the Mini zoo, Haddo, Port Blair, Andaman and Nicobar Islands (CZA, 2004), whereas the Lion-tailed Macaque are sexually dimorphic species. Such improper maintenance of studbooks can lead to loss of precious data on the behavior of the captive populations.

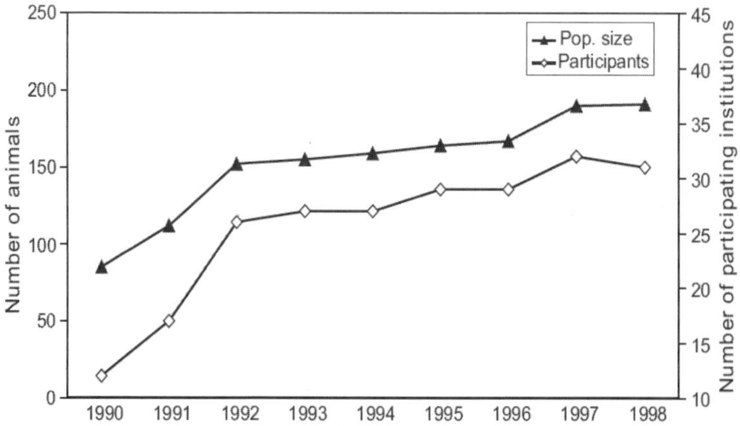

Fig. 8.1. Trends in number of participating institutions in EEP (Kaumanns *et al.*, 2001).

A random visit to the Arignar Anna Zoological Park (AAZP), Vandaloor, Chennai, in Tamil Nadu during the period of report writing (June, 2006) showed that there was no monitoring on the population of 5 individuals seen. In contrast, behavioral changes in captivity have been recorded and monitored day and night in the captive populations of North America. Recent studies (Singh *et al.*, 2006) on their reproductive biology have also been a compilation of observations in the wild in South India with the data available for around 50 years from North American and European zoos, since no reproductive data from Indian zoos were available.

(2) *Long term viability of captive populations*

In the SSP and EESP breeding programs, the reproductive successes have been better than the Indian programs. But the productivity in terms of surviving infants have been showing decreasing trends. It has been reported by all the zoos that it is difficult to keep Lion-tailed Macaque for captive breeding due to their low reproductive success in general. Even in the wild, the reproductive successes have been showing a decreasing trend, because of fragmentation of habitat. As the patch size decreases, the birth rate decreases and the age at first birth increases (Kumar, 1987). Due to the decrease in the fragment size, the size of the group increases in proximity. Thus, sexual competition results due to high female numbers and reproduction are suppressed ultimately.

Apart from the physical number of individuals forming a social group, the genetic makeup of the individuals also play a vital role in the long-term viability of a population. Population Viability Analysis (PVA) is a technique using computer simulations to model the vulnerabilities of small populations under threat of extinction. It is a powerful tool for management of rare species. The concept of Minimum Viable Population (MVP) is the number of individuals "having a 99% chance of remaining extant for 1,000 years despite the foreseeable effects of demographic, environmental and genetic stochasticity and natural catastrophes" (Shaffer, 1981). However, MVPs were not widely used in the later periods, but the principle that the smaller the populations become, the more are the chances of becoming vulnerable to extinction (Shaffer, 1990) has been accepted widely. Thus, PVA is considered and used as an important tool in management of wild and captive populations (Lindenmayer *et al.*, 1993).

The major concern in captive breeding is the inbreeding depression. Thus, the zoos introduce individuals from other zoos to maintain the heterozygosity and gene flow in a population. In the wild, male LTMs on sexual maturity migrate from one group to other, acting as a natural barrier for inbreeding and maintenance of gene flow between subpopulations. But this has to be brought about by translocations in captive breeding. Many of the captive bred populations of mammalian species have been observed to suffer from inbreeding depression, loss of genetic variability through genetic drift (Snyder *et al.*, 1996). These defects can be passed on to the wild by reintroducing, but in the case of LTM, these defects have led to failure of breeding programs showing declining reproductive rates.

The data has shown that females in EESP have largely varying reproductive rates, which implies that only some females are successful. This might lead to artificial selection. Due to the absence of translocations during the first generation, there are increased incidences of homozygosity in EESP populations. The same concept also applies to the Indian population, though there were occasional translocations. It is also because of the behavioral changes and responses of a female to a migrant male. It has been observed that females prefer migrant males for mating and social contacts in the wild (Kumar *et al.*, 2001). The same can be applied in the captive environment, by translocating new males during the breeding season. In India, there were two instances of adding two males to a breeding group from exhibiting zoos, but the older males were not removed and the reproductive success has not been reported yet. Non-removal of males that were already in the group will alter the male female sex ratio that is not conducive for breeding.

Thus, it can be seen that there are two major factors affecting the viability of the populations, (1) the social group showing male migration, thus introducing of new genes to maintain the genetic diversity and (2) the size and social environment of a social group.

The papers (Kaumanns *et al.*, 2006) show that there is no seasonality in breeding in captivity, whereas there is a birth peak in the wild. So, vasectomization has been carried out. The reports of LTM breeding from SSP show that there were incidences of vasectomised males to prevent unwanted pregnancies. In terms of conservation, vasectomy is seriously questionable when the species is on the brink of extinction suffering from habitat fragmentation and low reproductive potential. Such males can be isolated to a separate social condition to a group of males and used for migration and translocation during the breeding program. However, it should be noted that the surplus males isolated from a group should not be kept in solitary environment, because social behavior and contacts are important in primate mating. Additionally, vasectomization has also increased the cycle intervals in females, reducing their mating potential.

Another reason for the argument that males should not be lost is that this review shows the number of breeding females in the captive Lion-tailed Macaque populations all over the world is very low. Thus, it is very important to introduce variation at least from the male migrants. It is appreciated that a recent report of the Government of India (India Report, 2005) shows that there are proposals for translocations of 13 LTMs from San Diego Zoo to AAZP for breeding. However, it was found that the 13 will be of 2.11 compositions, i.e. 2 males and 11 females. If this proposal is accepted, this can augment the genetic variation among females in Indian captive population.

Endemic species reintroduction

Lion-tailed Macaques are endemic to Western Ghats of India. The SSP has carried out introduction programs in St.Catherine Island in North America. Though the release site is an island, the ethics of release of an endemic species into a new habitat is questionable. Instances of introduced species becoming invasive should also have been considered. However, it has been reported that they were recaptured, but details on that is not available. Other reports mention that there is a huge breeding group in the island that is breeding successfully compared to the other small groups in the zoo. This can be supportive information for translocations to other zoos for migrant males. In any case, reintroduction of endemic species should be done only in its native habitat. This issue has not been addressed in any of the reports from SSP, North America or CZA.

It has been also agreed upon that successful introductions are very rare even in spite of careful preparations for reintroductions (Kleiman *et al.*, 1994) and a review shows that only 16 of 145 reintroductions in different animal groups have been successful in establishing in the wild (Beck *et al.*, 1994). Still, extreme caution should be exercised before planning for reintroduction of such endemic species.

Additionally, the question of captive breeding and release of the captive bred individuals to augment the wild population should also be considered. The captive environment poses less selective forces. Thus the captive population might consist of more deleterious alleles. This can increase the risk of extinction in such fragile wild populations. Modeling of populations in the future is essential in such cases.

Recommendations

(1) *Number of participating institutions*

The endemic status of LTMs demands more attention and involvement from the Government of India. The number of participating institutions should be increased and more exhibits should be used for breeding due to the alarming decline of the wild populations. Though at times, one or two males are transported from exhibiting zoos to the breeding groups, these exhibiting zoos do not have monitoring programmes. The participating institutions should also monitor the captive-bred individuals in their enclosures for reproductive behavior and other behavioral changes. Unless observed, the reasons for failures cannot be clearly identified.

In addition to the participation, the conditions maintained in the zoos should be the best possible mimic of the natural habitat. The Lion-tailed Macaque inhabit the unique dense tropical forests of the Western Ghats. But the enclosures seemed drier and thinner in AAZP than that in the wild. The AAZP is a huge zoological park with many species in exhibits and breeding, and one of the properly maintained zoological parks in the country. Lion-tailed Macaque are primarily arboreal, spending 90% of their time in trees, however rare observations of climbing down to the ground for water and insects have been made in the wild. Thus, enclosures should have enough dense vegetation to provide shade and to mimic its natural habitat. The enclosures in AAZP had enough water and climbing setup, but were not dense as in the wild. Spending more time in trees can be correlated as a protective behavior from predators in the ground. Thus, when arboreal characters are not retained in captive populations, adaptation to the wild is expected to be poor in case of future reintroductions to the wild. Predation of reintroduced individuals is a huge problem in most of the reintroduction programs in particular primates (Britt *et al.*, 2003).

Knowledge sharing between research groups should be parallel with the training of the zoo keepers to understand the biology and behavior of the species. This can augment the capacity of the zoo keepers to record and maintain proper studbooks.

There are 15 unsexed Lion-tailed Macaque in Andaman and Nicobar islands as per the CZA inventory of zoo animals 2003–04, which is astonishing. Such improper maintenance of studbooks and inventories is regrettable in an endemic endangered species like Lion-tailed Macaque which is considered as one of the biological treasures of Western Ghats.

Studbook maintenance should be accurate and proper for other reasons also. Data in studbooks should be reviewed every year to understand the population in the past year. For example, Trivandrum zoo which is one of the participating institutions in India, has 9 males and 6 females (9.6.0), which have required proper attention, but one more male has been translocated from New Delhi Zoo in 2005. These are serious issues to be addressed as Lion-tailed Macaques have a female population that is twice the number of males in a social group.

(2) *Size of the captive bred population*

Size and social environment of the breeding group are also important. Similar to the importance of male migration, a typical breeding group should consist of many females with migrating males totaling to 20 individuals in a group (Singh *et al.*, 2000). LTMs follow social patterns similar to other primates; where the young females learn parenting from other mothers and infants in the social group. The captive populations consist of social groups of very low size with maximum size of 12 compared to the wild. Due to the different reproductive capacities of females, most of the young females would have grown in an environment with no infants or fewer infants. This leads to poor parenting in the next generation.

The inbreeding should be avoided as much as possible by frequent translocations between zoos. It is advisable to translocate males mimicking the male migration in the wild.

The Zoos Inventory show that there are individual males that are exhibited in solitary conditions in Indian zoos (Table 8.3). This situation can be changed by shifting the males to a single group, where social conditions prevail. These males can be translocated as a means to introduce male migrants into social groups with breeding females. Such translocations

Table 8.3. Situations of one or more LTM males in solitary or group living in Indian zoos.

Zoo/Institution	Number of males/females	Total at the end of year 2004
Rajiv Gandhi Zoological Park, Pune, Maharashtra	1 male	1 male
Alipur Zoological Gardens, Calcutta	1 male	1 male
Assam State Zoo cum Botanical Gardens	1 male	1 male
Tiger and Lion Safari, Shimoga, Karnataka	3 males	2 males (1 death)
Mahendra Choudary Zoological Park, Chhapbir, Punjab	2 males	2 males

are rare, so far only a few have been reported, like the recent single male translocation from New Delhi Zoo to Trivandrum Zoo in 2005.

The captive behavioral studies show that there is no intra-group aggression between males (Lindburg, 2000). So the risk of aggressive behavior can be positively ruled out. The scientific advice from experts is an imperative for selecting males for translocation exercises. So, continuous dedicated and combined efforts from scientific and zoo authorities is a must for maintaining a viable population that is not suffering from inbreeding depression.

The other option to save this species is use of assisted reproductive techniques in captive breeding procedures. The source of viable sperm has to be identified before using such techniques. Cryo banking of viable Lion-tailed Macaque sperms can be done for future research in this field (Morell and Hodges, 1998). Other options are artificial insemination (AI), embryo transfer (ET) and *in vitro* fertilization (IVF). Artificial Insemination (AI) with cryo preserved sperms can be tried in the captive populations to study the reproductive success. Trials on similar Macaque species have been carried out in Baltimore Zoo. The Pig-tailed Macaque (*Macaca nemestrina)* was chosen to be a model to try these techniques as they also belong to the *Macaca* genus. These techniques have been already applied in Rhesus Macaques (*Macaca mulatta*) and cynomolgus monkeys (*Macaca fascicularis*) (Morell and Hodges, 1998), with some level of success i.e. 1 out of 3 inseminated females gave birth to a surviving infant. A preliminary proposal for use of these techniques in LTM was found in the web, but unpublished. As expected, the reproducing Lion-tailed Macaque males had more sperm count (38-200 million per mL) and sperm mortality (54%). This involves collection of semen through electro ejaculation. Thus, health concerns arise keeping in mind a small captive population. Success in producing four embryos following implantation into a stimulated captive female was achieved in Baltimore Zoo (Cranfield *et al.*, 1992). These are presently stored in liquid nitrogen. Thus, cryo preservation of sperms of successfully breeding LTM males is essential for future research and development. To maintain genetic diversity, electro ejaculation can be performed from wild males and semen can be collected. These can be cryo preserved and aritificially inseminated in captive females. Due to the labour, expertise and fund-intensive characteristics of AI, these techniques have not been in practice in India. But emphasis is required to show that it is a potential method to overcome the reproductive failures in captive LTM populations.

Last but not the least, good management practices in the zoos is an imperative for a successful breeding program. Cooperation between zoos already exists, but the cooperative programs should be speeded up and moved to the front to prevent deaths that occur in captivity. These are some of the general but important considerations. The legislation and policy related to Indian zoos (CZA, Nov 2004) are elaborate and detailed enough, but efficient enforcement is made difficult because of the nature of administration. CZA is a central body, whereas the zoos in the states fall under both CZA and the State's forest department. But, there is no separate consideration for endemic species, which is also important. This federal system of governance slows down procedures and introduces difficulties in decision making.

Finally, awareness of the public about the endangered and endemic position of LTMs is very important to improve the social participation, behavior in the zoos and funding for captive breeding programmes. Zoo education programmes should be designed specifically for this species to create awareness among the public, in particular the school children.

Conclusion

Biodiversity in India faces multi-dimensional threats from social, economic and political perspectives. Thus, co-operation at the international level is necessary to save this species from extinction. Scientific research should be directly directed towards conservation. *Ex situ* conservation should be used only to supplement the *in situ* methods. Protection of the LTM habitats in the Western Ghats is important, because if the captive breeding is successful, there should be suitable habitats for the captive bred individuals to be released. Otherwise, the whole exercise of conservation breeding becomes meaningless. Enforcement of legislation to protect the habitat and establishing the proper scientific basis for design of breeding programmes will be the two building pillars for the survival of the Lion-tailed Macaque.

Specific to their captive breeding, the three reasons behind the failure of breeding programs, the low birth rates, high infant mortality, and slow population growth should be solved only by the combination of modern and classical scientific methods.

Thus this report acts as a review of data on the three major captive breeding programs of Lion-tailed Macaques world wide. The comparison of these three captive breeding programs shows that there are no clear data available on Indian Lion-tailed Macaque populations, the country to which it is endemic. More research is required to answer important questions, such as, Why does not the females give birth even after mating? More trial experiments should be conducted by reducing the number of solitary exhibits to increase the genetic variation in captive population. Future research can be extended to the possibilities of embryo transfer (ET) into females to ascertain the success of assisted reproductive techniques. Scientific data from the Indian captive population of Lion-tailed Macaques is yet to be drawn, so such areas could be potential research areas. Only concerted efforts can save this magnificent species from extinction.

References

Beck B B, Rappaport L G, Price M S, Wilson A (1994) Reintroduction of captive-born animals In: Olney P JS, Mace G M, Feistner A T C (Eds) *Creative conservation: Interactive Management of Wild and Captive Animals.* Chapman and Hall, London pp. 265–284.

Cranfield M R, Ialeggio D M, Berger N G, England B, Kempske S E (1992) The search for a reversible male birth control in the lion-tailed macaque as a model for other primate species. *Proc. Am. Assoc. Zoo and Wildlife Veterinarians.*

CZA (2004 web) Recognition of Zoo Rules (2004 Amendment), published online at http://www.cza.nic.in/rules.html

CZA (2004) Inventory of animals in Indian zoos (2003–2004), Central Zoo Authority, Ministry of Environment and Forests, Government of India.

Gledhill L (1996) A Master Plan for the Indian *Ex situ* population of Lion-tailed Macaques, *Macaca silenus*, Primate Info Net: *Old World Monkeys Taxon Advisory Group Newsletter* **3**(2) Winter, pp. 10–12.

India Report (2005) India Report for the 2005 Annual (60th) WAZA meeting, New York, USA, published by CZA, Government of India.

Kaumanns W, Krebs E, Singh M (2006) "An Endangered Species in Captivity : Husbandry and management of the Lion-tailed Macaque", from *mySCIENCE* **1**(1), pp. 43–71 at *http://myscience.uni-mysore.ac.in*

Kaumanns W, Schmid P, Schwitzer C, Husung A, Knogge C (2001) The European population of Lion-tailed Macaques (*Macaca silenus*): Status and problems, *Primate Report* **59**, pp. 65–75.

Kleiman D G, Stanley Price R M, Beck B B (1994) "Criteria for reintroductions" in: Olney P J S, Mace G M, Fiestne A T C (Eds.) *Creative Conservation: Interactive Management of Wild and Captive Animals.* Chapman and Hall, London, pp. 287–303.

Kumar A (1987) "The ecology and population dynamics of lion-tailed monkeys (*Macaca silenus*) in South India". Ph.D. dissertation, Cambridge University.

Kumar A (Ananda), Singh M, Kumara H N, Sharma A K, Bertsch C (2001) "Male migration in lion-tailed macaques", *Primate Report* **59**, pp. 5–17.

Lindburg D G (2000) "A century of involvement with Lion-tailed Macaques in North America", *Primate Report*, **58** (November), pp. 51–64.

Lindenmayer D B, Clark T W, Lacy R C and Thomas V C (1993) "Population Viability Analysis as a tool in Wildlife Conservation Policy: With reference to Australia", *Environmental Management* **17**(6), pp. 745–758.

Magin C D, Johnson T H, Groombridge B, Jenkins M, Smith H (1994) Species Extinctions, endangerment and captive breeding, pp. 3–30.

Morell J M, Hodges J K (1998) Cryopreservation of non-human primate sperm : priorities for future research, *Animal Reproduction Science* **53**, pp. 43–63.

Olney P J S, Ellis P (eds) (1991) 1990 *International Zoo Yearbook Volume* **30**, Zoological Society of London, London.

Shaffer M L (1981) Minimum population sizes for species conservation. *Bioscience* **31**, pp. 131–134.

Shaffer M L (1990) Population viability analysis. *Conservation Biology* **4**, pp. 39–40.

Singh M, Kumara H N, Ananda, Kumar M, Sharma A K, Defalco K (2000), "Status and conservation of lion-tailed macaque and other arboreal mammals in tropical rainforests of Sringeri forest range", Western Ghats, Karnataka, India. *Primate Report* **58**, pp. 5–16.

Singh M, Sharma A K, Krebs E, Kaumanns W (2006) "Reproductive biology of Lion-tailed macaques: An important key to the conservation of an endangered species", *Current Science* **90**(6), pp. 804–811.

Snyder N F R, Derrickson S R, Bessinger S R, Wiley J W, Smith T B, Toone W D, Miller B (1996) Limitations of captive breeding in endangered species recovery. *Conservation Biology* **10**, pp. 338–348.

9

INTEGRATED CONSERVATION AND DEVELOPMENT PROGRAM APPROACH IN PROTECTED AREA MANAGEMENT: CASE STUDY OF ANNAPURNA CONSERVATION AREA PROJECT IN NEPAL

ASHOK PRASAD OJHA, KOH KHENG LIAN and JOSEPH CHUN

Nepal has established a wide spectrum of protected areas with the primary aim of conserving biodiversity. Various problems relating to the management of protected areas have emerged over the years. The Integrated Conservation and Development Program (ICDP) approach in protected areas is an important tool to link conservation and development with the aim of resolving various problems associated with protected area management regime.

This study investigates the impact of the ICDP approach on biodiversity conservation, local participation, local socio-economy and livelihood. This was carried out in the Annapurna Conservation Area (ACA), Nepal. A social survey using various tools such as Participatory Rural Appraisal (PRA), semi-structured interviews, and questionnaires was conducted.

The results indicated that ICDP had a significant positive impact in study sites compared to before its intervention. The study suggests that forest condition was improved, there was significant reduction in the use of fuel wood, local people had easier access to resources, the wildlife population increased, people's participation had increased, and the local economy improved. The ICDP approach was successful in involving the majority of the local population in conservation. People were found to be very positive towards conservation initiatives and park authority. The ICDP approach was successful in getting local support and collaboration in conservation initiatives. As a result, there was significant development in the local institutional arena and local capacity building.

Acronyms

ACA	Annapurna Conservation Area
ACAP	Annapurna Conservation Area Project
CAMC	Conservation Area Management Committee
CBO	Community Based Organization
CBS	Central Bureau of Statistics
ICDP	Integrated Conservation and Development Program

Kg	Kilogram
KMTNC	King Mahendra Trust for Nature Conservation
NBS	National Biodiversity Strategy
PA	Protected Area
PRA	Participatory Rural Appraisal
SLC	School Leaving Certificate
TMsC	Tourism Management sub-committee
UCO	Unit Conservation Office
UNDP	United Nations Development Program
VDC	Village Development Committee

Introduction

Two approaches have been used for biological diversity conservation in Nepal. The first is a conventional protection-conservation approach, where conservation policies only focus on the preservation of biodiversity and nature by keeping local communities strictly outside the realm. These policies have been based on conventional perception that people always have a negative impact on the natural environment and human interaction with the environment leads to its exploitation (Ghimire and Pimbert 1997). Most of the wildlife reserves and national parks in Nepal have been designed using this approach.

The second approach is the Integrated Conservation and Development Program. This links protected area management with the economic activities of local communities.

This study evaluates the impact of the first Integrated Conservation and Development (ICDP) approach-based conservation program in Nepal, with the aim of conserving the biological diversity and improving socio-economic conditions of the local inhabitants.

The aims of this study are:

1) To analyze the effectiveness of the integrated conservation and development program (ICDP) approach in local economy/rural livelihood, biodiversity conservation and overall local developmental arena.
2) To examine the willingness and level of local people's participation in ICDP activities.
3) To examine the support for ICDP biodiversity policies at the local and international level.

4) To recommend the appropriate 'enabling conditions' for ICDPs to succeed in the context of Nepal.

Methods

The research included collection of primary data in the study sites through questionnaire surveys, focal group interviews, informal meetings, formal meetings, direct observation and the use of various Participatory Rural Appraisal ("PRA") tools. The study was conducted in the Annapurna Conservation Area ("ACA"), where ICDPs have been implemented for more than a decade. The questionnaire survey was administered in two Village Development Committees ("VDC"). A semi–structured questionnaire was prepared for this purpose and a random sampling strategy was used. A focus group discussion was organized to discuss the ICDP approach and people's participation aspects. 12 persons were formally interviewed during the study period; 27 persons were informally interviewed during the study. The author observed programs and activities launched by Annapurna Conservation Area Project ("ACAP") in the field, people's participation level, functioning and maintenance of some projects such as provision of drinking water, toilets, bridges and village meetings to verify and validate the information collected from various sources.

Participatory tools such as matrices and focal group discussions were used to obtain insights and discover resource use patterns, changes in wildlife populations, and effects of the ICDP approach in the conservation and development arena. Information provided during PRA was cross-checked during interviews. PRA exercises took about 3 to 4 hours in a group of six to eight people. Different locally available materials such as stones and grains were used. Using wildlife photographs facilitated discussion regarding wildlife issues. For example, relevant photographs of wildlife species were used to discuss changes in wildlife population after ACAP intervention in the area.

Various socio-economic, bio-physical, policy and institutional parameters were selected to analyze the impact of the ICDP approach in the ACA Management system. These were:

Evidence of changes in wildlife population

Crop damage by wildlife
Livestock depredation by wildlife
Perceived change in wildlife population
Evidence of wildlife in the forest

People's participation in ICDP

Attitude towards ICDP
Attitudes of people towards park authority
Willingness to participate in ICDP
Participation and decision-making
Involvement in conservation
Mode of participation

Basic developmental improvements local economy

Change in economic status of local people
Tourism
Increase in forest area after ACAP initiatives
Alternative energy devices
Institutional development
Capacity building
Participation and decision-making
Use and access to park resources

The quantitative data were analyzed using Likert's Rating Scale. The statements with "agree" and "strongly agree" were coded 4 and 5 respectively. Whereas "neutral", "disagree" and "strongly disagree" were coded 3, 2 and 1.

Attitudes and perception data were examined using 5-point scale statements which respondents were asked to agree or disagree. To obtain an overall score for all respondents on some of the attitude scale, responses to each of the statements were added and an average score was estimated.

Limitations of the Study

1. Poor baseline information: ACAP was one of the major sources of baseline information on ICDP approach in ACA management but data and information on some aspects were not found updated by ACAP and the systematic documentation and database was poorly kept by related local institutions. This created some difficulties in analyzing the effectiveness and level of people's participation in ICDPs.
2. Information based on memory recall: The participatory matrix scoring was mainly based on the past retrieved memories of the participants. This method is not always accurate.

3. Short time period: The main limitation with this study was the short time span to conduct extensive field studies. The time period allocated for completing and submitting this study was 3 months (May-July, 2005). The short time period limited the number of household surveys and community consultations.

Results

The socio-economy and demography of respondents as outlined by the survey is as follows:

Educational status of respondents

About 50% of the respondents were illiterate, which is more than the national average of 45% (CBS, 2003). 41% of the respondents were literate, 6% of the respondents left the school before obtaining the School Leaving Certificate ("SLC")[1] and about 3% of the respondents had educational status above SLC.

Respondents by age group

16% of total respondents belonged to age group from 20–40 years, 42% were from 40–60 years age group and 42% of respondents were above 60 years old.

Village representation of respondents

The respondents were from different villages in Marpha and Jomsom Village Development Committees ("VDCs"). 4% of respondents were from Chairo village of Marpha VDC, 30% of respondents were from Marpha village of Marpha VDC, 16% of respondents belonged to Syang village of Marpha VDC, while 24% of respondents were from Syang village in Jomsom VDC and 26% belonged to Jomsom village in Jomsom VDC.

Respondents by family type

81% of respondents were from single families[2] while the remaining 19% were from joint families.[3]

[1]Graduates after 10 years of formal education.

[2]A type of separated family where parents and their own children only stay in one house.

[3]A type of combined family composed of parents, their children, children's' spouses and grandchildren in one household.

Respondents by gender

About 85% of respondents were male and the remaining 15% of respondents were female.

Drinking water availability and sources

36% out of 102 respondents had access to drinking water from private taps, while 63% had access to community taps and 1% used the river as a potable water source.

Access to toilets

About 49% of respondents in the study area had access to toilets and 51% of respondents did not have access to toilets. About 98% of respondents had toilets with safety tanks and 2% of respondents had toilets with pit (Pit Toilet).

Land classification

The survey found 347.45 ropani[4] of arable land, 57 ropani of private plantation, 10 ropani of pastureland, 18 ropani of cultivated land and 31 ropani of fruit orchard within ownership of the respondents.

Ownership of pastureland

50% of respondents had 5–10 ropani, while another 50% had only 0-5 ropani of the pastureland.

Ownership of private plantation

62% of the respondents had 5–10 ropani of private plantation land, while the remaining 38% had 0–5 ropani of the land.

68% of respondents had 0–5 ropani of fruit orchard, 13% had fruit orchard land area of 5–10 ropani, 6% had 10-15 ropani, 10% had 15–20 ropani and 3% of respondents had 30–35 ropani of fruit orchard.

51% of respondents had 0–5 ropani of arable land, 37 respondents had 5–10 ropani of arable land, and 8% had 10–15 ropani of arable land, 2% had 25–30 ropani of arable land and 2% of respondents had 35–40 ropani of arable land.

[4]A unit of measurement, one hectare of land equals 19.79 Ropani (1 ropani = 5476 sq ft.)

Ownership of land based on tenure

35 out of 102 respondents had land ownership based on tenure. 30 respondents had arable land ownership,[5] 4 had fruit orchard ownership[6] and 1 had ownership with other lands based on the tenure system.[7]

Household access to irrigation facility

85% of households had access to irrigation facility, while remaining 17% did not have access to irrigation.

Respondents usually cropped four types of agricultural crops and 35% of total arable land was allotted for barley cultivation, 24% of arable land was for maize production, 24% was for buck wheat and 17% of the total arable land was allotted for potato cultivation.

Food security

Out of a total of 102 respondents, 14% produced more food than they really needed, while 25% of respondents produced sufficient food and 61% of total respondents suffered from food deficiency.

In regard to the respondents, who suffered from food deficiency, 2 had food sufficiency for just four months, 4 respondents for three months, 4 respondents for nine months, 3 respondents for eight months, 5 respondents for two months, 3 respondents for four months, 5 respondents for seven months and 9 respondents for six months.

30% of the respondents' main means to cope up with food insecurity was by working in either governmental organizations or private organizations, 6% were found using other land for cropping, 29% were engaged in tourism, 8% were engaged in livestock rearing and 27% of the respondents had other means beside those mentioned above.

Agriculture contributed 18% to the total economy, tourism and business contributed 41% of total income of respondents, year round employment's contribution was 17%, contribution from daily wages was 9% and total contribution from other sources was 15%.

The social survey for institutional development before and after conservation intervention with the ICDP approach shows a very encouraging trend.

[5]Ownership of land that is fertile compared to other land and could be cultivated through out the year. especially land with irrigation facility.

[6]Ownership of land on which fruit trees are grown.

[7]Holding of certain area of land as a tenant for a specific period of time.

Institutional setup and peoples' perception towards the mechanism

The establishment of CAMCs (Conservation Area Management Committees) under the CAMR, 1997 (Conservation Area Management Regulation) was a major step towards local institutional development in the two study VDCs. A high proportion of residents in these sites were aware of the roles and responsibilities of this institution. All respondents knew that CAMC existed. Out of 102 respondents, 17.6% were fully aware of the roles and responsibilities, while 46% were partially aware. However, 36.32% were not aware of the roles and responsibilities of CAMC. The social survey showed that 45.1% respondents were involved in at least one local organization, which was very encouraging. 75.49 % out of 77 respondents believed that the ACAP approach had a significant role in the institutional development of local communities.

The institutional analysis using the PRA method also showed considerable institutional development in the forms of functional committees in the VDCs studied. CAMC, mothers group (women's group), Tourism Management Sub-Committee ("TMsC"), and local youth groups were some of the main functional committees in the study VDCs.

The evolution of various local institutions including CAMC over the last 11 years in VDCs was another significant development in the institutional arena. This can be attributed to changes in awareness and attitudes towards participation in various community programs and realization of the importance of collective actions and a shared value system among local communities.

Capacity building as indicated by the National Biodiversity Strategy (NBS) is capacity enhancement of local communities and staff through training, workshops, seminars and interactive programs, as a major sectoral strategy in protected area management.

The ICDP approach of ACAP focuses on building capacity of local communities and local organizations in order to make them able to deal with conservation and development programs in a more efficient and effective way. The study reveals that ACAP had provided various trainings to different local organizations and community members, including resource management, income generation, alternative energy, and tourism management. The social survey indicated that 45.1% of the total 102 respondents attended at least one training/workshop.

Based on the above-mentioned findings, it can be concluded that the impact of ICDP in the ACA is very encouraging in the area of capacity building. In fact, income generation trainings were less focused.

Infrastructures such as road access, trail access, drinking water, electricity, and sanitation are important factors for enhancing rural livelihood and improving the community's quality of life. Improvements in basic infrastructure such as good sanitation, drinking water and basic education help to improve human development and also help to reduce poverty by raising the level of human capacity (UNDP, 2002). Basic infrastructure developments set up through the ICDP approach were very visible and important benefits were received by local communities.

The social survey showed that places had better basic infrastructure developments after the ICDP approach intervention of ACAP. 69.83% of the respondents supported this, 49% of respondents indicated that they have good access to toilet and safe drinking water, more than 72% of respondents reported that basic infrastructure such as drinking water, sanitation, and village access had been improved following the ICDP approach intervention.

Use and access to park resources

Timber for construction, fuel wood and fodder are basic subsistence needs of local communities in ACA, as elsewhere in the country (Hough & Sherpa 1989).

The Conservation Area Management Regulation ("CAMR") 1997 provides legal authority to manage and utilize resources by local communities. Promulgation of this regulation, in general, had not prohibited subsistence use of their resources. This is very different from other PAs in Nepal, where resource management and protection are carried out by the government without much involvement of local community and local welfare oriented programs (Nepal, 2002) and where in reality local people do not play any role except in buffer zones, in most of the protected areas in Nepal (Sharma & Wells, 1996). Thus many communities living within or outside PAs in Nepal get access to park resources either illegally or under strict license (Sharma, 1990). Therefore this is one of the major strengths of the program.

The social survey indicated that respondents were found to collect 10,236 Bhari[8] of fuel wood from different sources, of which 9,645 Bhari (94%) were collected from ACA. The remaining 6% was purchased from market. The social survey also indicated that 44% of households required fodder for different purposes. The respondents were found to collect 8,621 Bhari (Approx. 215,525 kg) of fodder from different sources, of which

[8]Bhari is the traditional unit of measurement for fuel wood and forage in the studied areas and one bhari=25kg.

7,978 Bhari (Approx. 199,450 Kg) (99%) was collected from ACA; the remaining 1% was collected from market and household peripheries. The majority of the respondents pointed out that access to major resources such as fuel wood and fodder had improved after ACAP intervention.

Timber was used mostly for hotel and office construction. The majority of hotel and lodge owners in the study areas used timber from other VDCs within ACA.

Concerning access to timber use from ACA, 31.37% of respondents reported that they had easy access to timber from forests, while the remaining 68.62% did not. This may be because CAMCs had stopped issuing timber permits in the last 3–4 years, in order to regenerate the forest, so that it can sustain the timber demand for a long time. Another reason was that CAMC used to issue timber permits only for the optimum required quantity, they did not issue permits as demanded by the users. In fact, CAMCs used to issue permits by carrying out actual need assessments. The result is that those that really need the timber received permits.

Local economy and livelihood

Tourism, trade and agriculture are the main economic and livelihood sources of local people in the area. The cost-benefit ratio of conserving a protected area must ultimately be positive for local communities, if the area is to be sustained in the long run (McNeely, 1995). The current evidence shows that local communities in the area have received various direct and indirect benefits from the ICDP approach. Creation of economic opportunities to uplift local economy is one of the direct benefits.

The majority of the local people in the area were involved with the hotel and lodge business. The increasing number of tourists reflected the growing importance of this business in the last few years.[9] Agriculture and livestock rearing was the predominant occupation of local communities in the area. About 37% of respondents were directly engaged in this business. The major direct economic benefits received by local people in the study area from the ICDP approach of ACAP were from tourism management and promotion, inputs for agricultural development and through alternative income generation trainings.

ACAP's annual investment in ICDP, financed either through sharing of revenue from park entry fees or from the support of international donors, is

[9]It should be noted that in recent years, the number of tourists are declining due to the political situation.

one of the significant benefits to the local communities. Almost 68.2% of the total respondents believed that there was a direct or indirect relationships between conservation and development in their area. Unlike most national park entry fees, the entry fee of ACA does not go to the National Government Treasury but are retained by the King Mahendra Trust for Nature Conservation ("KMTNC") for conservation and development in ACA (Bajracharya, 2003). This provision is unique in Nepal and little known by international communities (Sharma & Wells, 1996).

Support for agricultural development was another economic benefit received by local communities in the areas studied. A major proportion of respondents believed that ACAP helped them in agriculture development through supporting irrigation and promoting other technologies, such as the introduction of green houses for vegetable farming.

Local communities had opportunities to participate in training programs, which provided them with skills to organize economic activities. About 35% of 102 respondents had participated in trainings. Among them 27% had participated in income generation trainings provided by ACAP and 81.56% of respondents believed that these trainings helped them to boost their economy.

The results are supported by official reports. It has been reported that more than 2,702 local residents from the study site benefited from training programs (including conservation awareness, tourism awareness, mobile camps) from 1993–2004 (KMTNC-ACAP, 2005). Of the total 168 training events, 24.39% were focused on economic activities, such as accounts management, food processing, vegetable seeds production, saving and credit management, skills development, alternative energy, community development, and porter and licensed guide training. Of the total respondents in the area, 68.56% reported that economic opportunities in villages had increased after ACAP intervention in the area.

Sources of livelihood

Tourism is the most promising business in the studied area; almost 13% of 102 respondents were directly involved in the tourism business. From the 1970's, the Government recognized the Annapurna region as a tourist destination. Since then, international trekking began in the Annapurna region. The tourists used to visit the area with their own management, as there were very few facilities available to them. The tourism business was not encouraging at that time and only very limited households were benefiting from tourism. It was estimated then that only 7% of the total tourist expenditure was retained in the local economy. (Gurung & Nepal, 2002). Most of the services used to be charged below cost, so at that time, tourism was developing at a cost to the local economy and ecology. It can be said that tourism in the Annapurna region before ACAP's intervention was not a significant contributor to the local economy.

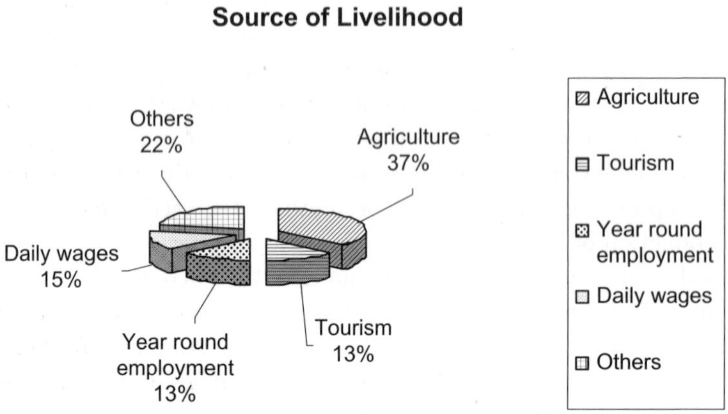

After ACAP's intervention, the area has become the most popular trekking destination in the Nepalese Himalayas (Banskota & Sharma, 1995). ACAP has implemented and completed different types of tourism programs with the aim of uplifting the living standard of local communities in the area. Tourism data indicated that there was an increasing trend in the annual number of visitors in ACA. However the last few years have seen a decrease in tourist numbers due to national problems which have created considerable insecurity in the country.

The social survey indicated that 20–25% of tourists' expenditure was retained to the local economy compared to 7% before ACAP intervention. The discussions clearly indicated that the ICDP approach in ACA management have had a significant positive on the economy and there was little doubt that the ICDP approach had brought increased economic opportunities to the area, where agriculture and livestock rearing were traditionally the main occupations of almost all households.

Forest management and conservation

A social survey on forest management and conservation was carried out in study sites to ascertain the perceived changes after ACAP intervention. The survey examined the extents of forest cover, legal authority for community based organizations (CBOs) and resource use conflicts. The survey indicated that community based restoration programs and devolution of forest management authority to local community decreased the use of fuel wood and increased awareness for conservation. The introduction of alternative energy devices has contributed to the effective conservation of forest compared to before ACAP intervention.

All respondents knew that they had community managed conservation forests in their area. It was found that 9.09% of respondents strongly agreed that forest conditions have improved after ACAP intervention, 54.54% agreed that forest condition had improved, while 27.26% showed neutral response and 9.09% believed that forest conditions had deteriorated after ACAP intervention.

The survey indicated that 55.29% of respondents believed that CAMC, as a community-based authority was responsible for forest management and conservation, while 37% indicated that the forest was managed and conserved through traditional systems and 5% of the respondents did not have an opinion.

Questions were asked to find out the perceptions of local communities towards the legal authority of CAMC. The result was that 15.6% of the respondents said CAMC is a forest product permit issuing authority, 23.5% viewed it as an authority which can ban cutting and collection of timber and live trees, 11.7% viewed it as an authority that could form sub-committees, 16% responded that it had the right to ban hunting, while 24.5 indicated it had legal rights but they were not sure of the specific rights and 7.8% of respondents indicated that CAMC does not have any legal authority.

Resource use conflict is reported in the majority of protected areas in the world and this is also true in ACA but the intensity and frequency of the conflict might vary with other protected areas. Based on the survey, 27.27% of respondents said that they frequently had conflicts with other communities for resource utilization, 9% said they sometimes faced conflict with other communities, while 63% did not have any conflicts.

Community plantations play a vital role in decreasing the anthropogenic pressure on natural forests for diverse use. Based on the survey, 81.8% of respondents reported that they were involved in community plantations, while 18% were not. Further, 75.45% of respondents believed that community plantation programs in their places were supported by ACAP, while 24.55% reported that community plantation programs were supported by organizations including Government authorities.

These responses are backed by the official data of ACAP. These showed that 38,495 seedlings of different tree species were produced in the ACAP Unit Conservation Office (UCO), Jomsom project nursery since its inception in Lower Mustang. Out of this, 10,917 seedlings of various species were used in community plantations, block plantations and private plantations (Source: KMTNC-ACAP, Unit Conservation Office (UCO), Jomsom, 2005).

Evidence from earlier research on fuel wood use by local communities and tourist lodges also indicated a decrease in fuel wood use. This was also supported by the increasing trend of adoption of various alternative energy devices (KMTNC-ACAP, 2001). ACAP has significantly contributed in introducing alternative energy devices. Plantation records on the northern sectors of ACA also substantiate reduced use of forests for the purpose of fuel wood harvesting. Collection of dead wood and conservation education has cumulatively contributed to reduced pressure on forests. ACAP has been promoting alternative energy devices. As of 2004, ACAP has promoted 34 back boilers, 26 smoke water heaters, 26 solar water heaters and installed 1 kerosene depot in the study areas.

Pasture land management

Grazing by livestock is an important issue for management of the national parks and protected areas. Livestock rearing is important to rural livelihood, therefore pasture land management plays a very important role in the PA management and sustainable rural livelihood cannot be expected without sustainable management systems. (Wangchuk, 2002). Effective implementation of the programs to improve quality of pastureland is one of the indicators of the success of protected areas management. Local families from the studied areas graze their livestock (yak, sheep and horse) in pasture lands inside ACA. The study found that 43 households were involved in livestock farming and all grazed livestock inside ACA. One household bought feed from the market and 24 households reported that ACAP had supported in pastureland management through various programs.

16.67% respondents believed that the overall condition of pastureland had improved after ACAP intervention, 14.7% reported that the overall condition had deteriorated after ACAP intervention and 68.6% believed that there was no change in the quality of pastureland. It was found that ACAP had not significantly implemented pastureland management programs in the study areas. This might be because livestock rearing in the study area was not the primary occupation of the local communities. Thus livestock did not have any significant negative impact in the management and conservation of resources. Therefore none of the organizations, including ACAP had focused on this issue. During discussions with local communities, it was found that considerable changes had taken place in livestock population, structure and management systems in ACA over the last few years. The big herds of cattle have been replaced by smaller herds.Labour shortage, a decreasing interest of young people in traditional livestock farming and increased involvement of local people in tourism related businesses had contributed to the reduction in livestock numbers.

Change in wildlife population

Changes in wildlife population, especially the Snow Leopard (*Uncia uncia*), Musk Deer (*Moschus moschiferous*) and Blue Sheep (*Pseudois nayaur*) after ICDP intervention of ACAP in the study areas were mainly analyzed through questionnaire survey, focus group interview and participatory wildlife matrix scoring.

Almost 81.38% of the total respondents knew that these wildlife (Snow Leopard, Blue Sheep, Musk Deer) were found in their area. The survey reflected that the wildlife in the study area has increased. 89.23% believed that the Blue Sheep population increased, 62.27% believed that the Musk Deer population has increased and 53.9% respondents believed that the Snow Leopard population has increased after ACAP intervention. One of the reasons behind these increases is the increasing level of awareness among the local community because the survey found that almost 89.2% respondents indicated that wildlife was not problematic to community.

This perceived increase in wildlife population in the study areas was supported by a participatory wildlife matrix scoring, based on field surveys, which indicated an increasing number of snow leopard, musk deer and blue sheep in the study areas.

ACA's records of legal actions after conservation initiatives showed that there were only seven illegal hunting cases filed (KMTNC-ACAP, 2004). Compared to reported cases from any army-guarded national park and wildlife reserve in Nepal, illegal hunting in ACA is comparatively very low.

There have been criticisms that conservation areas do not protect wildlife and there have also been reports that incidences of wildlife poaching are far greater in conservation areas than in national parks or wildlife reserves (Bajracharya, 2003) but the present study indicated that this is not the case in this area. The present study is also supported by some studies in ACA. A study of the Himalayan Tahr (*Hemitragus jemlahicus*) populations indicated that there is a population of the animal in the area (Gurung, 1995). A recent study in the area reported a 20% increase in the population over a five-year period (Shrestha and Ale, 2001) This study concludes that the wildlife population of Snow Leopard, Blue Sheep and Musk Deer is increasing after the ICDP approach through ACAP in the study areas.

However, perception towards the effectiveness of wildlife conservation was different among different respondents. A Likerts Scaling test showed that 46.32% respondents strongly agreed and 24.5% respondents agreed that the ICDP approach helped to conserve wildlife effectively, while 84.32% respondents either strongly agreed or agreed that wildlife hunting has been decreased after ACAP intervention in the study area. 30.4% respondents strongly agreed and 59% agreed that conservation awareness helped to effectively protect wild life species, while 38.25% respondents strongly agreed and 30.4% agreed that wildlife species were freely hunted before ACAP initiatives.

Involvement in ICDP

The study showed that local people participate in ICDPs in ACA for various reasons.

83.2% of the 102 respondents indicated that they participated for protecting their resources, 62.57% told that they participated for improving basic infrastructures, 34.2% reported that they got involved only for providing suggestions and comments to the park authority for improving their management system of ACA. Regarding the rationale for involvement in ICDP, a majority of respondents indicated that they participated because they regard ICDP approach of ACAP as an effective conservation and development measure.

Attitudes towards the ICDP approach

The results indicated that a majority of respondents in the studied areas held a positive attitude towards the ICDP approach. A high proportion of respondents (76.15%) either strongly agreed or agreed to the statement regarding the success of ICDP in conservation areas compared to before ACAP intervention. Similarly, a major proportion of respondents (72.57%) either strongly agreed or agreed that present developmental activities in their villages were due to the ICDP approach.

The social survey indicated that 74.50% of 102 respondents supported the present ICDP approach in ACA management, 18.62% did not support it, while 6.86% remained neutral.

- 87.2% respondents strongly agreed or agreed that they had easy access to fodder and fuel wood after ICDP approach intervention,
- 83.3% respondents believed that institutional management capabilities had been enhanced,
- 65.7% respondents reported the ICDP approach had effectively protected forests and wildlife,
- 65.7% respondents reported that basic infrastructural development in their villages had been increased, 79 respondents reported that ICDP had introduced alternative devices,
- 84.3% respondents reported that toilet and sanitation condition had improved,
- 83.3% respondents reported that local capacity had developed after ICDP approach intervention,
- 73.5% respondents reported enhancement in awareness level among local residents.
- 23.5% reported that technical support for agriculture, health and infrastructure had increased,
- 60.8% respondents either strongly agreed or agreed that economic opportunities had widened after ICDP approach intervention,
- 6.8% respondents strongly agreed or agreed that a local system of governance had been set up in the communities after ICDP approach intervention.
- 71.5% respondents either strongly agreed or agreed that the ICDP approach increased wildlife population and
- 68.6% either strongly agreed or agreed that forest conditions improved after the ICDP approach intervention of ACAP in their villages.

Attitudes of people towards the park authority

Protected areas are a genuine means to conserve biodiversity and ecosystems but designing a protected area without involving local people in the whole management does not usually work. There are numerous examples, where Protected Areas have resulted in conflict between local communities and state conservation authorities and deprived local communities of their rights to meet subsistence requirements, resulting in negative attitudes towards the park authority.

The results of the surveys of attitudes towards park-people relationship in the study areas indicated a highly positive relationship. This positive relationship has developed due to strategically located field offices of ACAP, regular conservation, education and awareness raising activities, empowerment of local people through diversified programs and a system of regular discussions with community peoples through various meetings, workshops and trainings.

A survey was carried out to determine local people's attitudes towards park authority. 84.83% of the respondents reported they were either always or often involved in decision–making, implementation, operation & monitoring-evaluation of ICDP. 86.30% believed that there was a good flow of information between park authority and communities either always or often, while 87.58% of the respondents indicated that the community members either always or often attended meetings and 86.21% pointed out that community members either always or often took part in planning and decision making.

The effectiveness of the ICDP approach in ACA management is illustrated by changes in attitudes of local people towards ACA authority. The results of the survey indicate a high positive relationship between local communities and ACA authority, in contrast to other protected areas in the world, where a lack of communication with local communities residing within and around the park was a key issue (Infield & Namara, 2001). More than 86.30% of the respondents agreed that there was good communication between ACA and local communities.

Local people's participation level and mode of participation in ICDP

Public participation in ICDP can be viewed as a process, extending from full government control to full community control. The lowest level of participation is compliance that, essentially, is the imposition of a decision on the community as seen in most of the National parks and wildlife reserves in Nepal. Here the community has no choice or involvement in the decision-making process. At the other end of the process is self-directed action, whereby the community is given full responsibility for decision-making and control of the process.

The social survey identified some of the aspects related to the ICDP approach in ACA and the level of local people's participation, mode of local participation in conservation and development activities after the appearance of ACAP in the studied areas. The social survey indicated that there was a very encouraging level of participation in

ICDP activities. The participation level varied from cooperation to consultation. 64% of total 102 respondents said that they took part in ACAP meetings, 75% reported that they took part in planning processes, 52.5% of the respondents reported that they suggested and commented on ACAP planning, 28% told that they participated in solving legal cases of ACAP, 87% of the respondents indicated that they participated in ACAP conservation and development activities, 89% of the respondents contributed by labor in ACAP initiated development works, while 31.35% of the respondents contributed in cash in ACAP developmental works.

The social survey found in that 33% of total 102 respondents expressed their views in meetings. When asked about the perceived changes in participation in ICDP after ACAP, intervention, 67.59% reported that there was significant increment in peoples' participation level in ICDPs, while 13.23% indicated that participation decreased after ACAP intervention and 19.18% of the respondents reported that participation neither increased nor decreased after ACAP intervention.

It can be interpreted that people's participation in conservation activities have increased after the ICDP approach of ACAP. This is because of the direct and indirect benefits people are receiving from conservation and development activities along with other activities the project has carried out in the area. It is clear that participation can only be achieved when local people are involved in decision-making, implementation, benefits development and evaluation of the activities targeted towards park management.

Costs of conservation

Despite benefiting from the conservation initiatives in ACA, a small fraction of respondents reported that there were some difficulties and costs arising from conservation initiatives. Major difficulties and costs of conservation have arisen from crop damage by wildlife and livestock depredation by wildlife.

The survey indicated that only 2% of 102 respondents reported that they were either experiencing the crop damage by wildlife regularly or frequently.

7% of 102 respondents reported that they were experiencing this problem in their villages. During the field study, participants indicated that the intensity and frequency of livestock depredation was decreasing every year. This might be because local people were slowly giving up the traditional livestock rearing profession due to emerging economic opportunities in the areas. All the respondents reported that the Snow Leopard was the main wildlife that killed their livestock.

Discussion

The aim of this study was to assess the effectiveness of the Integrated Conservation and Development Program approach in the Annapurna Conservation Area in biodiversity conservation, local development arena, economy, livelihood, and people's participation level. The management of Annapurna Conservation Area by a National non-governmental organization, King Mahendra Trust for Nature Conservation, with full support from Government is an exceptional arrangement in Protected Area Management in Nepal, and probably in South Asia. The results of this study indicate the importance of an Integrated Conservation and Development Program approach in achieving protected area management objectives and goals in an efficient and effective way.

The Integrated Conservation and Development Program (ICDP) had demonstrated improvements in ecological and social issues in the study areas. This approach in Annapurna Conservation Area is undoubtedly delivering conservation benefits to the Protected Area and local communities. Evidence obtained during the study reflects that wildlife poaching has decreased and wildlife population has increased.

The integration of programs designed to increase conservation awareness, plantations, and the introduction of alternative energy technologies has contributed to a reduction in fuel wood collection from the forest inside the conservation area and thus, enhanced the quality of the forest to support the increasing wildlife in the area.

Local communities were found to be empowered through appropriate authority, policies and legislation to conserve biodiversity. Attitudes towards the park authority were found to be very positive, reflected by changes in the behavior and attitudes of local communities in conservation.

A majority of residents perceived positive changes in conservation and development in their villages, thanks to this ICDP as applied in the Annapurna Conservation Area Project. They perceived positive changes in forest cover, wildlife population, participation level, basic developments, and local economy. This approach has empowered them with respect to sustainable use of natural resources, while at the same time it has helped to improve basic developmental needs in the villages by direct involvement of local peoples in planning and decision making.

The institutional development in Annapurna Conservation Area has had remarkable momentum due to this approach, especially the empowerment and capacity building of local institutions. Tourism is one of the important elements in Annapurna Conservation Area While tourism also has negative impacts, the negative impacts of tourism have been minimized with the implementation of the Integrated Conservation and Development Program approach in the Annapurna Conservation Area. Presently, tourism in Annapurna Conservation Area has contributed to a win-win situation, where the ecological and socio-economic environment has benefited (Nepal 2002).

The most significant result drawn from this study is that if a proper balance between conservation and development as well as benefits and activities is ensured, local communities can live in harmony and continue their traditional activities compatibly with the objectives and goals of protected areas management.

References

Bajracharya, S.B. (2003). *Community Involvement in Conservation: An Assessment of Impacts and Implications in the Annapurna Conservation Area, PhD Thesis*, University of Edinburgh, pp. 159–274.

Banskota, K. and Sharma, B. (1995). *Tourism for Mountain Community Development: Case Study Report on the Annapurna and Gorkha Regions of Nepal.* Kathmandu: International Centre for Integrated Mountain development: pp. 233.

CBS. (2003). *Population Census by VDCs*, Central Bureau of Statistics, HMGN, Kathmandu, Nepal.

Ghimire, K.B. and M.P. Pimbert (1997). *Social Change and Conservation: An Overview of Issues and Concepts — In: Social Change and Conservation. Environmental Politics and Impacts of National Parks and Protected Areas*, Eds. K. B. Ghimire and M. P. Pimbert, pp. 1–45, London UK: Earthscan Publications Limited.

Gurung, G. and Nepal, R.C. (2002). *Management of Ecotourism in the ACAP*, Paper presented for South Asian conference on Ecotourism in Gangtok, Sikkim, India, Jan. 2002.

Gurung, J.B. (1995). *Population, Habitat Selection and Conservation of Himalayan Tahr in the Annapurna Sanctuary, Nepal. MSc Thesis*, Agricultural University of Norway, Norway: pp. 58.

Hough, J. and M. Sherpa (1989). "*Bottom up vs Basic Needs — Integrating Conservation and Development in the Annapurna and Michiru Mountain*

Conservation Areas of Nepal and Malawi. (www.yale.edu- accessed on 12th July, 2005).

Infield, M. and A. Namara (2001).*Community Attitudes and Behavior towards Conservation: An Assessment of a Community Conservation Program around National Lake Mburo National Park, Uganda.* 35(1). pp. 48–60.

KMTNC/ACAP. (2005). *Compiled Report: 1987 to 2004. Unpublished Report.* Pokhara: KMTNC- ACAP.

KMTNC-ACAP. (2001). *Tourists Facilities Survey.* Unpublished Report. Pokhara: KMTNC-Annapurna Conservation Area.

KMTNC-ACAP. (2004). *Records of Legal Actions in the Annapurna Conservation Area. An Unpublished Report.* Pokhara: Legal Unit, KMTNC-Annapurna Conservation Area Project.

McNeely, J.A. (1995). Partnerships for Conservation: An Introduction. In: *Expanding Partnership in Conservation*, ed. J.A. McNeely, pp. 1–10. Washington DC, Island Press.

Nepal, S. (2002). *Linking Parks and People: Nepal's Experience in Resolving Conflicts in Parks and Protected Areas.* International Journal of Sustainable Development and World Ecology 9:75–90

Nepal, S.K. (2002). Involving Indigenous People in Protected Area Management: Comparative Perspectives from Nepal, Thailand and China. *Journal of Environmental Management* 30(6): 748–763.

Sharma, U.R. and Wells, M.P. (1996). Country Paper- Nepal. in: *Decentralization and Biodiversity Conservation*, Eds. E. Lutz, and J. Caaaldecott, pp. 65–76. New York; the World Bank.

Shrestha, R. and Ale, S.B. (2001). Species Diversity of Modi Khola Watershed. An Unpublished Report. Pokhara, Nepal: KMTNC-Annapurna Conservation Area Project.

UNDP. (2002). *Nepal Human Development Report*, Kathmandu, Nepal.

Wangchuk, S. (2002). *Grazing Management in National Parks and Protected Areas: Science, Socio-economics and Legislation (Tenure): (http://www.bhutanstudies.org.bt/journal/vol7/v7-3.pdf accessed on 27th June 2005).*

10

DEVELOPING A LANDSCAPE EVALUATION TOOL IN DEVELOPING COUNTRIES, A CASE STUDY IN BINTAN ISLAND, INDONESIA

GUNAWAN TANUWIDJAJA and LAI CHOO MALONE-LEE

Developing countries generally face great development pressure with many resulting environmental problems. The need for integrated ecological planning emerges from these concerns. However, there are constraints on the implementation of the requisite processes; these include the lack of political will, as well as lack of expertise, tools, information and funding. In this study, a feasible landscape evaluation method is proposed, which is suitable for application in these countries.

The research involves identifying the key elements of all suitable methods, which are later distilled and refined to develop the system referred to as Adaptive Landscape Evaluation Tool or "**ALiT**". The strengths of this system lie mainly in establishing the evaluation methods, the proposal for an adaptive list of data, and a system of scoring thresholds that embraces sustainable land development principles. The method was validated in a case study of Bintan Island in Indonesia.

The proposed evaluation method, **ALiT**, is comprehensive yet manageable and practical. It encompasses four elements as part of the development and evaluation process, namely, initial secondary data collection, rapid survey to verify critical data, multi-disciplinary analysis with Geographic Information System (GIS), and stakeholders' inputs. **ALiT** can be a very useful tool for reducing adverse impacts on the environment in the course of economic development in developing countries.

Introduction

In this millennium, humankind faces six major environmental problems that threaten livelihoods. The major problems are: population growth, resource depletion, global climate change, extinction of plants and animals, loss of wildlife habitats, increasing pollution, and high levels of poverty in many countries. Many experts believe that these problems are caused by unsustainable development in most countries (Miller, 2003, p. 1).

In many countries, the development of land is one of the contributors to unsustainable development. From the economic viewpoint, such unsustainable development is related to the quasi-public good characteristic of natural capital, including land (Thampapillai, 2002). In developing countries, population growth, economic development, and poverty also cause degradation and conversion of large natural areas. For example, since 1950, 97% of the forests of the Philippines and 84% of those in Madagascar have been degraded (Miller, 2003). In Indonesia, it was reported that around 3,180,243 ha of forests is lost annually;[1] while 147 mammals, 114 birds, 28 reptiles, 91 fishes and 28 invertebrates have become extinct because of deforestation.[2]

Unsustainable land development can be reduced through the adoption of suitable land development practices or systematic landscape evaluation before development takes place, to reduce environmental impacts. Since the 1960s, several landscape evaluation methods have been suggested by McHarg and other authors. However, difficulties are encountered in the application of the methods in developing countries. The problems are caused by the complex nature of the available methods, and lack of qualified personnel, data and relevant tools. This forms the background and rationale for the current research.

The purpose of the research is to develop an integrated landscape evaluation method for planners and government authorities of developing countries, to be applied particularly in regional planning. To validate the usefulness and applicability of the tool, a case study of Bintan Buyu, Bintan Island, Indonesia was undertaken. The proposal to develop this area embodies the usual conflicts faced by developing countries as they pursue economic progress. The case study was undertaken to demonstrate the applicability of the tool and also to draw lessons from its limitations, and how it can be adapted for use in other situations.

A simple landscape evaluation tool is needed in the context of developing countries. The tool must include all the important factors of development types and it should simplify the evaluation process while not affecting the validity and usefulness of the method in the given context. It must also involve a comprehensive multi-disciplinary approach including the participation of the stakeholders that can deliver the best land use choices and recommendations.

---

[1] http://www.indonesianforest.com/

[2] http://www.profauna.or.id/ indo/Fakta_satwa.html.

Methods

Bintan Island in Indonesia was chosen because of proposals which have been made to develop it in future, and the ecological importance of much of its area. The purpose of the study was to develop a landscape evaluation tool for developing countries due to the potential for wider application of the results of the research.

A general survey was conducted, and secondary data were compiled from local government agencies. Interviews were conducted with stakeholders to address the lack of data on many important factors. Data were also collected from the Indonesian National Geological Research Centre in Bandung that were crucial for the geological, hydrological and physiographic analyses. Arc View GIS 3.1 and AutoCAD 2000 were used in the creation of the secondary database and final analysis, while the landscape evaluation ranking was done with Microsoft Excel.

The full site survey was conducted in June 2006, with the help of two biologists. The final data analysis and the development of the landscape evaluation tool followed the final site survey. A simple review of the evaluation tool was conducted with the Bintan Regency Planning Agency. Further interviews with several planners from Indonesia who practise in Singapore and Indonesia were conducted to obtain more views on the method, which was then refined accordingly. Finally, the recommendations of the study were communicated to the Bintan Regency.

The Bintan Buyu area was chosen as the primary research area for two reasons. First, the site is the area designated as the new Regency Township of Bintan Regency. Thus, the development process would involve the pressures that are commonly faced by developing countries. The second reason is the presence of important habitats with a high level of biodiversity, and water resources which are located near Gunung Bintan.

Most developing countries face problems such as low incomes, huge economic disparity, poor health and lack of education, and governments face difficulties in their efforts to solve them (Todaro and Smith, 2003). These problems have also, directly or indirectly, led to environmental problems such as deforestation, overgrazing and water pollution that are common in these countries. Low levels of productivity, high population growth, and high dependency rate might worsen these conditions. The government agencies of developing countries also face problems such as weak leadership, unclear vision, lack of human resources, limited budget, and corruption. For these reasons, the governments often allow development projects to proceed without proper planning and environmental impact assessment, only for environmental and social problems to emerge after the developments encroach upon, and affect the carrying capacity of the environment.

In this context, it is necessary to develop a landscape evaluation tool which can assist decision making to minimise the negative environmental impacts of development. The tool developed in this research is named ALiT, the acronym for **A**daptive **L**andscape Evaluat**i**on **T**ool.

Landscape Evaluation Method

There are eight stages of **ALiT**:

1. Defining the Purpose, Scope, and Context
2. Criteria Selection
3. Secondary Data Collection
4. Setting the GIS Database
5. Rapid Survey of Critical Areas
6. Expert Consultation (Semi-Delphi)
7. GIS-based Analysis (including Ecological Factors, Other Natural Factors and Social Economic Factors)
8. Stakeholders' Evaluation.

The main purpose of landscape evaluation is to determine the best use for the land in the framework of development (economic and social). The second purpose is to predict the possible environmental impacts and prescribe mitigation measures. Other possible uses of this tool could be for comparing the feasibility of several areas for townships, tourist attractions and industrial areas. The tool accommodates several land uses; the land use types were adapted from Anderson (2000), McHarg (1992) and Singapore's Environmental Pollution Control Act (Code of Practices on Pollution Control, 2000). Residential and commercial are the main land uses that need to be evaluated first. The tool is suitable for the evaluation of urban brownsites, urban fringes and rural sites.

Criteria Selection

In this step, 16 general criteria of secondary data were selected to represent the environmental functions and economic considerations. These criteria were selected from the lists offered by McHarg (1992, 1998), White (1985), Ratcliffe (1977), Anderson (2000), Balmori and Benoit (2003). These criteria were divided into three groups that represent ecological factors, other natural factors and socio-economic factors.

Group 1 Data Set was selected to measure the ecological value of the land in order to conserve the important areas. The ecological value is represented by ecosystem type as the main factor of suitability for conservation. The other factors, such as patch size, rarity of plant species, ecological connectivity, riparian buffer, beaches buffer, and rarity of animal species were also included to supplement the information on the main factor. The rationale of the application is to minimise the destruction of ecologically important areas.

Group 2A Data Set was initially prepared to measure the suitability for development relating to construction, and social and environmental costs. These criteria are the existing land use, physiography (especially slope), geology, inland hydrology, marine hydrology, climate, and resources. They are important criteria because the area is prone to hazards such as flooding, typhoons and landslides.

Later, Group 2B was added to the criteria. It includes criteria related to physical factors such as accessibility and existing infrastructure. Factors such as distance from the main road, or other cities, as well as the regional international centre determine economic development potential, whereas the standard of utilities and infrastructure can reduce the construction cost.

Group 3 Data Set represents the social, political-legal and economic aspects of the area. Some of the factors could suggest that sites identified may be unsuitable for development as they are in areas with a high crime rate, or which are endemic for certain diseases. These parameters can also be used for determining policy options such as increasing the size of the police force in the area, creating new jobs and promoting hygiene. Finally, the human sensory aspect, comprising visual quality and other forms of pollution were also included in Group 3. These criteria are useful in determining suitable locations for tourism areas, hotels, resorts or high-end residential developments that require good visual quality.

It is pertinent to note that not all the required data and information will be available on all occasions. Thus, there would be instances where the evaluation process would have to be undertaken with the available data. This is the background of the adaptive approach for example, in this research, emphasis was placed on the need for critical biological data, which could be derived from a rapid biological assessment.

GIS Database Processing

There are several options for data processing. The more appropriate ones should take into account resources availability in the government agencies in developing countries.

The first tool proposed here is Arc View GIS 3.1 for database management and analysis. The software is selected because of its accuracy through Global Positioning System (GPS) and wider compatibility with other programmes. Although the price is high, and this can be a constraint, this is buffered by its impact in reducing environmental problems.

The second tool proposed for the analysis is McHarg's transparent layer maps (McHarg, 1998). This is the next best option to computer tools. The different scale of the factor maps can make this process difficult. This can be solved with the advanced capabilities inherent in some of the present-day photocopier machines.

If Arc View GIS 3.1 is adopted, then the next stage would involve purchasing a baseline vector map from a remote sensing service provider, such as *BAKOSURTANAL* (Indonesian National Coordinating Agency for Surveys and Mapping)[3] and CRISP (Centre of Remote Imaging, Sensing and Processing) NUS.[4] The rationale of this step is to make the compilation process and ground checking faster and more accurate.

Another alternative of the step is to redraw a cadastral/ topographical map with AutoCAD R14 or later version. The baseline map must be drawn in decimal degree unit and referenced to correct GPS coordinates, with reference to common usage in maps, GPS tools and navigation practices.

There are two choices of analysis method. The methods are grid system and shape-file (SHP) system. The grid is usually used for general evaluation purposes while the shape-file is usually used for more detail analysis. The next step for SHP processing is redrawing all factor maps in the SHP format. Thus for every factor, one SHP theme is produced. First, one polygon could be drawn following the areas' borders. Later, it could be divided and coded following the properties.

Rapid Survey

The ALiT proposes that high priority areas should be selected prior to conducting the survey. This is due to the budget and time constraints. The

---

[3]http://www.bakosurtanal.go.id.
[4]http://www.crisp.nus.edu.sg.

prioritised areas are areas with predicted high biodiversity, wetlands, areas earmarked for development in the near future, and areas with high scenic values.

Several survey methods were analysed in the study. Based on that, McHarg layer maps, photographs, video documentation, GPS tools as well as simple ecosystem tables were found to be the most appropriate. These tools were chosen owing to the flexibility of data collection they enable, and the shorter time the survey would take.

Expert Consultation (Semi Delphi)

Interpretation of the secondary data should be undertaken with assistance from a team of experts. For this study, a team comprising a Planner, Landscape Ecologist, Conservationist, Hydrologist, Civil Engineer, Geologist and an expert in the Social Economic Sciences was formed to conduct the ranking exercise.

Later in the study, all data should be interpreted and translated into a ranking criteria as described by McHarg (1992), White (1985), Ratcliffe (1977), Anderson (2000), Balmori and Benoit (2003). A similar multi-disciplinary approach was suggested by McHarg (1998). The difference lies in the scoring method for interpreting the factors which recognise different impacts for each criterion. The concept is later translated into two types of scoring scales. The first scale is 0 to 30 for major-weighting factors while the second one is 0 to 3 for minor-weighting factors.

Zero is chosen to represent high conservation value areas, while 30 or 3 are chosen to represent the suitable area for development. This is a new approach since normally, a higher score represents a higher value. The weighting method is adapted from the Environmental Impact Assessment (EIA) process representing high, medium and low impacts. For example, the ecosystem type is considered to be of major impact because the development of one important ecological area can affect all plants and animals in it. The rare species might be lost while the ecological functions of the area can be disrupted. On the other hand, the developing a remote area may not affect the environment in such a major way. This is an improvement from previous evaluation techniques.

GIS Analysis

GIS Analysis process was undertaken by translating the attributes in the database into scores. The most urgent factors are used for analysis for every specific use. Later the matrices application can be done with Excel or other database software. A GIS expert can help to carry out this step more efficiently.

Further, all SHP themes should be intersected with the Geo-Processing Wizard. This process will produce parcels of land with certain factors. After all themes have intersected, the final SHP theme could be used to evaluate the suitability. The most important one at this stage is the total score from the database to determine the conservation value and suitability value.

Because of the ecological approach that is emphasised in this method, the total scoring is divided into three. Total Group 1 score represents the ecological value, the Group 2 score represents the natural factors and the Group 3 score describes the socio–economic condition and validates other scores. The rationale of the classification is to balance the ecological values and economic values in creating a sustainable land use recommendation.

In order to determine the specific suitability of certain land uses, certain criteria that are directly impacted by land use change were chosen. Because of different number of data used and different land uses, the threshold must be adjusted according to the available factors. From the interpretation of the criteria, the threshold is described below.

The conservation threshold was determined by identifying areas in the site with least ecological value. In Bintan's case, the areas that were prescribed by the expert were scrubland, agriculture, mining, barren earth. The threshold was adjusted to 47% because of this. For comparison, Freeman (1999) recommended a score 14 of total 30 in the Leeds case or 46%.

Further, the suitability thresholds were suggested in the Bintan Case. The consideration of development suitability prescribes 33.33% as the limit of conservation and 66.66% for limit of development with special treatment.

Stakeholder Evaluation

Participatory planning and development have been suggested by many experts and applied in many projects in Indonesia. The difficulties in executing the approach are time limitation and possible conflict of interests among the parties. Sometimes, there is no political will to implement this in the development process. Furthermore, a low level of education also makes this process more difficult.

For better landscape evaluation, this method suggests a stakeholder meeting which should be attended by representatives of the relevant government agencies, private sector organisations, members of the local community and experts from the universities. This takes the form of a

Focus Group Discussion. This kind of approach has been implemented several projects such as the Indonesia Kecamatan Development Program,[5] Bandung City Discussion Forum,[6] and RW 11 Cibangkong Area, also in Bandung.[7] The participatory approach has been able to produce the best policies for the community. However, for landscape evaluation, trust and equity must be built first before conducting a meeting, especially for areas with a history of land conflicts.

The results of the meeting should be followed up by the relevant planning agency with a view to developing guidelines and a master plan of the area. This process requires the services of professional consultants to be engaged.

This landscape evaluation method has been developed by distilling the important landscape evaluation factors in the context of developing countries. Moreover, it considers the practicality of the application of the method with criteria selection and data prioritisation. It also harnesses current landscape evaluation methods with application of GIS.

The method also prescribes a rapid survey to validate the secondary data and to gather important additional data. It also reduces bias in the results by using three data sets for evaluation, which are: the ecological, natural and socio-economic factors. Two groups of participants were also involved in determining the suitability of the land: one is the team of experts and the second comprises representatives of the public sector, private sector and the people. This method was validated in Bintan Buyu, Bintan Island, Indonesia. The process is now described.

Application of ALiT (Adaptive Landscape Evaluation Tool) in Bintan Buyu

The purpose of this research was to review the suitability of the method for Bandar Seri Bintan. Bintan Planning Agency had recognised errors in

---

[5]http://web.worldbank.org/WBSITE/EXTERNAL/COUNTRIES/EASTASIAPACIFI-CEXT/EXTEAPREGTOPSOCDEV/0,,contentMDK:20477526~pagePK:34004173~piPK:34003707~theSitePK:502940,00.html.

[6]Reports of Bandung Urban Discussion Forum on Garbage Management Problems, January 2002 and Reports of Bandung Urban Discussion Forum in Housing Problems, August 2001, COMBINE.

[7]http://www.combine.or.id/program_04g.html and http://lnweb18.worldbank.org/ESSD/sdvext.nsf/60ByDocName/CaseStudy4-IndonesiaParticipatoryApproachesinBudgetingandPublicExpenditureManagementSDNote73March2003/$FILE/sdn73.pdf#search='cibangkong'.

the 2004 Master Plan for the townships, relating to incorrect planning scale and incomplete analysis. The agency was now preparing up a new Master Plan to replace it. So the results of the study would give valuable input into the new Master Plan of Bandar Seri Bintan. The second purpose of the evaluation is to reduce the environmental impact of the developments, especially in some areas with important ecological values. At the same time, areas with less ecological value were suggested to be evaluated and developed to fulfil the new requirements for township development. As Bintan Buyu area had already been determined as the new Regency Centre, comparison of several areas in the new feasibility analysis of Regency Township was not included in the scope of the research study.

The relevant land uses evaluated in Bintan Buyu, prescribed in the Municipal Plan, are:

- Conservation Area or LU-1
- Historical Building, Religious Facility, Social Facility or LU-2
- Commercial and Office and Government Office or LU-3.2
- Residential or LU-4.1 and LU-4.3
- Clean Water Treatment, Sewage Water Treatment, and Landfill or LU-6
- Open Space or LU-8.1
- Agriculture or LU-9
- Health Facility or LU-15
- Education Facility or LU-16

The secondary data collection was started after receiving permission for the study from the Political Bureau of Bintan Regency.

Data was collected from local government agencies, such as the Planning Agency (*BAPPEDA*), statistics agency (*BPS*), forestry agency (*Dinas Kehutanan*), public works agency (*Dinas PU*) and the Investment Coordination and Regional Promotion Board (*BKPMD*) of Bintan Regency based on the ALiT data checklist. Several national government agencies such as the National Geological Research Centre and National Water Resources Centre were approached for the data.

Arc View GIS 3.1 was selected as the database. The baseline map was collected from the Bintan Regency Public Works Department (1993). It was the most up-to-date data in all Bintan government agencies. Furthermore, all available data were collected, scanned, geo-referenced. The data was redrawn later into shape-files (SHP). The attributes were fed into the SHP data in a manual process. This involved a trial and error approach. For example, the attribute could be input as numeric or string data, depending on the analysis method used.

Results

A comprehensive site survey was conducted in June 2006. The survey was executed using transparent maps (simplified version, McHarg, 1998), photographic documentation, video documentation, GPS tool and simple tables. The methods proved to be effective in recording the site conditions. These also helped the researchers to draw a bigger picture of the site.

Simple tables were used to record the site conditions. The tables consisted of ecosystem type, plant and animal species name, existing land use, water body features and GPS coordinates or distance measurement from a particular reference point. They were realised that the survey could have been conducted more accurately with GPS tools and GIS Map.

Some points selected for biological assessment were derived from the secondary data. They were predicted to represent primary forest, secondary forest, mangrove, marshes, plantation and human settlement. Because of the time difference from secondary data collected, which were from the 1990's, and the present, many ecotypes of the area have been degraded or altered. This was the reason for the need to obtain the latest remote sensing image.

GIS Data analysis was started after all data were drawn in SHP format. The analysis was done by feeding in the score attribute and calculating the score based on the urgent factors of every form of land use. From the secondary data and survey, the ecological value and development suitability of the site were arrived at. Table 10.1 shows the criteria which were evaluated.

The general ecological condition of Bintan Island was compiled from the Natural Resources Inventory from the Bintan Planning Agency (Riau Islands Planning Agency, 2004). The data was later validated with the Rapid Biological Assessment (RBA) in Bintan Buyu. Major changes in the ecological condition of Bintan Buyu were apparent. For example, the 1993 Natural Resources Inventory Report reported the existence of several secondary forests, but during the survey the team found that the patches had been converted into agricultural lands.

From interviews with local people, Bintan Buyu had been occupied and utilized for agriculture since the 1950s. The designation of the catchment area was declared by the government without adequate public education and due process to purchase the land. So the local people still developed the land according to their customs and traditions. This phenomenon had created the rural- natural area patchwork that is evident in Bintan Buyu.

Table 10.1. Criteria evaluated in Bintan Buyu case study

No	Criteria	Availability
	Group 1	
A	**Vegetation**	
A1	Ecosystem Type	Yes
A2	Patch Size	No
A3	Rarity of Plant Species	No
A4	Ecological Connectivity	Yes
A5	Riparian & Beaches Buffer	Yes
B	**Animal**	
B1	Rarity of Animal Species	Partial
	Group 2A	
C	**Existing Land Use**	
C1	Compatibility with Existing Land Use	No
C2	Existing Facility Adequacy	Yes
D	**Physiography**	
D1	Elevation	Yes
D2	Slope	Yes
E	**Geology**	
E1	Foundation Strength	Yes
E2	Earthquake and Volcanic Zone	Yes
E3	Anticline, Syncline, and Fault	Yes
F	**Soils**	
F1	Foundation Strength	Yes
F2	Erosion by Soil Type	Yes
	Group 2A (Cont'd)	
G	**Inland Hydrology**	
G1	Recharge Areas	Yes
G2	Aquifer Productivity	Yes
G3	Surface Waters and Flood Zone	Partial
G4	Catchments Areas	Yes
H	**Marine Hydrology**	
H1	Marine Features	No
I	**Climate**	
I1	Hurricane, Typhoon	Yes
I2	Annual Rainfall	Yes
I3	Rainy Days in 1 year, Dampness, Average Temperature, Monthly Temperature, Wind Pattern, Sun Angle and Trajectory	Yes
J	**Resources**	
J1	Mining Resources	Not applicable because of designated as Catchments Area
	Group 2B	
N	**Location**	
N1	Distance from the Main Road	Yes

(*Continued*)

Table 10.1. (*Continued*)

No	Criteria	Availability
N2	Average Distance from Other City within Area	N/A
N3	Average Distance from Regional and International Centre	N/A
O	**Circulation**	
O1	Inland Transport	Yes
O2	Water Transport	Yes
P	**Utilities**	
P1	Availability of Facility	Yes
	Group 3	
K	**Politics and Legal Factors**	
K1	Municipal Government Economic and Human Resources Capacity	Yes
K2	Future Government Plan	Yes
L	**Economic Factors**	
L1	Land Market Pricing	Yes
L2	Land Ownership	No
L3	Regional Gross Domestic Product Trend	Yes, only in general
M	**Social Factors**	
M1	Higher Education Level	Yes, only in general
M2	Middle Education Level	Yes, only in general
M3	Criminalities Level	No
M4	Population Density	Yes, only in general
M5	Number of Patients and Deaths by Diarrhoea	Yes, only in general
M6	Number of Patients and Deaths by Dengue/Malaria	Yes, only in general
Q	**Human Sensory**	
Q1	Visual Quality	Yes
Q2	Other Pollution	Yes

Bintan Buyu's ecological condition is composed of several eco-types. First, the highest ecological areas are coral reef, mangrove, primary forest, and fresh water swamp forest. These areas must be conserved as much as possible because of their important functions, such as being habitats for plant and animal life and nursery ground for marine life and birds, providing coastline protection, serving as a waste assimilator, and being water purification systems. In the economic sector, there are non-forest products such as medicinal raw materials, rattan, and bamboo (Whitten *et al.*, 2000).

The second group has the second highest ecological value. It comprises secondary forests, marshes and abandoned plantations. Whereas these components fulfil certain ecological functions, they are still in the process of recovery after human interventions. For example, the team found Banded Leaf Monkeys in the abandoned plantations. This shows that the area is a relatively good habitat but it has generally less ecological value than it did in its original condition.

The third group has the second lowest ecological value areas, and includes the agricultural land and scrubland. This group can be considered as being a partially man-made environment. Comparatively low biodiversity was found in the area, comprising mainly insects, such as dragonflies, and birds. From the survey, the team found several abandoned agricultural lands and scrub areas that can be rehabilitated to preserve the high conservation value areas. At the same time, some patches of scrub must be preserved to provide habitat for several common species which were found.

Finally in the survey, the team found some very low ecological value areas, and these can be recommended for development. They are the mining or ex-mining areas and barren earth. Although some animals were found in these places, the ecological value of the areas was considered to be very low.

Another important feature of the ecological condition is the connectivity between the eco-types. From the survey, the ecosystem patches were found to be segregated. This has been caused by extensive unsustainable agricultural and mining practices. This phenomenon also might have resulted in the absence of large mammals or predators.

Lastly, two important species were found in the area: the Banded Kingfisher (*Lacedo pulchella*) and Banded Langur (*Presbytis siamensis rhionis*) are two examples of the endangered species in the conservation area.

Several parameters could be used to determine the economic potential of the area; three most important ones were selected. They are Land Market Pricing, Land Ownership and Regional Domestic Product Trend. There were no official data on the Bintan Buyu area for the first and second criteria because of its designation as a catchment area. Information obtained from the Bintan Land Administration Agency showed that no official certificates had been produced for the land parcels so far.

The community and the Bintan Forestry Agency own most of the land. While the land prices vary from Rp. 1,000 to Rp. 4,000. per sq. metre, it was found that the sites near the main routes fetched the higher prices. Now, because of the new Township Plan, most of the land has been parcelled and sold illegally to investors.

Bintan's Regional Domestic Product was recorded to have increased by 7.90% from 2000 to 2004. The main contributors to Bintan's Domestic Product were the industry, mining, trade and tourism sectors, while the biggest increases were reached by the energy, construction, agriculture and transportation and communications sectors (BKPMD, 2005).

It was recognised that for a new area to be developed, it must have a viable population. For this reason, the population analysis should be conducted before the determination of the development area. In 2004, it had been recorded that the population of the Bintan Regency had reached 115,675 persons.

The educational level in Bintan has been quite good. In 2004, it had been recorded that only 13% of the population above 10 years old age were illiterate. Some 2.60% of the population had 1–3 year diplomas or university degrees and 17.52% of the population had graduated from senior high schools and vocational schools. For developing countries, this may be considered as a moderate level of education. Besides education, the size of the labour force can also determine the suitability of an area for development. In 2004, 56.64% of the Bintan population had been recorded to form the labour force. However, there is an observed lack of skilled labour in Bintan. This can be observed from the figure that only 3.35% of the population of Bintan graduated from senior high or vocational schools and 1.35% graduated with a diploma or university degree.

Owing to the time limitation, data on the Bintan crimes rates were not collected. Other social statistics that were recorded were the diarrhoea and dengue fever occurrences. In 2004, in the six districts in Bintan, 2,985 people have suffered from diarrhoea and 34 people have suffered from dengue. This shows that the community has a low hygiene level and the area could be a source of endemic diseases. Specifically, in the Bintan Buyu area, it had been recorded that the population in April 2006 was 2,065 persons. By analysing the available data, it was determined that the population is increasing at 0.88% annually.

The Bintan Buyu people are mainly farmers, and fishermen (this information was obtained from an interview with the Head of the village). There were no detailed statistics on the level of education, crime and disease occurrence, and local revenue. However, it has been observed that, so far, there were no major crimes and related trends or similar social problems in the area. From an interview with the Head of the village, it was identified that the Bukit Batu people could not accept the presence of outsiders in the area. This might create social segregation in the future.

The inland public transport systems available in Bintan are taxis, public buses, motorcycle taxis. The motorcycle taxis are normally used for shorter distances inside the township, while buses serve the Tanjung Pinang to Tanjung Uban route. There is a school bus provided by the Bintan Regency to transport school children in the island. It was recognised that the transportation network has not been developed much (BKPMD, 2005). Besides public transport, the rural people normally use the motorcycle as the main means of transportation., Some cars and truck were also found during the survey and interviews.

Inland water transport is not common in Bintan, although speedboats, boats and a ferry services are widely used to connect Bintan with other islands. It is recorded there are four main ports and several private ports for transporting passengers and cargo regionally, nationally and internationally. These are the Sri Bintan Pura Port and Sri Payung Batu Anam (in Tanjung Pinang), Sri Bayintan Port and Tanjung Uban Port, which are the main ports in Bintan (BKPMD, 2005).

The transportation in Bintan Buyu is mainly by public buses and motorcycles. There is no ferry terminal at the site. Accessibility is one of the problems that might be faced by the township residents.

The utilities currently available in Bintan are clean water treatment and supply system, open landfill for waste, electrical power supply, and both land and mobile telephone transmission services. It was observed that the utilities are concentrated in towns such as Tanjung Pinang, whereas remote areas such as Bintan Buyu are not being served yet. Bintan Buyu, in particular, does not have a clean water supply network, sewage treatment and pipeline network, and telephone lines. There is also no major landfill site and incinerator facility in the location, although a small recycling facility or factory was found to the North of Bintan Buyu, in Kijang area. However, there are electrical power lines, individual water provision (mainly from wells) and a new cellular telephone service.

The limited provision of some of the basic utilities might create problems in the future after the township begins to operate.

Although visual quality could be categorised as a subjective parameter, it is important for several projects such as tourist attractions and hotels. For Bintan Buyu, only a small area could be considered beautiful and suitable for development. This area comprises the abandoned agricultural land and scrubland in northern part of the site. On the other hand, several areas were recognised to have unaesthetic views such as former mining sites and bare earth.

The existing mining operations in Bintan Buyu apparently could be the only source of land and water pollution. The tailing water is not treated before it is discharged. This can pose threats to the ecologically important areas and catchment area for drinking water. The other pollution identified was the traffic noise, which is generally at medium levels.

In 2005, Bintan Planning Agency and P.T. Metaforma Planning Consultant produced the Master Plan for Bandar Seri Bintan (Seri Bintan Township). Apparently there was a mistake in the calculation of the total area. The plan was designed to cater for a population of 30,000. In fact, the area of Bintan Buyu is less than that figure. The total area of Bintan Buyu is actually 61 sq. km (excluding the coral reef of approximately 8 sq. km.) Following this figure, the population should have been set at around 15,000 persons. This is one of the suggestions related to population proposed by this research.

Singapore had planned new townships with 45,000 to 80,000 families based on a high-rise concept.[8] This could be achieved with medium height residential buildings (12 stories high), townhouses and landed properties. However, this concept cannot be transplanted in the Bintan Buyu area because of lack of funding in the government and private sectors. Another alternative for setting a population plan is McHarg's (1998) suggestion that limits the growth of a city to the carrying capacity of the city. This concept was prescribed for Wilmington and Dover, considering sewage disposal capacity, ground water supply and reservoir water supply of the area. McHarg suggested a population of 33,100 people for the city. This concept could be applied in Bintan by considering the water resources of the area. The lowest monthly rainfall in Bintan was 197 mm^3 (BKPMD, 2005). Riau Islands Planning Agency *et al.* (2005), reported that the area of the Bintan River catchment comprises 4,628 ha and 57% runoff coefficient. Based on that, the average flow rate of the rivers are 129,260 m^3/day. Assuming that only 30% of the flow is utilised for the municipal water supply, 38,778 m^3/day of clean water will be produced. It will accommodate a population of 57,877. This is based on the rationale and method from Nathanson (2003, p. 252). McHarg (1998) also recommends limiting development to less ecologically important areas. In Bintan's case, these areas are the scrub, agriculture and probably some of plantation areas. This approach might also result in reduction of the population size that could be accommodated.

[8] http://www.ura.gov.sg/dgp_reports/punggl21/home.html.

Considering all the recommendations above, it was decided to combine the third and fourth alternatives. The authors recommend that the plan should be for a population of 57,000 persons (19,000 housing units) and the development should be limited to the less ecologically important areas. They also suggest that the mixed-use neighbourhood unit concept with a 400 to 600 metres radius be adopted. This is adapted from Perry's Neighbourhood Unit Concept (Urban Planning Guide, 1986).

Discussion

Bintan Buyu area is a degraded environment because of agricultural activities, human settlement and mining. Most of the ecosystem patches were found to be fragmented. Some good biological areas exist, for example, coral reef, mangrove, primary forest and fresh water swamp forest. These areas need to be conserved to protect the biodiversity. The conservation area could be converted for eco-tourism and recreation for city dwellers without disturbing the area. This strategy has been implemented in some of the other areas in Bintan such as the highly successful Banyan Tree Resorts and Hotels development.

Since this area has been designated as the new regency township, that the Bintan government should build the town in less sensitive biological areas, which comprise the agricultural land, scrubland and barren earth or the former mining sites. The most appropriate area is located north of the site comprising $10.5\,km^2$.

Assuming that only 50% of the area is going to be designated as residential to accommodate 57,000 persons, then the residential density would be

$$= 57{,}000 \text{ persons}/5.25\,\text{km}^2$$
$$\approx 10{,}857 \text{ persons}/\text{km}^2 \approx 109 \text{ persons/ha}$$

The development type suggested for the site is mixed-use low- to medium-rise buildings. This is because of the limited economic capacity of Bintan. Public utilities and integrated waste management systems (including waste water and solid waste treatment, and drainage system) should be provided for the area.

Ecological connectivity is a key factor for successful conservation in the area. Thus, the various areas of scrubland, forest, and riparian area that serve these functions should be preserved and improved; this is related to forest rehabilitation and monitoring.

The other natural factors of Bintan Buyu did not pose as obstacles to the development. However, inherent threats such as flooding must receive proper attention in the planning process. The site plan must not disturb the existing water balance and physiography. The existing master plan cannot be implemented because of the wetlands in the centre of the area. If the plan was implemented, not only would the construction and landfill costs be very high, but also, the ecological functions would be disrupted, and biodiversity loss and flooding might also occur in the area. Another recommendation is that polluting activities, such as general industry and mining should be reduced. It would not be feasible to implement water catchments because of the polluting activities which are currently taking place. It would be necessary to develop other water sources to replace the Kangboi River as the raw water supply.

Several of the findings of this research support, explain and illustrate the application of the ALiT approach. From the interviews with officers of the Bintan Government Planning Agency and other agencies with the experience in evaluating the agricultural potential of land in Bintan over several years, the evaluation method was found to be feasible. It seemed that the agencies had roughly applied this survey method. Currently, they have a team of several qualified professionals to prepare the master plan and land evaluation process.

Implementation of the evaluation method in the context of developing countries, as found in Bintan, will face several obstacles. These include the high expense of the GIS software and remote sensing data. For example, Arc View GIS 3.1 costs around US$7,500, while SPOT (remote sensing) images for just half of Bintan cost US$3,000. The lack of important data was found to be another challenge in the case of Bintan. In the research, the response to this was to use the adaptive list in the ALiT tool. In future, effort should be devoted to comprehensive data collection and management to support decision-making in the development process. Indonesia and other developing countries have little experience in these particular regards, for example, at present, in Bintan, statistical data are gathered only for bureaucratic and political purposes. In processing the data, problems were encountered from the differences in the accuracy of the maps, and in the various techniques of data gathering which were applied. For these reasons, corrections and assumptions had to be introduced when translating the factors map into the GIS shape-files (SHP files). Each factor had to be verified in the more detailed ground survey. The rapid biological survey conducted by the team and one of the authors was one such survey which was undertaken to verify each of the factors. However, costs can limit the time and personnel which can be marshalled to undertake the ground survey. One strategy which was adopted during the research study was to focus on monitoring points in areas earmarked for development in the near future and areas which are predicted to have high ecological value. The combination of ground survey and detailed remote sensing data was useful in these regards. The factors influencing the land use varied, and this was identified to have resulted in difficulties in considering the total and suitability thresholds. In subsequent applications of the method, research with a multidisciplinary approach could be used to validate the scoring.

The evaluation method developed in this research can be an important factor in studies relating to land development. Future users of this tool are the officers of government agencies, and consultants. On each development proposal, a team comprising a civil engineer, planner and ecologist should be formed to prescribe an ecologically balanced recommendation that paves the way for the attainment of sustainability in the implementation of the scheme. In such a multi-disciplinary team, the ecologist should be able to comprehend the nature of the local ecology; the civil engineer should be able to evaluate the soil types and hydrology regimes; and the planner should be able to determine and understand the social and economic potential as well as the constraints in the location. This approach would produce better recommendations than a narrow, single discipline approach.

The tool developed in this study made it possible for the team to mark different impacts, and correlations among evaluation factors and land uses. For example, stronger soil formation is important for physical development of residential and other buildings in urban areas, while for agriculture, soil fertility is the key factor. This was found during the scoring calculations of the different land uses. Another finding from the Bintan case study was the bias which could be introduced into the results from the accumulation of the scores of all the factors. It was also found that a screening of areas which are prone to potential hazards should be undertaken before calculating the suitability thresholds in order to reduce the environmental destruction and costs from the occurrence of natural disasters. The cases of Hurricane Katrina in New Orleans, and the tsunami and earthquake in Aceh and Jogjakarta, and more recently, in Pangandaran in Southern Java signify the need for this. It is pertinent to note that not all the results of the analyses can be simply followed. In most cases, there must be a compromise solution to the development decision. For example, in the case of Bintan Buyu, one could suggest that several plantation areas be developed and it could be argued that the area could be sacrificed for township development to protect the more important patches of forest.

Conclusion

A sound landscape evaluation process is an important step in the sustainable development of land. However, the cost of the necessary evaluation tools, and lack of data and time constrain the implementation of such processes in the developing countries. The assessment of the existing methods revealed the importance of selecting the evaluation criteria, proper weighting, rapid survey, and a multi-disciplinary and stakeholders approach in solving the problem. ALiT (**A**daptive **L**andscape Evaluat**i**on **T**ool) has been developed specifically for such contexts.

The strengths of ALiT can be found at five stages. The first is the criteria (factors) selection. This is important to achieve effective comprehensive evaluation. The second is the expert consultation for setting scores and weights for the factors. This step actually avoids the bias which may be associated with subjective analysis. The third is the rapid survey to collect the data that is not already available. In the study, the process was conducted by a multi-disciplinary team to which validated the secondary data. The fourth is the stakeholders' evaluation. Although this was not done during the current research, the participation of the people, public and private sectors will contribute to the successful realisation of the development. Finally, the method reduces result bias by using three data sets in the evaluation, which are ecological, natural and socio-economic factors.

ALiT is useful in the decision making process on implementation of sustainable land development. ALiT can be implemented in developing countries to reduce environmental destruction and attain economic development in sustainable development; this has been demonstrated in Bintan Island, Indonesia. Several possible improvements can be made to the tool, for example, through research on the relationships between the factors and land use. For instance, that housing and agriculture need different soil types could be accommodated with different ratings. It should be noted that the implementation of ALiT does not eliminate further detailed research on the various aspects of development suitability. For example, the loading test is still required to evaluate the strength of the soil formation in the particular area after site selection. This also applies to ecology, hydrology, and pollution control measures.

References

Anderson L.T. (2000), Planning the Built Environment, Planner Press, Chicago, USA.

Balmori D. and Benoit G. (eds.) (2003), The LAND Code: Guidelines for Environmentally Sustainable Land Development, Yale School of Forestry and Environmental Studies, USA.

Bintan Regency Public Works Department, 1993, GIS Map.

BKPMD Kab. Kepulauan Riau (Riau Islands Regency Investment Coordination and Regional Promotion Board), Potential Investment Research Report (Pengkajian Peluang Investasi Potensi/ Unggulan Daerah), 2005.

Code of Practice on Pollution Control (2000 Edition) (with amendments in Feb 2001, Jun 2002 and Feb 2004).

Freeman C. (1999), "Development of A Simple Method for Site Survey and Assessment in Urban Areas", *Landscape and Urban Planning:* 44, pp.1–11.

McHarg I. (1992), Design With Nature, John Wileys & Sons, Inc, New York.

McHarg I. (1998), Steiner Frederick R. (ed) To Heal the Earth, Selected Writings of Ian L. McHarg, Island Press, Washington D.C.

Miller G.T. (2003), Environmental Science, Working With Earth, 10th edition, Brooks/Cole Thomson Learning USA.

Nathanson J.A. (2003), Basic Environmental Technology: Water Supply, Waste Management, and Pollution Control, 4th ed, Prentice Hall, New Jersey, USA, p. 252

Ratcliffe D.A. (1977), A Nature Conservation Reviews, vols. 1 and 2, Cambridge University Press, Cambridge.

Riau Islands Planning Agency, Revision of 2006–2015 General Regional Plan Riau Islands Regency (Rancangan Umum Tata Ruang Wilayah Kab. Kepulauan Riau), 2004.

Riau Islands Planning Agency and CV Geometric Tehnik (2005), Research and Mapping of Water Recharge and Catchments Potentials in Riau Island Regency (Kajian dan Pemetaan Potensi Daerah Resapan dan Tangkapan Air Hujan di Kab. Kepulauan Riau).

Thampapillai D. (2002), Environmental Economics, Concepts, Methods and Policies, Oxford University Press, Melbourne.

Todaro M.P., and Smith S.C.(eds.) (2003), Economic Development, Pearson, Harlow England, pp. 46–71.

Urban Planning Guide (1986) (ASCE Manuals and Reports on Engineering Practice; no 49), American Society of Civil Engineers, New York.

White E.T. (1985), Site Analysis: Diagramming Information for Architectural Design, Architectural Media, USA.

Whitten T., Damanik S.J., Anwar J., and Hisyam N. (2000), Ecology of Sumatra, Periplus, Singapore.

11

STUDY OF PARKS AND NATURE RESERVES' FISCAL MANAGEMENT MODELS

NG CHUN, JUSTIN and LAI CHOO MALONE-LEE

Fiscal management is an integral component in the management of parks and nature reserves. Budgetary challenges faced by public sector agencies in Singapore and other countries have resulted in the introduction of innovative fiscal instruments in the management of parks and nature reserves. The inherent social elements in the provision of parks and nature reserves means that the introduction of such tools and instruments has to be carefully considered against the backdrop of the benefits attributed to parks and nature reserves. This is particularly so in the case of the intangible social benefits of parks and nature reserves, which cannot be simply weighed against any economic benefit. In this chapter, the fiscal aspects of parks and nature reserves management in Singapore are examined.

Fiscal management models are discussed with particular emphasis on established fiscal instruments and mechanisms that are based on broad principles of cost-recovery, cost-reduction and income-generation. The applications of these fiscal instruments and mechanisms are analyzed with respect to Singapore's experience in the provision of parks and nature reserves. In response to various challenges that park managers have to face in the management of parks and nature reserves, recommendations are made to enhance the fiscal management model towards long-term fiscal sustainability.

Introduction

In 40 years of nation-building, the Singapore government had always emphasized a pragmatic approach to the conservation of her natural areas and heritage. During the early years of independence from the 1960s to 1980s, Singapore was confronted with the simultaneous challenges of needing to house a growing population and to ensure a sustainable economy based heavily on the manufacturing sector. This "pragmatism" was visible in the development blueprint of a then rapidly-industrializing Singapore. Vast areas of land were cleared to develop public housing estates, factories and the requisite supporting public infrastructure. A significant proportion of Singapore's rural areas and natural biodiversity were thus lost, somewhat ironically given that Singapore simultaneously embarked on a national programme to green the city through the roadside tree planting, and landscaped parks and open spaces (Yuen, 1996).

Since the 1990s, Singapore has enjoyed, for the most part, a consistent level of economic progress. This "pragmatism" in public policies had seen a gradual but visible shift in emphasis from economic development to the protection and conservation of the remaining natural heritage. Examples of this shift can be observed in the conservation of natural bird sanctuaries at Sungei Buloh as a gazetted Nature Reserve (MND, 10 Nov 2001), and the deferral of development works at Pulau Ubin following the discovery of bio-diversity abundance at Chek Jawa (MND, 14 Jan 2002).

However, the recent shift in emphasis from a focus on fulfilling economic needs to environmental conservation fails to mask the fact that there is still the perennial conflict between economic development and the conservation of the natural environment, a conflict which can only be resolved by the fragile act of "balancing". How then does one "balance" the mostly quantifiable benefits of economic development against the intrinsic and largely intangible worth of the natural environment?

This question of "balance" is made even more complicated by the need to understand that the conflict between economics and environment is never quite static. As a country progresses through different phases of development, the dynamics of this conflict change with time to reflect the evolving needs and aspirations of the people. For example, in Singapore, between the 1960s to 1980s, societal perceptions deemed economic survival to be of greater significance than environmental conservation, while the latter only began to gain prominence and public recognition in the 1990s.

Parks and nature reserves provide an important social function. As a matter of public policy, whether a particular park or nature area should be preserved or sacrificed for development purposes, should ideally be based on the concept of social benefits. The conventional approach would be to compare the aggregated social benefits with the costs of providing these benefits (Dixon and Sherman, 1990). However, owing to the shortcomings of current valuation techniques, the cost-benefit analysis (CBA) approach presents numerous obstacles to attempts to frame a method to measure the net social benefits. The debate and argument is still continuing over the appropriateness of the existing valuation techniques, and whether their results can ever approximate the true worth of parks and nature reserves.

Moving forward, it might be of greater significance to be able instead to focus on the tangible costs of funding and providing these social benefits. The important question then is not whether parks or nature reserves should be conserved vis-à-vis economic developments but rather, when the relevant decision has been made, how to ensure long-term financial sustainability in managing and maintaining these parks and nature reserves.

This study seeks to:

i) Identify and synthesize best practices from various funding and fiscal management models that have been applied to the management of national parks and nature reserves in other countries.
ii) Analyze the applications of these fiscal management models and strategies to the management of parks and nature reserves in Singapore for long-term fiscal sustainability.

The study focuses on the funding and fiscal management models of parks and nature reserves through the various life-cycle phases, i.e. planning, development and management. The other aspects of the total management framework are also critical to the successful administration and management of parks and nature reserves, and are highlighted in the latter case study as they are also relevant to the follow-up discussions on financial sustainability.

In the context of this study, it is difficult to rigidly define the terms "parks", "nature reserves", and the related "open space", "nature areas" or "conservation areas". The review of relevant literature yielded many implied definitions. For example, while "parks" and "nature reserves" both provide greenery amenities and nature-based recreation opportunities, the difference between them might lie in the size, proximity to built-up areas, the degree of artificiality or even whether there is any significant biodiversity.

Fiscal Management of Parks and Nature Reserves

The need for government intervention in the provision of parks and nature reserves is commonly justified on the grounds of market failure (Turner, 2002). In the conventional approach, these estates are fully funded and operated by either the government or a government-linked entity. This chapter examines some fiscal management models, and the diverse funding structures and fiscal mechanisms. An illustrated summary of these fiscal mechanisms is used to highlight the extent and suitability of their applications.

Fiscal management models

There is a growing awareness amongst park authorities and departments of the need to scrutinize the means of improving the management efficiency of parks and nature reserves (Bennett, 2000). To this end, Garvin and Berens (1997) identified three emerging trends in park development and management models from an analysis of 15 case studies. These models are primarily differentiated by the level of private sector involvement.

The first model is one where the management of parks and nature reserves is fully undertaken by the public sector. In the case of parks, the state or municipal government typically identifies the needs of the community and commences the design and physical development of the park. Upon completion, the park is placed under the direct management of the parks department. As for nature reserves or conservation areas, these tend to require relatively minor input during the design and development phases. However, once identified and designated as conservation areas through legislation, these areas are usually placed under functional units within the parks departments.

In the second model, the public sector maintains ownership and responsibility for the park, but promotes joint ventures with the private sector to undertake selective development or redevelopment projects through donations or fund-raising. While the completed park is fully managed by the public sector, engagements with the private sector are encouraged through special programmes, improvements or donations to supplement the public-sector budget. The main difficulty in this model is in maintaining the involvement and support of the private sector entity(ies). Often, once a private corporation exhausts all the goodwill associated with the development of parks or nature reserves, the improvements and programmes are discontinued as donations dry up. Furthermore, under a public/private co-venture model (PPCM), there may be a detrimental over-reliance on funding from corporate injections and less emphasis on ensuring fiscal sustainability compared to a market-oriented civic model (MCM).

The third model, MCM, emphasizes the establishment of sustainable long-term partnerships between the public and private sectors for both park development and management. This model is characterized by the utilization of sophisticated market mechanisms such as non-profit development corporations or a business improvement district (BID).[1] The use of such institutional mechanisms brings private-sector responsiveness and risk-taking to the fore, and usually results in a proliferation of revenue-producing functions to provide for the improvement and management of the parks and opens spaces. Under this model, some of the more controversial mechanisms are the taxation of properties surrounding urban parks as levies for the resultant increase in their commercial and property values. The implicit assumption behind this market-oriented civic model is that the potential commercial opportunities in and immediately around urban parks are vastly under-utilized by the public-sector managers and could be more efficiently managed with private-sector approaches.

Private for-profit model

It is pertinent to note that arguments have also been put forth in favour of a pure private for-profit model that will maximize the efficiency of parks and nature reserves management. Proponents of private-sector provision contend that the involvement of the private sector typically results in higher diversity of pricing and service standards (Eagles, 2002), and may even be more responsive to societal needs.

However, a clear distinction must be made between ownership and management. While private ownership is ruled out owing to the high risks of market failure, parks and nature reserves can still be publicly owned but privately managed (Bennett, 2000). In the context of this study, this equal combination of public ownership and private management should be considered as a logical extension of the market-oriented civic model (MCM).

Application of models

Owing to the social element of parks and nature reserves, although the three models involve different levels of private sector engagement, each of them still positions the public-sector agency as the key stakeholder in the provision, development and management of parks and nature reserves. Not only does this allow the public-sector agency to retain its core social

---

[1] The use of business improvement districts (BID) were highlighted in Garvin and Berens' (1997) case study on Bryant Park, New York. Essentially, BIDs are self-financed legal entities that allow property owners and business leaders to provide common services beyond those that the city can provide. In this particular case, the Bryant Park Restoration Council (BPRC) financed the restoration and maintenance of Bryant Park from a collective tax levied on property owners within the designated city block.

and political responsibilities, but it also translates to less resistance to any inertia in adopting and restructuring the management of parks and nature reserves to these models.

Distinction between parks and nature reserves

It must be noted that some fiscal models and strategies are more relevant to urban or neighbourhood parks while others are more applicable to nature reserves or conservation areas. Some of the conventional approaches of distinguishing between parks and nature reserves are now considered in order to facilitate the discussion of the application of the fiscal management models to both parks and nature reserves.

Artificiality and urban-ness

Perhaps the most popular criterion in differentiating parks and nature reserves is the degree of artificiality in the physical form. Parks are generally perceived to be man-made, where the landscape and terrain are carefully sculpted, and the trees and plants are arranged based on a pre-conceived design. This is further underlined by the fact that most parks are found in built up areas and involve the use of modern development and construction techniques. However, park agencies are increasingly attempting to retain existing natural features in the development plans of urban parks, in order to avoid a blank-canvas approach to developing parks. Ironically, it is sometimes more costly to retain natural features like mature trees, terrains, existing ponds or streams in urban parks (Gobster, 2001).

Purpose and function

The artificiality criterion is also linked to the purpose or function of the parks. It is a common expectation that urban parks generally serve a recreational function. Whereas nature areas or nature reserves also fulfil this recreational function to some extent, the ease of accessibility associated with urban parks means that convenience as an outlets for recreation is one of the key considerations in the provision of an urban park.

Bio-diversity abundance

Nature reserves play a larger societal role than just fulfilling the recreational needs of the people. A significant consideration in the conservation and protection of nature areas is the proliferation of bio-diversity, or ecological merits. Evidence of bio-diversity abundance is a logical criterion in attempting a clear distinction between urban parks and nature reserves. An urban park generally contains much less bio-diversity since a large component is artificial. Furthermore, the long-term sustainability of any bio-diversity, including one that is artificially introduced to the park, may be threatened.

Status of legal protection

Due to the need for long-term preservation, nature areas and to a larger extent, nature reserves, are accorded legal protection, thus enabling the natural bio-diversity in the nature reserves to be sustained for longer periods.

Physical size

Perhaps the least-used criterion, but often held perception, is that nature reserves and nature areas are generally larger in size than urban parks. While that may be true for the majority of cases, there may be exceptions to this generalization, for example, a large urban park may approximate to the size of a nature reserve or nature area when developed as part of an urban renewal strategy on an expansive brown-field site (De Souza, 2004).

It is difficult to further refine these characteristics and features of parks and nature reserves into definitive categories. Even if the various criteria are used in combinations, the subjectivity of assessing these criteria will elicit variances in opinions and viewpoints.

Fiscal strategies in parks and nature reserves management

In this section, the funding mechanisms and instruments commonly used in the management of parks and nature reserves are considered. These include single-source government budgets and common fiscal strategies in terms of cost-recovery, cost-reduction and income-generation mechanisms. The merits and limitations of the strategies and their applications to conventional urban parks and nature reserves or conservation areas are analyzed with specific reference to the three management models.

Full government funding

While the government's budget is still the conventional means of funding the management of parks and nature reserves in most situations, it has been widely criticized as being inflexible in adapting to circumstances that justify modified approaches. MacAndrews (1998) pointed out that the fundamental drawback of single-source government funding is that the budget provisions for parks and especially nature reserves are often allocated independently of the specific needs and requirements presented in the long-term management plans. Funding for parks and nature reserves typically comprises only a small component of the total government budget allocation and is therefore more susceptible to budget adjustments under fiscal pressures.

However, full government funding has its advantages in the development and management of large-scale parks and nature reserves as high capital costs and long capitalization periods associated with these types of projects may render them less attractive to private-sector investments. Furthermore, government provisions of parks and nature reserves usually result in high economies of scale in terms of organizational set-up, manpower resources and administrative support.

Cost-recovery strategies

Costs can be minimized through cost-recovery or cost-reduction. Cost-recovery strategies are methods and mechanisms of recovering all the costs involved in developing and managing parks and nature reserves from users, partners and stakeholders (Eagles, 1995). Essentially, cost-recovery strategies are based on the principle that some monetary worth can be attributed to the usage or goodwill of parks and nature reserves, which can be used to offset costs.

Visitor or user fees

Visitor or user fees are currently adopted in many large nature reserves in relation to recreation and tourism. It allows for an easily-administered and convenient way of recovering part or all of the costs of developing and managing the park or reserve. However, typical user fee mechanisms can only be applied to nature reserves or parks with sufficient ecological, historical and cultural elements of interest. Thus, the application of visitor or user fees to urban or neighbourhood parks may not be practical.

A key concern in the application of visitor or user fees to nature reserves or large-scale parks is the matter of equity. More and Stevens (2000) raised this issue in the context of a local survey to establish whether the introduction of user fees would exclude low-income people from taking part in the recreation provided by parks and nature reserves. While the study was only undertaken in a specific locality, it does beg the question of whether the charging of user-fees contravenes the social functions and public purpose of parks and nature reserves. Willis (2003) proposed a potential solution in the adoption of discriminatory pricing of user fees based on different intensities of demand exhibited by two groups — *economically inactive* (the elderly, children or students) and *economically active* (all other visitors) to mitigate the exclusion effect. Indeed, such a discriminatory pricing approach may increase total revenue relative to a unitary price if the *economically active* are charged a significantly higher entry than the *economically inactive.* In theory, such discriminatory pricing can be a useful. However, the inherent difficulty is in identifying park visitors with different intensities of demand and to enforce and collect the selective fees. The associated costs of administering these measures will have to be considered when employing discriminatory pricing.

Instead of differentiating user fees by the demand intensities of visitors, Buckley (2003) suggests that visitor or user fees could be broken down into different usage components. For example, within the context of a simple unitary entrance fee, admission charges could be imposed for special interest areas within the parks or nature reserves, such as camping sites. This diversification of the fees and charges to the specific usage not only ensures equity but also, it allows managers of parks and nature reserves to employ user fees as effective visitor management tools.

Visitor or user fees could be creatively and effectively applied to the management of large parks or nature reserves under the PPCM or MCM model. However, due care must be exercised when using such fees. The recovery of costs should not be positioned as the only objective so as to ensure that the public mission in providing the parks and nature reserves is not compromised.

Sponsorships and donations

The solicitation of corporate sponsorships and donations has been increasingly viewed as a major source of developmental and management funding for parks and nature reserves. When applied in the appropriate context, sponsorships and donations provide much-needed fiscal boosts in response to budgetary cuts and pressure from increasing costs. Furthermore, corporate entities sponsoring worthy causes gain too from a significant outpouring of public goodwill. However, Barton (2000) warns that over-reliance on corporate sponsors may result in a loss of image, control, and possibly even ownership, compromising the intended social ideals of providing parks and nature reserves. This may be especially so in situations where the corporate sponsors or donors have interests which may be in conflict with how a particular park or nature reserve is managed. For example, a company, having adopted or sponsored a nature reserve, may insist on holding its corporate or public relations events in the reserves, without showing due sensitivity to the potential disturbances these activities may cause.

Taxation mechanism

A relatively radical cost-recovery mechanism is the collective taxation of adjacent properties for the increase in commercial activities and property values that proximity to urban parks may yield. While user fees or sponsorships may be intuitively inferred as the recovery of costs accrued to some direct usage or associated goodwill, taxation mechanisms are somewhat differentiated by the fact that the occupants of the adjacent properties being taxed may have very little direct usage of the parks.

Yet, these property-owners benefit from a spillover effect of park visitors that are in turn likely to visit their own commercial properties for retail, food and beverages, and so on. Taxation mechanisms thus provide a channel for the transfer of these indirect monetary benefits back to the park agency for allocation to improvement and development works. However, the introduction of taxation necessitates the continual and collaborative efforts of both the public agencies and private businesses to sustain the feedback system and to stay attuned to market sensitivities.

Cost-reduction strategies

Apart from recovery of cost from users and other stakeholders, cost-reduction strategies can also be utilized to minimize the internal costs of managing parks or nature reserves. Both cost-containment strategies are commonly applied in tandem so that the arguments are equally compelling from both internal and external perspectives.

Outsourcing of services

The direct reduction of staffing costs through outsourcing of some services is one of the most immediate ways to minimize the internal costs of managing parks and nature reserves, and can be logically introduced in all the three fiscal management models. A typical approach involves the government agency identifying a service or set of services that could be rationally contracted out to the private sector. The units providing these services are either downsized or entirely eliminated. Private firms are then invited to bid for contractual provisions of these services. However, one of the main difficulties in such outsourcing is the abrupt loss of the knowledge base and skill sets which the departing staff possessed, possibly leading to a discernable dip in the quality and effectiveness of services. Bruce (2001) proposed an alternative approach where, upon identification of the services to be outsourced, existing government employees were encouraged to form partnerships with private firms and bid for the contracts. The criteria for the awarding of these contracts could incorporate a period of monopoly to allow these service providers to acclimatize themselves to market-efficient operations. When the initial contract expires, another round of tendering for the services is undertaken. The main advantage of this alternative approach is that it allows the residual expertise of the staff to be retained and transfused to the new private set-up.

Administrative restructuring

In contrast to the reduction of costs through outsourcing, the other common cost-reduction mechanism involves the restructuring of the agency to reflect the enhanced roles and portfolios that are added to the agency's core functions. Administrative or organizational restructuring has the potential to result in more efficient fiscal management systems. Compared to outsourcing, the restructuring of the administrative or organizational systems allows the park agencies greater leeway to fine-tune their functional responsibilities in response to societal demands.

Income-generation strategies

Income-generation instruments may take the form of licenses or concessions that add value to the economic worth of a park. As parks and nature reserves primarily serve a social role, the commercial potential of parks and nature reserves are significantly less than that of other alternative land uses. However, there are still various opportunities to maximize the income-generating potential of parks and nature reserves from third party licensees.

Income from tourism and recreation

The development of tourism and recreation-related facilities and services in parks and nature reserves is considered the most reliable source of generating income from parks and nature reserves (Cardona, 2003). Typical facilities and infrastructure include restaurants, cafeteria and eateries, souvenir shops, bicycle kiosks and car-parks. Examples of services and activities are recreational programmes, guiding services for hiking and nature treks, and weddings.

Under a PSM, concessionaires licensed by the public agency for a period of time may provide many visitor services, such as accommodation, tours, and consumer products, but the income from these concessions is typically returned to a central government for management and disbursement. This may be effective when there are sufficient government budgets but park managers are constrained in their ability to respond to changes in visitor numbers (Eagles, 2002).

An emerging trend in park-related tourism is the development of more sophisticated commercial entities that integrate the tourism and recreation components with the intrinsic appeal of greenery and nature (Eagles, 2004). For example, large parks or nature reserves are retained as the central theme in any integrative proposals, while the commercial viability would be improved by the inclusion of hotels, museums, resorts, and retail or commercial facilities. Arguably, the growing sophistication of these facilities and services requires marketing and promotional expertise beyond the domain of the public sector. Although basic recreational facilities and services can be provided under the PSM, it is under the provision of the PPCM and MCM that a more conducive environment is created. The private sector expertise can be engaged in the creation of more sophisticated facilities and services, and the management of such integrated facilities.

Income from resource extraction

There has been considerable interest in the extraction of resources as a viable option for generating revenues in conservation areas and nature reserves. In its various manifestations, such as 'eco-mining' or 'bio-prospecting', the sustainable extractions of natural resources from identified areas within conservation areas could supplement direct tourism-related income in the pursuit of fiscal sustainability (Gurung, 2003).

The key consideration in adopting resource extraction is in ensuring that these activities do not cause any adverse impacts on the ecological sustainability of the nature reserves or conservation areas. Introducing resource extraction activities under a pure PSM incorporates the caution and sensitivities necessary in such projects, but may not be feasible due to a lack of technical expertise. On the other hand, the scale and intensities of resource extraction projects require a substantial initial outlay of capital, and on infrastructure. The occasional involvement of the private sector based on a PPCM may result in a less than committed partnership over the longer term, and that could jeopardize these types of projects.

It is essential that implementation of sustainable resource extraction incorporates the three-fold challenges of balancing sensitivities towards ecological preservation, adoption of technically-proficient methods, and ensuring long-term commitments by all stakeholders. It is conceivable that resource extractions as a fiscal instrument is most suited to the MCM.

To minimize the risk of excessive extraction activities to improve the bottom line, legislation could be introduced to dictate the maximum limits of resource extraction that would be allowed. Alternatively, Inamdar and de Merode (1999) proposed that a more market-efficient fail-safe mechanism could be the attachment of an accreditation scheme or eco-label to resource extraction practices. Essentially, this allows any potential overuse to be reflected through market responses when consumers signal their disdain by boycotting the specific extracted resources (Sasidharan, *et al.*, 2002).

Trust funds

A trust fund is a mechanism that encapsulates many of the fiscal management instruments discussed in the previous sections. Trust funds are, a fundamental form of institutional set-up that incorporates governance structures involving various stakeholders, clear and transparent operational procedures and sound fiscal management practices (Norris, 1999). Typically, a trust fund comprises three vital components (Inamdar and de Merode, 1999), namely, a set of assets capitalized by government grants or corporate donations is set aside as the trust fund, a set of laws or constitution is drawn-up to stipulate why and how the funds are spent, and, lastly a board of trustees comprising a cross-section of the stakeholders representing the government, private donors, local communities and technical experts to administer the allocation and disbursement of funds.

Trust funds are commonly adopted as an institutional mechanism for managing nature reserves or conservation areas as the threats to the ecological and bio-diversity sustainability is a long-term one and requires sustained and continual commitments (Norris, 1999). As the initial capital is typically expected to sustain the long-term needs of the protected areas, the board of trustees performs a crucial role in ensuring that sound fiscal management and good governance is practised. However, while most current environmental trust funds have been instituted for nature reserves and conservation areas, their application is not solely restricted to ecological and bio-diversity concerns. There have also been examples of trust fund mechanisms that have been adopted for urban parks that exhibit significant cultural or historical values. These mechanisms may differ from environmental trusts in employing more direct income from commercial opportunities.

Case Study

Following the discussion of the fiscal management models, strategies, and instruments, a case study of Singapore's experience in the provision and management of parks and nature reserves is now presented. Existing fiscal instruments are discussed, and related to the present social, political, economic and cultural context in Singapore through trend analysis to facilitate an assessment of the extent of their practical usage.

History of NParks

The National Parks Board (NParks) is a statutory board within the Ministry of National Development (MND) which is responsible for providing and managing quality parks, greenery, and related services to meet the needs of residents and overseas visitors. The Parks and Recreation Department (PRD), the precursor of NParks, was set up in June 1967 as a government department formed to achieve the objectives of greening the city. The emphasis was on the planting of roadside trees throughout Singapore to provide shade and greenery in the rapidly developing urban environment, as well as the development and management of parks and open spaces.

In June 1990, the Singapore Botanic Gardens (SBG), Fort Canning Park (FCP) and both Bukit Timah Nature Reserves (BTNR) and Central Catchment Nature Reserves (CCNR) were moved out of the PRD's management and placed under the former National Parks Board. The formation of the National Parks Board was intended to achieve a higher degree of expertise in the development, management and promotion of these national parks and nature reserves as valuable resources for recreation, conservation, research and education. In July 1996, the new National Parks Board (NParks) was re-constituted with the merger of PRD and the former National Parks Board. This allowed NParks to undertake all the different aspects of providing and managing greenery within a single organization, be it in the form of parks, playgrounds, roadside greenery or nature reserves and nature areas.

Distinction between nature reserves and nature areas

NParks currently manages over 1,800 hectares of parks, playgrounds, park connectors and open spaces, and over 3,326 hectares of nature reserves, comprising Bukit Timah Nature Reserves (BTNR), Central Catchment Nature Reserves (CCNR), Sungei Buloh Wetland Reserves (SBWR) and Labrador Nature Reserves (LNR).

At this juncture, it is necessary to highlight the differences between Nature Reserves and Nature Areas in terms of protection intentions and management responsibilities. The Urban Redevelopment Authority (URA), which is the planning authority in Singapore, differentiates nature reserves from nature areas primarily based on their ecological merits and state of legal protection.[2] A simple summary of the differentiation is now presented.

Nature Reserves

> *These are nature areas protected by law and managed by NParks. They contain key indigenous ecosystems found in Singapore, such as marshes, mangroves, lowland forests, coastal, mudflats and grasslands. For Nature Reserves, ecological management measures are conducted to ensure that these areas are sustainable in the long run. For example, reforestation is carried out for degraded areas and exotic species are weeded out. Besides maintaining ecological processes, Nature Reserves have varied uses ranging from education, scientific research to recreation.*

Nature Areas

> *These are areas where the natural flora and fauna remain relatively undisturbed by human activity. These areas will be recognized on the Special & Detailed Control Plan[3] for as long as the land is not required for development.*

---

[2]Urban Redevelopment Authority (URA) Press Release, *URA Launches Draft Master Plan,* 28 February 2003.

[3]Non-statutory, supplementary land-use plan detailing special uses and micro-controls that are not reflected in the Master Plan.

However, beyond this simple distinction between nature reserves and nature areas, one of the implicit considerations is that the nature reserves are managed by NParks and enjoy legal protection under the National Parks Act whereas nature areas could remain as state land under the Singapore Land Authority (SLA). The long-term status of nature areas is much less certain as they may be given up when the land is required for development.

Most of the identified 18 nature areas are on state land and not under NParks' direct management. However, an exception is the Pulau Ubin nature area which encompasses the Chek Jawa wetlands (NParks, 29 Apr 2005). While the discovery of bio-diversity abundance at Chek Jawa has resulted in a halt to long-proposed reclamation plans in Pulau Ubin, it was still insufficient for Pulau Ubin to be gazetted as a nature reserve, although the bio-diversity is arguably of a higher quality than that found elsewhere.

The difficulties in rationalizing land-use decisions are very relevant in Singapore. Although the URA has attempted to formulate a criterion for differentiating and establishing nature reserves from nature areas, it will still be subject to interpretation by decision-makers considering various other concerns, including those of an economic nature. In this sense, the efforts to conserve these nature areas have also been aided by the NParks' attempts to rationally incorporate the identified nature areas into future developments of regional parks[4] to serve the recreational needs of the people. These include the incorporation of Kranji marshes and mangroves into a future nature trail; the proposal to include the Sungei Cina mangroves in the future Woodlands Regional Park, and the conversion of the Dairy Farm nature area into an interim park.

Scope of parks and nature reserves management

The functional differences in the management of nature reserves and urban parks have been reflected in the organizational set-up of NParks; there is a distinction between the management of regional and urban parks, on one hand and nature reserves and conservation areas on the other. The management of urban greenery; regional and neighbourhood parks and park connectors, are under the *Parks Division*, while the nature reserves are managed by the *Conservation Division*, with a functional unit responsible for each nature reserve. This illustrates the more varied

[4]Long-term developments of regional and urban parks as identified in URA's (2002) Parks and Waterbodies Plan.

operational functions required in the management of nature reserves, besides recreational outreach and management: the need to maintain the ecological sustainability in the reserves probably alludes to the necessity for a more structured organizational set-up in managing nature reserves. This also points to the possibility that park managers have a greater degree of autonomy and freedom in altering their management decisions to suit their particular needs. In the context of fiscal management, this will afford them the flexibility of adopting unique fiscal instruments to their particular areas.

Fiscal Management

NParks, is a statutory board under the Ministry of National Development (MND). Its operations are fully funded by the government by way of grants. Any rental income from licensees and concessionaires are channeled back into the government's consolidated funds. However, the agency has the opportunity to use alternative income streams, especially from sponsorships and donations, to fund special developments and targeted programmes.

From the discussion above, it can be considered that NParks is operating on a fiscal management model lying between a PSM and a PPM. It can be deduced that NParks is thus subject to a substantial degree of political and public scrutiny over the use of government funding despite being an autonomous statutory board. For the sake of comparison, a distinction could be made between NParks' management of parks and nature reserves, and Sentosa Development Corporation's (SDC) management of the greenery and nature areas on Sentosa Island. The positioning of Sentosa as a recreational and leisure attraction (Goh, 1995) and SDC's ability to retain its income from admissions and license fees gives it greater fiscal autonomy in managing the natural estates within.

Although it is fully-funded by the government, NParks has managed to introduce a variety of fiscal instruments to aid it in the management of parks and nature reserves. As will be discussed in the subsequent sections, it should be noted that in some of the cases, the starting principle for introducing these instruments was not to improve the agency's fiscal sustainability but more with regard to other social and political rationales.

Cost recovery strategies

The charging of admission fees is one of the main means of subsidizing the high costs of providing parks and nature reserves. NParks currently has admission charges at only two of its locations; Sungei Buloh Wetland

Reserves (SBWR) and National Orchid Gardens (NOG) within the SBG. NOG is positioned as a stand-alone tourist attraction owing to the high costs of maintaining the orchid collections and related display features. In this sense, the introduction of admission fees at NOG is a healthy step towards the fiscal sustainability of the attraction.

On the other hand, the introduction of admission charges at SBWR was justified on a different basis. As SBWR is a feeding ground for migratory birds, there is a need to maintain the natural attraction it holds for the birds, i.e. the quiet and undisturbed setting. Ironically, in its previous use as prawn farms, the lack of outside visitors was a huge factor in attracting migratory birds (Ng, 1998). However, upon conversion to a wetlands park in 1993, there was a critical need to strike a balance between catering to the needs of visitors and maintaining the undisturbed sanctity of the park. Admission charges were thus introduced as a mechanism for crowd monitoring and control to limit the stress on the natural environment. The fee was deliberately set at a low $1 to signal this intention. It is evident that the recovery of costs was not a major consideration in the introduction of admission charges at SBWR.

Figure 11.1 shows that there was a significant decrease in the income earned by NParks from the combined admission charges it collected over the five-year period, 2000–2004. If this was to be used as an indicator of performance, it would have reflected rather poorly on the fiscal performance of the park managers. But, the decision to selectively introduce admission

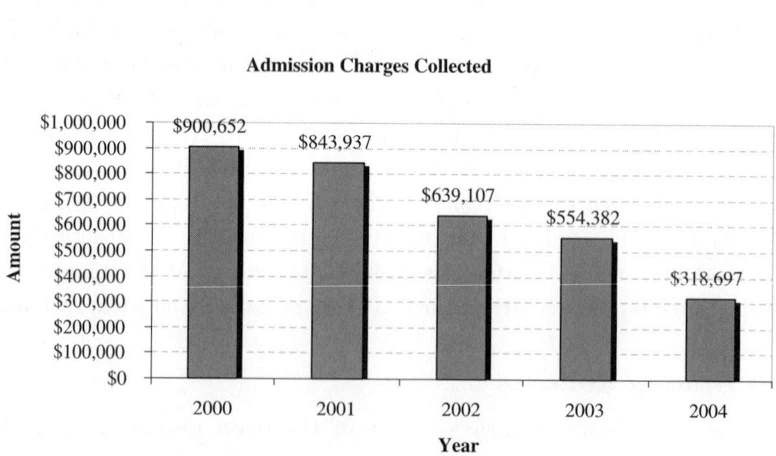

Fig. 11.1. NParks' 5-year admission charges collections.

charges at only the two locations was taken with due consideration of the other social ideals in park provisions.

As for cost-recovery through sponsorships and donations, there has been a gradual increase in terms of sponsorships and adoptions of parks and nature reserves by private corporations as well as institutional bodies. Examples include Hong Kong and Shanghai Banking Corporations' adoption of Chek Jawa (NParks, 16 Oct 2004) and the Tree-Top Walk in BTNR (NParks, 26 Nov 2004), as well as Shaw Foundations' donations towards the redevelopment of the Symphony Lake in SBG (NParks, 27 Apr 2005).

Figure 11.2 shows the donations collected by NParks over 2000–04. It can be observed that amount of money donated to NParks hovered around a consistent level for three years before the Garden City Fund (GCF) was launched in 2003 as a charity vehicle to consolidate donations to NParks. After the record amount of collections in 2003, the donations subsequently fell by half in 2004.

The principle of the Adopt-a-Park scheme was to increase the sense of ownership of residents, community and stakeholders involved in the maintenance of parks and gardens. Even when the (GCF) was launched in 2002 as a platform through which the public can take a personal stake in the sustenance of the Garden City efforts, the broad social intention was still evident that increased ownership and stake-holding was the more important driving force rather than the recovery of development or management costs (NParks, 1 July 2004).

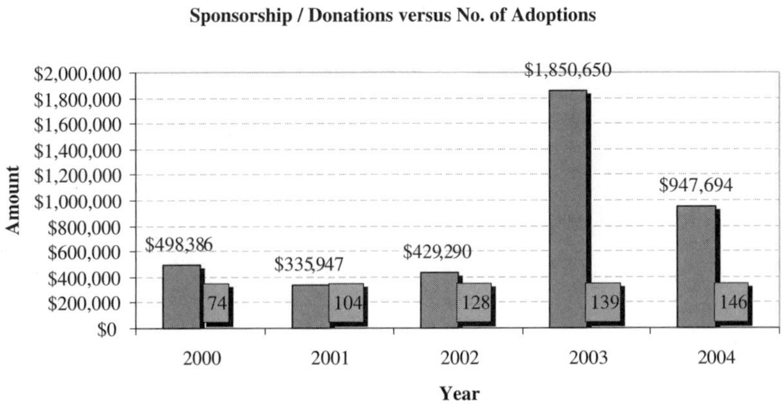

Fig. 11.2. NParks' 5-year donations collections and no. of park adoptions.

A point to consider is that if ownership and stake-holding are the more critical considerations in the Adopt-a-Park and GCF schemes rather than income generation, then the true barometer should be the number of corporations and institutions involved in the programmes and in partnerships with NParks rather than the dollar amount to measure their success.

Cost reduction strategies

With respect to reductions in the number of staff, either through natural attrition or outsourcing, it was highlighted above that one of the main stumbling blocks to such cost containment measures is the abrupt loss of knowledge and expertise. At NParks, in parallel with a trend of reduction of staff numbers (Fig. 11.3), there is a commitment to ensure the continued employability of the outgoing staff. Attention should be paid to the setting up of the *Industry Development Division* to ensure that the landscaping industry is upgraded and the service and skills level of the workers in the industry are relevant.

Figure 11.4 shows savings in terms of staff costs. From 2000 to 2003, when the staff strength was in the 900 or high 800 range, there were overall increases in staff costs, with the exception of 2004 when the staff numbers dropped by over 10%. The trend of increasing staff costs could be the result of public-sector employment policies then where wage increases were applied to all staff rather than on the basis of performance. This does point

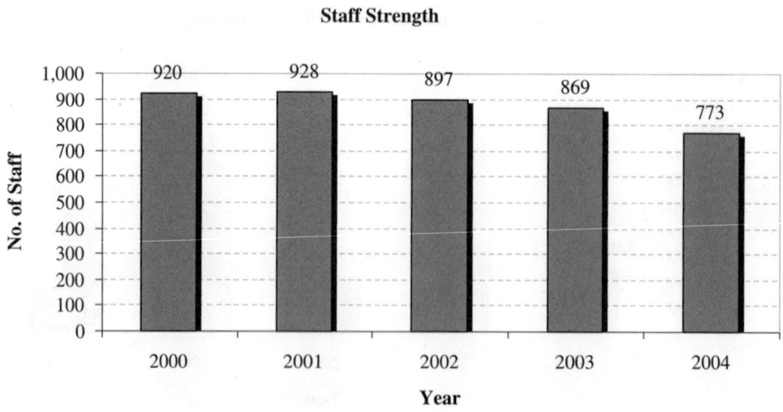

Fig. 11.3. NParks' 5-year staff strength.

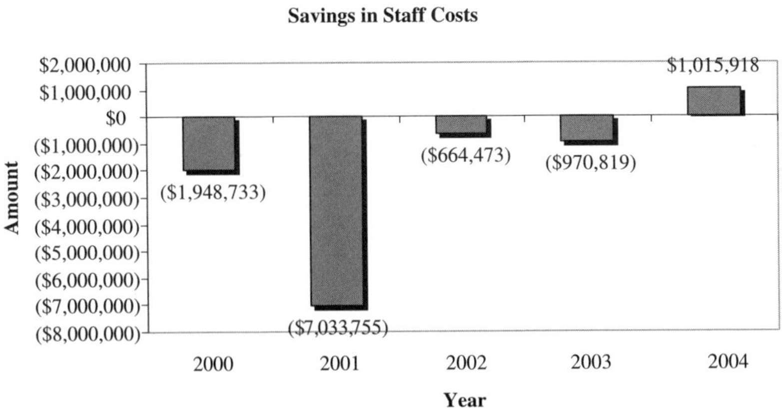

Fig. 11.4. NParks' 5-year staff costs savings.

to the importance of constantly keeping track of the cost implications of administrative and organizational set-up of NParks.

Income generation strategies

The introduction of commercial facilities such as eateries, cafes, restaurants and leisure services in parks and nature reserves should be done in the context of improving service levels and bringing convenience to park users and visitors. Based on this principle, the volume and types of commercial licenses and services would be kept within a certain defined range. The amount of rental income from these commercial licenses (Fig. 11.5) would

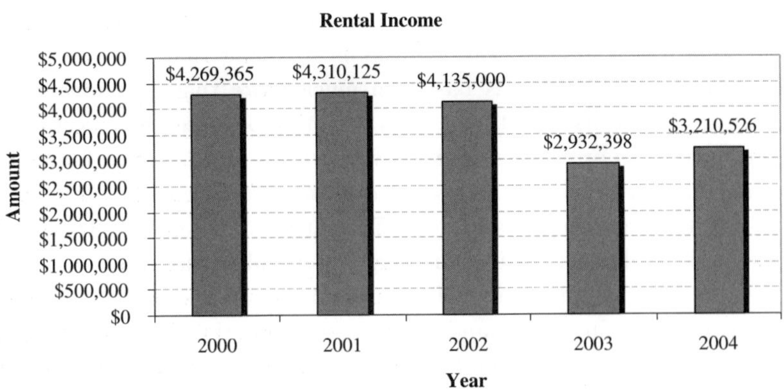

Fig. 11.5. NParks' 5-year rental income.

be welcome if retained by NParks instead of being returned to the central government funds. However, the ability to retain these revenues might encourage the development of larger and higher value commercial facilities that may disrupt the natural ambience of the parks and nature reserves.

The development of integrated resorts in the Marina Downtown areas and Sentosa is correspondingly generating a high level of public interest from the public and private operators on the possibility of integrating commercial facilities spanning retail, hospitality, healthcare and leisure, restaurants, and recreation within and next to parks and nature areas. The sensitive adoption of these proposals will be critical in ensuring that the general public does not lose environment to economics, albeit in terms of ambience rather than in absolute land areas.

With regards to resource extraction, it would be timely to highlight that in Singapore's context, where it is widely acknowledged that existing nature reserves and nature areas are already in a fragile state, any further exploitation in terms of resource extraction on a commercial scale will potentially cause irreversible damages. In that respect, the generation of resource extraction income has very limited potential for utilization in Singapore. However, if done sensitively to further research and provide educational benefits, allowing the selective and sustainable extraction of valuable bio-resources in the tropical rainforests within the nature reserves and nature areas would be a welcome boost to Singapore's positioning as a regional biological research centre.

The future of parks and nature reserves in Singapore

In Singapore, most of the existing fiscal instruments have been introduced not with the sole intention of alleviating NParks' cost pressures but for other social goals. In this way, NParks is signaling that the provision of parks and nature reserves is based on much more than just income and expenditure concerns. Yet, the gradual evolution of these fiscal mechanisms has given rise to a promising situation where there is an existing foundation of financially sustainable practices and systems that could be employed. Moreover, the organization and its staff are aware of the need to continually apply good fiscal practices in managing parks and nature reserves.

Under NParks, the future of parks and nature reserves management in Singapore is in safe hands, and the potential is clear that its fiscal management instruments could be further adapted with a fiscal management model which is close to the MCM. While it may not be realistic to adopt this fiscal model now, especially some of the more radical instruments associated with the model such as taxation of surrounding properties, it would be useful to begin to incorporate some private-sector thinking in the public-sector management of parks and nature reserves in Singapore.

Besides the various cost-recovery, cost-reduction and income-generating strategies, it would appear ideal that a trust fund mechanism be considered for NParks' management of nature reserves and even nature areas. Currently, four nature reserves (BTNR, CCNR, SBWR and LNR) are legally protected, with NParks acting as trustee, but the essential formation of a cross-section of private sector, public sector and community stakeholders within a fiscal framework would ensure a more socially relevant and committed representation working towards the long-term preservation of the reserves and nature areas.

Conclusion

One of the possible topics of debate to arise from this study is the perceived cost-pressures associated with each of the three fiscal management models. The pertinent question is whether the pressures for fiscal sustainability, i.e. to contain costs and generate income, are equally felt across the PSM, PCM and MCM. One would expect that the further an agency moves from a pure public sector model in parks and nature reserves provision, the higher is the corresponding pressures to attain fiscal sustainability. However, as this study has been based on a freeze-frame analysis of the various fiscal models and instruments, the possible effect of the critical factor of time has largely been omitted in the discussions.

It should be noted that the emphasis on the societal goals behind the provision of parks and nature reserves means that the choice of fiscal model and the fiscal instruments probably matter less than how they are applied. Moreover, fiscal sustainability in parks management denotes an ideal state that may never be fully realized. Thus, it might be more apt to view fiscal sustainability not as an end in itself, but to view the adoption of sustainable fiscal measures as viable and responsive means of tackling the many budgetary challenges that park agencies face.

Budgetary challenges to park agencies can come in many forms. For example, public sector restructuring and reforms may in turn exert fiscal pressures on social agencies such as park departments. Within the public sector, adoption of new budgeting systems such as performance-based budgeting may inconvenience social agencies relative to economic agencies (Andrews, 2005). Park agencies may also need to tighten their budgetary allocations to accommodate net stock increases due to new park developments.

So far, the topic of fiscal management and sustainability has been discussed within the context of parks and nature reserves provision. An interesting issue that merits further discussion pertains to the inherent difficulties in public policy and decision-making when attempting to compare social benefits of parks and nature reserves to other public sector social functions, for example, in the areas of public health, education, arts and sports promotion, and community and social development. Again, the fundamental principle may apply that in the absence of a method that could be used to effectively compare across a wide spectrum of social benefits, the best that social agencies can do is to improve the fiscal sustainability within their immediate management and subsidiary operations.

Among the challenges that public sector managers face in the provision and management of parks and nature reserves is the ever-increasing opportunity for public scrutiny of the various facets of their work. An emerging trend that Eagles (2004) observed in the management of protected areas is that the increasing access to information technology means that potential, current and past park-users and visitors are better informed and knowledgeable about what leisure and recreational opportunities exist in parks and nature reserves, the state of the management structure and the consequences of implemented management actions. This makes decision making in the management of parks complex. This points to a need for public-sector agencies to adopt a more consultative approach to the management of parks and nature reserves, and the public to intensify their participatory involvement in decision-making concerning parks and nature reserves.

Limitations of study

In the course of this study, there was a lack of numerical data that could be utilized for quantitative studies. While annualized aggregated income and expenditure figures were used in the case study analysis, specific and detailed figures, such as annualized operating costs of individual parks, and revenues from the different types of commercial licenses and concessions, could have been used to enhance the robustness of the quantitative element in support of the discussions in the study.

Application of study

The case study demonstrates that Singapore's experience in the provision and management of parks and nature reserves can be applied in countries which share the same social and economic challenges, such as countries with small geographical areas. However, this application must be done with due sensitivity to the particular context, including the collective social maturity and how the general citizenry view parks, nature reserves and other components of the natural heritage.

This study raised a number of questions on the many facets of fiscal management. This study is not a definitive framework or guide to fiscal management strategies; instead, it can serve as an introduction to the importance of fiscal sustainability, and the impact of fiscal strategies on public sector management of parks and nature reserves.

Recommendations for future research

Urban parks and nature reserves may be distinguished from each other by their public functions and objectives. While this study has incorporated most of the fiscal management models and mechanisms irrespective of their application to parks or nature reserves, it is suggested that future research may incorporate a more rigid scoping for either urban parks or nature reserves to take into consideration their unique characteristics and functions.

As can be observed from Singapore's experience, there is an increasing attention to maximize the income-generation potential of parks and nature reserves, through the introduction of more, and varied, commercial facilities. A common misgiving is that the over-zealous introduction of such facilities in parks and nature reserves may alter the inherent sense of serenity and nature that one would associate with parks and nature reserves. The public perceptions with regard to the commercialization of parks, and the intensity and mix of commercial amenities and services within parks and nature reserves and the impact on fiscal sustainability could be a topic for further research. It is hoped that such a study may provide some objective points towards answering the question: "How much is too much?"

References

Andrews, M., (2005). Performance-Based Budgeting Reform: Progress, Problems, and Pointers. In Shah, A. (eds) *Fiscal Management*, The World Bank: Washington, D.C.

Barton, R., (2000) The Implications of Sponsorship for State Park Management, *The George Wright Forum*, Vol. 17, 40–47.

Bennett, J., (2000) Privatising Parks — Why the Private Sector can Enhance Nature Protection, *Policy (Centre for Independent Studies, Australia)*, 3–7.

Bruce, C., (2001) Contracting Out at Parks Canada: Employee Takeovers. In Anderson, T.L., James, A. (eds) *The Politics and Economics of Park Management*, Rowman & Littlefield Publishers: Lanham, MD.

Buckley, R., (2003) Pay to Play in Parks: An Australian Policy Perspective on Visitor Fees in Public Protected Areas, *Journal of Sustainable Tourism*, Vol. 11, 56–73.

Cardona, F.G., (2003) Financial Strategy for Colombia's National Parks System (SPNN), Presented at *Vth World Parks Congress: Sustainable Finance Stream*, Durban, South Africa.

De Souza, C.A., (2004) The Greening of Brownfields in American Cities, *Journal of Environmental Planning and Management*, Vol. 47, 579–600.

Dixon, J.A., Sherman, P.B., (1990) *Economics of Protected Areas: A New Look at Benefits and Costs*, Earthscan Publications: London, UK.

Eagles, P.F.J., (1995) Tourism and Canadian Parks: Fiscal Relationships, *Managing Leisure*, Vol. 1, 16–27.

Eagles, P.F.J., (2002) Trends in Park Tourism: Economics, Finance and Management, *Journal of Sustainable Tourism*, Vol. 10, 132–153.

Eagles, P.F.J., (2004) Trends Affecting Tourism in Protected Areas, *Working Paper of the Finnish Forest Research Institute Vol. 2*, 18–26.

Garvin, A., and Berens, G., (1997) Urban Parks and Open Spaces, Urban Land Institute: Washington D.C.

Gobster, P.H., (2001) Visions of Nature: Conflict and Compatibility in Urban Park Restoration, *Landscape and Urban Planning*, Vol. 56, 35–51.

Goh, C.Y. (1995) *Impact of Tourism on the Environment: A Case Study of Sentosa Island.* Unpublished undergraduate dissertation, School of Building & Estate Management, National University of Singapore.

Gurung, G., (2003) Securing Financial Sustainability for Protected Area Management: A Case Study of Annapurna Conservation Area Project, Nepal, Presented at *Vth World Parks Congress: Sustainable Finance Stream*, Durban, South Africa.

Inamdar, A., de Merode, E., (1999) Towards Financial Sustainability for Protected Areas: Learning from Business Approaches, *WWF Sustainable Development Series, The Environment and Development Group*, Oxford, UK.

MacAndrews, C., (1998) Improving the Management of Indonesia's National Parks: Lessons from two Case Studies, *Bulletin of Indonesian Economic Studies*, Vol. 34(1), 121–137.

Ministry of National Development (MND) Press Release, (2002, 14 January) *Deferment of Reclamation Works at Pulau Ubin.*

Ministry of National Development (MND) Speech, (2001, 10 November) *Mr Mah Bow Tan, Minister for National Development at the launch of the Upgraded Facilities at Sungei Buloh Nature Park.*

More, T.A., Stevens, T., (2000) Do User Fees Exclude Low-income users from Resource-based Recreation? *Journal of Leisure Research*, Vol. 32, 341–357.

National Parks Board (NParks) Press Release, (2004, 1 July) *Heritage Trees Dedicated to Donors of Garden City Fund.*

National Parks Board (NParks) Press Release, (2004, 16 October) *HSBC Adopts Chek Jawa.*

National Parks Board (NParks) Press Release, (2004, 26 November) *NParks Relaxes Rule on Camping Permits.*

National Parks Board (NParks) Press Release, (2005, 27 April) *Official Launch of New Shaw Foundation Symphony Stage & Singapore Press Holdings' Gift of Music Series.*

National Parks Board (NParks) Press Release, (2005, 29 April) *Launch of Pulau Ubin Conservation and Management Plans — Ubin-HSBC Volunteer Hub, Chek Jawa, Jelutong Campsite & Bike Park.*

National Parks Board (NParks) Website, http://www.nparks.gov.sg/about/about.shtml.

Norris, R., Curtis, R., (1999) *Funding Protected Area Conservation in the Wider Caribbean: A Guide for Managers and Conservation Organizations*, United Nations Environment Programme and The Nature Conservancy.

Sasidharan, V., Sirakaya, E., Kerstetter, D., (2002) Developing Countries and Tourism Ecolabels, *Tourism Management*, Vol. 23, 161–174.

Singapore Government Press Release, (2005, 18 April) Ministerial Statement by Prime Minister Lee Hsien Loong, *Proposal to Develop Integrated Resorts.*

Tranel, M.J., Hall, A., (2003) Parks as Battlegrounds. In Harmon, D., Putney, A.D., (eds) *The Full Value of Parks: From Economics to the Intangible*, Rowman & Littlefield Publishers: Lanham, MD.

Turner, R.W., (2002) Market Failures and the Rationale for National Parks, *Journal of Economic Education*, Vol. 33, 347–356.

Urban Redevelopment Authority (URA), (2002) *Parks and Waterbodies Plan.*

Urban Redevelopment Authority (URA), (2003) *Master Plan 2003.*

Willis, K.G., (2003) Pricing Public Parks, *Journal of Environmental Planning and Management*, Vol. 46, 3–17.

Yuen, B., (1996) Creating the Garden City: The Singapore Experience, *Urban Studies*, Vol. 33, 955–970.

12

PROBLEM OF MARINE INVASIVE SPECIES AND THEIR IMPACT ON SINGAPORE WATERS

MANOJ KUMAR PUTHIA, PETER NG KEE LIN
and RUTH O'RIORDAN

The introduction of invasive marine species into new environments by ships' ballast water, ships' hulls and via other means has been identified as one of the greatest threats to the biodiversity of the world's oceans. This study analyzes the present status of invasive marine species in Singapore waters and relates this to global context.

Most countries are being affected by the adverse impact of invasive marine species. Developed countries are responding to this threat by introducing national legislation, adopting new technologies, and by researching into the subject. International organizations are playing their part by planning to impose effective legislation to prevent the international spread of invasive marine species. Numerous research projects are underway to identify the alien species, to determine their potential impacts, and to discover effective ballast water treatment technologies.

Studies reveal that no detailed research has been done in Singapore to ascertain the status of alien species in its waters. Officially, only one species of alien mussel and a few species of toxic dinoflagellates have been found in Singapore waters thus far. These species have already shown their adverse impacts in other parts of the world and they are potentially harmful in Singapore as well. There is currently, no regulation of the discharge of ballast water or the introduction of alien species into Singapore waters. Studies show that Singapore is particularly vulnerable to this problem. Thus, there is a need to establish regulations to prevent the introduction of alien marine species. However, local baseline information on the native marine species is inadequate and more research work is required.

Introduction

An "invasive species" is defined as a species that is non-native (or alien) to the ecosystem under consideration and whose introduction causes or is likely to cause environmental or economic harm or harm to human health (National Invasive Species Council, 2002). The impact of invasive species on aquatic ecosystems over the last 25 years has in some cases been overwhelming. Most of the invasive species are the result of human activities. It is instructive to examine some examples of marine invasion by alien species, which shows the severity of problem.

- In 1982 a species of comb jellyfish was found in the Black Sea; it was previously unknown to this region. It was identified as *Mnemiopsis leidyi* from America (Vinogradov *et al.*, 1989).The jellyfish increased in number and by 1988 their standing stock was calculated as thousands of metric tonnes. These comb jellies consume zooplankton with great speed and by 1989 anchovy (*Engraulis encrasicolus*) landings were reduced by ten times from their previous level (Harbison & Volovik, 1994). The introduction was due to ballast water discharge by a ship coming from a port in the Americas in 1979 or 1980 (Carlton, 1996).
- *Potamocorbula amurensis* (a bivalve) was introduced into San Francisco Bay by the ballast water of a bulk cargo vessel from a Chinese river port. By 1988, the standing stock of this invasive species reached 10,000 individuals per square metre (Carlton *et al.*, 1990). The impact of the invasion was so severe that the spring, phytoplankton bloom in the North Bay had almost disappeared (Alpine & Cloern, 1992).

Carlton (1996) explained that biological invasions in the sea consist of natural range expansions (i.e. the often predictable ebb and flow of organisms due to changing climates or currents) and human-mediated introductions. Invasions due to human activities are usually unpredictable and natural barriers of space and time do not play a big role in them.

Of the various types of human mediated introductions, the ballast water of ships is the most important identified vector responsible for the unintended transfer of aquatic organisms (Carlton & Geller, 1993). The introduction of non-native aquatic species by ballast water is capable of seriously damaging the ecological systems and resource based industries. The impact of ballast water introduced invasive aquatic species, for example, the zebra mussel in the Great Lakes (Griffith *et al.*, 1991) and toxic dinoflagellate in Australia (Hallegraeff & Bolch, 1992) has amplified awareness of the invasive species problem, and this has increased the requirement for effective solutions. The introduction of larger and faster modern ships and well-organized cargo loading and unloading facilities and procedures has further intensified the transfer of ballast water between ports. An increasing number of ships are transporting more cargo and between more regions, faster than ever before. As a consequence, the world's aquatic communities are becoming more homogenous with technological progression.

Although, all components of the environment are facing invasion by alien species, the marine environment is somewhat different in the sense that it is easily penetrable. Most of the marine organisms have planktonic stages in their life cycles and these are easily transferable from one place to another. Human activities, primarily the global movements of organisms associated with ocean-going vessels and with commercial fisheries products, have led to the redistribution of a large number of marine organisms over the past five centuries (Carlton, 1989). The modern scale and rate of new human-mediated invasions in the oceans are difficult to investigate due to lack of communication among scientists working on different groups of organisms, different habitats and different regions (Carlton 1989).

During studies in laboratories at the University of Oregon's Institute of Marine Biology, Carlton (1989) found that almost all major and most minor marine phyla survive voyages and often arrive in vast quantities. Carlton and Geller (1993) collected over 200 species of living zooplankton and phytoplankton from the ballast water of 70 cargo ships arriving from Japan in Coos Bay, Oregon.

Objectives

Singapore is one of the busiest ports in the world. Singapore is a small island country and the marine biodiversity of its sea is rich. For several reasons, Singapore's economy depends on its marine resources. There are concerns about the introduction of alien species in the country's waters. This study has the following objectives:

1. To study the global status of the problem of invasive marine species.
2. To study the status of the problem of invasive marine species in Singapore waters.
3. To suggest the possible solutions to the problem.

Methods

In this study, past work done by marine biologists was examined and the available literature was studied. Various cases of alien invasion that have occurred in the recent past in different parts of the world were probed. This investigation was undertaken with a view to find out what happens after the invasion of alien species occurs and also to study the effectiveness of various guidelines set up by countries around the world.

Government websites and those of reputed and recognized organizations were accessed between 1st November 2002 and 31st March 2003. Opinions of marine biologists on the subject were sought though e-mail communication.

Although, there are various modes of introductions of alien species, ship mediated introductions are most important among them. Due to the introduction of larger and faster ships and increased dependence on shipping, ship mediated introductions are of main concern. In this study, primarily ship mediated introductions of alien species are considered.

Different words are used to describe the species occurring in the ecosystems in which they are not indigenous, such as "non-indigenous", "non-native", "exotic", "alien", "pest", "new", and "foreign". A definition in the *Interim Guiding Principles for the Prevention, Introduction and Mitigation of Impacts of Alien Species* which was developed under the framework of the CBD (Convention on Biological Diversity), Subsidiary Body on Scientific, Technical and Technological Advice, is:

> "Alien species is a species occurring outside its normal distribution".

In biological terms, a species is considered to be native in its past or present natural range (the habitats and ecosystems where it lives or has lived) or within its natural dispersal potential (the area it can reach using its own legs, wings or wind/water-borne dispersal systems, or even if it is rarely found there) (IUCN, 2000). Where members of a species occur outside their normal distribution, they are considered to be alien in this new location. Some kind of human intervention is involved in introducing the species in new areas.

The scope of the term "species" also needs to be explained. Species should be interpreted to include subspecies and lower taxa, as well as any part, gametes, seeds, eggs or propagule of such species that might survive and subsequently reproduce (IUCN, 2000). The term "invasive" has no standard meaning. It is used interchangeably with terms like "pest" or "weed". These could be applied to both native and alien species. The common feature of such terms is the concept that these species have adverse impacts in the form of damage inflicted on the receiving species, site or ecosystem (IUCN, 2000). The Convention on Biological

Diversity says:

> "Alien invasive species is an alien species which threatens ecosystems, habitats or species".

Another definition given by IUCN is:

> "Invasive species means an alien species which becomes established in natural and semi-natural ecosystems or habitat, is an agent of change, and threatens native biological diversity".

There are various modes of introduction of alien species like ships' ballast water, ship fouling, aquaculture, public aquarium release, and private release by individuals. Ballast water and sediment constitute the most important vector for trans-oceanic and inter-oceanic movements of shallow-water coastal and marine organisms and approximately, 10 billion tonnes of ballast water are transferred each year (IUCN, 2000). A single ship is able to carry several hundred to over 100,000 tonnes of ballast water. About 3000 species are estimated to be transported in ballast water each day (IUCN, 2000). Even after a journey of several months of duration, several species of animals, plants and bacteria can survive in a viable form in ballast water. When ballast water carrying these species is discharged into the waters of port states, some of the harmful species may establish there and may pose threats to indigenous human, animal and plant life.

Ballast is important in ships carrying small or no cargo. The stability of ships is principally provided by the cargo when they are fully loaded. If a ship is empty or lightly loaded, the buoyancy and, therefore, the instability of a ship increase. By addition of weight to the lower part of a ship in the form of ballast, positive stability is attained by drawing the centre of gravity below the centre of buoyancy. Ballast water that is pumped in is coarsely filtered by a grid intended to exclude large objects and with a second screen having holes of 5 to 10 mm diameter fixed in the "mud box" for the elimination of smaller objects. There is no micro-filtration of ballast water to stop the intake of small marine organisms and sediment. Ships have the capacity to carry large volumes of ballast water, the amount varying from ship to ship. The volume of the ballast water is decided by the captain of the ship according to the buoyancy requirement at any particular time.

There are 4 main phases in the translocation of organisms: take-up; transit; release; and establishment. When water ballast is pumped on-board a ship, organisms present in the water may also be taken onboard. Additionally, organisms may also be present in any sediment that is taken up with the ballast water. Depending on the size of the vessel and, therefore, on the volume of the ballast water intake, such organisms can range in size from micro-organisms such as bacteria and viruses to a fish.

Some ships have deposits of mud in their ballast tanks of up to 10 cm in depth that is a good medium for some organisms to grow and many species may be transported in ballast tank sediment (Schormann *et al.*, 1990). Other than providing a food source, sediments create habitats for potentially viable resting life stages of organisms such as toxic dinoflagellate spores.

The release of ballast water unavoidably liberates organisms and sediment that have been carried in the tanks from source locations. Factors that determine the distribution of a species are the higher and lower limits of its temperature tolerance, the time period in which the ambient temperature is favorable for reproduction, and presence of a suitable environment (Walford & Wickfund, 1973). Other factors include salinity of water, habitat, predators and food sources. If the physical properties of the new environment are similar to those of the original one, the introduced species may well survive and multiply (Walters, 1996). A United States study found over 400 species of plankton and bottom living organisms in the ballast tanks of Japanese ships over a period of four years (EPA, 1995).

More than twelve marine and brackish water phyla have been recognized in ballast water. The most regularly occurring are nematodes (round-worms), rotifers, polychaetes (bristle-worms) and crustaceans. Other organisms that are found are ciliates (protozoans), fish and larvae of mussels and snails (Jansson, 1994). A study of foreign ballast water in more than 70 vessels arriving at Chesapeake Bay (United States) found that at least 90% of the vessels carried live organisms. Additionally, schools of juvenile fish were found in the ballast tanks of ships entering Chesapeake Bay from the United Kingdom, the Netherlands and Israel (Chesapeake Bay Commission, 1995).

There are two main types of marine fouling organisms. There are micro-organisms such as bacteria, diatoms and algae that form the "primary slime film". Secondly, macro-fouling species such as molluscs, sea squirts, sponges, sea anemones, bryozoans, tube worms, polychaetes and barnacles. The risk of translocation of exotic marine species via hull fouling may be as great as that associated with ballast water discharge. Organisms may be introduced by means of fouling of hulls and other parts of the ships, for example, propellers, and underwater discharge and suction openings and their gratings (Walters, 1996). Fouled wharf pylons offer a source of larvae of sessile fouling organisms in close proximity to berthed ships. These larvae may attach onto vessel hulls forming and adding to hull fouling communities. Mobile organisms may nestle in fouling encrustations and could be transferred from one place to another. Additionally, when a ship is berthed, it may bump against the wharf, dislodging fouling organisms from its hull into the water. Organisms and larvae originating from encrustations on a fouled vessel may settle on nearby pylons or other ship's hulls (Walters, 1996).

Organisms may also be released from other areas of the ship such as pipes, anchor chains and wells (Walters, 1996). Introduction of organisms by fouling may be facilitated by underwater or in-water cleaning of hulls. Throughout in-water cleaning, divers use semi-hydraulic equipment to remove fouling encrustations from the hulls of berthed vessels. The removed fouling is allowed to sink to the seabed where a number of species could continue to exist and successfully establish new populations. The process of hull cleaning may also lead to the release of spores and gametes from crushed organisms.

A large number of fouling organisms produce eggs or spores and spread them in the water column for external fertilization or they are released as brooded larvae. Planktonic larvae can remain viable in the water from a few hours to several weeks or months. By this manner, fouling species may be transferred by ballast water during the larval stage of their life cycle.

Apart from the introduction of invasive species, fouling causes an increase in the surface roughness of ship and greater frictional resistance. A slime layer of merely 1 mm thickness can cause a 15% loss in ship speed (Lewthwaite, 1985). This increases operating costs through higher fuel consumption and the inability of vessels to meet time limits. Evaluation shows that for a fouled 50,000 DWT tanker, the average speed loss at full power would be 2 knots, which if the ship was sailing at 13 knots would result in a 39% increase in fuel consumption (Ship Care & Maritime Management, 1991). This, in 1995, amounted to an extra cost to that ship in the order of US$ 340,000 in fuel alone (Donaldson, 1995).

Concern over impacts of species transfers has motivated a large amount of research on ballast water and aquatic invasive species globally since the early 1990s. The need for innovation in ballast water management has brought together national governments, international organizations, the shipping industry, research institutions, resource-based industries, policy specialists, and treatment vendors, along with their respective research priorities (Mays, 2001). Some efforts have been made to collect much of this research information into centralized locations (GloBallast, 2001; AIRD, 2001).

Mays (2001) assessed the amount of funding that has been used on ballast water research projects. He evaluated the collective direction and impact of that investment. Various organizations are attempting to attract global attention to this problem. As part of the Global Invasive Species Programme, development of a *Global Invasive Species Database and Early Warning System* is underway. This is an easy to use database and it provides wide-ranging information on alien species, their taxonomy and ecology, their native and invaded distributions, reports on control and eradication, and contacts and references for further information. The 'early warning system' will predict potential new invasion sites by a comparison of habitat types (IUCN Invasive Species Specialist Group.[1])

The International Maritime Organization (IMO), in cooperation with the Global Environmental Facility (GEF) and the United Nations Development Programme (UNDP), has started a new program called *Global Ballast Water Management Programme* or "GloBallast". This programme is providing US$ 10.2 million to help developing countries to decrease the transfer of marine alien species and protect their coastal and marine resources from marine alien species (Ballast Water News, Issue 1, 2000).

More than US$34 million has been used on the 108 projects which report funding information, resulting in an average of US$318,923 per project. Analysis shows that the United States is the country at the forefront of ballast water research with nearly 50% of the research projects. The U.S. is also leading the field in posted collaborative research projects with other countries and regions, and is partnered in 85% of projects involving multiple countries/regions.

Mays (2001) reported that the number and variety of research organizations and institutions undertaking ballast water research is wide. Parties being impacted by the introduction of aquatic invasive species are represented in these research studies. However, such representation is not evenly distributed among the parties concerned. Universities, government and independent research organizations make up the bulk (75%) of all research.

[1]URL: http://www.issg.org

The International Maritime Organization (IMO) and government departments such as the U.S. and Canadian Coast Guards, Australian Quarantine and Inspection Service (AQIS) and the Research Council of Norway are playing lead roles in providing funds to their respective regional programs (Mays, 2001).

The mpacts of the invasive marine species can be broadly categorized into ecological, economic, and human health impacts.

Ecological impacts

Invasive alien species have the potential to seriously disrupt ecosystem relationships and environmental processes, threatening native species and the integrity of the natural environment, human health, aquaculture sites, tourism and enjoyment of coastal amenities (Walters, 1996). Table 12.1 presents examples of the ecological impacts caused by alien invasive species.

Economic impacts

The Zebra mussel, for example, has caused instabilities in aquatic ecosystems including declines and extinctions of native mussel populations and infestations have also resulted in clogged municipal and industrial water intake pipes, such as the intakes to thermal and hydro-electric stations. It was anticipated that by the year 2000, the costs of measures against the mussel were in the order of US$5 billion (De Klemm, 1994). Others estimate that the cost of keeping water intakes from the American Great lakes free of mussels will be in the range of US$2-4 billion per year (Lodge, 1993). According to the Commonwealth Science and Industrial Research Organization (CSIRO) of Australia, the economic impacts of zebra mussel were (only on remedial engineering, cleaning pipes and water systems heavily encrusted with the mussel) estimated to be around US$600 million a year.

Apart from the expenditure required to mitigate and control the impacts, more than US$52 million has been spent on ballast water research projects worldwide (an average of US$ 319,000 on each individual research project) (Mays, 2001). The number of international and domestic ballast water research projects has been increasing over the last decade. A significant increase in the number of projects occurred following National Invasive Species Act (NISA) authorization in 1996 and in anticipation of the national ballast management program becoming mandatory. From an international point of view, the IMO's plan to introduce a convention

Table 12.1. Ecological impacts of exotic marine species (Adapted from Report on Ballast Water and Hull Fouling in Victoria.)

Reduction of biodiversity	As indigenous species decline in response to aggressive invading species, populations become increasingly homogenous. The loss or reduction of native species' populations may occur due to predation by exotic species; competition for food and habitat; and the introduction of new diseases.
Genetic changes to populations	Inside a species, populations from separate locations may be genetically dissimilar to each other. Invasion of a species between ports may result in a change in the genetic structure of the population of the destination. This may change the sensitivity of the population to disease, or affect its ability to adapt to future changes in the environment (Walters, 1996)
Physical effects on the environment	Invasive species may change the physical environment, altering the physical properties of substrata or the sediment chemistry, or affecting light penetration. These changes consecutively may hamper or enhance the survival and reproduction of other species in the community.
Eutrophication	Exotic species may disturb the ecological processes that help to maintain the balance of nutrients in a marine system. When excessive levels of nitrate and phosphate are present in the water, frequent algal blooms may occur (eutrophication). Overgrowth of algae depletes oxygen levels in the water, which may cause the death of fish and larger aquatic plants. Nutrient enrichment is one of the most serious threats to the aquatic ecosystems of the world (EPA, 1995). Other impacts of algal blooms include blocking of pipes and the production of odors and toxins.

on ballast water also appeared to encourage research. Mays (2001) notes that the countries leading the research are those which are facing severe ecological and environmental impacts caused by the introduction of ship-mediated aquatic invasive species.

Considering Singapore's dependence on shipping, it is important for the country to pay greater attention to this problem. The impact of aquatic invasive species has been estimated to cost the U.S. economy billions annually (ANS Task Force, 2001). To address the issue, many individual states in the U.S. are adopting state-based mandatory and voluntary ballast management programs. Bearing in mind the small populations of Australia, Canada and New Zealand, the level of participation in research on the subject in those countries is remarkable.

The economic and industrial structures of these countries indicate that they are dependent on the shipping industry. About 98% of all Australian commodities is transported by coastal shipping (AQIS, 2000). Introduced marine organisms such as toxic dinoflagellates have the potential to threaten the seafood industry. In 1992, France banned scallops from Japan due to the presence of paralytic shellfish poisoning (PSP) toxins from toxic dinoflagellates. Japan has also banned abalone from Spain because of PSP toxins (Report on Ballast Water and Hull Fouling in Victoria). In the U.S. state of Massachusetts, in 1988, 16 whales died after eating Atlantic mackerel that contained small traces of dinoflagellate toxins (Beckmann, 1988); this affected the tourism industry. Similarly, in the U.K., the toxic dinoflagellate *Alexandrium tamarense* has led to bird kills through the ingestion of mussels and sand eels (Campbell, 1994).

Health impacts

Even if some of the impacts of alien marine species are ignored by some countries, human health impacts cannot be ignored. Invasive marine organisms can cause more immediate hazards; they may cause death or illness in humans and marine animals. Long-distance dispersion of human pathogens is possible through ballast water. Ballast water may be important in the worldwide distribution of microorganisms, as well as for the epidemiology of waterborne diseases affecting plants and animals (Ruiz *et al.*, 2000).

Singapore is an island nation located south of Peninsular Malaysia. The total area of Singapore is approximately 650 km^2 and there are about 60 offshore islands. Singapore is highly maritime by its nature and this city state is greatly dependent on its marine assets including its strategic location on a main international sea-lane for commercial shipping. The coastal area has played an important role in Singapore's economy and development. Due to good planning and the innovative approach of the government, Singapore has set an example of phenomenal growth in a comparatively short time. The top fifteen of the ports in the Asia Pacific region (in terms of total ballast water discharge) represented only 6% of all identified ports (251), but accounted for 44% of all overseas arrivals and

30% of the total ballast water discharge. With regard to total discharge, Singapore is second in the list.

Chou and Lam (1989) observed that no studies have been undertaken on the ecological significance and implications of introduced species in Singapore. They noted that matters of species invasion and their effects on the environment have not been given attention, but demand systemic investigation. Ng *et al.* (1993) observed that there was no documentation on introduced species in Singapore and no detailed local studies had been undertaken. An alien mussel, *Mytilopsis sallei*, which has shown devastating impacts in few other countries, has been found in the Singapore waters (Ng & Sivasothi, 1999; CSIRO, 2000). Holmes and Teo (2002) discovered a toxic dinoflagellate, *Gymnodinium catenatum*, in the Singapore waters and believed that "Singapore has one of the busiest ports in the world, which makes Singapore a potential node for the import and export of nuisance marine species".

In Singapore, *Mytilopsis sallei* is an introduced bivalve from Central America and the Caribbean (Ng & Sivasothi, 1999). In many parts of Singapore, these exotic bivalves have colonized the walls of intertidal monsoon drains (Tan & Chou, 2000). They are similar to the zebra mussel (*Dreissena polymorpha*) from Europe which invaded waterways in the U.S., and created problems by attaching themselves to the internal walls of cooling systems and choking them. They can tolerate a wide range of salinities and are found in many estuarine habitats. *Mytilopsis sallei* were possibly introduced to this part of the world through ballast water of ships coming from the Caribbean and Central America (Goh *et al.*, 2002). These mussels are known for their ability to survive in fluctuating adverse conditions and may outnumber the native species for food and space.

The Centre for Research on Introduced Marine Pests (CRIMP) (CSIRO, 2000) also noted that the black-striped mussel (*Mytilopsis* spp.) has been found in Singapore, Taiwan, Japan and Fiji; and that this mussel is believed to have invaded an Indian port in the 1960s. The most probable vector for its spread is on the hulls of commercial or recreational vessels. This mussel attains a maximum length of around 25 mm and is highly productive; tens of thousands of eggs are released at each spawning. Its productivity can reach $100\,kg/m^2$per year, and it has the capacity to outnumber native marine communities (CSIRO, 2000).

Holmes *et al.* (2002) identified the Paralytic shellfish poisoning (PSP) toxin producing dinoflagellate, *Gymnodinium catenatum*, in Singapore waters. This dinoflagellate has been found in several phytoplankton samples since 1997 (Holmes & Teo, unpublished data). Dinoflagellates are microscopic, mostly unicellular algae that inhabit fresh and marine waters. They constitute an important part of aquatic phytoplankton. Among microalgae, dinoflagellates are the most important toxin-producing group. They produce a large variety of toxins that accumulate through marine food chains. Events associated with the poisonings from dinoflagellates are called harmful algal blooms, or "red tides".

Seafood is an important component of the diet of Singaporeans. Previously, *Gymnodinium catenatum* PSP contamination of shellfish was recorded mostly in temperate waters, for example in southern Japan, the Atlantic coast of Spain and Portugal, the Mediterranean coast of North Africa, southern areas of Australia, and the Pacific coast of Mexico, New Zealand and South America. During their life cycle, *Gymnodinium catenatum* produce cysts that are highly resistant to environmental stress and this may have helped the global spread of species through the ballast water of ships (Hallegraeff, 1998).

Another non-toxic species of dinoflagellate, *Gymnodinium impudicum,* was also discovered in Singapore waters in 2000 (Holmes & Teo, unpublished data). This was the first record of that organism in South-east Asia.

Three types of human poisonings associated with dinoflagellate species found in Singapore waters have been recorded in the country. The most important of these is PSP (Holmes *et al.*, 2002). Shellfish containing $0.8\,\mu g$ or more PSP toxin/g edible tissue is considered unfit for human consumption and in the Philippines this limit is set at only $0.408\,\mu g$ in order to protect children (Holmes & Lewis, 2002). Public health can be protected by adopting rigorous testing programmes but shellfish growers and seafood industries suffer a loss due to inability to harvest or sell contaminated products (Holmes & Teo, 2002). In the 1980s, PSP occurred in Singapore, and five people suffered toxicity, of which two died (Tan & Lee, 1987).

A second type of toxicity from dinoflagellates is Diarrhetic shellfish poisoning (DSP). Fish that contains more than 0.2 DSP toxins/g shellfish meat is considered unfit for human consumption (Holmes & Lewis, 2002). Holmes, *et al.* (1998) reported the presence of DSP toxin in tropical shellfish in Singapore waters and predicted that the disease might occur in Singapore and South-east Asia.

The third type of toxicity from the dinoflagellates is ciguatera. It happens due to the consumption of contaminated finfish. More than

25,000 people are affected by this poisoning every year (Lewis, 2001). Dinoflagellates *Gambierdiscus* produce ciguatoxins that are responsible for the poisoning. One species, *Gambierdiscus yasumotoi*, is a new species that has been found in Singapore waters (Holmes, 1998).

Legislation

The Maritime and Port Authority of Singapore (MPA) safeguards the country's maritime interests in the international arena. MPA is responsible for the overall development of the shipping, port, navigational safety, management of vessel movements and management of the marine environment in the port. It also controls the licensing and regulation of marine and port services and facilities. The introduction of invasive marine species is not covered in the regulations. There is no restriction on the discharge of ballast water in Singapore waters. At the Singapore ports, vessels have to follow the guidelines for discharging clean ballast specified by the "Clean Seas Guide for Oil Tankers", published jointly by the International Chamber of Shipping (ICS) and the Oil Companies International Marine Forum (OCIMF). The Guide provides that only clean ballast or segregated ballast may be discharged within 50 miles of the nearest land or within a special area. Ballast water discharged from segregated ballast tanks should be visually checked prior to, and during, discharge for the presence of any oil.

In the port information section of "Year 2003 Singapore Tide Tables and Port Information" published by Hydrographic Department of the MPA, ballast water related regulations or procedures are not mentioned. Some legislation and other measures have been initiated to prevent the entry of introduced species. Under the Fisheries (Piranhas) Rules 1971, the importing of piranhas is banned, and the rearing or release of fish into natural waterways is illegal (Chou & Lam, 1989). These laws cover only few freshwater species on the island and fail to protect the marine environment.

There are signs of increased awareness among ship operators. The Singapore-based "World Tankers Management" is the first tanker operator in Singapore to have its fleet certified under Lloyd's Register's new Ballast water management plan (BWMP) (Lloyd Register, 2002). This indicates the ship owner's approach to meeting the highest standards of ballast water management, without waiting for mandatory requirements in legislation. Although, it seems that the concern of ship owners is mainly the safety of their ships during ballast management procedures, it shows that they are ready to comply voluntarily with any future regulations.

Lloyd's Register is recognized as the organization that can provide ballast water management expertise and solutions. In addition to certification and advisory services, it provides industry-wide training and has many ballast water management workshops around the world. Ship owners of some other Asian developing countries such as India are showing their concern. The Great Eastern Shipping Co. Ltd of India became the largest private-sector shipping company in the country to receive Lloyd's Register's Ship Right BWMP descriptive note; it has received certification for its tanker and gas carrier fleets, a total of 17 ships.

Researchers in Singapore are concerned with this problem. A study was carried out by the Environmental Technology Institute (ETI) in collaboration with the MPA and the National University of Singapore (NUS) on the evaluation of ballast water treatment technologies (*Business Times*, 29th March, 1999). The research started with a preliminary evaluation of various technology choices that were suitable for shipboard operations. Three technologies were selected for pilot-scale investigations: media filtration, screen filtration and biocides applications. A dock-side pilot plant, constructed at Sembawang, has been operational from February, 1999. The study gave promising results for screen filtration systems but, it still requires testing on ships.

Impacts

The black-striped mussel that has already invaded Singapore was introduced into India in the late 1960s, since then it has had impacts similar to those of the zebra mussel in North America (CSIRO, 2000). It is difficult to predict impacts but preliminary predictions are that it may cause fouling of wharves, marinas, recreational and inshore vessels, seawater systems (such as mariculture pumping facilities, vessel ballast and cooling systems) and marine farms (such as those for pearl oysters). Ecological impacts are more complicated to forecast, but are expected to be extensive. In onshore, low estuarine habitats, introduced populations of black-striped mussels may form virtual monocultures of 10–15 cm thickness.

It is difficult to project potential economic impacts of species invasion in Singapore waters. An indication of potential disaster can be perceived from an examination of their impacts in other countries. When black-striped mussel was first found in Australia in 1999, the following actions were taken (CSIRO, 2000):

> "The Northern Territory Government invoked Natural Disaster legislation and has quarantined the Cullens Bay, Tipperary and Frances Bay marinas. National Black-Striped Mussel Task Force (NBSMTF) was established with a Scientific Subcommittee to advise on response options, including treatment of infected sites and the tracking and cleaning of infected vessels. Chemical treatment of the Cullen Bay marina using chlorine was begun immediately. Port and marina records indicate some 400 vessels which have visited the three infected areas of the Port of Darwin area at sea, either returning to Darwin round for other Australian international ports. Joint state commonwealth efforts are underway to track these vessels and treat any that are found to be infected with the mussel. Methods developed to control zebra mussels in North America are being adapted by Northern Territory and CRIMP scientists and applied to the black-striped mussel. Response to the invasion has been coordinated initially by the Northern Territory Government; the Scientific Subcommittee of the NBSMTF includes representatives from CSIRO, the Northern Territory Museum and the West Australian Museum. As the effort has broadened to include attempts to track and treat potentially infected vessels in the domestic recreational and fishing fleet, an ad hoc Commonwealth/state coordination process has been set up".

As mentioned earlier, another alien species found in Singapore is the toxic Dinoflagellate, *Gymnodinium catenatum*. Singapore has already experienced its toxicity (Tan & Lee, 1987). Algal blooms of Dinoflagellates may also affect the tourism industry.

Potential impacts of pathogenic microorganisms are more alarming. Invasive marine organisms can cause direct hazards; and they may cause death or illness in humans and marine animals. As most of the pathogenic microorganisms are not found naturally in the sea water, their introduction may have adverse impacts. Due to fast pace of shipping, this may pose an immediate threat to human health.

Solutions

It is recognized that once introduced and established in an open aquatic system, exotic species are extremely difficult to eliminate (Carlton, 1993).

Therefore, the best solution is the prevention of the introduction of invasive species into the native waters. Various organizations have been seeking approaches to reduce the impacts. The IMO has been working to find such solutions since the mid-1970s. In 1997, the IMO assembly adopted the *Guidelines for the Control and Management of Ships' Ballast Water to Minimize the Transfer of Harmful Aquatic Organisms and Pathogens.* The guidelines are to assist Governments and relevant authorities, ship masters, operators and owners, port authorities and other interested parties, in minimizing the risk of introducing harmful aquatic organisms and pathogens from ships' ballast water and associated sediments while protecting ships' safety. Many of the stakeholders are participating in, and funding research projects. For example, the shipping industries are contributing to the search for knowledge, regulation and treatment. For instance, under California's Ballast Water Management and Control Program, qualifying voyages must pay a fee which funds the program (Faulkner, 2000) and AQIS's Ballast Management Program was partially funded by a levy on shipping (AQIS, 2000).

Governments are working with research institutions, private and non-profit institutions and universities to address ballast water management. This has enabled research on different aspects of the matter to be undertaken, and resources from several sources to be exploited to support research and implementation.

It would be more effective if researchers worked in co-operation. For example, in the Great Lakes Ballast Technology Demonstration Project, researchers from government, universities and industry work together (Cangelosi, 1997). Such collaborative work leads to quality research as experts in the experimental field are able to draw on the resources, knowledge and experience of the shipping industry (Cangelosi, 1997). This allows both parties to benefit from the results of the research (Mays, 2001). The relationship between the shipping industry, government and researchers will continue to be bolstered by further research, particularly on-board testing of prevention and treatment systems (Mays, 2001). The shipping industry's input is important in making the transition between voluntary and regulatory ballast management procedures more acceptable.

Non-experimental research that gives biological and ecological baseline data of aquatic invasive species has also been conducted, but it is decreasing in prominence relative to other forms of research. Baseline monitoring is vital for determining the extent of introductions, their history, distribution and impacts (Mays, 2001). It is also helpful for interpreting the features that make up a "successful" invader, and in the risk evaluation of ballast water transport.

Full-scale design studies of ballast treatment systems and operating vessel trials are also vital in expediting the search for successful prevention methods. A big fraction of operating vessel trials and in-port ship sampling procedures still involves research related to biological impacts rather than prevention or treatment (Mays, 2001).

The pressing need for a solution to the aquatic invasive species problem is further evident in the types of research being conducted. Current research is focusing on the prevention/treatment aspects of ballast water, rather than on the assessment, management and policy issues. As the issue of invasive species is comparatively new, baseline information will help to make policies for the prevention and control of this problem.

There is a need for an accurate scientific and technical knowledge base to help government and authorities to make the right decisions on introductions, to assign limited resources, and to implement effective control measures (IUCN, 2000). Countries such as the U.S., Australia, New Zealand, and UK have adequate information to evaluate the situation. Singapore still lacks sound baseline information on its marine biodiversity. Complete statistics would help in building up the political will for new legislation. It would be helpful to fill the gaps in basic information and knowledge on local biodiversity, including taxonomic information on local species, their status and distribution and the extent to which they are threatened by alien invasive species (IUCN, 2000).

To prevent the introduction of exotic species in oceans, a series of national and international conventions have been introduced that largely urge a "no introduction policy". These include the 1973 "code of practice" of the International Council for the Exploration of the Sea (ICES), the 1973 position paper of the American Fisheries Society, the 1984 recommendations of the Council of Europe, and the International Union for Conservation of Nature and Natural Resource's (IUCN) "Translocation of Living Organisms" position statement in 1987 (Carlton, 1989).

An article in the "Convention on the Law of the Sea" states:

> 'States shall take all measures necessary to prevent, reduce and control... the intentional or accidental introduction of species, alien or new, to a particular part of the marine environment, which may cause significant or harmful changes thereto'.

Article 22 of The United Nations Convention on the Law of the International Watercourses is devoted to the problem of the introduction of alien species into international watercourses. It provides as follows:

> 'Watercourse states shall take all measures necessary to prevent the introduction of species, alien or new, into an international watercourse which may have effects detrimental to the ecosystem of watercourse resulting in significant harm to other watercourse States'

A number of major port states, such as Australia, Canada and the USA have introduced regulations intended to prevent ships that come into their waters from discharging ballast water that contains 'non-native' aquatic species. In compliance with these regulations the master of a ship is expected to show that he has taken steps to prevent such discharge by a method that is acceptable to the relevant port state.

Although, currently there are no internationally agreed regulations covering ballast water management, in 1997 the IMO published 'Resolution A 868 (20) Guidelines' for the Control and Management of Ship's Ballast Water to minimize the transfer of harmful aquatic organisms and pathogens. Those countries requiring ships to exchange or treat ballast at sea prior to entering port, coastal or state waters have generally adopted these guidelines. IMO guidelines require a ballast water management plan (BWMP) specific to each ship, and identify the procedures to be implemented for the effective management of ballast water on board the ship.

To make invasive species and ballast water regulations effective, it is necessary to provide an extensive amount of information related to this subject. There is a need for a strategic plan for research on the invasive species problem. At global level, the 1995 'FAO Code of Conduct for Responsible Fisheries' explains principles and standards for responsible fisheries practices that are planned to ensure the effective conservation, management and development of living aquatic resources, with appropriate consideration for the ecosystem and biodiversity. Moreover, to develop national ballast water management regimes, the U.S., Australia, the European Union and New Zealand are making efforts to decide on the most feasible international ballast water prevention, treatment and management solutions (Mays, 2001). All these countries are major supporters of ballast water management from an international viewpoint with the support of the IMO.

In-water hull cleaning has been banned by the Victorian Channels Authority since 1997. The ban is applicable to ships over 200 gross tonnes in the waters of Melbourne, Geelong, Hastings and Portland. It also permits the cleaning of sea chests, sea suction grids and other apertures provided the debris does not enter the water column. In a dry-docked ship the hull is typically cleaned by high pressure water jets and grit blasters (Walters, 1996). In many countries, it is a legal obligation that ships be dry-docked for cleaning, repair and maintenance of hulls at least once every five years. The operators of these ships are also required to dispose of hull cleaning waste at appropriate on-shore sites.

The procedure of implementing state, national and international ballast water and aquatic invasive species regulations is difficult. According to the IUCN (2000), 'planning and decision making procedures on alien species issues are complex and may require the involvement of governmental and non-governmental stakeholders in different sectors and at all levels'.

Existing options to avoid the spread of aquatic organisms through ballast water include exchanging the ballast water in the deep ocean where there is less marine life and non-deep water organisms are less likely to survive. Other methods include various types of treatments of ballast water to destroy the living organisms. These include filtration, thermal, chemical and radiation treatments of ballast water. The other options are as follows: (i) non-release of ballast water; (ii) taking on only clean ballast; (iii) treating the ballast water en route (such as by heating, with chemicals like chlorine, or ultraviolet radiation); (iv) depositing the ballast water in special reception tanks at port; and (v) biological treatment through the addition of predatory or parasitic organisms to ballast water.

At present, reballasting at sea seems to be the best-available solution to reduce the risk of transfer of invasive aquatic organisms (GloBallast, 2002). This technique is being practiced by some of the countries but it is not completely effective in removing organisms from ballast water (GloBallast, 2002). Many authors argue that reballasting at sea may itself help the dispersal of harmful species, and that island states located 'down-stream' of mid-ocean reballasting areas may be at particular risk from this practice (GloBallast, 2002). Most of the research work for ballast water is concentrated on developing an effective and economic method to treat the water.

The maneuvering of a ship can be risky, and ballast water treatment methods should not interfere with the safe and economical operation of the ship. Any control measure that is developed must meet a number of criteria

(GloBallast, 2002), including: it must be safe; it must be environmentally acceptable; it must be cost-effective; and t must work.

The shipping and scientific communities believe that shipboard treatment technologies may offer a medium to long-term solution for managing the ballast water problem. Still, the efforts to develop a cost-effective shipboard treatment technology faces enormous challenges, mainly due to the huge amount of water involved and the limited space available onboard ship.

To prevent hull fouling, tributylin (TBT) anti-fouling paints are used; these, in themselves, pose threats to the environment due to their toxicity. TBT is one of the most toxic substances ever deliberately introduced into the ocean (Schrantz, 1990). In many countries, deleterious effects of TBT on oysters and snails have been recorded. As a result, most countries have banned the use of TBT paints on small vessels (as they are more numerous in estuaries) and on marine and estuarine structures such as piers and buoys. Japan, New Zealand and the USA have imposed a ban on the use of TBT. The IMO wishes to phase out the use of TBT paints around the world within 10 years. Researchers are trying to discover a more eco-friendly substitute.

Few countries have imposed a ban on in-water hull cleaning. This ban reduces the risk of alien species introductions, but it does not eliminate the risk of transfer, as fouling organisms may still transfer from ships' hulls to nearby wharves and pylons; drop to the sea floor; or release spores, eggs or larvae into the water column. The disposal of fouling material from shipyards should be properly managed and there should be monitoring of the levels of fouling on incoming vessels.

Various governments and reputable organizations appear to be doing their best to make available the information regarding invasive marine species. For example, there is an overwhelming number of websites on the subject. Several research projects are underway in different countries. In Singapore, most of the work has been carried out on freshwater introductions (Alfred, 1961; Johnson, 1973; Ng *et al.*, 1993; Chou & Lam, 1989). The authorities concerned need to respond before it is too late to act. The government and researchers should involve the public during the planning process. Such participation has educational benefits and it leads to a rise in public awareness.

To address the problem of invasive marine species, many initiatives and activities are being undertaken throughout the world. These activities involve many disciplines such as marine science, marine resource management, human health, engineering, ship design, ship building, shipping operations, research and development, economics, law and education, training and communications (GloBallast, 2002). In Australia CSIRO has a separate unit, the Centre for Research on Introduced Marine Pests (CRIMP). The CSIRO's national port survey program started in 1995 and its purpose is to provide baseline information on the status of introduced species in Australia's trading ports. This program is supported by port authorities, state agencies and the Australian Ballastwater Management Council.

Conclusion

There is sufficient evidence and data to confirm that invasive species pose a great threat to marine ecology, the economy and human health. Due to increased shipping, ballast water has become one of the key sources of dispersal of non-indigenous marine species. Most developed countries such as the US, Australia, New Zealand and UK, are making efforts to deal with the problem. Many affected countries are spending large amounts of money to mitigate the impacts caused by invasive marine species in their waters. A number of research projects are underway around the world. The IMO has responded by supporting research projects and it has proposed a new convention on ballast water. Other organizations, including non-governmental ones, are actively working to deal with the impacts.

At the time this study was undertaken, there had been no significant research work in Singapore to identify the invasive marine species and to assess their potential ecological, economic and health impacts. Singapore's waters have already been invaded; the presence of some invasive marine species such as alien mussels and toxic dinoflagellates has been noticed. Singapore is one of the busiest ports of the world and more than 400 ships arrive in, or leave, the country's waters every day. Each year, millions of gallons of ballast water are discharged, and this makes Singapore vulnerable to the potential impact of invasive marine species. No regulations are imposed in the country to control ship-mediated invasion of exotic marine species in its waters or regarding the discharge of ballast water by ships.

Some awareness has been shown by a few ship owners regarding ballast water management. Some research projects on ballast water treatment technology are in progress. Researchers, government agencies and shippers should respond to prevent the invasion of alien marine species in

Singapore waters. Singapore should act proactively as several countries have unilaterally adopted compulsory regulations to minimize the risk through ships entering their ports.

References

Alfred, E.R. (1961). Singapore freshwater fishes. *Malayan Nature Journal*, **15**: 1–19.

Alpine, A.E. and Cloern, J.E. (1992). Trophic interactions and direct physical effects control phytoplankton biomass and production in an estuary. *Limnology & Oceanography*, **37**: 946–955.

ANS Task Force (2001) What Are Aquatic Nuisance Species and Their Impacts? www.anstaskforce.gov/moreimpacts.htmAQIS (2000).

The Aquatic Invasions Research Directory (AIRD) (2001). Smithsonian Environmental Research Center. http://invasions.si.edu/aird.htm

Australian Quarantine and Inspection Service (2000) Australian Ballast Water Program Overview — AQIS.
http://www.aqis.gov.au/docs/ballast/ bprogram1/htm

Ballast Water News , Global Ballast Water Management Programme. Issue 1, 2000. http://globallast.imo.org/index.asp?page=newsletter.asp&menu=true

Beckmann, R. (1988). Blooms and gloom. *Ecos*, **57**: 19–23.

Campbell, D. (1994). Translocation of exotic marine organisms in ballast water-Why worry? *In*: Ballast Water Symposium Proceedings, 11–13 May, 1994, AGPS Canberra.

Cangelosi, A.A. (1997). Battling biological pollution. *NEMW Economic Review*, May/June; Northeast-Midwest Institute, Washington D.C.

Carlton, J.T. (1989). Man's role in changing the face of the ocean: Biological invasions and implications for conservation of near-shore environments. *Conservation Biology*, **3**: 265–273.

Carlton, J.T. (1993). Biological invasions and biodiversity in the sea: The ecological and human impacts of nonindigenous marine and estuarine organisms. *In*: Proceedings of the Conference and Workshop, Nonindigenous Estuarine and Marine Organisms (NEMO) Seattle, Washington, April 1993, U.S. Department of Commerce.

Carlton, J.T. (1996). Marine bioinvasions: The alteration of marine ecosystems by nonindigenous species. *Oceanography*, **9**: 36–43.

Carlton, J.T. and Geller, J.B. (1993). Ecological roulette: Biological invasions and the global transport of nonindigenous marine organisms. *Science*, **261**: 78–82.

Carlton, J.T., Thompson J.K., Schemel, L.E., and Nichols, F.H. (1990). Remarkable invasion of San Francisco Bay (California, USA) by the Asian clam *Potamocorbula amurensis*. I. Introduction and dispersal. *Marine Ecology Progress Series,* **66**: 81–94.

Chesapeake Bay Commission (1995). The introduction of nonindigenous species to the Chesapeake Bay via ballast water. Strategies to Decrease the Risks of

Future Introductions through Ballast Water Management, Chesapeake Bay Commission.

Chou, L.M. and Lam, T.J. (1989). Introduction of exotic aquatic species in Singapore, 91–97 pp. *In*: S.S. De Silva (Ed.) Exotic Aquatic Organisms in Asia. Proceedings of the Workshop on Introduction of Exotic Aquatic Organisms in Asia. Asian Fish. Society Special Publication, 3, 154 p. Asian Fisheries Society, Manila, Philippines.

Commonwealth Scientific and Industrial Research Organisation (CSIRO) 2000. http://www.marine.csiro.au

De Klemm, C. (1994). The introduction of exotic species and the law. *In*: Introduced species in European coastal waters. Boudouresque, C.F., Briand, F. and Nolan, C. (Eds.), European Commission Publication, Luxembourg: 85–92.

Donaldson, K. (1995). Figure calculated using bunker fuel prices for 1991 and 1995, ABARE.

Environment Protection Authority (1995). Protecting water quality in Port Phillip Bay, State Environment Protection Policy (Waters of Victoria) draft Schedule F6 (Waters of Port Phillip Bay) and draft Policy Impact Assessment, Publication 434, February 1995.

Faulkner, M. (2000). California's Ballast Water Management and Control Program Progress Report — September 2000. Ballast Exchange, vol. 3, Fall 2000.

Global Ballast Water Management Programme (Globallast), International Maritime Organization. http://globallast.imo.org

Goh, B.P.L., Sodhi, N.S., Daiqin, L., Hung, H.S. and Ng, P.K.L. (2002). A guide to urban creatures. Singapore Science Centre.

Griffiths, R.W., Schloesser, D.W., Leach, J.H. and Kovalak, W.P. (1991). Distribution and dispersal of the zebra mussel (*Dreissena polymorpha*) in the Great Lakes region. *Canadian Journal of Fisheries & Aquatic Sciences*, **48**: 1381–88.

Hallegraeff, G.M. & Bolch, C.J. (1992). Transport of diatom and dinoflagellate resting spores in ship's ballast water: Implications for plankton biogeography and aquaculture. *Journal of Planktonic Research*, **14**: 1067–1084.

Hallegraeff, G.M. (1998). Transport of toxic dinoflagellates via ships' ballast water: Bioeconomic risk assessment and efficacy of possible ballast water management strategies. *Marine Ecology Progress Series*, **168**: 297–309.

Harbison, G.R. and Volovik, S.P. (1994). The ctenophore *Mnemiopsis leidyi*, in the Black Sea: A holoplanktonic organism transported in the ballast water of ships. *In*: Proceedings of the Conference and the Workshop on Nonindigenous Estuarine and Marine Organisms (NEMO).

Holmes, M.J. (1998). *Gambierdiscus yasumotoi* sp. nov. (Dinophyceae), a toxic benthic dinoflagellate from southeastern Asia. *Journal of Phycology*, **34**: 661–668.

Holmes, M.J. and Lewis, R.J. (2002). Toxin-producing dinoflagellates. *In*: Menez, A. (Ed.), Perspectives in Molecular Toxinology. Wiley and Sons, Chichester. 39–65 pp.

Holmes, M.J. and Teo, L.M. (2002). Toxic marine dinoflagellates in Singapore waters that cause seafood poisonings. *Clinical and Experimental Pharmacology and Physiology*, **29**: 829–836.

Holmes, M.J., Bolch, C.J.S., Green, D.H., Cembella, A.D., and Teo, S.L.M. (2002). Singapore isolates of the dinoflagellates *Gymnodinium catenatum* (Dinophyceae) produce a unique profile of paralytic shellfish poisoning toxins. *Journal of Phycology*, **38**: 96–106.

Holmes, M.J., Lee, F.C., Teo, S.L.M., and Khoo, H.W. (1998). A survey of benthic dinoflagellates on Singapore reefs. *In*: Reguera, B., Blanco, J., Fernández, M.L. and Wyatt, T. (Eds.). Harmful Algae. Xunta de Glacia and International Oceanographic Commission of UNESCO, Santiago de Compostela. 50–51 pp.

IUCN-The World Conservation Union. (2000). A guide to designing legal and institutional frameworks on alien invasive species. Environmental Policy and Law Paper No. 40.

Jansson, K. (1994). Alien species in the marine environment- introductions to the Baltic Sea and the Swedish West Coast, Swedish Environmental Protection Agency.

Johnson, D.S. (1973). Freshwater life. *In*: Animal Life and Nature in Singapore, (Ed.) S. H. Chuang. Singapore University Press, Singapore, 103–127 pp.

Lewis, R.J. (2001). The changing face of ciguatera. *Toxicon*, **39**: 97–106.

Lewthwaite (1985). Cited in Gordon, D.P. and Mawatari, S.F. (1992), Atlas of Marine-Fouling Bryozoa of New Zealand Ports and Harbours, Miscellaneous Publications, NZ Oceanographic Institute, **107**: 1–52 pp.

Lloyd's Register (2002). http://www.lr.org/code/home.htm

Lodge, D.M. (1993). Biological invasions: Lessons for ecology. *Tree*, **8**: 133–137.

Mays, N. (2001). Ballast water research and funding: A history and analysis. Northeast Midwest Institute, Washington, D.C.

National Invasive Species Council (2002). http://www.invasivespecies.gov/council/main.shtml

NISA (National Invasive Species Act, 1996). http://www.nemw.org/nisa_summary.html

Ng, P.K.L. and Sivasothi, N. (1999). A guide to the mangroves of Singapore. II, Animal diversity. Singapore Science centre.

Ng, P.K.L., Chou, L.M., and Lam, T.J. (1993). The status and impact of introduced freshwater animals in Singapore. *Biological Conservation*, **64**: 19–24.

Report on Ballast Water and Hull Fouling in Victoria, Parliament of Victoria http://www.parliament.vic.gov.au/enrc/inquiries/old/enrc/ballast/Ballast-28.html

Ruiz, G.M., Rawlings, T.K., Dobbs, F.C., Drake, L.A., Mullady, T., Huq, A., and Colwell, R.R. (2000). Global spread of microorganisms by ships. *Nature*, **408**: 49–50.

Schormann, J., Carlton, J.T., and Dochoda, M.R. (1990). The ship as a vector in biotic invasions. IMAS, Marine Technology and the Environment.

Schrantz, J. (1990). TBT marine paint under EPA scrutiny. Industrial Finishing, Hitchcock Publishing Company, **66**: 58p.

Ship Care and Maritime Management (1991). Fuel savings from hull cleaning. *Ship Care and Maritime Management*, **23**, Oct/Nov 1991.

Tan, C.T.T. and Lee, E.J.D. (1987). Paralytic shellfish poisoning in Singapore. *In*: Gopalkrishnakone, P., Tan, C.K. (Eds.). Progress in Venom and Toxin Research. National University of Singapore, 429–432 pp.

Tan, K.S. and Chou, L.M. (2000). A guide to common seashells of Singapore. Singapore Science centre.

Vinogradov, M.E., Shushkina, E.A., Musayeva, E.I., and Sorokin, P.Y. (1989). A newly acclimated species in the Black Sea: The ctenophore *Mnemiopsis leidyi* (Ctenophora: Lobata). *Oceanology*, **29**: 220–224.

Walford, L. and Wicklund, R. (1973). Contribution to a world-wide inventory of exotic marine and anadromous organisms. FAO Fisheries Technical Paper No. 121, Food and Agricultural Organisation of the United Nations, Rome.

Walters, S. (1996). Ballast water, hull fouling and exotic marine organism introductions via ships- A Victorian study. Publication 494, Environment Protection Authority (Victoria).

13

A SOCIO-ECONOMIC ANALYSIS OF FISHERY SUSTAINABILITY IN SINGAPORE

GINA KUM and RUTH O'RIODAN

It is generally agreed that the global captured fishery is facing a crisis as a result of over-harvesting, which has taken place in the past few decades. Likewise, capture fisheries in the Southeast Asia region have been stagnating in recent years. Culture fisheries, on the other hand, are increasingly recognised as a potential sustainable source of production, both regionally and internationally. By examining both the capture and culture fish sources in Singapore and regionally and forecasting the future demand for fish in Singapore, the study seeks to determine whether the path of development and management of the fishery sector in Singapore would promote or undermine sustainable production, both locally and regionally. This study revealed some interesting insights into the patterns of consumption of animal protein, including fish, in Singapore. It has predicted that Singapore's dependency on imports from the three regional countries, Malaysia, Thailand and Indonesia will not be a major contributing factor that undermines a sustainable fishery in the region. The inadequacies of national laws and policies for the management of fisheries in Singapore were also analysed.

1. Introduction

There is a general consensus that the global captured fishery is facing a crisis, primarily as a result of over-harvesting which has taken place in the past few decades. Two infamous cases of fishery collapses, Peruvian anchoveta in 1972 and the Canadian Grand Banks Cod in 1992 have illustrated the possibility of a stock collapse of a particular species.[1] Recent research by Myers and Worm (2003) suggests that ninety percent of large predatory fish, such as the swordfish, marlin, and sharks, has been cleared from the seas in the past fifty years or so. The authors used data going back 47 years from four continental shelves and nine oceanic systems, ranging from the tropics to the Antarctic. Whether off the coast

[1]Schiermeier (2002) pp. 662–665 and Myers *et al.* (1996), pp. 629–640. http://www.oneworld.net/external/?url=http%3A%2F%2Fens-news.com%2Fens%2Fapr2003%2F-2003-04-28-05.asp

of Newfoundland, Canada, or in the Gulf of Thailand, the findings were alarming. Furthermore, other commercially important species, such as tuna, cod, halibut, skate and flounder are now generally far smaller in size.

Due to global concern, the World Development Summit agreement was reached during the Johannesburg summit held in 2002, to protect the world's fishing stocks. The aim is to restore most of the world's major fisheries to levels that can produce the maximum sustainable yield by the year 2015.[2] The main recommendations include permanent non-fishing zones to protect breeding areas by 2012, temporary fishing bans in areas experiencing serious depletion and action against illegal fishing.[3]

While the global catch may be at its highest, with the highest on record at 94.9 million tonnes in 2000, as compared to about 20 million tonnes in 1950,[4] forecasts of upper limits to capture fisheries are being increasingly substantiated by the evidence of recent years.

According to Grainger and Garcia (1996), the gap between demand and supply towards 2010 are not likely to be fulfilled by marine capture fishery production. Using trend analysis, they showed that most marine fishery resources were increasingly or fully fished and predicted that world fishery resources might be fully fished in 1999.

World culture production on the other hand, continues to expand. In the Southeast Asian region, the fisheries sector plays an important role in the region from a socio-economic point of view. It provides an important source of animal protein food and creates income and jobs.[5] The major production comes from Indonesia, Thailand and the Philippines for the 1991–1999 period. On the whole, fisheries in the Southeast Asian region have shown upward trends in capture and culture fish production. Since the introduction of modern fishing technology in the 1960s, fish production in the region has steadily increased. However, with short-term management objectives for increasing production levels rather than the long-term comprehensive and sustainable management of fisheries resources, the rate of increase in capture fishery production is now declining.[6]

Aquaculture is becoming an increasingly important and dependable source of fish in the region. But, bearing in mind the presence of

[2]http://www.johannesburgsummit.org/html/basic_info/basicinfo.html
[3]http://news.bbc.co.uk/1/hi/world/africa/2221116.stml
[4]See FAO (2002a) and FAO (2002b).
[5]ASEAN-SEAFDEC (2001a), pp. 18–42.
[6]ASEAN-SEAFDEC (2001a), pp. 1–2.

unsustainable culturing practices in some countries in the Southeast Asian region, increased culture production does not necessarily suggest sustainable use of aquatic resources and the environment. There are many aspects and concerns to take into consideration, for example, the possible negative environmental impact of intensive farming systems and resource competition with other economic activities.[7] With increasing pressure from a possible global fisheries crisis, Singapore has a role to play in preventing driving the stock level down to the detriment of fishing communities. Embracing the concept of sustainability and following the outlook of the global and regional fishery status, this study attempts to examine fishery sustainability in Singapore.

This study takes an integrated approach in examining the two sources of fisheries: capture and culture. The overall aim of this study was to determine whether the consumption patterns of fish, the current status of the fishery sector and the path of development and management of the fishery sector in Singapore, would promote or undermine sustainable production both locally and regionally.

2. Methods

For the analytical part of this study, secondary data, published and unpublished, were collected from relevant government authorities such as the local AVA, MFRD and Singapore Statistics, the regional SEAFDEC and the global FAO. Collection of such data was used to identify trends of local and regional fish production from capturing and culturing, trends of local and regional consumption of fishery products and the relationship to the per capita income and population growth. Projections for the local future demand of fish were also conducted through simple linear regression and time series trend.

The development and current condition of capturing and culturing practices in Malaysia, Thailand, and Indonesia were studied. Such information was then compared to Singapore especially in aquaculture practices.

3. Fisheries in Singapore

The role of the fisheries sector is of very limited significance to Singapore's economy. There are no data on the fishery sector itself, but it is categorised

---

[7]ASEAN-SEAFDEC (2001b), pp. 123–157.

under the sector of agriculture and fishing for the years 1991 to 1995 and under the sector agriculture, fishing and quarrying for 1996 to 2001. These sectors represent less than 0.3% of the total GDP of all industries in Singapore. With limited sea resources and land for agriculture, Singapore's agricultural developments take place mainly in allocated areas, called Agrotechnology Parks on land and Marine Parks at sea of which the latter is for marine fish farming.[8] Singapore has two fishing ports, the Jurong fishing port and Senoko fishing port, both operating under Agri-Food and Veterinary Authority (AVA), providing a 24-hour daily service to foreign and local fishing vessels for discharging and trans-shipment of fish.[9]

As a member of the Southeast Asian Fisheries Development Centre (SEAFDEC), Singapore manages the Marine Fisheries Research Department (MFRD) through the AVA. It is the regional centre for Research and Development and aids in the transfer of technology to the local and regional processing industry and government agencies through training programs, publications and information services.[10]

The Marine Aquaculture Centre (MAC), comprising of a hatchery and a deep netcage fish farm, was set up in 2000 at St. John's Island and is responsible for the development of marine fish production, technology on faming systems such as the coastal and deep net-cages systems, and on fish breeding and nutrition.[11]

i) *Capture fishery*

As of 2003, the local fishing fleet consists of only 134 vessels; 66 offshore and 68 inshore fishing vessels.[12] Inshore fishing vessels include those using palisade traps, drift nets, beach seine nets, and fish pots while offshore fishing vessels include otter trawls and longlines.[13] There is a trend towards a reduced fishing boat fleet and the capture fish production; it can be assumed that dependency on the local fleet for local consumption of fish has reduced over the years. The continual need to meet the local consumption demand for fish may have hence shifted to depend more on importing or culturing.

---

[8]FAO Country Profile and Mapping Information System.
[9]AVA website http://www.ava.gov.sg
[10]See AVA Annual Report 2000/01 p. 52/53 and SEAFDEC website http://www.seafdec.org
[11]AVA Annual Report 2000/01 pp. 52/53.
[12]AVA website http://www.ava.gov.sg
[13]SEAFDEC (2002).

ii) *Culture fishery*

Marine fish farming in the coastal waters of the Johor Straits began as early as in the 1930s, when fish were caught by palisade or fishing traps and shrimp fry were caught in brackish water ponds by the traditional Chinese method of tidal entrapment. In the early 1970s, the Primary Production Department, PPD (now AVA) introduced the idea of placing fixed cage nets at palisade traps, for the operators to hold an assortment of small but economically important fish species for rearing to marketable sizes. The department too introduced the Marine Fish farming scheme to regulate the farming activities in the coastal waters, setting aside specific areas in the east and west Johor Straits for such activities.[14] Sea-based floating netcage farms, located in the northeastern and northwestern part of Singapore account for the bulk of cultured fish production in Singapore, producing both finfish and shellfish. There is also a handful of land-based freshwater and marine fish farms. Marine water culture dominates as the leading type of production in Singapore due to the limited land and freshwater resources of the island state. Brackish water culture began only in 1996. Freshwater culture began a year later in 1997.

Floating net-cage farming is mainly for the culture of popular and high value finfish species like the grouper (*Epinephelus tauvina*), seabass (*Lates calcarifer*), golden snapper (*Lutjanus johni*), milkfish (*Chanos chanos*), Pompano (*Trachinotos blochii*) and shellfish like green mussels (*Perna viridis*) as well as mangrove crabs (*Scylla serrata*). Landbased farms products include shrimps like the tiger shrimp *(Penaeus monodon)* and banana shrimp (*P. merguiensis)*.[15]

In addition, Singapore's first commercial Deep Sea Fish Farm was established in 1998, off St. John's Island. The Deep Sea Fish farm together with a hatchery forms the Marine Aquaculture Centre (MAC) of which most facilities were completed by the end of year 2001.[16] The establishment of technology for deep cage fish farming has led to the successful culturing of species like seabass, kingfish, trevally and threadfin.[17] The centre works with a vision to catalyse the advancement and expansion of marine fish aquaculture in Singapore and regionally.

---

[14]Chou and Lee (1997), pp. 115–117.
[15]AVA website http://www.ava.gov.sg
[16]AVA Annual Report 2001/02 p. 52.
[17]AVA Annual Report 2001/02 p. 31.

4. Sources of Fish

The source of fish for local consumption comes from both local production and import. With a small amount of local production, Singapore depends heavily on imports to meet local consumption needs. The major sources of supplies of imported fish come from Malaysia, Indonesia and Thailand.[18]

i) *Capture and culture fisheries in Malaysia*

Malaysia has a coastline of 4,810 km and with the declaration of the Exclusive Economic Zone (EEZ), has control over the marine resources within 450,000 square nautical miles. The Malaysian fishing industry is dominated by small-scale sector, scattered along its coast and involves multi-gear fisheries ranging from the traditional fishing stakes, gill nets and barrier nets, to modern gear, like trawl nets and purse seines.[19]

The bulk of the capture production in Malaysia comes from the marine area, with a small percentage from inland water. The marine capture fisheries are categorised as both coastal and deep sea fisheries. Contribution from inland waters is insignificant as Malaysia has no large river systems, but there are several sizeable reservoirs that are being developed for inland fisheries. In 1977, a study on the statistical data for the landings of fish for 1964 to 1975, revealed declining catch or little growth.[20] According to the study, any growth was due to the increased landing of trash fish that was a result of increased fishing efforts of trawlers. Furthermore, with the increasing amount of export, the study claimed that the amount of fish available for local consumption fell over the years, while the prices soared. And essentially, the welfare of the locals suffered. However, it is noted that the different regions of Malaysia have varying levels of growth in production. Production from Peninsular Malaysia rises slowly because the catch from the West Coast has stagnated, while the Sabah production is rising steadily. Both Sarawak and Labuan have stagnated but could be due to the lack of facilities and climatic changes rather than full exploitation. This is because of their small fishing fleets and harvests.[21]

The aquaculture industry consists of freshwater, brackish water and marine water culture. The major culture systems in use include ponds,

[18]AVA statistics from Annual report 2000/01 pg. 54.
[19]Information from the FAO Country Profiles and Mapping Information System.
[20]CAP (1977).
[21]Malaysian Agriculture Directory & Index 8th and 9th ed., Chap. 4.

tanks, raft and poles, freshwater cages and marine net-cages. More than 15 species of finfish, molluscs (mainly mussels and cockles), and crustaceans (mainly prawns) are being cultivated.[22]

The National Agriculture Policy (1992 to 2010), which was later published as the National Agriculture Policy (1998 to 2010) of Malaysia, initiated the development of the capture fishery industry towards a fully modern and highly commercialised industry for the exploitation of available fish resources on a sustainable basis. As the inshore waters are seen to be fully fished, Malaysia will develop and expand its deep sea fishing through modern fisheries infrastructure and a comprehensive human resource development program.[23] Deep sea fishing is expected to contribute substantially, estimated at about 430,000 tonnes annually.[24] While the Department of Fisheries aims to increase national fish production, it is also adopting a cautious and conservative approach to maximise exploitation of the deep sea resources.[25]

On the side of aquaculture, the National Agriculture Policy drew up a development plan for an increase of production to 200,000 tonnes by the year 2010. A master plan detailing the various action programmes aims to facilitate the development of the industry including the identification and zoning of culture sites, provision of technical support services, setting up of management measures to overlook environmental issues, and establishment of a fish health management and quarantine system.[26]

ii) ***Capture and culture fisheries in Thailand***

Thailand has a 2,624 km length of coastline. Marine fisheries have a significant socio-economic role in Thailand, both as a food source and a generator of income.[27]

Capture fisheries production in Thailand comes from freshwater fisheries and marine fisheries, consisting of approximately 10% and 90% of the total production respectively. Marine capture fishing activity takes place in the Gulf of Thailand and in the Andaman Sea.[28] The marine capture fisheries can be broadly divided into commercial and small-scale

---

[22] Nagaraj & Singh (1998) and FAO Country Profiles and Mapping Information System.
[23] MOA (1993) pp. 46 & 47.
[24] MOA (1999) p. 80.
[25] Mazlan (1997) pp. 15–20.
[26] Nagaraj & Singh (1998).
[27] Information from the FAO Country Profiles and Mapping Information System.
[28] Chullasorn & Chotiyaputta (1997) pp. 125–137.

sub-sectors. Small-scale vessels typically employ gillnets, small push nets, lift nets or other modern small-scale gear, or alternatively set traps, bag nets and other stationary gear in estuaries or protected inshore waters. As for commercial fisheries, trawlers and purse seines are the most important components of the fleet. Push nets and gillnets also take significant catches.[29]

The continual advances in fishing technology have increased the captured production over the years, resulting in the depletion of stocks particularly in the Gulf of Thailand. According to Chullasorn and Chotiyaputta (1997), the most important species namely Indo-Pacific mackerel, anchovies, round scad and sardines have been reported as fully exploited.

Freshwater culture is the oldest practice and is widely practised in most parts of the country, particularly in the central and north-eastern areas. The most cultivated species for freshwater aquaculture are tilapias, catfish, local carp, gourami, and freshwater prawn, while the larvi-culture of seabass, grouper and mullet is practised in marine and brackish water. Freshwater culture is divided into four groups according to their facility, namely, pond, paddy field, ditch and cage culture. The culture systems vary from extensive to intensive, and from monoculture to polyculture or integrated fish culture. Coastal culture is mostly large commercial intensive farming systems using cages or ponds.[30] In addition, fish culture integrated with pig, poultry and rice farming, among others is also practiced to some extent.[31]

With the objective of sustaining fisheries development and management, the National Fishery Development Plan envisaged that total fisheries production should be maintained at not less than 3.5 million tonnes per annum through sustainable fisheries development means.

Thailand has undertaken efforts to manage the marine and coastal fisheries including rehabilitation of fisheries resources while reducing excessive fishing and minimizing by-catch. Measures introduced include the banning of certain fishing gear, restricting fishing seasons and areas, increasing artificial fish spawning and nursing grounds, prohibiting destructive fishing practices, reducing excess fishing capacity, regulating

[29]ICLARM http://www.worldfishcenter.org/demandsupply/inception_reportapr02
[30]Prathak (1998), Tokrisana (1993) and FAO Country Profiles and Mapping Information System.
[31]ICLARM www.worldfishcenter.org/demandsupply/ inception_reportapr02

mesh sizes, introducing responsible fishing operation, improving post-harvest technology and strengthening institutional legal and financial frameworks.[32] On the side of aquaculture, the plan aims to increase aquaculture production by 5% annually from the current production levels. Strategies include the strengthening of aquaculture techniques and management, promoting cost-effective and environmentally-friendly aquaculture, upgrading production quality and hygiene, and expanding markets for aquaculture products.[33]

iii) *Capture and culture fisheries in Indonesia*

Indonesia is an archipelagic state, which consists of large and small islands. It possesses a long coastline of 81,000 km with an EEZ of 3.1 million km^2. The relatively richer part of the archipelago in the west, produces about two-thirds of the total fish catch and attract large amounts of fishing effort, particularly in the areas relatively close to shore. This part includes the Sunda Shelf area and the large islands of Sumatra, Java and Kalimantan.[34]

Capture fisheries are being practiced both in marine and inland waters. The former includes coastal or small-scale fisheries and offshore and commercial fisheries while the latter includes lakes, rivers, reservoirs and dams. The fishing fleet in marine waters can be categorized as small-scale and commercial scale, but capture fisheries in inland waters are mostly small scale, artisanal, and labour-intensive. The majority of Indonesia's fishing fleet is small-scale, with limited capacity to sail offshore. The dominant fishing gears are hook and line, gillnet, traps, lift nets, seine net and purse seine. Others use pole and line, trolling gear and mini-seines for skipjack and tunas and small pelagics.[35]

The aquaculture industry comprises of freshwater, brackish water and marine culture. The aquaculture industry was developed in Indonesia to be the main source of fish supply where 25.67 million hectares has been allocated for development. Small-scale farms dominate the industry and are carried out in brackish water ponds, paddy fields, freshwater ponds, rivers, irrigation canals, water reservoirs and lakes. The majority are engaged in fresh water pond aquaculture (65 percent), paddy field aquaculture

[32]FAO Country Profiles and Mapping Information System.
[33]Tokrisna (1993) and FAO Country Profiles and Mapping Information System.
[34]FAO's Country Profiles and Mapping Information System.
[35]ICLARM http://www.worldfishcenter.org/demandsupply/ inception_reportapr02

and brackish-water aquaculture, with a few using cage aquaculture. In freshwater aquaculture, monoculture of tilapia and common carp in floating cage systems and monoculture of common carp in running water systems and ricefields are the most widely practiced. Most aquaculture activities are located in Java, while the rest are spread out around the country, especially in Sumatra, Sulawesi, Kalimantan and Lesser Sunda Islands.[36]

In 1998, the Directorate General of Fisheries (DGF) implemented a rescue program for national economy recovery. With the availability of under-exploited fish stocks, the fisheries sector was to be further developed to assist in the recovery of the Indonesian economy. Other development programs for marine capture fisheries deal with the efficiency and productivity of fishing vessels, fish quality and development and improvement of capture fisheries' infrastructure and facilities.[37]

Under aquaculture development, the Directorate General of Aquaculture (DGA), under the Ministry of Marine Affairs and Fisheries, is mandated to formulate the policies and standardization of culture techniques. The policies include the development of an integrated aquaculture zone for marine, brackishwater and freshwater for enhancement of growth and sustainability of aquaculture business, regulations on environmental protection and the amount of production.[38]

5. Trends in Consumption Patterns

It is generally agreed that people worldwide eat more fish than any other type of animal protein. Fish is the primary choice of protein for 950 million people in the world, and is an important part of the diet of many more.[39] The importance of fish is no different in Singapore. The main sources of protein meat for Singaporeans include fish, chicken, pork, mutton, beef and duck. The per capita consumption of fish takes up an average of 30% of the protein diet of Singaporeans. Fish and chicken top the list of choice in Singapore's protein diet.

[36]Supardan (1993), ICLARM http://www.worldfishcenter.org/demandsupply/inception_reportapr02 and FAO Country Profile and Mapping Information System.
[37]ICLARM http://www.worldfishcenter.org/demandsupply/inception_reportapr02
[38]Supardan (1993) and ICLARM "Strategies and Options for Increasing and Sustaining Fisheries and Aquaculture Production to Benefit the Poor Households in Asia" at http://www.worldfishcenter.org/demandsupply/inception_reportapr02
[39]ICTSD-IUCN (1999) pg. 2.

According to Yeo (1996), the increasing affluence in Singapore is identifiable in food consumption patterns such as the trend to adopting a healthier lifestyle in consuming more "white" meat as contrasted with "red" meat.[40] The per capita consumption of fish has dipped over the years, from 31.5 kg in year 1991 to a decade low of 23.3 kg in 2001. Chicken overtook fish as the preferred choice of animal protein since 1994 onwards, with a per capita consumption of 35.9 kg in 2002.

Looking at the retail prices of fish and other animal protein in Singapore, fish is generally more expensive than chicken, pork and duck, but cheaper than beef and mutton. However, the prices of fish are calculated based on the average of four types of fish, naming **Pomfret, Seabream, Spanish Mackerel** and **Gold Banded Scad**.[41] This may not reflect accurately the prices of all types of fish. From an economic pricing point of view, the fall in the price of chicken, as the substitute "white" meat, can be an explanatory factor why chicken took over fish as the preferred choice of animal protein since 1994. On the other hand, the rate of inflation since 1998 for fish has fallen. The per capita consumption of fish continues to fall, picking up only slightly in 2002. Hence, it is not conclusive whether the higher prices of fish and the simultaneous fall in prices of chicken may be the reason for the decline in the per capita consumption of fish. Non-economic explanations such as changes in lifestyle choices, eating habits and tastes can also influence the trend on fish consumption.

6. Future Projection in the Demand For Fish

The factors that can be expected to affect fish consumption can be both economic and non-economic. For the latter, in microeconomic theory, the demand and supply model explains the intersection of quantity demanded equilibrium. The law of demand stipulates the lower the price, with other factors that affect consumption held constant, the larger the quantity consumers' wish to purchase. The disposable income of consumers is another economic factor that affects consumption level. Most demand analysis concentrates on price elasticity, income elasticity and cross price elasticity.[42] Non-economic reasons include individual preferences and varying and changing tastes.

---

[40]Yeo (1996) pp. 8–13.
[41]Information provided by AVA's officer Ms Sharon Foo.
[42]Browning and Zupan (1996).

On the other hand, the supply affects the amount of fish available for purchase, and an increased amount of production will depress the price of fish. The increase in price may not lead to an immediate increase in the supply of fish as the nature of production in the fishery sector depends on many factors, especially for the capture source. However, it may positively influence the production in the following years as, by theory at least, higher prices would induce producers from the capture and culture source to increase production.

This section aims to forecast the future potential demand for fish in Singapore. The forecast analysis is modelled closely to the FAO's World Fish Demand Model (WFDM) whereby the calculation for prediction of future demand used a mixture of Per Capita Gross Domestic Product (GDP), time trend, 10-year and 20-year averages.[43] For this study, the independent variables that will be used as predictors for the consumption level in Singapore are per capita GDP and time.

There are several reasons why this study, similar to WFRD, is a forecast of potential demand and not consumption. Consumption can be affected by many reasons under the demand theory and hence it is hard to determine the extent to which forecast demand may be realized because it depends on the production level as well. With the lack of information on the price evolution and their effects, it is only possible to forecast potential demand.

The forecast analysis of this study is, however, not a mere blind duplicate of the WFRD model. The author recognised that the simplicity of the model can be highly assumptive and risky with respect to its accuracy. But, simplicity can also be a virtue itself. The author does not seek to make precise forecast of consumption of fish through this study but seeks to show the general trend of the historical consumption in Singapore and the future potential demand and whether this would exert a negative impact on local and regional fish sustainability.

Several considerations were taken into account. Data on the retail price of fish in Singapore are an average of only four different species and hence may not be a good representation of all species of fish consumed. And to use cross elasticity is highly data demanding and complicated and the list of substitute and complement goods to fish that need to be determined is not an easy one.

Like the WFDM, per capita GDP is used as the best proxy for disposable income of consumers. While it is recognised that an increase in

[43]Ye (1999).

GDP need not necessarily mean an increase in consumers' income, it is the next best available data that can be collected. The long-term forecast on the GDP used for this study is obtained through extraction from a combination of several papers from the Singapore Ministry of Trade and Industry (MTI), Asian Development Bank's (ADB) *Asia Development Outlook* 2003, the World Bank's *Global Economic Prospects* and the Developing Countries 2003 and International Monetary Fund's (IMF) *World Economic Outlook*.[44]

The forecast for the GDP in 2003 is taken as 2.3% as predicted by the Asian Development Bank.[45] This is in line with the Singapore's MTI outlook of growth, first maintaining it between 2 to 5%, which was later revised to 0.5% to 2.5%.[46] The following years of annual growth for 2005 to 2010 is taken to be at 4.0%. This uses the long-term trend growth rate that is estimated through econometric tests in the paper Singapore's *Engine of Growth: A demand side perspective* by MTI.[47] There is a limitation to the GDP forecast as the long-term growth trend is dependent only on the demand side, ignoring supply factors. However, with a lack of information on long-term GDP projections, the forecast of potential demand for fish has to proceed with the best available information. The forecast for the population is obtained from the Human Development Network from the World Bank Group.[48]

Based on the estimated 2.3% growth in the GDP in 2003 and 4% annual growth thereafter, and the population projections, GDP projection in 2005 and 2010 are S$177,980 and S$216,540 respectively. With projected population of 4.422 million in 2005 and 4.669 million in 2010, the per Capital GDP will be at S$40,249 in 2005 and S$46,378 in 2010.

Simple regression analysis with application of linear Method of Least Squares is undertaken here.[49]

Method of Least Square for Simple Linear Regression determine the best fit line as $Y^* = -0.0008X + 57.359$ as seen in Fig. 13.1.

---

[44]http://www.worldbank.org/prospects/gep2003/index.htm & http://www.imf.org/external/pubs/ft/weo/ 2001/02

[45]ADB (2003).

[46]" Performance of Singapore Economy in 2002 and Outlook for 2003" by MTI at http://www.channelnewsasia.com/cna/econsurvey/2002/fullyr.pdf and "Economic Survey of Singapore: First

Quarter 2003" at www.mti.gov.sg/public/PDF/CMT/EDA_2003Q1_Ch2.pdf?sid=44

[47]Tan *et al.* (2003).

[48]http://devdata.worldbank.org/hnpstats/DPselection.asp

[49]Allen (1997).

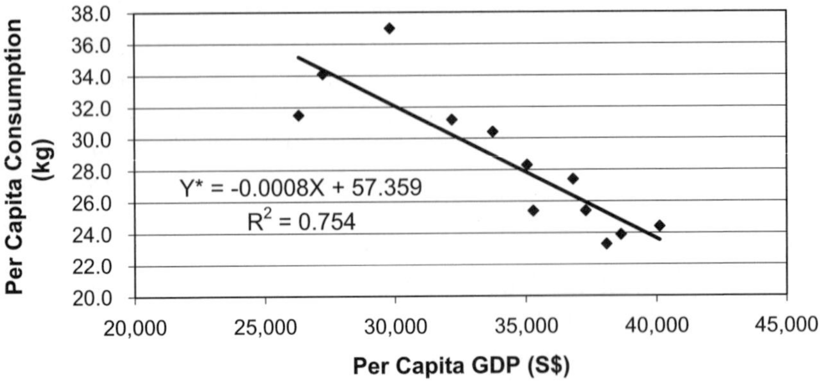

Fig. 13.1. Linear regression for per capita consumption.

where Y* is the Estimated Per Capita Consumption

X is the Per Capita Consumption of Fish

The co-ordinates points plotted on the graph are the actual Per Capita Consumption, denoted as Y against the actual Per Capita GDP, denoted as X.

Per capita consumption is predicted to fall to 25.2 kg by the year 2005 and to 20.3 kg in 2010. The projected decline in the per capita consumption in 2010 would overall reduce the total consumption from 111,434 tonnes in 2005 to 94,780 tonnes in 2010, despite the rise in human population.

Time series data are also explored to determine any strong correlation for historical consumption and future demand. In this case, the change in time itself is not the cause of the change of the dependent variable (per capita consumption), but may aggregate the effects of other factors such as the prices of fish and disposable incomes of consumers. A strong assumption is made here that the future behaviour does not change and depends only on past behaviour.

Method of Least Square using time series model gives the plot the best trend line in Fig. 13.2 as

$Y* = -1.0985t + 35.7$[50]

where Y^* is the Estimated Per Capita Consumption for the past years t is the time (Year)

[50]A manual calculation to check the best fit, through Method of Least Square, was conducted.

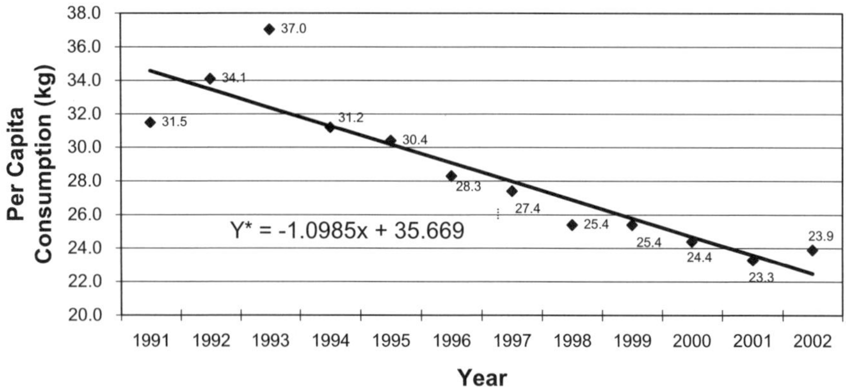

Fig. 13.2. Time trend regression for per capita consumption.

The co-ordinates points plotted on the graph are the actual Per Capita Consumption, denoted as Y.

This represents the best linear fit to the actual demand over the past 12 years and hence will be used to forecast the per capita consumption for 2005 and 2010.[51] This also shows the forecast of potential demand for the period 2003 to 2010. Based on the time series trend, the future potential demand is a continual decline whereby the potential demand will drop to 13.7 kg per capita in 2010, from 23.9 kg per capita in 2002.

Although this calculation gives an indication to potential future demand, there are some caveats. Past behaviour does not guarantee to produce a similar future trend. Furthermore, fish constitute an important source of protein meat for Singapore and it is more likely that the amount consumed will stabilise at recent levels. It is likely that the forecast would be more accurate if it were possible to obtain reliable longer-term data, stretching back to 1960, for calculation of 10-year averages.

Bearing in mind the shortcomings of the forecasts, it is possible to predict that the impact of such a reduction in the future demand of fish would reduce the dependency on importing from Malaysia, Thailand and Indonesia. This is without considering that the actual per capita consumption of fish of recent years appeared to have reached a plateau at a lower average consumption of 24 kg per capita. Furthermore, considering the

[51]Logarithmic Regression was also tested to check possibility of a better fit. However, a smaller co-efficient of correlation R and F-Statistic was found and hence the function was rejected.

fact that Singapore's stated intention is to expand local culture production to provide for 40% of the total local consumption of fish by 2010, Singapore's dependency on importing from the neighbouring three countries will be further reduced.

7. Constraints and Prospects of Aquaculture in Singapore

i) *Lack of Resources*

One of the major constraints in Singapore is the lack of resources. The lack of land and freshwater resources limits the development of brackish water and fresh water culture farms. Sea area for culturing is also limited in Singapore because port and other activities take precedence, and the number of farm units in coastal areas is limited by the scarce coastal resources.[52]

Furthermore, farming along the coast of Johor Straits and Pulau Ubin island faces the possibility of pollution from oil spills due to shipping traffic. A chemical spill due to a tanker collision near Pulau Ubin island in August 2000 lead to the AVA suspending harvest and sale of the fish from the farms in the affected area and lifting the suspension only after close monitoring and satisfaction that the fish were safe for consumption.

With the limited sea area for farming in Singapore, one way of increasing the production of fish by aquaculture is to increase the levels to those achievable by industrial fish farming systems such as the deep-cages in open waters. Such open water systems can be located far from the coastal activities and make use of the vertical water column rather than sea area. As suggested by Chou and Lee (1997), this would be a more sustainable option for aquaculture development in Singapore.

ii) *Technical Constraints*

At present, fish spawning and the production of broodstock of major species is not a problem in Singapore. Fish broodstock, that are either wild species or reared from fry, are raised in netted cages in coastal waters. But production of healthy and wholesome fish products from aquaculture will demand a secure supply of good quality seed.

At least 100 million fry would be required to produce 40,000 tonnes of fish. In Singapore, private hatcheries are the main source of fry stock. As the

[52]See Chou and Lee (1997), p. 126.

local demand for fry is too low, it is difficult to encourage investment in the development of hatchery technologies. Currently, the Marine Aquaculture Centre, MAC that has a hatchery, could help commercial hatcheries to produce fry. The opening of MAC can be seen as a promising initiative to catalyse the technological advancement and development of the aquaculture industry.

The prevention of disease is another area that needs to be addressed. Singapore regulates the movement of live animals, conducts disease surveillances and diagnostics and cooperates with regional countries to prevent any possible spread of diseases. Oxytetracycline and Vitamin C are commonly used for disease control.[53]

Further steps should be made through practice of good quarantine and sanitation control, implementing fast and precise tools for the detection of pathogens and improving the management of fish farms. AVA should also conduct training on diseases and their prevention to farms through trained specialists.

iii) *Environmental Constraints*

Another aspect to take into consideration is the possible negative environmental impact of aquaculture. The potential of developing the aquaculture industry in Singapore should not just be a focus to increase production as a self-sustaining food source, but also to develop an industry that is environmentally sound and sustainable. The marine cage aquaculture system, which is primarily practiced in Singapore, is known to have a high potential for causing environmental degradation as these systems are in relatively open waters and the untreated waste is discharged directly.[54] As a further step, pollution problems should be addressed through establishment of legal guidelines and regulation with strict enforcement. Pollution monitoring and environmental impact assessment (EIA) should be mandated to assess the pollution potential in the selection of sites for setting up of culture farms.

iv) *Regulation and Policies Direction*

Although there are certain guidelines that are enforced by AVA on the types of feed prohibited for use, there is currently no specific law and regulation

[53]See ASEAN-SEAFDEC (2001b), pp. 48–59.
[54]http://www.start.or.th/got/gotwww/living.html

to provide proper guidance to the direction of aquaculture development. To work towards achieving expanding aquaculture on a sustainable basis, Singapore first needs to set clear roles and directions for the sector and adopt a set of laws and guidelines. Close collaboration with SEAFDEC on the implementation of the RCCRF on the management of aquaculture sector is also desired.

v) *Alternatives*

Extension into deeper regions of the marine waters has been looked upon as a possible alternative for the expansion of capture fish production. Through the application of new technologies, deeper waters of the continental slopes and plateaux can be explored. Such technologies that have been suggested include the use of submarines, remotely operated vehicles (ROV) and bottom crawlers using air-bubble of electric field "net" with sophisticated pumping system. The submarines will allow filtering of fish species to retain target species, while releasing any by-catch or unwanted species. And the use of ROV is to allow direct observation of the settling and hauling of the gears.[55] But for feasible application for the capture fishing community, such alternatives must be cost-effective and practical.

Moreover, such alternatives can be looked upon as merely increases in fishing effort. Such efforts would still ultimately be limited by the available stock in the sea. Moreover, top predator fish found usually in deep-sea water because of being at the top of the ocean food-chain will consist of a smaller population as compared to the smaller predators. For sustainable use of the top predator fish species, they need to be fairly fast growing, widespread in location and exist in areas where there are abundant smaller fishes to feed on.[56]

8. Conclusions and Recommendations

Through the assessment of data and analysis of the trends in the consumption of the different types of protein meat for the period 1991 to 2002 in Singapore, chicken overtook fish as the preferred choice of "white" meat since 1994. Disposable income, as reflected by the per capita GDP in this study, does not have a strong correlation to the choice of "white" or "red" meat. It is generally assumed that an affluent country would choose

[55] Ibarahim (1997) pp. 21–28.
[56] Merrett & Haedrich (1997).

healthier options of meat. However, since 1994, despite the rising GDP, the per capita consumption of "red" meat such as beef and mutton remains rather stable whereas the per capita consumption of fish and pork took a downward trend.

It is fair to predict that Singapore's dependency on imports from the three regional countries (Indonesia, Malaysia and Thailand) into 2010 will not be a major contributing factor that undermines a sustainable fishery in the region. The factors are:

a) Singapore had the second highest per capita consumption, after Malaysia, for the period 1991 to 1994 and fell to the third place, after Malaysia and Thailand for the period 1995 to 1997. But due to the small population size in Singapore, the total consumption of fish is significantly lower than the other three countries. Using the data in 1997, Singapore's total consumption is merely one-tenth of Malaysia's, one-twentieth of Thailand's and one-thirtieth of Indonesia's.
b) Not only has the amount of per capita consumption in Singapore been declining since 1994, the amount of imports from the three regional countries has, also been on a downward trend since 1996.
c) Linear regression using the projected per capita GDP, predicted a potential per capita demand of 25.2 kg and 20.3 Kg in 2005 and 2010 respectively. Assuming that the projection is fairly accurate, per capita consumption level in Singapore would stabilise in the range of 20 to 25 kg per annum. The total consumption would not vary too greatly, or may even decline in 2010 as projected.
d) The dependency on imports will be further eased when taking into account Singapore's plan to be self-sustainable by providing 40% of the total local consumption of fish through aquaculture by 2010.
e) The overview on the development of the fishery sector in Malaysia, Thailand and Indonesia shows that the general objective is to increase production of fish in a sustainable manner. If this may be conducted successfully, the sustainability of fishery in the region can be further assured.

But, what may be of concern is the fact that the higher prices of fish in Singapore may induce the export of good quality fish to Singapore, creating social inequality in the region. However, further studies need to be conducted due to the limitation of data available including the lack of retail and wholesale prices. There is also ground for further studies on the possible effects on trade and the distribution patterns of fish for consumption due to the huge disparity in the per capita income of Singapore and the region.

In reviewing the international, regional and national fisheries management policies, law and instruments of Malaysia, Thailand and Indonesia, the inadequacy of Singapore's law and regulation and management system is reflected. While the Fisheries Act gives considerable control over the capture fishing industry, the State Lands Act protects the removal of corals or shells without a permit and Prevention of Pollution of the Sea Act and the Merchant Shipping Oil Pollution Act regulate marine pollution, there is no specific law to protect Singapore's marine ecosystem as a whole and to protect the social well-being of stakeholders (fishermen, farmers, consumers) in the fishing industry.

References

ADB. 2003. *Asian Development Outlook 2003: Special Chapter Competi- tiveness in Developing Asia.* Asian Development Bank. Oxford University Press. New York.

Allen, M.P. 1997. *Understanding Regression Analysis.* Plenum Press, New York.

ASEAN-SEAFDEC. 2001a. *ASEAN-SEFDEC Conference on Sustainable Fisheries for Food Security in the New Millennium: "Fish for the People"* Technical Document 19–24 November 2001. Bangkok. ASEAN, SEAFDEC, FAO and DOF. Thailand.

ASEAN-SEAFDEC. 2001b. *Outcomes of the National Seminars for ASEAN-SEAFDEC Conference on Sustainable Fisheries for Food Security in the New Millennium.* ASEAN, SEAFDEC, FAO and DOF. Thailand.

AVA. 2001. *AVA Annual Report 2000/01.* Agri-food and Veterinary Authority. Singapore.

AVA. 2002. *AVA Annual Report 2001/2002.* Agri-food and Veterinary Authority, Singapore.

Browning, E.K. & M. Zupan. 1996. *Microeconomic Theory and Applications.* 5th ed. Harper Collins College.

CAP. 1977. *Crisis in Malaysian Fishery.* Consumers' Association of Penang, Pulau Pinang.

Chou, R. & H.B. Lee. 1997. *Commercial Marine Fish Farming in Singapore.* Aquaculture Research. No. 28.

Chullasorn, S. & C. Chotiyaputta. 1997. "Fishing Status in Thailand" in SEAFDEC. 1997. *Proceeding of the Regional Workshop on Responsible Fishing* 24–27 June 1997. SEAFDEC. Bangkok.

FAO, 2002a. FAO Yearbook Fisheries Statistics: Capture Production, v.90/1. FAO, Rome

FAO, 2002b. The State of the World Fisheries and Aquaculture. FAO, Rome

FAO Country Profile and Mapping Information System. Retrieved Mar 15, 2003 from http://www.fao.org/fi/fcp/en/MYS/profile.htm

Grainger, R.J.R. & S.M. Garcia. 1996. *Chronicles of marine fishery landings (1950–1994): Trend analysis and fisheries potential.* FAO, Rome. FAO Fisheries Technical Paper No. 359, FAO, Rome.

Ibarahim, R. 1997. "Fisheries and the Environment" in B.J. Sihik, F.M. Yusoff, M.S. Mohd Zaki & T. Petr (eds.). *Fisheries and the Environment Beyond 2000.* Universiti Putra Malaysia. Malaysia

ICLARM "Strategies and Options for Increasing and Sustaining Fisheries and Aquaculture Production to Benefit the Poor Households in Asia" at http://www.worldfishcenter.org/demandsupply/inception_reportapr02

ICTSD-IUCN. 1999. *Fish for Thought: Fisheries, International Trade and Sustainable Development.* Natural Resource, International Trade and Sustainable Development Series No. 1. International Centre for Trade and Sustainable Development, Switzerland and the World Conservation Union Publications, U.K.

Malaysian Agriculture Directory & Index. 99/00. 8th ed. Agriquest Sdn Bhd, Malaysia.

Malaysian Agriculture Directory & Index. 01/02. 9th ed. Agriquest Sdn Bhd, Malaysia.

Mazlan, B.J. 1997. "Malaysian Fisheries Policy beyond 2000" in B.J. Sidik, F. Yusoff, M.S. Mohd Zaki & T. Petr. (eds.) 1997. *Fisheries and the Environment: Beyond 2000.* Universiti Putra Malaysia, Malaysia.

Merrett, N. R. & R.L. Haedrich (1997). *Deep-Sea Demersal Fish and Fisheries* Chapman & Hall. London.

MOA. 1993. *The National Agriculture Policy 1992-2010.* Ministry of Agriculture, Malaysia.

MOA. 1999. *The National Agriculture Policy 1998-2010.* Ministry of Agriculture, Malaysia.

Myers, R.A., N.J. Barrowman, J.M. Hoenig & Z. Qu. 1996. *The Collapse of Cod in Eastern Canada: the Edvidence from Tagging Data* ICESJ. Sci. No. 53.

Myers, R. & B. Worm (2003). *Rapid Worldwide Depletion of Predatory Fish Communities.* Nature Journal. V.423.

Nagaraj G. & T. Singh. (eds.) 1998. *Aquaculture Practices in Malaysia.* Malaysian Fisheries Society.

Prathak, T. 1998. "Thailand" in *Improving Management of Aquaculture in Asia.* Asian Productivity Organisation, Tokyo.

Schiermeier, Q. 2002. *How Many More Fish in the Sea?* Nature Journal, V.419. pp. 662–665.

SEAFDEC. 2002c. *Fishery Bulletin for the South China Sea Area 1998.* SEAFDEC. Thailand.

Supardan, A. 1993. "Aquaculture Development in Indonesia" in I.C. Liao, C. Shyu & N. Chao (eds.) *Aquaculture in Asia.* Asian Productivity Organisation, Taiwan Fisheries Research Institute, Taiwan.

Tan, K.Y., F. Wu, M.H. Toh, K.W. Seah & J.P. Thia. 2003. *Singapore's Engine of Growth: A Demand Side Perspective.* Ministry of Trade and Development. Singapore.

Tokrisna, R. 1993. "Aquaculture in Thailand" in I.C. Liao, C. Shyu & N. Chao (eds.) *Aquaculture in Asia.* Asian Productivity Organisation, Taiwan Fisheries Research Institute, Taiwan.

Ye, Y. 1999. *Historical Consumption and Future Demand for Fish and Fishery Products: Exploratory Calculation for the Years 2015/2030.* FAO Fisheries Circular No. 946. FAO, Rome.

Yeo P. K., 1996. *Trends in Food Consumption*, Singapore Statistics Newsletter, Singapore Department of Statistics. Singapore.

14

AQUACULTURE MANAGEMENT IN THUA THIEN HUE PROVINCE, VIETNAM: ENVIRONMENTAL CONSIDERATIONS

TRAN THI MINH HANG and CHOU LOKE MING

The Tam Giang lagoon, one of the biggest in Asia, is located in Thua Thien Hue province, Vietnam. Its area is about 22,000 ha with a length of 70 km along the coast. Tam Giang lagoon has great potential for aquaculture development which brings better incomes for local people. Aquaculture has significantly contributed to poverty reduction in Thua Thien Hue Province. However, there has not been an appropriate plan for local aquaculture, resulting in an uncontrolled increasing number of aquaculture ponds and net enclosures. That poses negative impacts on the environment and natural resources. This study examines the present status of aquaculture in Tam Giang lagoon and analyses the environmental problems that local aquaculture faces. From this synthesis, recommendations are formulated. The objectives are to reorganize local aquaculture in more appropriate ways; address technological, organizational and regulatory issues in order to improve environment protection and sustainable aquaculture development.

I. Introduction

There are many important lagoons along the coast of Central Vietnam such as Tam Giang Cau Hai Lagoon System in Thua Thien-Hue Province (referred to as Tam Giang Lagoon in this report), Son Khe and Tra O in Quang Ngai Province, Thi Nai in Binh Dinh province and Cu Mong and O Loan in Khanh Hoa province. The total combined area of these lagoons is about 28,000 ha of which Tam Giang Lagoon itself occupies nearly 22,000 ha making it the largest coastal lagoon system in Southeast Asia.

The Tam Giang Lagoon stretches along the coast for about 68 km. On the eastern side, the lagoon is separated from the sea by sandy dunes with two openings, Thuan An and Tu Hien. The western side of the lagoon consists of rice fields and river estuaries. Tam Giang lagoon also receives freshwater from several rivers, such as Huong, Bo and Dai Giang. As a result, the lagoon water is a constant mixture of fresh and salt waters with salinity fluctuations which are seasonal (rainy or dry seasons) and partial (depending on the distance to the estuaries or openings to the sea) (Nguyen X.H. *et al.*, 2000). With unique topography, the lagoon is rich in natural resources. According to Provincial Department of Fishery (DOFI), Tam Giang lagoon is home to about 742 species of 391 genera and 208 families (excluding about 100 species just identified up to genus).

At present, the Tam Giang lagoon system shares borders with 32 communes of 5 coastal districts and plays an important role in the local socio-economic development. About 300,000 inhabitants, nearly one-third of the province's population, have settled around the lagoon and earn their livelihood directly or indirectly exploiting natural resources in many ways, in and around the lagoon. (Truong *et al.*, 2000).

The lagoon is a shallow, brackish water area which has not only valuable resources but also high potential for aquaculture and fisheries. The average depth is 2 meters with a maximum of 3 to 4 meters in channels. Salinity and bottom topography are favorable for the construction of aquaculture ponds and net enclosures in most of the area, with concentrations nearest the Thuan, an estuary which has the highest salinity.

At present, the natural resources of the lagoon are showing signs of stress mainly as a result of over-exploitation. Moreover, there has been a considerable local population boom. Hence, aquaculture is considered as an alternative to improve local people's income, reduce poverty and decrease exploitation pressure on the lagoon. This is also a strategy for the development of aquatic products in the locality supported by the local government. As a result, there have been rapid aquaculture developments which attract more and more people.

With comparative advantages of market, socio-economic and natural conditions, coastal aquaculture has been developing rapidly in Vietnam during the last decades. Aquaculture is considered as a leading economic sector in the economic transformation and modernization of agriculture and rural developments in the period to 2010. The orientation is clearly expressed in many legal documents.

After the Government promulgated the resolution No.09/NQ-CP regarding structural changes in economic production, the coastal aquaculture area and activities dramatically increased. In 1999 there were 290,000 ha used for coastal shrimp aquaculture, but in 2001 the shrimp farmed area increased up to 478,800 ha. As stated in the Master Plan for Aquaculture Development (DOFI, 2001a), shrimp farming area will continue to expand, and would reach 700,000 ha by 2010. Along with area expansion, farming activities have been intensified considerably. All these efforts have contributed to fishery exports which ranked third in terms of export value in 2001 and second in 2002.

However, although much progress has been made on research and development activities in Viet Nam for coastal aquaculture, considerably less progress has been made to understand, evaluate and mitigate the negative environmental impacts associated with poorly sited and managed brackish water farms. Experience has shown that severe environment consequences for coastal ecosystems can result from unplanned and poorly managed coastal aquaculture. The potentially high financial returns from such developments often come at the expense of economic externalities that are hidden in the loss of breeding grounds, biodiversity, wetlands and negative impacts on other coastal resource users.

Negative impacts from other sectors, such as water pollution from agriculture and industry, can also adversely affect the sustainability of aquaculture. Such problems are particularly acute in the southern part of Viet Nam where, for example, in the Minh Hai province 60,000 ha (55%) of the mangrove forest area were lost to coastal aquaculture development during the 1980s. Inappropriate siting and poor operational management of brackish-water ponds have contributed to further problems, such as the shrimp disease outbreaks which devastated extensive shrimp farms in the Mekong delta in 1996 and uneconomically low pond yields (on average, Vietnamese shrimp pond yields are among the lowest in the Asian region). Such problems have led to shrimp ponds being left fallow in some areas, while some farmers build large ponds to compensate for low productivity, this leading to increasing negative environmental impacts (UNDP/MOFI, 2000).

Before 2004, regulations on fisheries activities have been promulgated such as Regulation of aquatic resources protection and development 1989, Decree No. 103/2000/QD-TTg dated 25 August 2000 by the Prime Minister to encourage aquaculture. However, fishery activities have become more complex. As a result, the Law of Fisheries should be modified and updated to address new issues. For example, there is an urgent demand of Vietnam fisheries in order to develop and integrate their activities within the global economy.

The Law of Fisheries 2003 has been formulated and came into enforce in July 2004. The law provides many guidelines and policies in aquaculture development, protection of aquatic resources, exploitation activities, aquatic product import and export, international cooperation in fisheries activities, among other issues. The Law of fisheries, affirms the key role of fisheries in Vietnam's economic system. It creates safe legal corridors for development of aquaculture and exploitation of aquatic products, leading to a promising future for Vietnam's fisheries (Nguyen D.G., 2004).

Following the national trend, aquaculture in Thua Thien Hue has also been developing rapidly recently. Tam Giang lagoon system is abundant in aquatic resources offers high potential for aquaculture development. Aquaculture is defined as a high priority investment sector aimed at becoming a key economic activity of the province.

Up to 2002, Thua Thien Hue has a total of 3851 ha of aquaculture area on and along lagoon which has increased 30 fold since 1991. Many related jobs have also been created as a result. Aquaculture generates more income and improved living standard in the area. This contributes greatly to the poverty alleviation plan of the province. Aquaculture really becomes a promising and highly attractive livelihood for local people.

Considering the benefits and profits from aquaculture, Thua Thien Hue People's Committee has made a plan for aquaculture development in the period 2001–2010 with some main objectives:

- Encouraging using uncultivated lands, water surface near lagoons and fallow areas near the seaside for aquaculture.
- Increasing the area of aquaculture from 3851 ha in 2003 to 7100–6500 ha in 2010 with a targeted production of about 7000 tons/year.
- Diversifing the breeding species, type of aquaculture (emphasize on intensive culture and industrial culture).

However, there has not been an appropriate plan for local aquaculture, thereby resulting in an uncontrolled increase in number of aquaculture ponds and net enclosures. These developments also raised concerns of negative impacts on the environment and resources. Some of the obvious consequences include the appearance of a milky tide in some regions and wide spread of disease over the whole lagoon (Ton, 2004). Up to October 2004, of the 3,755 ha of total shrimp culture area, 1,360 ha were affected by diseases resulting in mass mortality of cultivated shrimps (DOFI, 2004b). Unplanned aquaculture development also destroys nursery habitat of various aquatic species and pollute waters partial in aquaculture areas and vicinities.

The cumulative effects of these impacts significantly influences the sustainable development of aquaculture of Thua Thien Hue Province and pose serious threats to the Tam Giang lagoon's biodiversity and environment. These problems need to be addressed and managed in an environmentally friendly manner if sustainable development is to be achieved. It urgently calls for an overall management plan for aquaculture development of Thua Thien Hue Province.

This study examines the present status and development trends of aquaculture in Tam Giang lagoon; identifes the socio-economic and environmental impacts of aquaculture development in the lagoon such as decreasing biodiversity, the destruction of nursery habitats, water pollution, and diseases appearing. The study proposes recommendations to reorganize local aquaculture in a more appropriate way, address technological, organizational and regulatory issues in order to improve environment protection and enhance sustainability of aquaculture.

II. Methodology

Fieldwork and interviews were employed in this study. Interviews with local people were conducted to collect information such as the trend of daily aquatic species caught (decrease or increase), diseases arisen recently and the way they deal with diseases, and the pollution of culture ponds. Recommendations formulated took into consideration the opinions and expectation of local people, local government (commune, village, etc.) as expressed through discussions and interviews.

i) *Overview of aquaculture status in Thua Thien Hue Province*

The official strategy for aquaculture development in Thua Thien Hua Province is based on a semi-intensive culture as part of their livelihood portfolio with state-run companies supplying seed and technical support. As a result of some successful experiments, the chosen type of aquaculture in the lagoon reflects a flexible polyculture. Common polyculture patterns include: shrimp (*Penaeus monodon*)/crab (*Scylla serrata*)/fish (*Siganus guttatus*)/seaweed (*Gracilaria tenuistipitata*); shrimp/seaweed and shrimp/crab/fish. Grouper and grass carp (*Ctenopharyngodon idellus*) are cultured in cages. It is observed that low intensity polyculture has resulted in increased economic benefits, due to low investments for seed and feed, high product quality and better prices than in high-density culture. In fact, intensive aquaculture cultivation companies have had to abandon their ponds or rent them to fishers (Ton, 2001b).

The rapid development of aquaculture in Tam Giang Lagoon is not only in terms of the increased cultivation area, but also in the range of products, methods of cultivation, output and productivity. There are different types of aquaculture products chosen to best suit local strategy and strengths. Many methods of cultivation have also been tested in order to pick the most suitable one. The total aquaculture area has increased by 423.63% from 830 ha in 1995 to 3,516.2 ha in 2003. Meanwhile, total output increased by just above sixteen times, resulting in a productivity increase of nearly 396.19%. However, aquaculture production in the lagoon is regarded as unstable and unsustainable. As a result, further research is required to address the associated problems that will be critically analyzed later in the report.

ii) *Values of Tam Giang Lagoon's biodiversity*

The lagoon has a diverse range of phytoplankton and aquatic seagrass. Research has identified 589 species of basic aquatic organisms. The following figures show a breakdown of the composition and density of phytoplankton in different areas of Tam Giang lagoon.

Among these, many species have high economic values such as *Gracilaria tenuistipitata*, *Enteromorpha* and *Chaeetomorpha* seaweed, Valisneria and Najas seagrass, *Penaeus monodon* and *Metapenaeus ensis* shrimp; *Scylla serrata* crab among others.

Tam Giang lagoon is also home for many birds including 73 species, 53 genuses and 29 families. There are 39 migratory species, 30 species listed in Strictly Protected Birds of EU and one species named in Vietnam's Red Book (*Limnodromus semipalmatus*).

The lagoon also acts as a shock absorber to reduce the impacts of flood for the area and a water buffer zone to prevent salinity intrusions. In

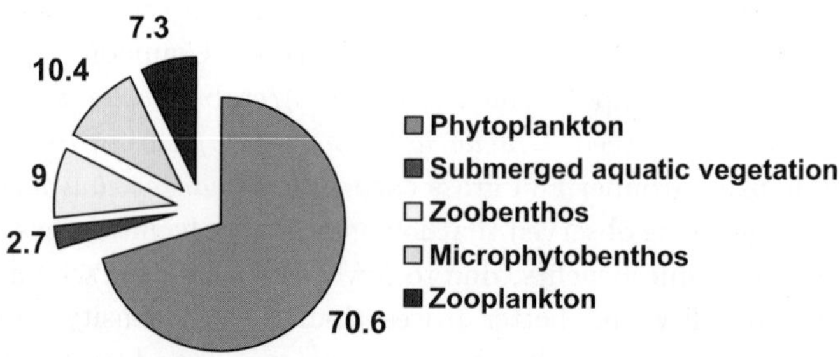

Fig. 14.1. Composition of aquatic species of Tam Giang lagoon.

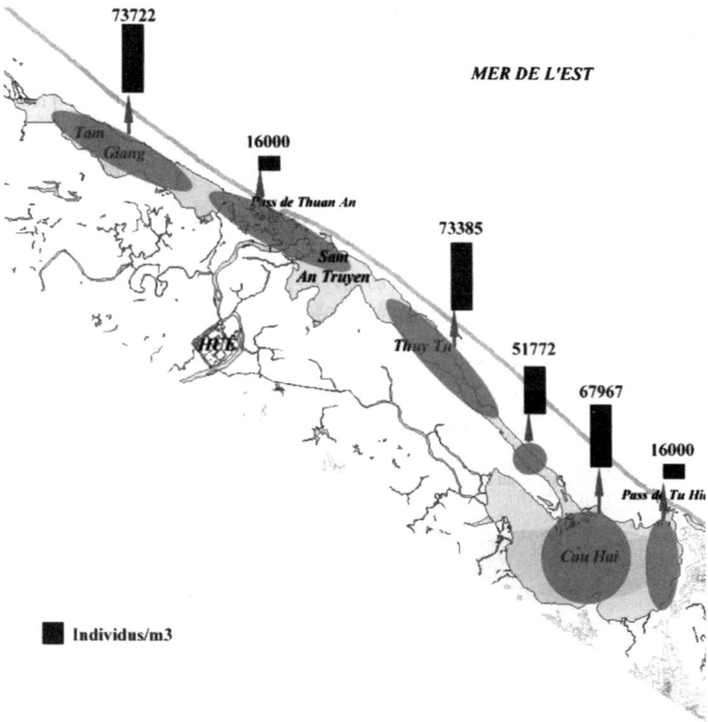

Fig. 14.2. Average density of phytoplankton in Tam Giang Lagoon (Source Ton. T. P, 2001).

addition, the lagoon system has potential for attractive ecological tourism which is one of the provincial development strategies. The provincial government is also planning to upgrade Tan Thuan seaport supports the whole region. The ecologically important and unique nature of the lagoon system is also internationally recognized, and part of it proposed as a RAMSAR site by the government of Vietnam (IMOLA, 2000).

iii) *The birth of aquaculture in Tam Giang lagoon*

Aquaculture development in Tam Giang lagoon dates back to the late 1970s when the Aquatic Branch of Thua Thien–Hue province started to encourage fishermen to develop their understanding of the lagoon environment to culture aquatic species from natural sources in the lagoon.

With the purpose of studying the lagoon and identifying appropriate and valuable species to culture, the Aquacultural Research Station was established in 1977 (Ong, 1991). With the success of *Gracilaria* experiments, the station was transformed into an aquacultural enterprise, the Agar Company with 95 ha of earth ponds. The company went on operating successfully and profitably with products ranging from shrimp to crab and fish, and on hatchery production of tiger shrimp (*Penaeus monodon*).

Following the Agar Company success, other firms entered the market and also succeeded such as Seaproduct Export Company and Vatex, a joint-venture shrimp specialisation company between Australia and Lobana Company. In 1990, the semi-intensive shrimp culture in Phu Xuan of the Seaproduct Export Company yielded 1.4 to 1.8 tons/ha.

As a result, the lagoon experienced a boom in aquaculture development, especially in tiger shrimp cultivation. In only 5 years, from 1994 to 1999, the area of pond and net enclosures increased from 137 to 473 ha in Phu Tan. The total area of aquaculture in the whole lagoon reached 1000 ha by 1999, 1700 by mid-2000, 1850 ha by the end of 2000, and to 2700 ha in 2001 according to the Department of Fisheries statistics. Programmes in support of aquaculture development in terms of technical training, loans and grants were also offered from 1992. Aquaculture hence became an important sector and cultural characteristic of the region (Ton, 2001b).

iv) *Models of aquaculture*

Early culture methods applied by fishers is considered as "fattening culture" by keeping and feeding shrimp, crab and fish in bamboo cages and net pens to marketable size. The development of earth ponds is constrained due to the limitation of suitable land and investment capacity. Therefore investment in earth pond construction was mainly made by state companies such as Agar Company, Hue Shrimp Company, Phu Xuan Shrimp Company and Aquatic Extension Centre. The provincial and local governments also encourage pond development through capital support, sites and tax incentives.

Extensive culture began in earth ponds built in inter-tidal areas and surrounded by earth dikes with small gates. Natural stocking is used — thus all the natural species found within the ponds are retained in large quantities for culture. Supplementary stocking of certain targeted species such as tiger shrimp, crab, mullet (*Mugil*) and rabbitfish (*Siganus*) caught throughout the lagoon, increases productivity. In this way, fishers depend largely on nature for their fry and postlarvae, food and culture environment although the naturally occurring food is occasionally supplemented with trash fish. Ponds are stocked at very low densities: shrimp at 0.03 per m^2, crab at 0.02 per m^2 and mullet at 0.0011–0.0015 per m^2. Extensive aquaculture takes advantage of the natural production in the pond. The fishers and the farmers who initially participated in training courses organized by Department of Fishery use this model of culture.

When hatchery-produced postlarvae shrimp and shrimp feed became easily available on the market and when wild postlarvae crab could be harvested in the lagoon in sufficient quantities for their culture demand, fishers began to increase the density of cultured species. Stocking density increased to 1 to 2 shrimps per m^2 and 1 crab per m^2. Natural food (i.e. naturally produced within the pond) was still considered important, however, prepared feed was used intensively (Le, *et al.*, 2000). It consisted of two forms: boiled fish ground and cast into dry grains for juveniles and boiled trash fish chopped up into small pieces for older individuals. Production in improved extensive culture depends mainly on stocked species. To reduce the risk of losing the whole crop to disease, the fishers applied polyculture. According to Le, *et al.*, 2000 Common combinations are:

- Shrimp and *Gracilaria*
- Shrimp, crab, fish and *Gracilaria*
- Shrimp, crab and fish

Knowledge of local climate and weather conditions and of the changeability of the lagoon environment helped the fishers develop a suitable seasonal calendar for aquaculture.

In 1995, the local government started to offer licenses for fishers to set nets along the boundary of their fish corral grounds. In the regulations, culture had to start within one year of issuing the license and the area had to be surrounded by an earth dike otherwise the license would be revoked. But in reality, due to lack of capital for investing in a dike and lack of knowledge of aquaculture techniques the fishers continued to culture using only nets.

Net enclosures developed into a system of multiple nets. Within the main nets surrounding the entire area, a smaller area was enclosed with nets of finer mesh size for the culture of tiger shrimp. The area outside the small culture area was devoted to fishing by corrals. This net enclosure aquaculture system was initiated by fixed gear fishers with their fishing corral located in the Phu Tan area. Selective exploitation is applied — harvesting large sized individuals and re-releasing smaller ones.

A considerable number of fishers have set up net enclosures around their fishing grounds without official approval.

v) *Socio-economic benefits*

The number of households engaged in aquaculture increased rapidly with the fast development of aquaculture. In 1996, the number of households was 867, with an average area of 0.79 ha per household. In 2001, it increased to 4207 households, with a slightly lower average cultivation area of 0.7 ha per household. Moreover, development of aquaculture has led to development of relevant services such as food supply, fry production, and trading of antibiotics. The percentage of households with profits vary from 64% to 75% during the period from 1996–2001. In 2001, the total value of aquaculture product is estimated at VND 112 billions (US $ 7 millions) Profit margins are around 25%.

As a result of higher income, the living standard of local people is also improved. Infrastructure is properly built up and maintained. Education and healthcare service are also getting better. The percentage of children receiving proper education is more than 80%. Aquaculture has also improved the awareness of community about natural resource protection and sustainable development. Now, people have to learn new sustainable techniques and improve their skills in order to keep production profitable.

III. Existing Problems

Aquaculture development has resulted in the privatization of fishing grounds — the transfer of community property to family or group ownership and the exclusion of mobile fishers from their traditional fishing grounds. As previously mentioned, there is a problem of illegal net enclosure activities. Moreover, the faster the net enclosures developed, the narrower the fishing ground of mobile gear fishers became which made their lives increasingly difficult. This has led to conflicts between the two groups of fishers. Weakness in the local authorities and management made the problems worse. Aquaculture in the area was developed to assist fixed gear fishers but did not take into account the problems of mobile gear fishers. As a result, a significant number of mobile fishers were forced to make a change in their livelihoods and either fish in remote areas or work as hired labour. In a pilot scheme in Quang Thai, Mobile gear fishers in Quang Thai demonstrated that they can successfully engage in aquaculture if offered certain priorities and assistance. Despite that, there have not been specific regulations or guidelines to assist mobile gear fishers to engage in aquaculture (Le, *et al.*, 2000).

i) *Environmental degradation*

A aquaculture has decreased water flow due to the crowded earth ponds and many net enclosures. The ponds built near the lagoon shore during the early stages of aquaculture are showing signs of degradation caused by very slow water exchange and production is poor. These polluted ponds cover an area of about 50 to 60 ha and are concentrated near the port and Dien Truong Bridge.

There is also a negative impact on the habitat of aquatic lagoon species. For example in Phu Tan, the widespread claim of lagoon areas for the construction of ponds has removed habitat for growth of aquatic species. This has also contributed to the decline in fishing catches in the lagoon.

ii) *Aquatic animal disease*

Aquatic disease and epidemics, a result of poor water quality, presents a serious constraint to the development of sustainable aquaculture in Phu Tan and in the whole lagoon area. For instance, though aquaculture has just developed in Quang Thai, diseases appeared in grass carp in the first culture season. The most common diseases are red body, yellow head, muscle necrosis, softened shell, black spot, black gill, blind eyes, organisms adhesion, carapace deformation, and leg and tail rot affecting shrimp; swollen gill and red shell affecting crab; red spot on fish; and white canopy on *Gracilaria* (Ton. and Le 1999). The resulting economic losses led to reduction in reinvestment and spending on health care and children education and family's living facilities.

iii) *Lack of planning in aquaculture*

Due to the lack of proper planning and design, the number of nets is not optimal, often over capacity. Consequently, water exchanges decreased as well as water quality. Most recently, small-sized net enclosures are being reinforced by earth dikes and converted into earth ponds.

IV. Current management of aquaculture in Thua Thien Hue Province

i) *Policies and regulations*

Regulations on fisheries in Viet Nam are not complete and aquacultural laws are still in its infancy. Currently, the system adopts several international Codes and Guidelines as follows:

- FAO Code Of Conduct For Responsible Fisheries 1995.
- Resolution on Sustainable Fisheries for Food Security for the ASEAN Region.
- Plan of Action on Sustainable Fisheries for Food Security for the ASEAN Region 2001. Section 6 of the Plan clearly pointed out overexploitation of fisheries.

The current fisheries laws include:

- Ordinance on Fisheries Resources Protection and Development 1989 by the National Committee of the Republic Socialist of Vietnam.
- Decree on Implementation of the Ordinance on Fisheries Resources Protection and Development 1990 by the Ministerial Committee.
- Circular on Guides for the Decree dated April 25th 1989 and Decree 195-HDBT dated June 2nd 1990 on Fisheries Resources Protection and Development.
- Master Plan of the Fisheries Sector to 2010: "Policies and Strategies for Development" — Ministry of Fisheries — DANIDA (Denmark).
- Master Plan of the Fisheries Sector to 2010: "Development Program on the Rationalization of Fisheries Exploitation" Ministry of Fisheries — Ministry of Planning and Projection.

In addition, the Fisheries Laws, effective from July 1st 2004, is a significant addition to the legislation framework although.

ii) *Relevant regulations*

Although less emphasis is placed on relevant regulations, these are often important factors contributing to the success or failure of the Master Plan on Aquaculture and Fisheries. For example, section 47 of the Amended Land Laws 1999 is concerned with the use of water surface for aquaculture. Section 15 of the Resource Laws, effective from January 1st 1999, specifies the protection of water quality for agriculture and aquaculture. Section 27 is concerned with the use of water resource for salt production, aquaculture and fisheries. Section 14 of the Environmental Protection Laws mentions that the use of agricultural and forestry land for aquaculture has to follow the master plan, land improvement plan to ensure sustainability and the use of antibiotics, pesticides, animal protection medicines and other products must comply with national laws.

iii) *Strategies*

Facing the unplanned expansion of aquaculture in the lagoon area, in 2001, the local authority started to develop a Master Plan for Aquaculture to 2010. However, the plan is rather simple and mostly focuses on short term methods of reducing the negative impacts of aquaculture on the environment. One of the main themes is the policy of removing net enclosures in certain communes. The main reason is the uncontrolled waste by net enclosures which is considered to be damaging both natural resources and aquaculture environment.

Nonetheless, the plan does not specify any particular short term and long term methods to achieve such objective. In fact, the total net enclosure area did not decrease but continued to increase from 2001 to 2003. By estimates, the total net enclosure area went up by 200 ha during the period. Although some communes, such as Phu Loc, had implemented effective methods to remove net enclosures, other communes including Phu Vang still face considerable problems of negative impacts caused by unplanned net enclosure expansion (DOFI, 2001a).

iv) *Institutional arrangement*

The highest government body in charge of Fisheries and Aquaculture development is the Ministry of Fisheries. In each province, there is a Provincial Department of Fisheries. This is the advisory body for the Provincial People Committee — The highest local authority. The total number of members of the Department varies from 15 to 20. The officials are specialised in a number a subjects including fisheries, aquaculture, fisheries production, engineering, fisheries management, economics, planning, human resources and laws.

The Centre of Aquatic Resource Protecting and Development including the Aquatic Resource Supervisory Board is responsible for fisheries and aquaculture management through registration, licensing, quarantine and monitoring activities. The Centre consists of the Administrative Department, the Quarantine Department and the Aquatic Resource Monitoring Team and Fishing Boat Registration Team.

At the communal level, fisheries officials are responsible for the local fisheries and aquaculture activities. They report directly to the District Department of Agriculture and Rural Development (DDARD). Currently, the number of fisheries officials is listed in the following table

Fisheries officials in different districts

DDARD	Phong Dien	Quang Dien	Huong Tra	Phu Vang	Phu Loc
Number of Fisheries Officials	2	3	1	4	2

The number of officials responsible for local fisheries and aquaculture activities however is rather small. Another common problem is the lack of expertise in the management of aquaculture and fisheries. As a result, the management capacity of fisheries and aquaculture at the local and operational level is very limited. It also takes time and resources to train a local specialist who can work closely with local fishers. An effective method of planning, managing and monitoring of aquaculture is recommended to solve this problem both in the short term and long term.

V. Potential Environmental Impacts of Aquaculture in Thua Thien Hue Province

As previously mentioned, aquaculture has rapidly developed in many communes in Tam Giang Lagoon. However, it is observed that the legal framework for aquaculture has not been completed and nor ready for long-term sustainable development. The significant short-term economic benefits are clear but the longer term benefits or losses are there to be fully assessed.

Aquaculture development itself has inevitable impacts upon the environment. For example, the expansion of net enclosure will decrease the natural areas. Moreover, along with the unplanned intensification of aquaculture, disease and pests are becoming more common. One of the serious problems is the conventional concept of fishers that "It's lucky or unlucky", so negative impacts, different environmental conditions and diseases are natural and cannot be avoided. Certainly, valuable experience is gained and transferred; it may be too late before the environment is seriously damaged.

Other communes in Viet Nam have experienced the consequences of unplanned aquacultural development such as mass mortalities of shrimps and other aquatic faunas in O Loan Lagoon, Phu Yen Province in 1998. The causes included: poor waterway systems leading to poor water exchange, water pollution due to uncontrolled waste, ineffective allocation of net enclosures and illegal encroachment of the lagoon area. The economic loss is substantial with 88% of the aquaculturists suffering great losses and only 2% with minimal profits. What is more important is the degradation of the lagoon aquatic resources which may take years for recovery.

i) ***Water exchange and circulation***

Tam Giang Lagoon is a semi-closed system and where tidal amplitude fluctuates to a maximum of 50 cm. Due to the region's particular climate, the water flow is very weak in the dry season but can cause floods in certain areas in the rainy season (Le, *et al.*, 2000). As a result, in the dry season, the flushing rate of the lagoon is low thus contributing to pollution and aquatic animal diseases.

Another problem is that tidal currents can not reach aquaculture ponds located further inland and thus worsens the water quality used for aquaculture. The intensification of aquaculture in the area is likely to produce a higher concentration of discharged substances than the lagoon environment has the capacity to absorb (Le, *et al.,* 2000).

While the lagoon provides suitable locations for cage farming many distinct physical features of the system may compound the ecological effects of the waste released from cage farms. It is observed that current tides and seawater flows are not sufficient to dilute and aid transport of fish farm waste. Tam Giang lagoon's cage farming system has a restricted entrance which limits the exchange of water between the lagoon and more open coastal waters. In addition, shallow entrance sills prevent the penetration of seawater at depth; thus, deep water within the lagoon may become isolated for a period of time up to a number of years (Lazier, 1963; Gade and Edwards, 1980). This has implications for the dilution of soluble waste from fish farms and the response of the lagoon ecosystem.

The high density of net enclosures in the lagoon may also pose threats to the quality of water and cause aquatic animal diseases. Following is an illustration of the high density of net enclosures in the lagoon. According to Ton, (2001a), the rapid uncontrolled expansion of net enclosures has resulted in a poor waterway system. As a result, these factors should be considered in the region's aquacultural planning to minimize losses and ensure sustainability.

ii) ***Seed supply***

The lagoon relies on two hatchery centers in Thuan An and two small hatchery stations in Thuan An and Loc Vinh. However, in the recent years, seed supply has been insufficient to satisfy aquaculturists' demand. As a result, a large amount of seed has been bought from the southern provinces such as Da Nang and Nha Trang. It is believed that the imported shrimp seed was the cause of recent disease outbreaks (Ton and Le 2001). It is critical that the local authority set up a temporary supervisory board to check on the situation. In the long-term, seed centers or nursery grounds should be developed and maintained properly.

The lack of sufficient seed supply may lead to a possible threat of the natural aquatic resources, which is the over-exploitation of shrimp and crab seed. In extensive pattern aquaculture (polyculture and continuous stocking and harvesting), natural seed sources are important to aquaculturists. It can be deduced that the natural seed supply makes a considerable contribution to solve the seed shortage. There must be control measures on the exploitation of natural seeds before the aquatic resources are seriously damaged.

A large amount of juvenile aquatic species is exploited to meet the demand for feed in aquaculture, which also decreases the lagoon's resources (Ton, 2001c). Short-term and long-term (what?) actions should be taken to control the situation.

iii) *Encroaching the lagoon*

As mentioned above, most of the lagoon's bottom is shallow and flat. Under these favorable conditions, the total area of aquaculture ponds increased quickly. As a consequence, seagrass beds — a habitat of fry and juvenile aquatic species — have been destroyed and the lagoon shape has changed considerably. At present, most shallow areas close to the lagoon edge are used for aquaculture, which reduces the lagoon area and affects the distribution of juvenile aquatic species, including native species and those migrating into the lagoon.

Expansion of the aquaculture activities is narrowing the natural area of the lagoon. For instance, in Phu Tan the area of natural lagoon was halved during the period between 1987 and 1998 (Ton and Le 2001).

Particularly, in Phu Tan, there are 20 groups of aquaculturists using net enclosures with a total area of 138 ha. Most of the lagoon water surface was divided into small areas with net pen and net enclosure as illustrated in the following map.

iv) *Destroying habitats, reducing biodiversity*

As a consequence of pond construction, many seagrass and freshwater macrophyte beds which are the habitats of fry and juvenile aquatic species,

MAP OF AQUACULTURE SYSTEM AND WATER WAY IN PHU TAN AREA

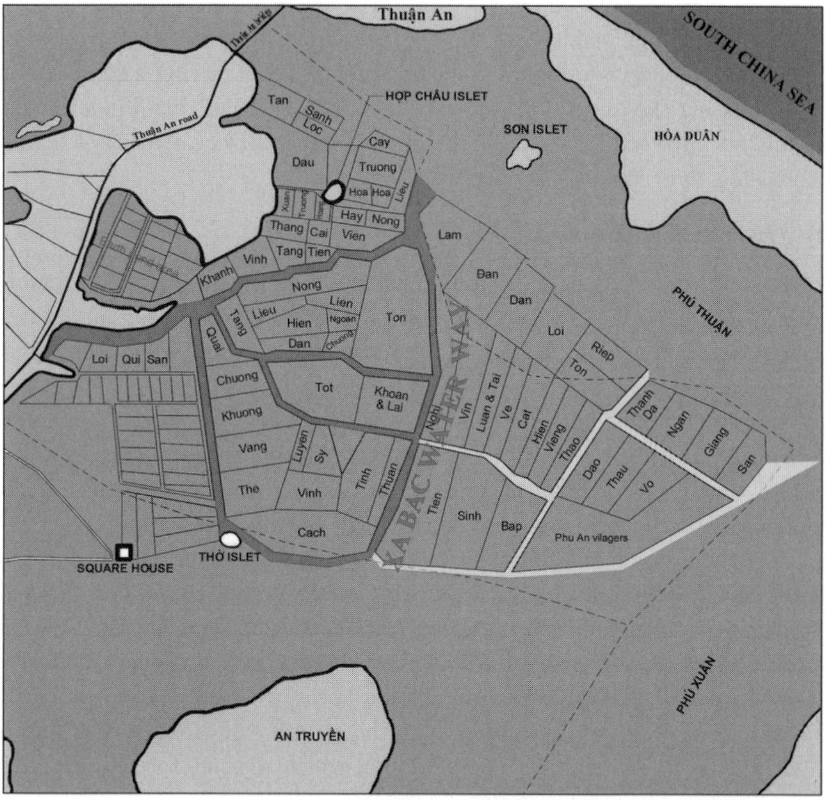

DRAWN BY TAN DUONG VILLAGERS WITH THE ADJUSTMENT OF THE RESEARCHERS OF PHU TAN GROUP - IDRC PROJECT

Fig. 14.3. In Sam-An Truyen area, seagrass ground of Cymodocea rotundata, Halophila beccari is almost replaced by aquaculture ponds.

have been destroyed and the shoreline has changed dramatically. For example, the seagrass *Cymodocea rotundata* is almost destroyed in An Xuan area and is degraded in Sam-An Truyen. The seagrass Najas indica and Halophila beccari around Vinh Thanh area is also affected severely. The overexploitation of seagrass which is used as food in aquaculture is destroying seagrass beds near the O Lau river estuary (Ton and Le 2003).

The ideal habitat is shallow water areas 30–50 cm deep which has dense beds of aquatic plants. However, with these habitats gone, fish and shrimp juveniles have to live in deeper waters where they lack food and safe shelter and where they are vulnerable to many predators. This increases the mortality rate among juveniles-one of many reasons for the decrease of catch in the lagoon. Catch rates in Ha Dao area, reputed as the highest in Phu Tan, have decreased by 4–5 times over the last 2 to 3 years.

Although net enclosures do not destroy seagrass and freshwater macrophyte beds, they encroach the lagoon area, obstructing water currents and impeding the natural fry migration. With water currents hindered, fish fry and shrimp seed have a difficult time reaching nursery areas thus further decreasing their chances of survival.

v) ***Overexploitation***

As aquaculture occupies the common fishing area, the majority poor mobile gear fishers have to go to deeper waters to fish. Moreover, as large adult shrimp and fish are not abundant anymore, these fishers have been forced to replace old gear (such as hook and line and gillnet) with new fishing gear to catch small juvenile fish and shrimp before they get a chance to reproduce. With little or no reproducing individuals, many species are at risk of disappearing from the lagoon. If exploitation of juveniles is not controlled, the natural lagoon resources will continue to degrade.

Another important finding related to the aquaculture expansion was of the new harvest and use of the fresh water macrophyte (called *"Rong"*) for fish pen culture in the northern lagoon system. The local people are concerned of the scarcity of *Rong* because they can no longer collect it nearby. Instead, they have to go further away into the lagoon or in the rivers to find it. It is also noted that the scarcity of *Rong* is also due to other destructive fishing activities, among which were motorized drag nets and eel rakes. There is also a decline in fish catch per unit effort of all major types of fishing at the north and middle lagoon. The rate is estimated at 23–45 percent during the year 2000 and 2003.

Scientists reported that there are many types of harmful and toxic phytoplankton present in Tam Giang Lagoon. If there are excess nutrients (from aquaculture effluents), they may grow and bloom rapidly (eutrophication) leading to a water bloom or algae bloom, thus Dissolved Oxygen (DO). Some phytoplanktons have the ability to produce toxic compounds effecting aquatic species. There are 30 harmful phytoplanktons that occupy 5.9% of the total phytoplanktons in Tam Giang Lagoon. They belong to different phylum: Cyanophyta: 1 species, Heterokontophyta: 2 species, Dinophyta: 27 species.

In 2000–2003, *Alexandrium* and *Prorocentrum cordatum* bloomed along the lagoon. Scientists have identified high density areas of them as shown in the following pictures (pictures?). For the first time in history, there was a "Milk Tide"- a strange bloom occurring in the Tam Giang lagoon (Ton, 2004). It appeared in June 20, 2004 in the lagoon area near to Vinh Ha commune. The layer of bottom lagoon water turned a milky colour. The dissolved oxygen in the bloom area is low 1-2mg/L. Various kinds of fish and eels were killed. The bloom disappeared after a heavy rain in the middle of July, 2004. The water sample was sent to Japan for testing but the strange species that caused this tide has not been identified. It is very small with diameter of 2 micrometers. This phenomena should be thoroughly studied and seriously treated if sustainable development is to be achieved. Although the water pollution and problem of aquaculture effluent are not that severe, preventative actions must be taken and proper planning as well as monitoring must be put in effect.

Fig. 14.4. Milk Tide in Tam Giang lagoon.

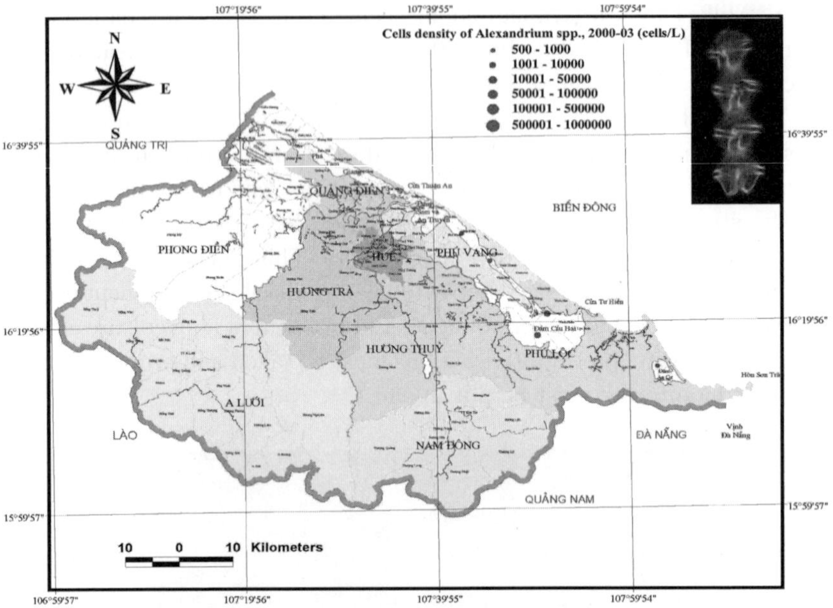

Fig. 14.5. High densities areas of *Alexandrium* spp.: 2000–2003.

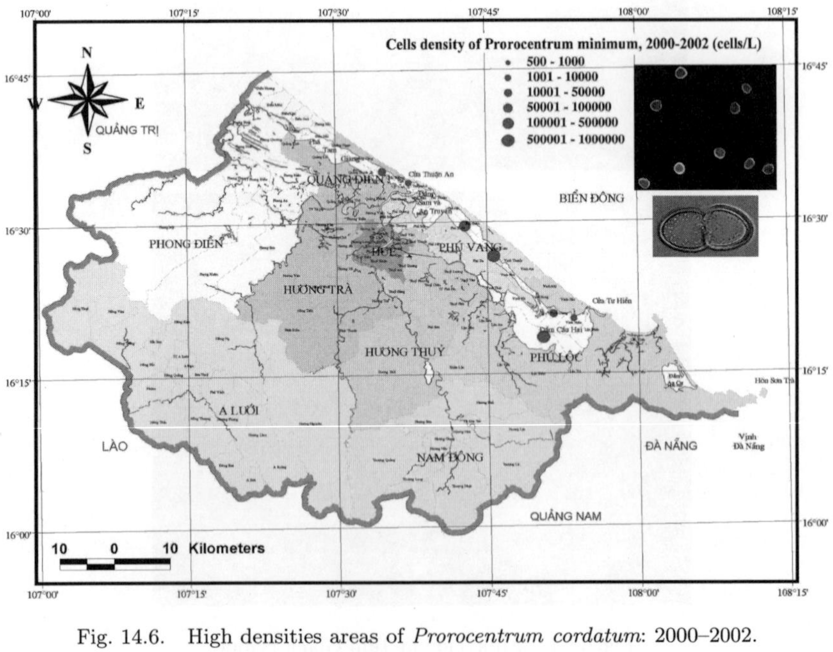

Fig. 14.6. High densities areas of *Prorocentrum cordatum*: 2000–2002.

vi) *Diseases*

Aquaculture development in Tam Giang Lagoon has faced rather serious problems of animal health since 1994. They tend to spread over a large area of the lagoon in March, April, July and August of each year. For example, during the 1995 shrimp disease epidemic, 300 million fry and 900 million full grown shrimps died. This resulted in huge capital losses for the farmers, and many have been unable to repay their bank loans. A survey showed that one household lost all its capital of VND 38 million (~ 3000USD) after stocking 20,000 PL15 in a 5,000 m^2 pond in 1998. Other household received only 15 million VND (~1200 USD) in returns after investing 52 million VND (~4500USD) in a 3,000 m^2 pond (Ton, 2001; Anon, 1999a).

In 1985, the "colourless" disease appeared in *Gracilaria* which the local fishers call "white canopy" disease. The tips of the fronds become white in colour and the thallus perishes. Mullets, rabbitfish and other species of fish living in the *Gracilaria* ponds suffer from ulcers (Ton. and Le, 1999). The diseased *Gracilaria* also affected the fingerlings of the shrimp used in aquaculture.

Between 1993 and 1994 shrimp disease affecting limited areas in the lagoon began to spread and from 1995 to 1997 shrimp disease broke out as an epidemic. Presently, disease is one of the major constraints to fish and shrimp culture in especially after the Century Flood of 1999 when the environment significantly changed. The common diseases in fry are luminescent disease, carapace deformation, red spot and external fouling. Interventions applied by fishers to reduce the impact of disease on affected animals consist of changing the water to improve water quality and providing enough food to help the animals overcome the disease. Some fry producers have used antibiotics and chemicals for treatment of disease.

vii) *Environmental effects on disease*

Aquaculturists used their accumulated knowledge and experience to rank environmental factors based on their effects on diseases in aquaculture.

Salinity (‰) is considered to have the highest impact on disease while "ro" water (Ro) (what's this?) and tide (Td) are second. These factors are followed by temperature (T°) then seasons (Ss) management (Mg, which aquaculturists referred to as care and management), location (Lo) and water colour (WC). The high density of net enclosures (Hd) and aquatic flora (Af) are considered to have the least impact on disease.

According to aquaculturists, aquatic flora has little impact unless combined with the depth and the quality of substrate of the net enclosure (Le, 2000).

viii) *Socio-economic impacts of disease*

In the Project at Tan Duong village located in Tam Giang lagoon, the survey results of 31 aquaculturists indicate varying economic loss as a result of disease. Almost one third of aquaculturists experienced an 80% loss of shrimp and the average loss of shrimp was 71%. Crab losses were fewer with 7 aquaculturists experiencing a 30% loss and another 7 aquaculturists a 50% loss. Average crab loss was 42%. Although the maximum loss experienced by aquaculturists was 60 million VND (~ 5000 USD), this is a big economic loss as they have little capital. Most of them borrow money from local creditors at high interest rates to invest for each season.

Most aquaculture households experience unfavorable social impacts of disease through effects on the debt status, constraint to aquaculture development, health care, children's education and the balance of labor division

Some of the positive impacts of disease in aquaculture include:

- The increased knowledge and experience collected from crop failures due to disease.
- A greater effort on the part of aquaculturists to pay back loans which increases self-reliance.
- A search for improved techniques on aquatic animal health care, aquaculture management and strategies for preventing and treating disease based on local experience.
- A search for better strategies for the prevention and treatment of disease on the part of DOFI and its extension department.

Aquaculture is the key livelihood of most of village families, thus aquaculturists usually maintain the activity, although they face disease and other risks. They understand that although nature has provided favourable conditions for the development of aquaculture, it can also be severe. The aquaculturists believe they must stay firm and collect experiences from both failures and successes in every culture season. Eventually these will contribute to improved methods on health management, aquaculture management and the development of household economics.

VI. Recommendations

According to Shehadeh, and Pedini (1997) the following issues must be addressed to ensure sustainable aquacultural developments:

- the establishment of an enabling environment,
- adoption of an integrated planning and management framework,
- increasing the efficiency of resource use and productivity in general at the farm level,
- reducing externalities and avoiding irreversible damage,
- establishing reliable databases and effective information management systems,
- maximizing positive sustainability attributes.

i) *Waterway systems*

The high income potential of net enclosures led to rapid uncontrolled application in many communes within the lagoon area and resulted in a poor waterway system (Ton, 2001a). It is essential that the waterway system has to be properly built and maintained. This will:

- Ensure waterway traffic safety.
- Ensure sufficient space for aquatic animal growths.
- Provide more fishing space for mobile gear fishers.
- Ensure proper water exchange between net enclosures and earth pond areas.

It is not difficult to design an effective waterway system for any aquacultural areas. The key challenge is how to apply and maintain the system and gain acceptance from aquaculturists. The answer lies in a co-management system identified and tested by Professor Ton That Phap (2001).

The co-management system has to be supported by education, negotiation and initiatives from the local government and the fishers as partners (Ton, 2001a). It is in the interest of the fishers to express their problems and to co-operate to find the solutions. The role of the government officials is to co-ordinate, supervise and provide guidelines and technical assistance. The pilot project in Phu Tan produced impressive results and set an example of co-management in a coastal community in Vietnam (Ton T.P., 2001a). It is believed that the system will achieve similar results in other communes within the lagoon area.

ii) *Mangrove replanting*

Mangrove replanting and protection is one important issue that must be addressed. The development of aquaculture sites, roads, ports and residential areas has reduced the mangrove area. As a result, the aquatic resource is threatened, which in turn led to lower productivity and increased pollution. It is a priority that a "green belt" of mangrove has to be identified, replanted and maintained. A key lesson from the Sumatra example is the timing of the planting: seeds must be planted in the rainy season to improve survival rates.

Another potential cause of mangrove replanting failure is concerned with land ownership as experienced in Sumatra. In the case of the Tam Giang lagoon, similar issues are found. There is an issue of "uncultivated areas", especially in the very sensitive mangrove areas, which are illegally occupied and exploited (DOFI, 2004b). As a result, these areas should be protected under clear legislations.

Disease control and raising awareness of aquatic animal health management

The encouraging results of the "Raising Awareness of Aquatic Animal Health Management Project in Trung Lang, Quang Thai Commune" prove that the method can be used effectively to control diseases and thus ensure aquaculture productivity and sustainability (Ton, 2001). Using participatory discussions and workshops, local fish farmers will become aware of the causes of common aquatic epidemic diseases, preventative methods, and the importance of environmental management in aquacultural development. As shown in the Trung Lang Project, fish farmers effectively learnt from the discussions and workshops and changed their fish farming methods to reduce the threats of diseases and achieve higher productivity (Ton, 2001).

iii) *Build and manage protected areas*

To ensure sustainable aquaculture and biodiversity, it is critical that the master plan categorizes different areas of the lagoon for different development purposes, areas with diverse compositions and high density of seeds. Moreover, most of these sites have large areas of seagrass. Some concentrated seagrass areas amount up to 1,762 ha. These are important habitats of aquatic resources, particularly the important seeds. There is a positive correlation between the concentration of seagrass and the amount of aquatic seeds.

These areas are different in terms of biodiversity and the sensitivity to negative impacts. Each area requires different management and should only be used for suitable activities. For example, the normal areas are suitable for commercial fishery and mass aquacultural production. Selective and controlled aquaculture activities can be carried out in the sensitive areas after all the preventative measures have been put in place. The very sensitive area consists mainly of critical natural capital and therefore should be strictly protected.

For proper protection and management, these areas should be classified as protected areas with the following objectives:

- To reduce negative impacts of aquaculture and fishery on rich biodiversity areas.
- To protect the nursery function of mangroves, seagrass beds and other nursery habitat for shrimp and fish species.
- To increase productivity and ensure sustainable aquaculture and fishery.

According to the Mekong Protected Areas field study conducted in Tam Giang Lagoon, there is evidence of lower productivity caused by lack of protected area intervention. As a result, it suggested that protected areas should be built and properly maintained. The following map shows the locations of the proposed protected areas in the lagoon:

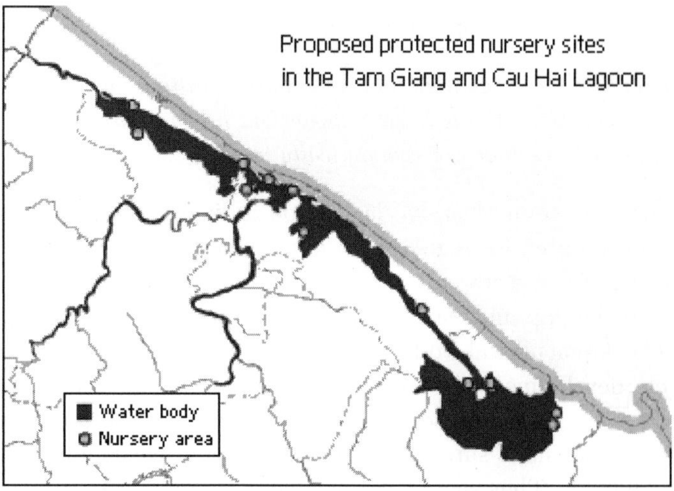

Fig. 14.7. Proposed protected nursery sites in Tam Giang Lagoon.
Source: Mekong Protected Areas Field Study in Thua Thien Hue Province 2003.

iv) *Legal guidelines*

The role of aquaculture has significantly increased from the perspective of both producers and consumers. Recently, consumer perception of food production has been negatively affected due to cases such as mad cow diseases or genetically modified food. As a result, consumers are increasingly interested in how farmed products are actually produced. They are also concerned with animal welfare and environmental impacts of the food production process. At the same time, driven by profits and competition, producers may not offer products that are safe or environmental friendly. It is therefore critical for law makers to ensure sustainable mutual benefits of all relevant stakeholders in the aquacultural sector. Specific aquacultural regulations and guidelines should be carefully considered and put into effect.

According to Cullinan, and Van Houtte (1997), in the long-term, the government needs to develop legal and institutional instruments to:

- Recognize aquaculture as a distinct agricultural sector;
- Integrate aquaculture concerns into resource use and development planning;
- Improve food safety and quality to safeguard consumers and meet the standards of importers;
- Improve the management of aquaculture, particularly where it has the potential to be socially or environmentally unsustainable.

In the short and medium term, it is suggested that Vietnam considers the Code of Conduct for European Aquaculture:

> "*Aquaculture should give the best, most healthy and nutritious product possible with the lowest impact on the environment possible using procedures that are economically viable.*"

Passing new comprehensive legislation is likely to take a number of years, while the need for regulation and management is urgent due to the rapid growth of the sector. As a result, a voluntary and non-biding Code would fill in the gap and establish a foundation for the formal legislation system. The Aquacultural Code of conduce will ensure the transparent and responsible development and management of the aquacultural sector for the benefit of consumers and producers ultimately. It is important that the Code should make sure that producers have a clear opinion of what "good conduct" is. They should also be able to communicate their interests, opinions and concerns. In addition, the Code should address the concern of final consumers and all other stake holders. The Code should at least cover topics such as:

- the interactions with the environment.
- the welfare of the livestock.
- the product supplied to the consumer.

The adoption of the Code will be a cornerstone for further actions and developments of the aquaculture sector in Vietnam in general and in the Tam Giang Lagoon area in particular. It would improve the competitiveness of the region's aquacultural products for export. As one of the largest importers of aquacultural products, the European Union currently imposes detailed conditions on the handling, slaughter, inspection, processing, packaging, identification and storage of fishery products, and applies stringent controls over the animal health conditions applicable to the marketing of aquaculture animals and products. It also increasingly pays attention to the quality of the water resources and the consequential need to ensure conservation of and control over the pollution of the aquatic environment, and, on the other hand, to develop harmonized standards for fish food products marketed within the EU.

In fact, the recommendation to develop a National Code of Conduct was agreed on in the National Workshop on the Code of Conduct for Responsible Fisheries and its practical application to coastal aquaculture development in Viet Nam (FAO, 2004. FishCode Review No.12). The workshop suggested the National Code of Conduct be built based on the FAO Code of Conduct for Responsible Fisheries. However, what is important is not which framework to embrace but the practical concerns of the national and local aquacultural development in the search for equity and sustainability.

The effectiveness of voluntary codes of practice can be enhanced by creating incentives to encourage compliance or disincentives for non-compliance or failure to join the scheme (FAO/FishCode Review No.12). For example, in India projects involving intensive culture or the conversion of non-saline land to brackish water aquaculture will not be sanctioned, and refinance support will be provided to meet the requirements of the National Bank for Agriculture and Rural Development. Tax incentives are also provided in Malaysia for aquaculture enterprises (FAO/FishCode Review No.12).

In preparation for the new Code of Conduct or legislations, it is suggested that the MOFI and the departments and research institutes under it should immediately formulate projects/programmes to collect and analyze relevant information in support of the implementation and dissemination of the Code.

v) *Community-based planning*

The importance of self-management and participatory planning has increasingly been recognized where national and local authorities do not have the sufficient resources and capability to effectively develop, implement and enforce aquaculture guidelines and regulations. In the case of Vietnam, another major constraint is the lack of knowledge about good production practices, public understanding of the importance of a sustainable coastal ecosystem. There is also the issue of the failure of the commonly used top-down approach. As a result, participatory or community based planning should become the central focus of sustainable aquacultural developments in Tam Giang Lagoon in particular and in Viet Nam in general.

The effectiveness and applicability of the approach in the management of aquatic resource and sustainable livelihoods of such approach has been well proven in many pilot projects conducted in the area (Lessons in resource management from the Tam Giang lagoon, 2002). This is further confirmed by a statement by a well-respected Vietnamese senior researcher in the field, Dr. Truong V.T. (2004, 15)

> *"The participatory approach helped us to change the ways of thinking from conventional methods which resulted in the scientific data and technology. That is the things not the human. We now have the tools and are interested in working with people to understand them and to come up with the necessary actions"*

One fundamental problem of the local aquacultural development that is already mentioned previously concerns with land use. For example, most of the productive lagoon area in Thuan An entails exclusive occupied rights by the more powerful villager groups without any legal permission for expansion of net enclosures. The current planning and legislation system does not seem to be capable of effectively resolving the problems. It is simply a failure of communication and finding root cause solutions. It is likely to take time to resolve such difficult problem, but participatory planning can reduce the negative impacts and allow room for negotiation until formal legal enforcement becomes effective.

In the Quang Thai pilot project (Truong, 2004), the participatory planning was able to bring out the legitimacy of the natural resource management at the local level. At the same time, it sustained people's livelihood because it recognized the current access based on customary rights. The resulting community actions were to regulate the exploitation and reduce pressure on lagoon use thereby minimizing resource degradation. It is believed that similar results will be achieved if the approach is employed by other communes of the Tam Giang Lagoon.

The planning in the project was also successful in rearranging pens without any force or actual financial incentives. As a result, competition was no longer necessary nor possible. Mutual benefits and developments became the focus. Good practices were also established. For example, the conventional practice of illegally occupying land then asking for permission was abandoned. Instead, a process of gear location registration was successfully implemented (Truong, 2004).

Planning also improved management capacity and effectiveness. Responsibilities and tasks were clearly defined and successfully carried out. The emphasis is on empowerment, self-reliance and self-management. As a result, the planning was able to create a base for development of consensus for registration, licensing and taxation and future formal regulations.

The key elements of the participatory planning that ensures its effectiveness include:

- For common understanding among the stakeholders of the context, problems, and planning approach.
- Active participation of all relevant stakeholders.
- Learning-by-doing.
- Empowerment and clearly defined responsibilities and tasks.
- Sufficient and timely support from relevant authorities.

VII. Conclusion

This study has provided a detailed analysis of aquaculture development and its impacts upon the environment in Tam Giang Lagoon. The lagoon is considered to be suitable for aquaculture development. The national as well as local government has formed strategies that emphasize aquaculture development as an effective tool to alleviate poverty and improve living standards.

Aquaculture in the lagoon has developed more rapidly than the local management capacity. It has brought about significant changes in terms of economic gains and social status. However, due to the lack of a proper legal framework for planning and managing, there are considerable potential threats including:

- Poor water exchange leading to poor water quality.
- Decline in seed supply leading to degradation of aquatic resources.
- Encroachment of the lagoon.
- Threats to habitat, nursery ground and damaging biodiversity.
- Overexploitation of natural aquatic resources.
- Water pollution.
- Aquatic animal diseases.

In the author's opinion, the total area for aquaculture in Thua Thien Hue is sufficient and should not be increased to ensure sustainable development. What is more important is the strategy to improve productivity and the technological, organizational and legal infrastructure. As a result, the report suggests several proven effective methods to reduce the negative impacts of aquaculture development as follows:

- Disease control and Raising Awareness of Aquatic Animal Health Management through participatory discussion and planning.
- Building and managing protected areas for long term seed supply and biodiversity protection.
- Mangrove replanting to provide an enabling environment for sustainable development.
- Improvement of waterway system through co-management methods.
- Suggestions on the adoption of certain international legal guidelines.
- Community-based approach to planning and management.

These are working methods that have been tested in certain communes of the lagoon system. As these communes share very much the same ecosystem characteristics, similar successful results should be achieved when applied in the whole lagoon area. Among these methods, the most important one is the community-based and co-management approach. As the local government still lacks expertise and the accumulated knowledge of fishers is still limited, this approach will ensure sustainable aquacultural development with minimized environmental impacts and economic conflicts. This is the key concept of new aquacultural management and development in this area.

References

Cullinan C. and A. Van Houtte (1997), *"Development of Regulatory Frameworks"*, Review of the state of world aquaculture, FAO Working Paper, FAO Rome, Italy.

DOFI, (2001a), *Master plan for Aquaculture of Thua Thien Hue 2001–2010.*

DOFI, (2001b), *Master plan for Fishery of Thua Thien Hue coastal zone 2001–2010.*

DOFI, (2004), *Report about shrimp diseases situation in 2004 and preventive program for 2005, Thua Thien Hue.*

FAO, (2004), *"Report of the National Workshop on the Code of Conduct for Responsible Fisheries and its practical application to coastal aquaculture development in Viet Nam"*, FAO Fish Code Review No.12.

Gade H.G. and Edwards A., (1980), *Deep water renewable in fjords*, In: Fjord oceanography (eds. Freeland H.J., Farmer D.M. and Levings C.D.), Plenum Publishing Corp., New York.

IMOLA, (2000), *Project Document of Project Integrated Management of Lagoon Activities in Thua Thien Hue province (IMOLA).*

Lazier J.R.N., (1963), *Some aspects of the oceanographic structure of the Jervis Inlet system*, Department of Physics, University of British Columbia, Canada.

Le T.N.T., (2000), *Sustainable development of aquaculture in tam giang lagoon*, p. 27–38 In: Lessons in resource management from the Tam Giang (eds.V.J. Brzeski and G.F. Newkirk), Dalhousie University, Halifax, Canada.

Le V.M , Ton T.P and Hoang N.D, (2000), *Aquaculture — Its introduction and development*, p. 115–133 In: Lessons from the Lagoon (eds. V.J. Brzeski and G.F. Newkirk), Dalhousie University, Halifax, Canada.

Nguyen D.G., (2004), Fundamental issues of Law of Fisheries, Vietnam Justice Publisher.

Nguyen X.H, Tran V.T and Ton T.P, (2000), *A historical perspective on the management of lagoon resources*, p. 39–54 In: Lessons from the Lagoon (eds. V.J. Brzeski and G.F. Newkirk), Dalhousie University, Halifax, Canada.

Ong, V.D. (1991), *The status and development of Gracilaria of Seaweed Culture and Processing Center (SCPC) Hue*, Seaweed Culture and Processing Centre Report, p. 19–21.

Shehadeh, Z.H. and M. Pedini (1997), *"Issues and Challenges"*, Review of the state of world aquaculture, FAO working paper, FAO Rome, Italy.

Ton T.P. and Le T.N.T., (1999), *Aquatic animal health assessment in Tam Giang Lagoon, Vietnam*, Presented at the DFID/FAO/NACA/GOB Asia Regional Scoping Workshop "Primary Aquatic Animal Health Care in Rural, Small-Scale, Aquaculture Development" held in Dhaka, Bangladesh, Sept. 27–30, 1999.

Ton T.P. and Le T.N.T. (2001), *Emerging disease problems tam giang lagoon aquatic systems health assessment*, Hue University of Sciences Hue City, Viet Nam.

Ton T.C. (2001), *"Raising Awareness of Aquatic Animal Health Management"*, Lessons in Resource Management from the Tam Giang Lagoon. (eds.V.J. Brzeski and G.F. Newkirk), Dalhousie University, Halifax, Canada.

Ton T.P. (2001a), *"Co-management in the Planning of a Waterway System for Aquaculture"*, Lessons in Resource Management from the Tam Giang Lagoon. (eds.V.J. Brzeski and G.F. Newkirk), Dalhousie University, Halifax, Canada.

Ton T.P., (2001b), *Development of Aquaculture in Tam Giang Lagoon, Thua Thien-Hue, Viet Nam*, Hue College of Science.

Ton T.P., (2001c), *Report on biodiversity in Tam Giang lagoon, Thua Thien Hue, Vietnam*, Integrated Coastal Zone Management (ICZM) Project.

Ton T.P. and Le V.M., (2003), Knowledge for sustainable development, Hue University of Science.

Ton T.P., (2004), "*Harmful Algal Bloom in Tam Giang lagoon*", Hue University of Sciences Hue City, Viet Nam.

Truong V.T. (2004) "*Participatory Local Planning For Resource Governance: A Case In Tam Giang Lagoon, Vietnam*", CBCRM Project, Hue University of Forestry and Agriculture, Hue, Vietnam.

Truong V.T. and Veronika J.Brzeski, (2000), *Biological resource management in Tam Giang lagoon*, p. 7–14 In: Lessons from the Lagoon (eds.V.J. Brzeski and G.F. Newkirk), Dalhousie University, Halifax, Canada.

UNDP/MOFI, (2000), Project Documentt of Project VIE/97/030/A/01/NEX Environmental Management in Coastal Aquaculture, www.undp.org.vn/undp/ prog/prodocs/env/vie97030, (accessed April 2004).

15

RELATIONSHIP BETWEEN INCOME AND ENVIRONMENTAL DEGRADATION A CASE STUDY OF THAILAND

RAWADEE JARUNGRATTANAPONG
and JESUTHASON THAMPAPILLAI

This study critically examines the theory underlying the environmental Kuznets curve (EKC) with reference to Thailand. It investigates the EKC using cross-regional panel data on emissions of four major air pollutants (PM10, NO_2, SO_2, and CO) and data on deforestation. There are two different relationships in the EKC of Thailand: an inverted U-shaped relationship for PM10 and CO, and a monotonically increasing relationship for SO_2 and NO_2. The relationship between per capita income-PM10 and per capita income-CO appear to support the EKC thesis; but these results could be biased due to the effect of the 1997 financial crisis, which substantially reduced these pollutions. However, environmental measures, both command-and-control (e.g. improvement of fuel quality, and improvement of emission standards) and economic instruments (introducing a price differentiation between leaded and unleaded gasoline, and demand-side management in the power sector), have played a major role to improve air quality in recent times. Unlike PM10 and CO, NO_2 and SO_2 violate the EKC theory by showing a rising relationship with per capita income. A plausible explanation for the monotonically increasing relationship for NO_2, SO_2 is the usage of energy resources in Thailand that are high sulfur and nitrogen content energy resources. Coal, which emits acidic gas in the form of SO_x and NO_x followed by SPM (suspended particulate matter) into air after burning, is a major fuel in electricity generation and other industries in Thailand. The increased coal consumption could explain why the relationship between income-SO_2 and income-NO_2 are a monotonically increasing relationship, even though environmental policies have been adopted to improve air quality. Furthermore, this study finds no evidence of an inverted-U shaped EKC for deforestation in Thailand. Population density is a factor that has an effect on deforestation and this study illustrates that the 1989 logging ban in Thailand had little effect on forest conservation. On the basis of this study, it is reasonable to conclude that a monotonically rising relationship between income and environmental degradation is more valid in the context of Thailand, implying that Thailand is on a rising path of EKC so far.

I. Introduction

Can we maintain high economic growth without reducing environmental quality? What is the relationship between economic growth and environmental quality? Policymakers have shown considerable interest in these questions. These have been a source of great controversy for a long time. Some people believe that higher levels of economic growth require larger input of material, energy and large quantities of waste discharge. They think that environmental quality could not go along with industrial development. Environmental quality could only be achieved by escaping from industrialization. On the other hand, some people argue that the way to maintain environmental quality is through economic growth.

The Environmental Kuznets Curve is named after Simon Kuznets who hypothesized that the relationship between a measure of inequality in the distribution of income and the level of income is an inverted U-shaped curve. Kuznets (1955) proposed an inverted U-shaped relationship between income growth and income inequality, which is known as the Kuznets curve.[1] The general form of the original Kuznets Curve is shown in Fig. 15.1.

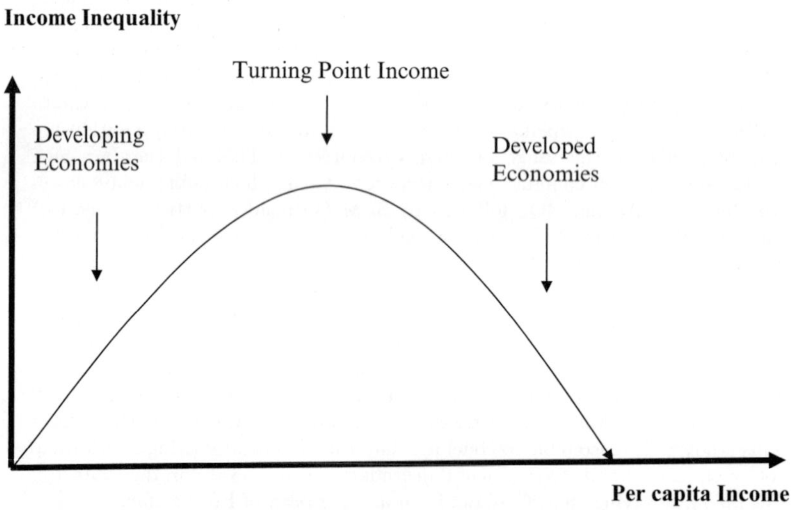

Fig. 15.1. The Kuznets curve.

[1] The original Kuznets curve that Simon Kuznets proposed is a U-shaped relationship between income growth and income equality.

Kuznets's hypothesis states that at the early stage of economic development income inequality increases while at the later stage it decreases. It was supposed that this change was alongside the structural change from agricultural to industrial economy. When a new industry is introduced into a matured (or stagnant) economy that may be an agricultural economy or industrial economy, the income gap will widen due to the gap between the new and old industries. However, when the economy moves towards a new equilibrium through a shift of resource allocation, the income gap will narrow. This may be explained in economic terms.

When an agricultural economy transforms into an industrial economy, it is necessary to give incentives to industrialize. Thus higher income inequality is inevitable at the early stage of economic developments. When the industrial sector succeeds in absorbing a large part of the labour force and the inequality is adjusted, the income gap will be narrowed (Ikemoto and Uehara, 2000).

The environmental Kuznets curve (EKC)

The environmental Kuznets curve theory proposed that there is an inverted U-shaped relationship between various indicators of environmental degradation and economic development. Kuznets's name was apparently attached to the curve by Grossman and Krueger (1995), who found its pattern was analogous to Kuznets's inverted-U relationship between income inequality and economic development. They estimated EKCs for SO_2, dark matter and suspended particles (SPM) using a cross-country panel of data on urban air pollution levels to estimate the point at which atmospheric concentration of SPM and SO_2 become decreasing functions of income. Figure 15.2 shows the typical shape of the environmental Kuznets curve.

The shape of EKCs can explain that at low income levels in the first stage of industrialization, economic growth is associated with increasing pollution levels or concentrations (for instance, the rate of land clearance, resource use and waste etc) because people are more interested in benefit and income than quality of environment, and environmental regulation is accordingly weak. As income rises, pollution gradually levels off and, structural change towards information-intensive industries and services, linked with increased environmental awareness, enforcement of environmental regulations, better technology and higher environmental expenditures, result in a leveling off and a gradual decline of environmental degradation (Vogel, 1999).

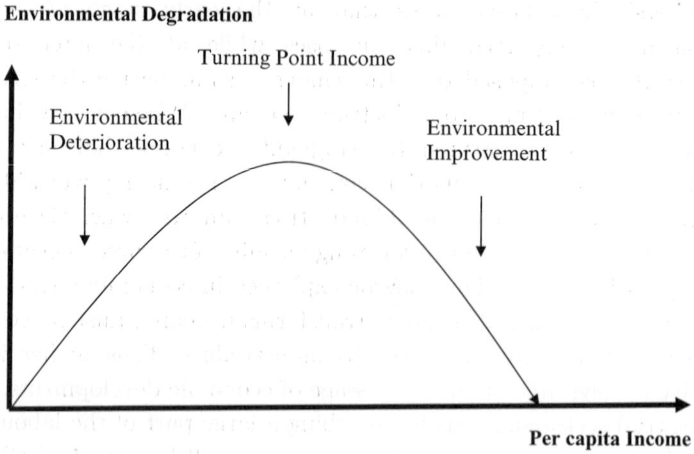

Fig. 15.2. Environmental Kuznets curve.

The patterns of relationship between income and environmental degradation

There are many different types of relationships between income and environmental degradation that can be distinguished between monotonic and non-monotonic curves. Monotonic can be either increasing pollution with rising income, or decreasing pollution with rising income. Non-monotonic patterns have been suggested in namely inverted-U curve. However, the patterns of relationship depend on the types of pollutants and the models that have been used for estimation (Panayotou, 2000).

Selden and Song (1994) suggest four arguments in shape of an inverted-U curve for air pollutants. These factors are: (i) positive income elasticities for environmental quality; (ii) changes in the composition of production and consumption; (iii) increasing levels of education and environmental awareness; and (iv) more open political systems. They reflected both market forces and changes in environmental regulation. Thus, environmental improvement requires that economies would pass through "stages of development" with at least some aspects of environmental quality decay in the beginning before improvements.

Panayotou (2000) suggests that a monotonic increase of environmental degradation requires for strict environmental regulations and even limits on economic growth to ensure a sustainable scale of economic activity within the ecological life-support system. A monotonic decrease of environmental degradation along a country's development path suggests that policies that accelerate economic growth lead to rapid environmental improvements and no explicit environmental policies are needed. Lastly, it is the inverted U-shaped of environmental degradation or the Environmental Kuznets Curve. There are many issues in this case: (1) what level of per capita income is the turning point? (2) how much damage would have taken place and can it be reduced? (3) would any ecological thresholds be violated and irreversible damages take place before environmental degradation turns down, and how can they be avoided? (4) Is environmental improvement at higher income levels automatic, or does it require institutional and policy reforms? and (5) how to accelerate the development process so that poor countries can experience the same improved economic and environmental conditions enjoyed by developed countries?

The aims of this study are to investigate the relationship between economic growth and environmental degradation in Thailand and to try to answer the question about what factors (e.g. economic growth, population, and environmental policy) affect environmental conditions. However, the objective of this study is not to prove or disprove the EKC hypothesis. For testing the EKC hypothesis, long-term data is needed to analyze the pattern of economy structure. Long term environmental data is not available for Thailand. Therefore, the expected results of this study show the pattern of income-environment relationship in Thailand with the limited data that is available. The data limitation also confines the study to air pollution and forest resource.

II. Methods

i) *Air pollutant data*

This study uses the entire available panel data reported by Thailand's Pollution Control Department (PCD) urban air quality in different cities in Thailand for the period 1996–2000. The PCD has been estimating emissions for Bangkok and other regions of Thailand. The Department of Energy Development (DEDP) has also applied emissions coefficient to estimated fuel use by sector. There are, however, significant differences in these data.

Emission inventories are prepared and reported by different organizations. Sometimes, these data differ largely from each other (World Bank, 2002). For example, the existing inventories in the Bangkok Metropolitan Region (BMR) do not account for the pollution that is generated outside its airspace, contributions from open burning of agricultural residues or open cooking. Furthermore, the emission factors used for vehicles may not be accurate, as these emission factors have not been developed for vehicles and conditions in Thailand. With so much uncertainty, the concentration, reported by the World Bank, is an appropriate alternative for this study.

The PCD has established networks for monitoring air quality in five regions of Thailand including, the Central, North, Northeast, East, and the South (See Fig. 15.3). The monitoring network of the PCD consists of 52 sites nationwide, which are linked to PCD's central computer system located in Bangkok.

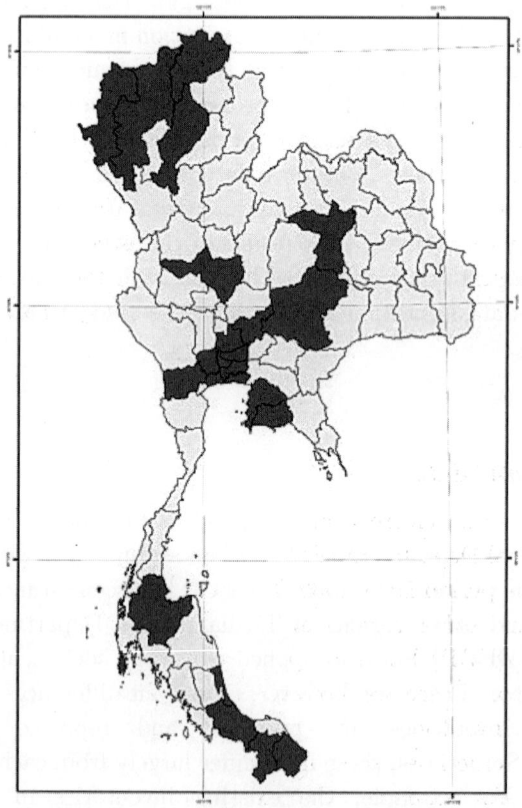

Fig. 15.3. Air quality monitoring station location.
Source: http://www.aqnis.pcd.go.th/station/allstation.htm (visited on May 25, 2010).

Thailand commenced air quality monitoring in 1983. The monitoring network has progressively evolved and expanded to cover a variety of pollutants. Air pollution monitoring assesses compliance with air quality standards. The majority of sites in all five regions monitor particulates, CO, NO_2, SO_2, and ground-level ozone. However, not all stations monitor all these pollutants, and some monitor other pollutants in addition to those mentioned here.

Air quality data reported by PCD are among the most commonly used indicators of air pollution in cities and other densely populated areas. Key air pollutants are dust, small particulate matter (PM 10), sulfur dioxide (SO_2), lead (Pb), carbon monoxide (CO), nitrogen oxides (NOx), hydrocarbons (HC), and ground-level ozone (O_3). Particularly, ambient PM has the most serious health impacts and still exceeds standards along traffic corridors such as Bangkok and suburbs. Other transboundary air pollutants such as ozone depleting substances (ODS), greenhouse gases (GHGs), and some persistent organic pollutants (POPs) have caused long-term impacts on the regional and global environments (World Bank, 2002).

However, the objective of this study is to investigate the relationship between income and environmental degradation, it does not examine health impacts. Therefore, air quality indicators, which were used in this study, should represent a change in environmental quality following economic change. The air indicators analyzed in this study use the average of 1-hour concentration to reflect environmental quality except for PM10 that used the average of 24-hour concentration.

ii) *Forest data*

The statistics of forest and woodlands are taken from the Royal Forest Department (RFD). The RFD's statistical record is based on reports since 1961. Unfortunately, the RFD did not collect data every year because of budget limitations. Thus, this study has to estimate data by extrapolating from an average growth rate of forest data.

iii) *Other data*

The gross provincial product on real per capita GPP (1998 prices) is a proxy for provincial income that air pollution monitors are located. This data is taken from the office of the National Economic and Social Development Board (NESDB). Provincial population data is taken from the Department of Provincial Administration, Ministry of Interior.

III. Air Pollutant Analysis

The results of the estimated reduced-form relationship between environmental change and per capita income are presented in Table 15.1 based on the coefficient estimates from the panel regressions. Given that the error terms from the regression are likely to be heteroscedastic,[2] this analysis also used White's approach[3] to solve this problem. Each table relates to one type of environmental quality indicators: PM_{10}, NO_2, SO_2, and CO. The patterns of the estimated relationship of income-environmental indicators are present in Figs. 15.3–15.6. The figures were constructed by time trend and interaction term between GDP and population density, GDP-squared and population density, and GDP-cubed and population density by the sum of the estimated coefficients for time trend and products of GDP and population density, with population density and time trend set equal to their mean values in the sample.

This study tested the relationship between population density on its own and airborne concentrations. The author hypothesized that the higher population density, the higher airborne concentrations. Urban people in Thailand prefer to have their own cars for transportation if they can afford to do so. The principal reason is poor public transportation, particularly in Bangkok. Although the Skytrain (BTS) was an effort to displace bus use that encourages a less-polluting mass transit option, usage remains low (40 percent utilization capacity). One reason is that the full-system of Skytrain has not finished yet. The current system did not cover the main areas in Bangkok. The privately operated BTS is handicapped by a large debt burden, and this prevents it from offering incentives to decrease ticket price (World Bank, 2002).

At any given income level, technology and environmental policy played a favorable role in improved local air quality. For example, the Government's national energy policy promotes the use of natural gas, this policy contributed to this decline because the combustion of natural gas releases around half of the amount of CO_2, and a small fraction of the PM10 and SO_2 that produced by conventional fuels such as oil and coal (World Bank, 2002).

The regressions of PM10 in Table 15.1 show that the cubic term fits best. Comparing to cubic form, quadratic form and linear form in PM10

[2]The variance of each disturbance term u_i is not constant number equal to σ^2, $E(u_i^2) \neq \sigma^2, i = 1, 2, 3, \ldots, n$ (Gujarati, 2003).

[3]See more details in Gujarati (2003): p. 413–4.

Table 15.1. Regression results for PM10 concentration (fixed-effects specification).

Variables	With population density			Without population density		
	Cubic	Quadratic	Linear	Cubic	Quadratic	Linear
P	−0.0955	.0359	.0362	—	—	—
	(−.708)	(.195)	(.196)			
	(.4815)	(0.8459)	(.8454)			
PY	6.93×10^{-6}	4.4×10^{-7}	2.7×10^{-7}	**6.58×10^{-6}**	3.97×10^{-7}	2.23×10^{-7}
	(2.301)**	(.3849)	(.425)	**(2.079)****	(.3582)	(.3432)
	(.0244)	(.7015)	(.6724)	**(.0412)**	(.7213)	(.7325)
PY^2	-4.46×10^{-11}	-5.54×10^{-13}	—	**-4.15×10^{-11}**	-5.63×10^{-13}	—
	(−2.359)**	(−.167)		**(−2.061)****	(−.169)	
	(.0212)	(.8681)		**(.0430)**	(.8667)	
PY^3	8.47×10^{-17}	—	—	**7.87×10^{-17}**	—	—
	(2.516)**			**(2.189)****		
	(.0142)			**(.0319)**		
T	−6.743	−6.016	−6.0297	**−7.175**	−5.815	−5.827
	(−3.268)***	(−2.991)***	(−2.988)***	**(−4.332)*****	(−3.782)***	(−3.792)***
	(.0017)	(.0038)	(.0039)	**(.0000)**	(.0003)	(.0003)
R^2	0.5229	0.4897	0.4896	**0.5212**	0.4894	0.4893
Number of observations	93	93	93	**93**	93	93

Note: 1) The figures reported in parentheses below the coefficient estimates are t-statistics and p-values respectively.
2) The intercepts varied across monitoring stations.
3) P = population density, PY = interaction term between population density and GDP, PY^2 = interaction term between population density and GDP squared, PY^3 = interaction term between population density and GDP cubed, and t = time trend.
* = estimate is significantly different from zero at the 10 percent level.
** = estimate is significantly different from zero at the 5 percent level.
*** = estimate is significantly different from zero at the 1 percent level.
4) Bold type is a chosen model.
Source: Authors' calculation.

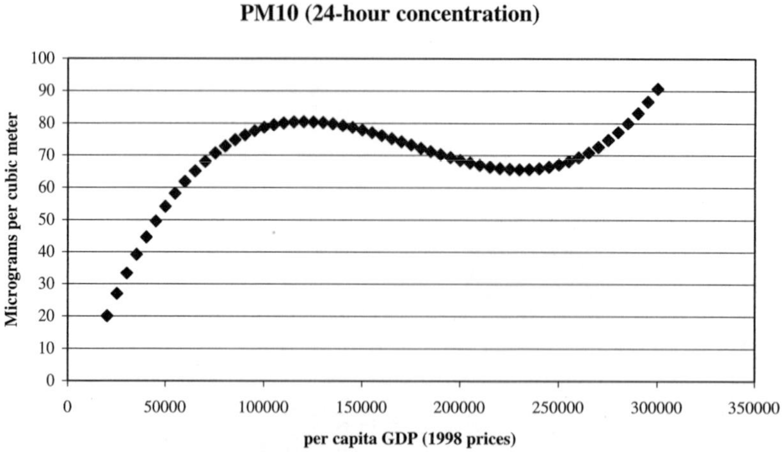

Fig. 15.4. Estimated PM10 concentration-income relationship in Thailand.
Source: The regression result from Table 15.1.

models without population, t-statistics and p-values show all variables in cubic term are significant, whereas quadratic form and linear form have only time trend that is significant. PM10 displays an inverted-U shaped relationship between PM10 and GDP only up to about 240,000 baht. (Fig. 15.4) The level of PM10 appears to rise with GDP at low levels of GDP, but finally reaches to a local peak (at 120,000 baht in the range of 20,000 baht to 240,000 baht), and then falls with GDP at higher levels of income. However, the estimated relationship turns up again at very high level of income, but there are small numbers of observations with income above 240,000 baht. It means that the shape of the curve in this range may not represent the relationship.

Comparing to Shafik's (1994) study, this turning point is higher than Shafik predicted. Shafik's turning point, US$3,280 at 1985 international prices, is equivalent to 165,856 baht at 1998 Thailand prices. But the turning point of this study is lower than Selden and Song (1994) predicted. Selden and Song's turning point, US$6,200 at the 1985 international prices, is equivalent to 313,508 baht in Thailand prices. However, the results of this study may be biased because the range of data in this study covered the post-1997 financial crisis which slowed down Thailand's economic growth. This phenomenon had an effect on energy consumption, especially in the transport and industry sectors.

For the regressions of NO_2 and SO_2, the linear form fits best. Compared to cubic form, quadratic and linear forms in NO_2 and SO_2 models without

Table 15.2. Regression results for NO_2 concentration (fixed-effects specification).

	With population density			Without population density		
Variables	Cubic	Quadratic	Linear	Cubic	Quadratic	Linear
P	−.001	−.007	−.007	—	—	—
	(−.069)	(−.484)	(−.485)			
	(.9455)	(.6299)	(.6291)			
PY	-1.30×10^{-7}	1.77×10^{-7}	2.66×10^{-7}	-1.34×10^{-7}	1.86×10^{-7}	**2.76×10^{-7}**
	(−.338)	(1.405)	(5.597)***	(−.367)	(1.464)	**(5.797)***
	(.7362)	(.1644)	(.0000)	(.7146)	(.1476)	**(.0000)**
PY^2	2.37×10^{-12}	2.90×10^{-13}	—	2.41×10^{-12}	2.92×10^{-13}	—
	(1.011)	(.695)		(1.100)	(0.706)	
	(.3154)	(.4894)		(.2750)	(.4828)	
PY^3	-4.0×10^{-18}	—	—	-4.07×10^{-18}	—	—
	(−.983)					
	(.3291)			(−1.070)		
				(.2883)		
T	.474	−.509	−.502	−.479	−.550	**−.543**
	(−2.513)**	(−2.639)**	(−2.644)**	(−2.979)***	(−3.458)***	**(−3.461)***
	(.0144)	(.0103)	(.0101)	(.0040)	(.0009)	**(.0009)**
R^2	0.8732	0.8717	0.8711	0.8732	0.8715	**0.8709**
Number of observations	92	92	92	92	92	**92**

Note: 1) The figures reported in parentheses below the coefficient estimates are t-statistics and p-values respectively.
2) The intercepts varied across monitoring stations.
3) P = population density, PY = interaction term between population density and GDP, PY^2 = interaction term between population density and GDP squared, PY^3 = interaction term between population density and GDP cubed, and t = time trend.
 * = estimate is significantly different from zero at the 10 percent level.
 ** = estimate is significantly different from zero at the 5 percent level.
*** = estimate is significantly different from zero at the 1 percent level.
4) Bold type is a chosen model.
Source: Authors' calculation.

Table 15.3. Regression results for SO_2 concentration (fixed-effects specification).

	With population density			Without population density		
Variables	Cubic	Quadratic	Linear	Cubic	Quadratic	Linear
P	−.036	−.032	−.032	—	—	—
	(−2.464) **	(−1.903)	(−1.707)			
	(.0162)	(.0611)	(.0923)			
PY	3.88×10^{-7}	2.11×10^{-7}	6.68×10^{-8}	2.57×10^{-7}	2.50×10^{-7}	$\mathbf{1.08 \times 10^{-7}}$
	(1.074)	(1.633)	(1.247)	(.681)	(1.923)*	**(2.144)****
	(.2866)	(.1070)	(.2164)	(.4984)	(.0584)	**(.0354)**
PY^2	-1.67×10^{-12}	-4.69×10^{-13}	—	-5.03×10^{-13}	-4.61×10^{-13}	—
	(−.757)	(−1.387)		(−.224)	(−1.311)	
	(.4514)	(.1697)		(.8236)	(.1940)	
PY^3	2.31×10^{-18}	—	—	8.02×10^{-20}	—	—
	(.609)			(.201)		
	(.5447)			(.9833)		
T	−.108	−.089	−.100	−.271	−.269	**−.279**
	(−1.181)	(−.932)	(−1.06)	(−2.55)**	(−2.27)**	**(−2.415)****
	(.2417)	(.3544)	(.2934)	(.0130)	(.0264)	**(.0183)**
R^2	0.8738	0.8729	0.8696	0.8633	0.8633	**0.8602**
Number of observations	93	93	93	93	93	**93**

Note: 1) The figures reported in parentheses below the coefficient estimates are t-statistics and p-values respectively.
2) The intercepts varied across monitoring stations.
3) P = population density, PY = interaction term between population density and GDP, PY^2 = interaction term between population density and GDP squared, PY^3 = interaction term between population density and GDP cubed, and t = time trend.
* = estimate is significantly different from zero at the 10 percent level.
** = estimate is significantly different from zero at the 5 percent level.
*** = estimate is significantly different from zero at the 1 percent level.
4) Bold type is a chosen model.
Source: Authors' calculation.

Table 15.4. Regression results for CO concentration (fixed-effects specification).

	With population density			Without population density			
Variables	Cubic	Quadratic	Linear	Cubic	Quadratic1 (with time trend)	Quadratic2 (without time trend)	Linear
P	−.0007	−.0011	−.001	—	—	—	—
	(−.957)	(−1.341)	(−1.331)				
	(.3417)	(.1842)	(.1873)				
PY	5.13×10^{-9}	2.72×10^{-8}	-5.06×10^{-10}	2.61×10^{-9}	2.86×10^{-8}	$\mathbf{2.93 \times 10^{-8}}$	9.08×10^{-10}
					(3.583)***	**(3.422)***	
	(.3270)	(3.549)***	(−.101)	(.167)	(.0006)		(.189)
	(.7446)	(.0007)	(.9195)	(.8677)		**(.0010)**	(.8500)
PY^2	6.01×10^{-14}	-9.0×10^{-14}	—	8.25×10^{-14}	-8.97×10^{14}	$\mathbf{-9.23 \times 10^{-14}}$	—
	(.578)	(−3.095)***		(.805)	(-3.047)***	(**−2.973**)***	
	(.5649)	(.0028)		(.4233)	(.0032)	**(.0040)**	
PY^3	-2.88×10^{-19}	—	—	-3.31×10^{-19}	—	—	—
	(−1.435)			(−1.680)*			
	(.1557)			(.0973)			
T	−.002	−.005	−.007	−.006	−.011	—	−.013
	(−.249)	(−.507)	(−.689)	(−.669)	(−1.389)		(−1.454)
	(.8035)	(.6137)	(.4931)	(.5054)	(.1694)		(.1502)
R^2	0.8636	0.8614	0.8465	0.8631	0.8599	**0.8575**	0.8450
Number of observations	94	94	94	94	94	**94**	94

Note: 1) The figures reported in parentheses below the coefficient estimates are *t*-statistics and *p*-values respectively.
2) The intercepts varied across monitoring stations
3) P = population density, PY = interaction term between population density and GDP, PY^2 = interaction term between population density and GDP squared, PY^3 = interaction term between population density and GDP cubed, and t = time trend.
* = estimate is significantly different from zero at the 10 percent level.
** = estimate is significantly different from zero at the 5 percent level.
*** = estimate is significantly different from zero at the 1 percent level.
4) Bold type is a chosen model.
Source: Authors' calculation.

population, t-statistics and p-values show the interaction term of income and population density, and time trend in linear term are significant at the 1 percent level and the 5 percent level for NO_2 and SO_2 models respectively, whereas the variables in cubic and quadratic forms are less significant. This study found a monotonically increasing relationship between NO_2 and income (Fig. 15.5) as well as between SO_2 and income (Fig. 15.6) at all levels of income in the sample range.

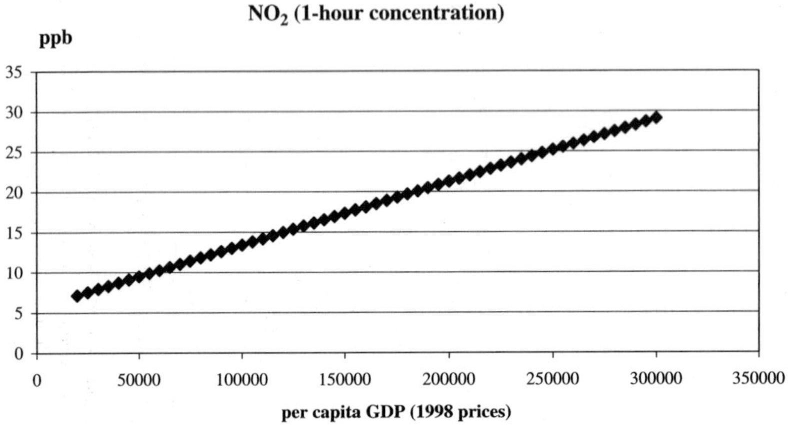

Fig. 15.5. Estimated NO_2 concentration-income relationship in Thailand.

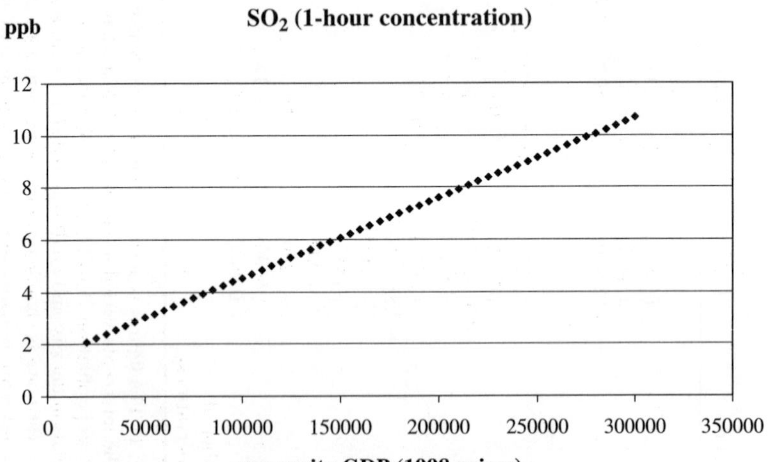

Fig. 15.6. Estimated SO_2 concentration-income relationship in Thailand.
Source: The regression results from Table 15.2 and 15.3.

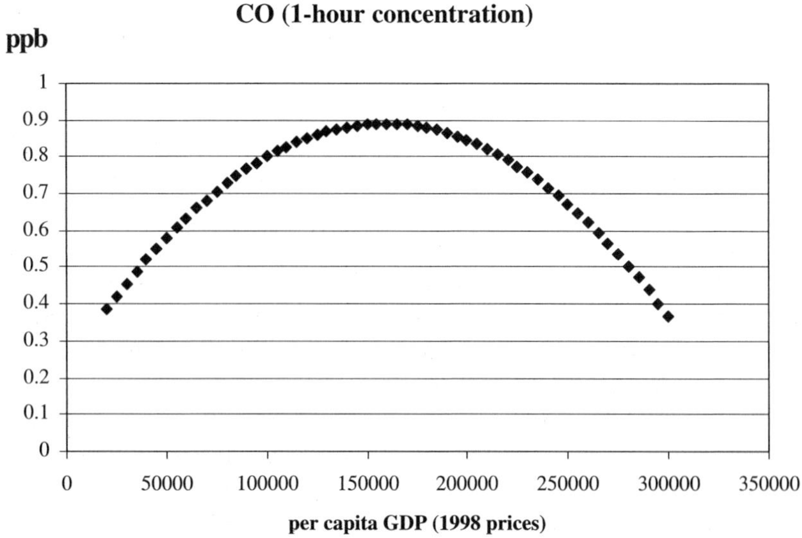

Fig. 15.7. Estimated CO concentration-income relationship in Thailand.

The regression of CO concentration shows the quadratic form without time trend (model 2) fits best. According to the quadratic form in CO models without time trend (model 2), t-statistics and p-values show all variables in quadratic term are significant, whereas, the time trend in model 1 is not significant (Table 15.4). This study found an inverted-U shaped relationship between CO concentration and GDP The level of CO appears to rise with GDP at low levels of GDP, but finally reaches to a peak (at 160,000 baht), and then falls with GDP at higher levels of income.

The results of the relationship between income-PM10 concentration and income-CO concentration did not either prove or disprove the EKC hypothesis because the range of data has limited in this study as it covered the post-1997 financial crisis which had an effect on energy consumption in Thailand. There are some explanations as to why the shape of income-PM10 and income-CO is an inverted-U relationship. The World Bank (2002) reported that the important reasons explaining the decline of PM10 concentrations in Thailand are control measures[4] and the Asian economic crisis of 1997.

[4]The negative coefficients of time trend in models reflect this evidence, implying that, at any given income level, the air quality is better than in the past due to factors not necessarily correlated with income (e.g. policy changes).

The Thai government has implemented many environmental policies to improve air quality: enacting the environmental law in 1992; completing the phase-out of leaded gasoline by 1995; improving fuel quality and engine specifications; curbing pollution from power plants; moving enterprises to cleaner production practices; tightening construction standards; improving public transport; and substantially reducing the use of ozone depleting substances. Examples of success stories include introduction of ULG, improvements in fuel quality, control of power plant sulfur emissions, regulations for motor vehicles, and the monitoring systems (World bank, 2002). Furthermore, responding to increasingly strict emissions standards, over 80 percent of Thai motorcycle production shifted from two-stoke engines to four-stroke engines.

The resulting economic slowdown of Thai economy due to Asian economic crisis of 1997 affected the energy consumption in 1998, which decreased at 8.19%. The decreasing of energy consumption in 1998 also contributed to a decline in airborne emissions in Thailand.

The results of the relationship between income-PM10 concentrations and income-CO may be biased due to the 1997 financial crisis. That means the income-PM10 concentration and the income-CO concentration relationship do not represent the relationship between income and air quality in Thailand. The monotonical relationship between income and air quality for SO_2 and NO_2 are more reasonable to represent the income-airborne concentration in Thailand. The statistical figures supported an inverted-U shaped in time trend variable, as a proxy of factors that are unrelated to income such as policy changes. The time trend variable was strongly significant and had a negative coefficient, implying that there are other factors that contributed to improve air quality in Thailand. One point worth noting is that environmental measures played an important role in this effect.

According to decomposition of the income-environment relationship, namely the scale effect, the composition effect, and the abatement effect, the scale effect has played an important effect in Thailand so far. Nevertheless, the environmental policies have played an important role to improve air quality in Thailand. In the absence of the financial crisis in 1997, the airborne concentrations might not have decreased noticeably, implying that the negative impacts of the scale effect still tend to overcome the positive impacts of the composition and abatement effects.

IV. Deforestation Analysis

None of the income terms are significant in any specification that is consistent with Shafik (1994). However, when using stepwise technique,[5] all variables were removed except population density that is significant at the 1 percent level. It can be concluded from these results that income and the 1989 Decree revoking all terrestrial forestry concession do not have any bearing on the annual deforestation changes in Thailand.

Furthermore, the effect of a logging ban mirrors Panayotou's (1990) study that "The ban has largely failed to reduce deforestation due to a lack of effective enforcement and forest management. In fact deforestation may have increased since the ban". The statistics on forest cover in Thailand also shows a decrease in forest cover after the logging ban. The econometric result reveals that population density is the key factor affecting deforestation, the coefficient on average population density is positive. Panayotou (1990) reviewed a study that examined the causes of deforestation in Northeast Thailand. The study found that "population density, poverty, and infrastructure (both dams and roads) as well as economic incentives (wood and crop prices) played a major role in deforestation". This reflects that there are many factors affecting deforestation in Thailand and population density is only one of them.

This study found an inverted-U shaped for PM10 and CO. Furthermore, this study found a monotonically increasing relationship between income-NO_2 and income-SO_2. The inverted-U shaped for PM10 and CO did not either prove or disprove whether the EKC exists in Thailand. The main reasons that contribute to decreased air pollution in Thailand were environmental measures and the financial crisis affecting energy consumption in Thailand. However, as indicated before, it is difficult to test the EKC with the limited data as employed here. Nevertheless, it was possible to test at least in a rudimentary way whether the per capita income values observed here correspond to either the ascending or descending

---

[5]In this method, variable entry and removal examines the variables in the block at each step for entry or removal. This is a forward stepwise procedure. The significance values in the output are based on fitting a single model. Therefore, the significance values are generally invalid when a stepwise method is used. All variables must pass the tolerance criterion to be entered in the equation, regardless of the entry method specified. The default tolerance level is 0.0001. Also, a variable is not entered if it would cause the tolerance of another variable already in the model to drop below the tolerance criterion (SPSS Inc., 1992).

Table 15.5. Comparing turning point in Shafik's (1994) study and Selden and Song's (1994) study with income range in this study. *Source*: Shafik (1994), Selden and Song (1994), and authors' calculation.

Indicator	Turning point in Shafik (1994)	Turning point in Selden and Song (1994)	Income range in this study	
			Min.	Max.
SPM	165,856	495,546	20,065	1,118,785
SO_2	185,577	450,037	20,065	1,118,785
NO_x	—	606,791		
CO	Not applicable Monotonically rising	313,509	20,065	1,118,785
Deforestation	126,414	—	22,649	84,428[6]

segments of the EKC of other studies. It was possible to do this comparison with Shafik's (1994) study and Selden and Song's (1994) study by making appropriate currency conversions.

In conclusion, the turning points on SPM, SO_2, and deforestation in Shafik's (1994) study occur at an income level of this study (Table 15.5). Also, the turning points on SPM, SO_2, NO_x, and CO in Selden and Song's (1994) study fell at an income level of this study. Because the turning points of Shafik's (1994) study and Selden and Song's (1994) study fell clearly within the per capita income range of this study. It is possible to tentatively suggest that the EKC does not hold for Thailand. That is say relationship between per capita income and pollution is monotonically increasing.

No evidence was found for an inverted-U shape for deforestation in Thailand. The factor affecting deforestation is population density. Moreover, if there is a turning point in Thailand, it could take more time to reach that point because even maximum income in this study is less than a turning point in Shafik's study (US$ 2,500 at 1985 international prices or 126,414 baht at 1998 Thailand prices).

V. Discussion

In October 2002, the Thai government announced the reorganization of its public sector after several years of preparatory work. The new Ministry of Natural Resource and Environment (MoNRE) was created to be responsible

[6]This figure is much smaller because it applies to the entire country. The figures above relate urban income.

for environmental management. In addition, a new Ministry of Energy (MoE) was also established.

i) *Air pollution policy in Thailand*

The Thai economy has shifted from a mainly agrarian economy into the manufacturing and service industries in the early 1990s. The rapid expansion of industry, without an appropriate implementation and enforcement to prevent environmental degradation, was followed by the rapidly increasing rates of air and water pollution, especially in the Bangkok Metropolitan Region (BMR).

In 1992, the United Nations Environment Program (UNEP) reported that the BMR was one of the most air-polluted cities in the world because of the traffic problem.[7] Thailand followed the 1992 National Environment Quality Enhancement and Protection Act (NEQA) with many pieces of environmental legislation; however, it has not created the legislative or institutional framework to successfully monitor and enforce emission regulations on point sources of pollution.

Many national and local agencies like the Pollution Control Department (PCD), National Energy Policy Office (NEPO), Land Transport Department (LTD), BMA, city of Chiang Mai, and others have formulated policies, programmes, and projects aiming to control air pollution. These efforts have contributed noticeably to improve air quality in Thailand. In the end of 2002, the World Bank (2002) reported that the air quality in Thailand overall has improved significantly in recent years and Bangkok air quality is now rated above Beijing, Manila, Jakarta and New Delhi. The measures that the Thai agencies have implemented can be categorized to three types, the command and control measures, the economic or market-based instruments, and other measures such as improving public transport and building overpasses in major traffic arteries.

ii) *Command and control measures*

The command and control strategy consists of relying on standards of various types to bring about improvements in environmental quality: ambient standards, emission standards, and technology standards. There are two main steps in these measures: monitoring and sanctioning. Monitoring refers to measuring the operation of polluters in comparison to whatever requirements are set in the relevant law. Sanctioning refers to the

[7]http://www.eia.doe.gov/emeu/cabs/thaienv.html (visited on June 23rd, 2003).

task of bringing to justice those who have violated the law. The objective of these measures is to get people to comply with an applicable law; hence some monitoring is essential. Moreover, command and control measures by themselves are no incentive for polluters to decrease pollutant emission.

In summary, measures related to air pollution control are based on:

- Improvement of fuel quality.
- Improvement of engine standards.
- Improvement of emission standards.
- Reducing garbage burning.

iii) *Economic instruments*

Economic incentives and disincentives are being employed to promote environmental conservation and efficient use of natural resources. Incentives contain preferential tax credits and accelerated depreciation allowances on pollution abatement and control equipment. Economic disincentives are often based on the polluter pays principle (PPP). Pollution fines are used to complement the enforcement of emission standards that depends on many factors, such as duration of violation, the quantity of effluent discharged, and exiting environmental conditions at that time.[8]

There are two success stories using economic instruments to improve air quality in Thailand: introducing a price differentiation between leaded and unleaded gasoline and demand-side management (DSM) in the power sector.

Most air pollution prevention in Thailand is based on regulatory mechanisms rather than on market forces or economic instruments. However there is a growing trend in economic instruments in Thailand. Thai people have a limited awareness of the importance of pricing resources to reflect their real economic value and social cost. Most Thai people think that natural resources are for free.

iv) *Forest policy in Thailand*

Annual rates of deforestation in Thailand are higher than neighboring countries: Burma, Cambodia, Laos, and Vietnam. The Thai government has attempted to form measures, laws, and policies that have been implemented for the purpose of forest conservation. One of the major obstacles of forest

[8] http://www.eia.gov/emeu/cabs/thaienv.html (visited on June 23, 2003)

conservation is the conflict between the State, which has been trying to conserve the forests by declaring forest reserves and conservation areas, and the landless rural people, who rely on the forests for their livelihood. This conflict has grown more serious over the years, especially with the significant increase in the rural population (Mingsarn and Pornpen, 2000).

There was general agreement that the impediment to protect and enhance the environmental quality of Thailand is the lack of capacity to implement policy and/or enforce regulation (Daniere and Takahashi, 2002;, Mingsarn and Pornpen, 2000). Deforestation in Thailand was due to the unsustainable practices employed by logging companies before the cancellation of logging concessions and post-ban illegal logging and upland encroachment by farmers.

There was a general agreement that community-based law was a better solution than law enforcement officers to detect illegal forest exploitation by outsiders (Brunner *et al.*, 1999). However, the progress of a community-based law in Thailand has been slow. A draft Community Forest Act was prepared in April 1996. In an effort to gain final enactment, a new constitution was ratified in July 1997. The effort to implement the Community Forest Act did not succeed because after the constitution came into affect, the new government came in following beginning of the 1997 financial crisis. In 1998, the cabinet in the new government canceled the 1997 resolutions and enacted a new one. Up to the present, the Community Forest Act has still been being drafted.

Environmental degradation in Thailand has been pushed by industrialising policies and the microeconomic agency in the early development of the Thai economy never addressed environmental issues. This reflected the pro-growth approach of the Thai government at that time (Rock, 2002). The 1961 First National Economic and Social Development Plan (NESDP) aimed to protect 50 percent of Thailand's areas as forests. The First Plan also stated that the target could be reduced to 40 percent, or about 128 million rai (20.48 million hectares), given the increase in population and the need to utilize land for other economic activities. Subsequently, the second NESDP, was declared in 1967, emphasized the protection of 40 percent of Thailand's areas as forests. Later, the first formal National Forest Policy was announced in 1985 that proposed 40 percent of land under forests, which was divided into 25 percent as Economic or Production Forest, and 15 percent as Conservation Forest. Finally, the seventh NESDP, 1992–1996, reversed conservation forest to 25 percent, and production forest to 15 percent.

It is difficult to uncritically accept the government's data of timber exports and imports, as information regarding illegal logging and smuggling is not accessible. There is a gap between what is reported by the exporter and the importer. This gap reflected illegal logging and smuggling between the two countries. (e.g. Cambodia to Thailand).

Although forest plantations have been incorporated, the rate of reforestation is largely below that of deforestation. Statistics show that the annual reforestation has declined from 52,277 hectares (0.3 million rai) in 1992 to about 7,300 hectares (0.05 million rai) in 1998 (World Bank, 2000).

There are several factors contributing to deforestation in Thailand. Sureeratna (2001) list the main contributing causes as follows:

- Agricultural expansion — both permanent and shifting cultivation,[9]
- Rural poverty, including disadvantaged and landless people,
- Population growth/over-immigration, resulting in population increase,
- Poorly planned and managed concessionaires of both legal and illegal logging operations,
- Poor coordination of policy planning and implementation between government agencies, involving forest resource management and conservation, and weak institutional capacity for the above activities, and
- Infrastructure development, particularly roads, dams and mining, and improved access into the previously remote natural forest areas.

VI. Economic Growth and Patterns of Environmental Change in Thailand

From the empirical results for air pollution, PM10 and CO, it seems that the environmental Kuznets curve exists in Thailand. However, the range of data, which is examined in this study, covers the 1997 financial crisis. In the absence of the 1997 crisis, the relationship between economic and PM10 concentrations might be a monotonically increasing relationship like ones in NO_2 and SO_2 concentrations.

There are some reasons to explain why an inverted-U shaped environmental Kuznets curve may not exist. Kadekodi and Agarwal

[9]Hill tribal people have often been blamed for destroying the watershed forests through shifting agriculture. However, shifting cultivation is not critical, as long as arable land is unlimited and population is low (Sureeratna, 2001).

(2001) summarized from their empirical study that the inverted U-shaped relationship between environmental resource and economic growth exists whenever the natural resource-based industry is more capital intensive than the manufacturing goods industry at lower levels of aggregate (K/L) ratio in the early phase of economic development. This seems to explain why no evidence was found for an inverted-U relationship between economic growth and airborne concentrations in Thailand.

For deforestation, this study found no relationship between economic growth and deforestation in Thailand. Not surprisingly for this outcome, as mentioned above that there are several causes of deforestation in Thailand which economic growth is not directly related. The finding of this study reflected that the population plays a major role in influencing deforestation. This mirrors Sureeratna's (2001) conclusion that agriculture expansion due to population growth, rural poverty, and population growth/over-immigration, which are impacts that have influenced deforestation in Thailand.

This study also conforms to policy's failure of a logging ban. Panayotou (1990) argued that the root causes of deforestation in Thailand include a fast-growing population, which leads to social pressure such as increased rural poverty and lack of alternative employment opportunities. The government encouraged landless people to utilize "deteriorated" forest land. This government response made this problem worse.

Income equality in Thailand is another reason that explains why there is no evidence found for income-environment relationship. Koop and Tole (2001) argued, "a country in which growth is not accompanied by the alleviation of poverty or inequality would be expected to have higher deforestation rates than one characterized by more balanced and equitable growth that improves the lives of the mass of the poor". It implied that if distributive policies marginalize the poor, environmental degradation may worsen despite the economy growing quickly (Steer, 1992). This assumption is true for Thailand. Thus, it is not surprising because deforestation rates in Thailand have increased rapidly as economic growth increases.

It seems the solution of deforestation is the community forest that is an alternative for forest management. However, there are other measures that can contribute to alleviate this problem. These measures include a land use zoning plan, allocation of concessions by public auction, a taxation system that rewards efficiency and penalized waste, and performance bonds coupled with a minimum level of state monitoring (Brunner *et al.*, 1999).

VII. Conclusion

This study investigated the relationship between economic growth and environmental quality in Thailand, where air pollution and deforestation represent environmental degradation. The results show that the PM_{10} and CO appear to support the environmental Kuznets Curve (EKC) thesis; but the results may be biased due to the effect of the 1997 financial crisis. Without the 1997 financial crisis, the relationship would be linear. However, the environmental policy also plays a major role in improving air quality in recent times. SO_2 and NO_2 violate the EKC thesis by showing a rising monotonically relationships with per capita income. A plausible explanation for a monotonically increasing relationship (NO_2, SO_2) is a reflection of energy resources in Thailand that contain high sulfur and nitrogen. Thailand shifted from imported oil to local energy resources, such as coal, natural gas, and biomass, after the oil crisis in the 1970's. Furthermore, coal burning emits acidic gas in the form of SOx and NOx followed by SPM (suspended particulate matter) into the air. The increasing use of coal, as shown, could explain why the relationship between income-SO_2 and income-NO_2 are a monotonically increasing relationship, even though the environmental policy has been adopted on air quality improvement. On the basis of this study, it is reasonable to summarize that a monotonically increasing relationship is more valid to represent the income-environment in Thailand, implying air pollution control in Thailand still requires stricter environmental regulations, increasing environmental public awareness, and new approaches, such as economic instruments, to shift from energy-intensive industry to technology-intensive industry.

For deforestation analysis, this study found no evidence of an inverted-U shaped EKC. By and large, only population density has an impact on deforestation. Furthermore, the study illustrates the 1989 logging ban in Thailand had little effect on forest conservation. Nevertheless, the results of this study should be interpreted cautiously since it seemed there are other factors that have an effect on deforestation, such as property rights, private land tenure institutions, employment opportunities and rural poverty, which is far beyond the scope of this study. Therefore, it does not mean that the solution of this problem is population control. Whenever issues of land tenure, employment opportunities and rural poverty cannot be resolved, deforestation continues to increase in Thailand.

It seems the solution of deforestation is community forest programmes that provides an alternative for forest management. However, in the meantime, the detailed content of the draft law on community forests is still subject to further debate. Furthermore, there are other alternatives measures that can mitigate deforestation, such as a land use zoning plan, allocation of concession by public auction, a taxation system, and performance bonds.

References

Brunner, J., Seymour, F., Badenoch, N., and Ratner, B. 1999. "Forest problems and law enforcement in southeast Asia: the role of local communities", *Resources Policy Brief*, World Resources Institute, Washington, D.C.

Daniere, A. and Takahashi, L.M. 2002. *Rethinking Environmental Management in the Pacific Rim: Exploring Local participation in Bangkok, Thailand,* Ashgate Publishing, Aldershot.

Grossman, G.M. and Krueger, A.B. 1995. "Economic growth and the environment", *Quarterly Journal of Economics*, 110(2): 353–77.

Gujarati, D.N. 2003. *Basic Econometrics*, fourth edition, International edition, McGraw-Hill, Singapore.

Ikemoto, Y. and Uehara, M. 2000. "Income inequality and Kuznets' hypothesis in Thailand", *Asian Economic Journal*, 14(41): 421–43.

Kadekodi, G.K. and Agarwal, M.M. 2001. "Why an inverted U-shaped environmental Kuznets curve may not exist" in *The Sustainability of Long-term Growth: Socioeconomic and Ecological Perspectives*, edited by Munasinghe M. et al., Edward Elgar Publishing, Cheltenham.

Koop, G. and Tole, L. 2001. "Deforestation, distribution and development", *Global Environmental Change*, 11: 193–202.

Kuznets, S. 1955. "Economic growth and income inequality", *American Economic Review*, 45(1): 1–28

Mingsarn Kaosa-ard and Pornpen Wijukprasert. 2000. *The State of Environment in Thailand: A Decade of Change.* Natural Resources and Environment Program, Thailand Development Research Institute, Bangkok.

Panayotou, T. 1990. "The economics of environmental degradation: problems, causes and responses", *Development Discussion Paper No. 335*, Harvard Institute for International Development, Harvard University.

Panayotou, T. 2000. "Economic growth and the environment", *CID Working Paper No. 56*, Environment and Development Paper No. 4, Center of International Development, Harvard University, July.

Rock, M.T. 2002. "Pollution management in Malaysia and Thailand" in *Pollution Control in East Asia: Lessons from the Newly Industrializing Economies,* Resources for the Future and Institute of Southeast Asian Studies, Washington D.C.

Selden, T.M. and Song, D. 1994. "Environmental quality and development: Is there a Kuznets curve for air pollution emission?", *Journal of Environmental Economics and Management*, 27: 147–62.

Shafik, N. "Economic development and environmental quality: an econometric analysis", *Oxford Economic Paper*, 46: 757–73.

SPSS Inc. 1992. SPSS 6.1: Base System, User's Guide, Part 2. Chicago.

Steer, A. 1992. "The environment for development", *Financial and Development*, 29: 18–21.

Sureeratna Lakanavichian. 2001. "Forest policy and history in Thailand", *Working Paper*, Research Center on Forest and People in Thailand, Danish Institute of Agricultural Science, December. (www.forestandpeople.org/publicat%5cup09.pdf visited on June 23rd, 2003).

Vogel, M.P. 1999. Environmental Kuznets Curves: A study on the economic theory and political economy of environmental quality improvements in the course of economic growth, Lecture note in economic mathematical system 469, Springer Verlag Berlin Heidelberg.

World Bank. 2000. *Thailand: Environment Monitor 2000*, The World Bank Office, Thailand.

World Bank. 2002. *Thailand: Environment Monitor 2002*, The World Bank Office, Thailand.

16

COST-BENEFIT ANALYSIS OF MINING IN PROTECTED FOREST: A CASE STUDY OF PT WEDA BAY NICKEL, CENTRAL HALMAHERA, INDONESIA

SILVIA IRAWAN and CHANG YOUNGHO

The Government of Indonesia (GoI) has demonstrated strong commitment in seeking to prudently manage the forest resources by issuing the Republic of Indonesia Law number 41/1999 in 1999. This is seen as a positive step in an improved commitment to environmentally sustainable use of forestry resources which have been exploited recklessly in past decades. This law aims to manage the forest resources of the country for current as well as future generations. It designates all forested areas into three classes according to their ecological function. However, this new forest classification has created significant issues in some areas where the newly designated functions of the forests are overlapping with other development activity which had previously been endorsed by other regulations. Some 155 mining companies had, in previous years, been granted Contract of Work rights to operate in areas that were later designated under this law as protected areas. This law compels the companies to suspend all activities as it prohibits open pit mining in protected forests areas. Some three quarters of these companies still continue to operate due to legal disputes. The GoI has to choose and foster the most desirable alternative and it needs assessment tools such as Cost Benefit Analysis which will enable it to make informed and justifiable determinations.

This study seeks to determine the course of action which is the most desirable for society as a whole. A Cost Benefit Analysis (CBA) is applied in the study. This study concludes that the benefits of the forest are expected to exceed US$ 12.32 million. Should the forest be preserved, , it will last for a potentially infinite period of time, and remain as a life support system for current as well as future generations. On the other hand, the benefits accrued from mining activities will last for only 20 years, and with limited additional benefits in terms of post-mining operations.

Introduction

In Indonesia, forest clearance has been an ongoing and contentious issue. The average rate of overall deforestation in Indonesia has reached a rate of 1.7 million ha/year over the past 12 years. The lowland forests of Sumatra and Kalimantan are predicted to vanish by 2010 if the current trend continues (World Bank, 2003). In response to this issue, the Government of Indonesia (GoI) has demonstrated strong commitment by seeking to prudently manage the forest resources by issuing the Republic of Indonesia Law number 41/1999 in 1999. This must be seen as a positive step in an improved commitment to environmentally sustainable use of forest resources which had been exploited recklessly in past decades.

The Law 41/1999, which aims to manage the forest resources of the country for the sake of current as well as future generations, designates all forested areas into three classifications according to their ecological functions.[1] These classifications are: protected, conservation, and production forests. In order to ensure that each of the designated forests continues to perform the function assigned to it, the law further specifies management guidelines for each of the classifications.

However, the new forest classification has created significant issues in some areas where the newly designated functions of the forests overlap with other development activity, which had previously been endorsed by other regulations. The situation is worsened as both activities cannot be mutually implemented in practice. For example, 155 mining companies had, in previous years, been granted a Contract of Work (CoW) to operate in areas that were later designated under the Law 41/1999 as protected areas.[2] The new law compels the companies to suspend all activities as clause 38/4 of the law prohibits open pit mining in protected forests areas. Of these 155 companies, some three quarters (124 companies) still continue operations due to legal disputes.[3]

In order to address these legal disputes, the government intends to exercise clause 19 of Law 41/1999 which allows it to alter the status of the forests in disputed areas with the approval of The House of Representatives (DPR-RI).[4] During the meeting with the Commissions III (on forestry affairs) and VIII (on mining affairs) of the DPR-RI in November 2003, the GoI proposed to alter the status of some protected forest areas in order to allow 13 companies to continue their mining activities. The 13 companies were selected from the 150 companies mainly due to the confirmed amount of mineral resources discovered in the disputed areas. Moreover, these

[1]Law 41/1999 Republic of Indonesia.
[2]Contract of Work (COW) grants exclusive rights over a defined area to explore, develop and mine any mineral deposit found, to process and refine all minerals extracted, and to market and sell the products.
[3]Presentation in the Commissions III and VIII of the House of Representative meeting on Discussion in Altering the Classification of Protected Area for Mining Purposes, 2003.
[4]Law 41/1999 Republic of Indonesia.

companies had started installing extensive and costly mining facilities. Should the DPR-RI approve the government's proposal, it would therefore agree to the clearance of 125,188 ha of Indonesia's protected forests for mining activity. While the total protected forests area in Indonesia is approximately 29 million hectares, this figure still represents a significant portion of this non-renewable resource.[5]

The justification put forward by the GoI for allowing the 13 companies to operate was mainly built around the economic benefits from the mining activity.[6] It was estimated that the Government would receive approximately US$ 501 million per year from taxes and royalties if the 13 companies were allowed to continue their activity. However, the mining sector, which contributed some 13.6% of the country's GDP in 2000, has inflicted substantial environmental damage in the country, including extensive land disturbance, contamination of rivers used for drinking water and food supplies, and loss of the forest cover and habitat (World Bank, 2003). The seriousness of the GoI in managing the nation's forest resources is therefore being tested. The test will be whether the government will allow a sizeable amount of protected forest areas to be sacrificed for the sake of the economic benefit from the operations of the 13 mining companies.

The conflict between development and preservation of nature is endemic in developing countries such as Indonesia where many people still live under internationally recognized poverty lines. One could argue that for the sake of future generations the government should keep the natural resources intact. However, should the country halt the developments for this reason, its future generations might inherit a poor and under-developed nation. Thus, it is imperative for the government to assess every development activity wisely and encourage those activities that provide more benefits to the society without significantly discounting the options and benefits for future generations.

This study attempts to assess net benefits which accrue to the society from both preservation and mining activity by analyzing the trade-off between real benefits to society, and the resource and opportunity costs that should be borne to obtain the benefits. Thus, the study aims to assist the decision makers to take informed decisions. In order to assess

[5]http://www.dephut.go.id/INFORMASI/INTAG/Eks_03/Bab_1.PDF

[6]Presentation in the Commissions III and VIII of the House of Representative meeting on Discussion in Altering the Classification of Protected Area for Mining Purposes, 2003.

which alternative is the most desirable to the society, Cost Benefit Analysis (CBA) is employed. Under this approach, the desirability of an alternative is essentially the excess of benefits over costs where all outcomes are identified and assessed in terms of society as a whole (Sinden and Thampapillai, 1995).

Net Present Value (NPV) tests are applied to obtain the economic benefits of the mining activity. As the benefit accrued from mining will be a flow of outcomes over time, the NPV-tests convert each annual net benefit over time to an equivalent value at a common time which is the present value. The first NPV-test (NPV_F) is conducted to obtain the benefits of mining activity without considering foreign ownership and environmental costs. The second analysis of NPV (NPV_I) is intended to obtain the real benefit accruing to GoI from mining activity as the government only gains from taxes and royalties. The study then calculates the net benefits of mining activity with consideration of the environmental costs (NPV_E).

Threshold value (TV) analysis is further exercised to obtain the minimum value of preserving the protected forest. The threshold value is commonly applied in the valuation of unique irreversible environments (Thampapillai, 2002). The study would recommend preservation over mining activity if the result of the TV analysis is found to be negligible.

Case study

The CBA conducted in this study focuses on the mining operations of one specific company in one specific location. PT Weda Bay Nickel was selected from the list of 13 companies which are being evaluated by the DPR-RI. One of the main reasons for this selection was the availability of data. PT Weda Bay's mining operation is situated in Central Halmahera, North Maluku province. Some 90% of the shares of the company are controlled by Weda Bay Co, a Canada-based mining company, The remaining 10% of shares are held by Aneka Tambang Co with 65% of these shares held by the GoI.

In 1998, a CoW was granted to PT Weda Bay by the GoI to explore, develop, and mine any mineral deposits in the 76,280 ha of land granted to it on Halmahera Island. The CoW obtained by PT Weda Bay covers the central part of the island at the convergence of the four main arms of Halmahera, and extends from Weda Bay in the south to Kau Bay in the north.[7] The resources in the deposits in the CoW area are 215 million dry tonnes, with 1,35% of nickel and 0.11% of cobalt.

The conflict occurred when Law 41/1999 designated nearly half of the area allocated by the CoW, 35,155 ha, as protected forest, including Mount Mirarong and Mount Tabobo. The forest area was later named Ake Kobe protected forest. During the meeting with the DPR-RI in 2003, the government proposed to alter 9,954 ha of the protected forest in this area for PT Weda Bay's operations.

The company proposes to operate in an open pit mining pattern which is prohibited by the law. The proposed method of operation is to employ a conventional hydraulic excavator and truck type operation with support equipment including bulldozers, graders, and front-end loaders.[8] It also proposes to discharge the tailings, using the Deep Sea Tailings Placement (DSTP) method. DSTP involves placing the mine tailings in the deep marine environment beneath the ocean, which is a biologically active zone.

Ake Kobe protected forest

Halmahera Island is covered by rainforest although the southern peninsula has a seasonal climate and low rainfall. The forest is usually classified as 'seasonal forest" and 'monsoon forest'. The four peninsulas are all mountainous with the highest point at 1,635 m in Gunung Gamkonora (Poulsen *et al.*, 1999). Prior to the issuance of Law 41/1999, numerous proposals to establish protected areas on Halmahera Island had been submitted to the GoI as it was the largest island in Indonesia with no forest protection. One of the proposals was submitted in 1995 by BirdLife International.

The interest shown by conservationists is mainly due to the biodiversity richness of the island. An analysis conducted in the Halmahera group islands (Halmahera, Bacan, Morotai, Obi and other smaller islands), based on the distribution of birds revealed that the islands is one of the biological diversity 'hotspots' of the world. The islands rank number ten out of a total of 218 biological diversity hotspots in the world (Poulsen *et al.*, 1999). Moreover, 23 endemic species of birds are found in Halmahera Island

---

[7]http://www.wedabay.com/s/Projects.asp?PropertyInfoID=893&View=All, visited on 5th of February 2004.

[8]http://www.wedabay.com/s/Projects.asp?PropertyInfoID=893&View=All, visited on 5th of February 2004.

(Poulsen *et al.*, 1999). Therefore, it was seen that the grant of protected forest status would ensure the protection of the endemic species of the island.

The government ultimately gazetted Ake Kobe forest as a protected area of the Halmahera Island. As designated by Law 41/1999, the island's forest provided several environmental services including supporting human life systems particularly in the regulation of hydrological cycles, controlling flooding and erosion, preventing intrusion of sea water and maintaining soil quality.[9]

Furthermore, the protected forest is inhabited by the indigenous people of Tugutil. The total population of the Tugutil is still unknown but they are believed to be primarily living in Wasile, Maba with small numbers also present in Kao and Tobelo sub-district (Poulsen *et al.*, 1999). The Tugutil are semi-nomadic forest dwellers who cultivate plots of land and harvest many forest products including dammar and gaharu (local vegetables) as well as derivatives of sago and palms. Some Tugutil tribes are also found in the area where nickel mining is proposed to be undertaken by PT Weda Bay Nickel (Poulsen *et al.*, 1999).

Impact on the ecosystem

The initial operations of PT Weda Bay will require the clearance of 9,954 hectares of Ake Kobe protected forest for installation of the open pit mine. PT Weda Bay has indicated that the initial operation will extract mineral deposits at Santa Monica, which are expected to comprise 33% of the total resources. The remaining deposits are scattered along the Coastal area, Kahuna Besar, Pintu, and Jira River. These areas have been assigned to the company under the CoW.[10]

Under the above scenario, the company will clear 9,954 ha of protected forest for its initial operation (6.23% of the total ecosystem). This comprises 35,155 ha (21.99% of the total forest ecosystem) as granted by the CoW. The operation will have a significant negative and permanent effect on the overall forest ecosystem as a nickel mining operation requires clearing of all surface vegetation in areas where there are mineral deposits (Poulsen *et al.*, 1999). The clearance of a patch of forest, which is commonly called "habitat fragmentation", will affect the integrity of the overall ecosystem, and reduce

[9]Law 41/1999 Republic of Indonesia.
[10]http://www.wedabay.com/s/Projects.asp?PropertyInfoID=893&View=All, visited on 5th of February 2004.

its ability to provide ecological services including supplying water resources, and preventing flooding and erosion.

Habitat fragmentation would also have severe effects on the broader ecosystem of the island. The following impacts might reasonably be expected on the species in the habitat area, which would ultimately affect the overall biodiversity of the island (Miller, 2002):

- Decrease in the sustainable population size of many species
- Increased surface area or "edge effects" which will make some species vulnerable to (1) predators, (2) competition from non-native and pest species, and (3) wind
- Creation of barriers that limit the ability of some species to disperse and colonize new areas, to find sufficient food, and to find mates.

This study does not simply focus on the amount or proportion of the forest area that would be lost. Instead, it is more prudent to view the Ake Kobe protected forest as one ecosystem as disturbance to any part of the forest will affect the entire ecosystem. As mentioned in Clause 1 of Law 41/1999, the forest is one ecosystem comprising mainly trees and their relationships with their surroundings. This unity of function and mutual supporting relationships cannot be separated.

Methodology

Cost benefit analysis

CBA is a policy tool that is commonly used by decision makers to choose between competing alternatives. Decision makers are continually confronted by numerous choices. However, owing to the scarcity of the resources they need to choose the course of action that would be most desirable to that society.

The CBA usually follows some common sequences and the whole process is based on scientific method of problem solving. The sequences include (Sinden and Thampapillai, 1995):

- Identifying the problem and defining the alternatives to resolve it.
- Identifying the social benefits and costs of each alternative (these refer to the outcomes accruing to society from a given alternative which may differ from the private point of view. In the social context, all environmental and other effects on society included).
- Valuing the benefits and costs of each alternative.

Tabulating the annual benefits and costs

- Calculating the net social benefits for each alternative (in this stage, the net benefits in each year of the project are converted to an equivalent net benefit at a common time which is the present. The present values of each annual benefit are summed up to provide a total of outcomes which is called a net present value).
- Comparing alternatives by their net social benefit.
- Testing the effect of changes in assumptions and data by performing sensitivity analyses.
- Making the final recommendation(s).

Based on the above outline of the CBA method, the following sequence of steps was adopted in the study in order to fulfill its primary objectives, taking into consideration time and data constraints.

- After examining the dispute between preservation and mining proposals, it was realized that both alternatives cannot co-exist in the Ake Kobe protected forest. Therefore, the government needs to decide on, and foster, those activities that are determined to be more desirable. This decision-making process would be assisted by assessing and weighing the net social benefits of both activities. To obtain the net benefits of mining activities, the study performs NPV tests which convert the annual net benefits of mining activity into an equivalent value at a common time.
- Three NPV tests were conducted in this study: (1) NPV_F was undertaken to obtain the benefits of mining activity without considering foreign ownership and environmental costs, (2) NPV_I was intended to obtain the real benefit accruing to GoI from mining activity as the government is only directly benefits from taxes and royalties, and (3) NPV_E which calculates the net benefits of mining activity with consideration of the environmental costs; this is considered to be the net social benefits of mining activity.
- In estimating the NPV, there are many uncertainties owing to the difficulties of accurately defining future benefits and costs. This may include disparities in data, projection, and estimates. Thus, a Sensitivity Analysis (SA) was undertaken to ascertain the impact of price and production volatility.
- As a third step, the study attempted to analyze the minimum value of the preservation that would make the mining activity unviable. It

employed the TV analysis, which is commonly adopted for the valuing of unique environments.

Net present value tests

The present value is the equivalent today of a future benefit or cost.[11] As every alternative has a flow of economic benefits and costs overtime, a NPV is therefore required to convert each annual net benefit to a common time, which is commonly adopted as the present, and sum up the flow of these benefits. The NPV also attempts to quantify society's preferences of an alternative which may prefer outcomes at one time rather than another through discounting (Sinden & Thampapillai, 1995).[12]

As the main purpose of CBA is to assist in selecting projects and policies that are efficient in terms of their use of resources, the criterion of the NPV test can be applied (Hanley and Clive, 1993). The criterion of project acceptance is if the NPV is greater than zero (NPV>0). According to Hanley and Clive (1993), if the NPV>0, the project can be said to represent an efficient resource allocation as it is deemed to result in an improvement in social welfare.

The NPV is calculated using the following equation (Sinden & Thampapillai, 1995):

$$\frac{\mathrm{NPV} = (\mathrm{B0} - \mathrm{C0}) + (\mathrm{B1} - \mathrm{C1}) + \cdots + (\mathrm{Bt} - \mathrm{Ct})}{(1+i)^1(1+i)^t} \tag{1}$$

where i = Discount Rate; B = Benefits; C = Costs; t = time t

Sensitivity analysis

'Sensitivity analysis is a recalculation of net benefits with different data together with the reinterpretation of the relative desirability of the alternatives'.[13] The value obtained from the NPV test is based on the assigned data used in the calculations. Changes of the data will lead to different results of the NPV test. Thus, there is some uncertainty in the study. In all ex ante analyses of CBA, the analyst must make predictions concerning various data. As these predictions are not made with perfect

[11]Sinden J.A. and D.J. Thampapillai, 1995, p. 115.

[12]Discounting is a process by which a future outcome is converted to its equivalent value at present using the percentage rate of compound interest (Sinden & Thampapilai, 1995).

[13]Sinden J.A. and D.J. Thampapillai, *Introduction to Benefit-Cost Analysis*, Longmans, Australia, 1995, p. 162.

foresight and knowledge, it is important to also conduct a sensitivity analysis (Hanley and Clive, 1993).

A sensitivity test will also assist the analyst to discover the parameters to which the NPV outcome is most sensitive (Hanley and Clive, 1993). These parameters include the discount rate, quantity and qualities of inputs and outputs, project life span, and market price. The following procedure for addressing this uncertainty is adopted in this study (Sinden and Thampapillai, 1995):

- Recalculating the net social benefits by inserting different data into the NPV formula such as various estimates of the price of the metal, and the quantity of output and input.
- Identifying the critical variables and describing the source of their uncertainty.
- Re-interpreting relative desirability with all the net social benefit data.

In this study, the result of sensitivity analysis is presented in the form of price and production elasticity. Elasticity is the percentage change in a key variable which follows a one percent change in another parameter (Sinden and Thampapillai, 1995). High elasticity means that the NPV is sensitive to the tested variable which requires the analyst to spend extra time to find the best estimate of the variable (Sinden and Thampapillai, 1995). The formula used to calculate price elasticity is:

$$\mathrm{e} = \frac{\text{change in NPV} \times 100}{\text{Initial NPV}} \div \frac{\text{change in price} \times 100}{\text{Initial value of price}} \tag{2}$$

The formula for calculating production elasticity is:

$$\mathrm{e} = \frac{\text{change in NPV} \times 100 \div \text{change in quantity of metal produced} \times 100}{\text{Initial NPV} \quad \text{Initial value of quantity of metal produced}} \tag{3}$$

Threshold value

In this situation, both preservation and mining activity have costs and benefits. In the context of this study, the social net benefit of mining activity is the NPV_E. On the other hand, as discussed above, the forests can also provide significant economic value were they to be preserved in their natural state. However, the economic value of preserving the forest, is not yet clearly known. To derive a reasonable figure for the economic value of preservation, the study employed the TV analysis.

The definition of TV is embedded in the simple calculus of CBA and can be explained in the context of this study as follows (Thampapillai, 2002):

Consider the two alternatives that exist at Ake Kobe protected forests, namely mining activity and forest preservation. The alternatives may be seen as mutually exclusive and cannot reasonably coexist.

Therefore let the $\mathrm{NPV_E}$ from mining activity be "$\mathrm{b_d}$", and the value of preservation in the initial year be "$\mathrm{b_{op}}$". In this context, the value of "$\mathrm{b_d}$" is represented by the net benefits accrued from mining activity; however, the value of "$\mathrm{b_{op}}$" is not known yet. The value of $\mathrm{b_{op}}$ usually comprises both environmental and economic services provided by the forest (i.e. erosion and flood control, carbon sequestration, and maintaining the hydrological cycle). The benefits of preservation are assumed to grow at "r" percent and the discount rate is "i" percent. Benefits of preservation are likely to grow over time (t). According to Fisher *et al.* (1972), there are several reasons to expect preservation benefits to grow at a positive annual rate, "r". First, it is due to the increasing relative scarcity. As the area of global tropical rain forests declines, each remaining hectare becomes more valuable and the willingness to pay (WTP) increases. Second, because the information on the importance of the ecosystem's structure and diversity increases and becomes more widely dispersed over time, people become better informed; thus, aggregated global WTP to preserve natural resources, such as rain forests, is likely to increase. Finally, as real incomes rise, demand for environmental goods and services may also increase.

The NPV of preservation thus can be defined as

$$\begin{aligned} &= \mathrm{b_{op}}[1 + \{(1+r)/(1+i)\} + \{(1+r)^2/(1+i)^2\} \\ &\quad + \cdots + \{(1+r)^t/(1+i)^t\}] \ldots \end{aligned} \tag{4}$$

If the NPV of preservation exceeds the NPV of commercial development, then:

$$\begin{aligned} &\mathrm{b_{op}}[1 + \{(1+r)/(1+i)\} + \{(1+r)^2/(1+i)^2\} \\ &\quad + \cdots + \{(1+r)^t/(1+i)^t\}] \geq \mathrm{b_d} \end{aligned} \tag{5}$$

$$\begin{aligned} \mathrm{b_{op}} &\geq \mathrm{b_d}/[1 + \{(1+r)/(1+i)\} + \{(1+r)^2/(1+i)^2\} \\ &\quad + \cdots + \{(1+r)^t/(1+i)^t\}] \end{aligned} \tag{6}$$

The value of b_{op} refers to the threshold value, and $b_d/(1+i)^1 + (1+i)^2 + \cdots + (1+i)^t$ represents the net present value of mining activity (NPV_E).

Furthermore, the $(1+r)^1+(1+r)^2+\ldots+(1+r)^t$ represents the Present Value (PV) of \$1 growing at r%.

A formal definition of the threshold value for environmental preservation in the context of such preservation being a mutually exclusive alternative is as follows (Thampapillai, 2002):

$$TV = \frac{NPV_E}{\text{PV of \$1 growing at } r\%}$$

The TV is defined as the minimum value of an environmental benefit (EB) stemming from the preservation of an environment in the initial year growing at a specific rate such that the present value of the EB is at least equal to the highest net present value of the alternative to preservation'.[14]

Preservation would be the preferred option if the TV obtained an insignificant result. The rationale underlying this premise is that "even a small value of environmental benefit is capable of offsetting the net benefit of mining". However, the use of the TV involves a subjective assessment of what is "large" or "small" (Thampapillai, 2002). In order to overcome this drawback, this study thus utilizes results of several studies that were conducted to value the environmental benefits of protected forests in other islands in Indonesia as benchmarks.

Estimation of the TV requires complete data on the alternative to preservation, and the environmental benefit. In the study, the alternative to the preservation of the Ake Kobe protected forest is considered only as the mining activity. Moreover, the growing rate of environmental benefit is usually determined by scientific enquiry or the use of pertinent data such as the rate at which demand for recreation, or biodiversity is expected to grow (Thampapillai, 2002). Fisher *et al.* (1972) argue that 'r' could be considered as the percent rate of growth in annual benefits. In order to obtain the value of "r", the study utilized data collected by Beukering *et al.* (2003) which were published in the report entitled "Economic Valuation of Leuser National Park on Sumatra, Indonesia" which analyzed the annual benefits of conserving that park for a 30-year period.

---

[14]Thampapillai, D.J. *Environmental Economics: Concepts, Methods, and Policies*, Oxford University Press, Australia, 2002, p. 87.

Results

Net present value tests

PT Weda Bay, on its official website, proposes two alternative scenarios for its proposed mining operations under different financial parameters . Under the first scenario, published in April 2002, to the company would produce 60,000 tons of nickel per year and 5,000 tons of cobalt per year. The revised scoping study for the first scenario, as published in March 2003 indicated that the company would produce 40,000 tons of nickel per year and 3,000 tons of cobalt per year. The second scenario presents an opportunity for a lower capital cost facility than the first scenario whilst still producing the quantities of nickel and cobalt required for a financially viable project.[15]

In order to calculate a NPV, the study built in several assumptions of the financial parameters of the project. The study could not simply adopt the parameters published by the company as there were not sufficient details in the information provided by the company. For instance, the operating costs excluding cobalt credits were listed as approximately US$0.78 per lb. It is not clear whether this figure includes administration, extraction, and processing costs. Considering this constraint, the study then employed slightly adjusted figures based on realistic assumptions. Some of the points which underlie the financial parameters as adopted by the study are now discussed:

Metal prices

There are various estimates of the outlook of the nickel price on the world market. The London Metal Exchange (LME) indicates that the current price of nickel has reached US$16 per lb. However, to obtain the most accurate result for the NPV, the study referred to some reasonable estimates and did not merely adopt the current market price, as the price fluctuates. The revised figure, therefore, refers to the estimate of nickel price as formulated by the Department of Natural Resources of Canada.[16] The price of nickel is predicted to range between from as low as US$2 per lb to US$4 per lb in 2005.

Cobalt, the co-product of nickel, is recovered by several processes including matte smelting, reduction roasting, ammonia leaching, and

---

[15]http://www.wedabay.com/s/NewsReleases.asp?ReportID=54546&_Type=News-, visited on 25 March 2004.
[16]http://www.nrcan.gc.ca/mms/pdf/nfo/nfo99/nick-e.pdf

pressure acid leaching. It is a specialist metal used in highly specialized application i.e. hybrid alloys in aircraft gas turbine engines, in cemented carbides for cutting and wear-resistant applications, and for various other uses where substitution is rare. Therefore, the demand for cobalt is no longer dependent on the price.[17] For instance, in 1996, the demand was rising rapidly at a time when the price increased substantially to US$30 per lb. On the other hand, a low cobalt price resulted in only a small reduction in cobalt production by mid-2002 due to a reduction in the demand for cobalt.[18] Therefore, the cobalt price used in this study is based on that provided by the Cobalt Development Institute, an independent association, in 2002. It prediced that the price would range from US$7 to US$9 per lb in 2004.[19]

Operating costs

The operation costs adopted in the study was at US$1.5 per lb. This figure includes administration, mining and processing costs. This figure is significantly different from the operating costs published by the company, at US$0.78 per lb for the first scenario and US$0.86 per lb for the second scenario. As the figures published by the company were not accompanied by a detailed breakdown, the study referred to an estimates made by Brookhunt as presented during the inaugural New Caledonian Nickel Conference in June 2002. It predicted that the total production cost in 2003, including mining and processing costs, would be US1.68 per lb.[20] Further complicating the calculation of the value of nickel, the company proposed to produce an intermediate sulphide product of nickel. This intermediate product would then be sent to refinery in another country. Therefore, the study assumes that the price of metal produced by PT Weda Bay should be lower than the price of the final nickel product on the world market. However, it was not possible to find sufficient information on the price of the intermediate nickel product. To overcome this difficulty, the study assumes that the intermediate nickel product has the same price as that of the final nickel product on the world market. The costs employed in

[17]http://www.gfms-metalsconsulting.com/Nickel%20&%20Cobalt.htm, visited on 23 January 2004.
[18]http://www.gfms-metalsconsulting.com/Nickel%20&%20Cobalt.htm, visited on 23 January 2004.
[19]http://www.formcap.com/Jan09-2003-CoTrends.pdf, visited on 23 January 2004.
[20]http://www.brookhunt.com/nickel/New%20Nickel%20Projects.PDF

the study should thus encompass mining and processing costs which make the operating costs at least equal to the costs of producing the final nickel product.

Capital costs and discount rate

To obtain both capital costs and the discount rate, the study referred to the figures published by PT Weda Bay. As for the discount rate, it will be held constant in each analysis, at 10%. The 10% discount rate is commonly applied in assessing government project in Indonesia.

Exploration costs

The exploration costs refer to the cost incurred by the company before the production stage. This includes the general survey, exploration, and feasibility study stages. Based on the annual reports published by PT Weda Bay in 2000–2003, the exploration costs adopted in this study is approximately US$200 million.[21]

Net Present Value without Considering the Ownership and Environmental Costs (NVPF) Using the financial parameters described above, the NPV_F obtained is estimated at US$369 million. The result reveals that under the low price estimation, the NPV_F obtained is negative. This means that the mining activity is economically unviable where the price of nickel and cobalt fall to US$2 and US$7 per lb respectively. Table 16.1 shows the NPV_F obtained using different estimates of price and quantity of metal produced.

Table 16.1. NPVF obtained with different financial parameters in million US$.

	Low price	Best price	High price
High production	−178.04	676.10	1530.25
Best production	−195.97	369.07	934.00
Low production	−215.54	60.41	336.36

High production: 60,000tpy nickel and 5,000tpy cobalt
Best production: 40,000tpy nickel and 3,000tpy cobalt
Low production: 20,000tpy nickel and 1,000tpy cobalt
Low price : US$2 of nickel and US$7 cobalt price
Best price : US$3 of nickel and US$8 cobalt price
High price : US$4 of nickel and US$9 of cobalt price

[21]http://www.wedabay.com/s/QwikReport.asp, visited on 20th May 2004.

The results obtained can be compared to the figures presented by the company. According to the press release issued by the company in 2002, the financial analysis of the first scenario of the project indicated that the internal rate of return on investment (IRR) is 16.2% and the Net Present Value (NPV_F) is US$380 million at a 10% discount rate. In the analysis, the estimated capital cost of the facility and associated support infrastructure is US$902 million. Furthermore, it was assumed that the average operating cost is US$0.78 per lb of nickel produced, exclusive of cobalt credits. Metal prices in this financial model are US$3 per lb and US$10 per lb for nickel and cobalt respectively. All the sulphide intermediate product is assumed to be sold.[22]

Sensitivity analysis

As shown in Table 16.1, several analyses were conducted to obtain the NPV_F using different estimates of price and quantity of metals. The study estimated that nickel and cobalt prices might fluctuate between US$2 and US$4, and US$7 and US$9 per pound respectively. Furthermore, the quantity of nickel produced is estimated to be 20,000 tons per year for the low estimation; 40,000 tons per year for the best estimation and 60,000 tons per year for the high estimation. However, the sensitivity analyses ignored the fluctuations of cobalt produced as cobalt is the by-product of nickel and cobalt production depends on the level of nickel produced.

The results of the sensitivity analyses reveal that the price elasticity of cobalt on NPV_F from high to best estimate and best to low estimate keeping all other parameters at best estimation are 0.87% and 0.85% respectively. Furthermore, the price elasticity of nickel on NPV_F from high to best estimate and best to low estimate keeping all other parameters at best estimation are 2.36% and 4.27%. The quantity elasticity on NPV_F ranges from a high to best estimation and best to low estimation keeping price at its best level are 1.36% and 1.67%. Table 16.2 presents the results of the sensitivity analyses.

From the results, the NPV is more sensitive to changes in the price of nickel than the changes in the price of cobalt or the quantity produced. The study therefore paid more attention to the nickel price fluctuations in order to obtain the best NPV_F. To this end, estimations by the Natural Resources Department of Canada are deemed to be appropriate to be used as references, as more accurate data could not be obtained.

[22]http://www.antam.com/News/Publications/QR/Adobe/2Q02QuarterlyReport.pdf

Table 16.2. Elasticity of different parameters.

Elasticity	High to best estimation	Best to low estimation
Nickel price elasticity	2.36%	4.27%
Cobalt price elasticity	0.87%	0.85%
Quantity elasticity	1.36%	1.67%

NPV accrued to Government of Indonesia (NPVI)

The NPV_I is the net benefit accruing to the GoI from taxes and royalties. Pursuant to the CoW granted by the GoI, the following tax liabilities should be fulfilled by the company:

- Deadrent: It refers to the annual amount of tax that should be paid by the mining company based on both the number of hectares covered by the CoW and the stage of the CoW. The dead rent is US$0.10 per hectare during the first year of the exploration stage, and US$3 during the operation stage.
- Royalties in respect of the company's production of Minerals: In the context of the study the royalty that is required to be paid is US$63 per ton of nickel produced and US$156 per ton of cobalt produced.
- Corporate Income Tax in respect to income received or accrued by the company: According to the Income Tax Law 1994, the company is liable to pay 30% or lower rate as set forth by the regulation on taxable income exceeding fifty million Rupiah.

Table 16.2 shows that the best estimate of NPV_I is US$323.3 million. In comparison, the expected income accrued to the country, as announced by the Government in the coordination meeting with Commission VIII and Commission III of the DPR-RI,[23] was reported at US$50 million per annum

Table 16.3. NPVI obtained with different financial parameter in million US$.

	Low price	Best price	High price
High production	87.85	521.51	1,354.51
Best production	55.95	323.33	867.87
Low production	24.04	124.78	380.79

[23]Presentation in the Commissions III and VIII of the House of Representative meeting on Discussion in Altering the Classification of Protected Area for Mining Purposes, 2003.

including the taxes and no-taxes income. Moreover, if this annual income is converted to present value, the NPV_I over 20 years with a 10% discount rate is US$380 million.

NPV with consideration of environmental costs (NPVE)

In calculating NPV_E, the study included the environmental costs that might reasonably incurred due to the proposal to discharge tailing using the STD method. In making this assumption, the many studies which were referred to showed that severe impacts on marine ecosystem is caused by applying the STD method. Weda Bay where the tailing is proposed to be discharged, sustains approximately 3,800 permanent fishermen and 240 non-permanent fishermen.[24] If the disposal of the tailings significantly harms the marine ecosystem, it will affect the income of these fishermen.

If the company does not employ the STD method, it would incur greater capital costs, but less environmental costs would be borne by the community. Therefore, this study argues that the environmental costs of discharging tailings into the marine environment could be equal to the cost saved if the company applies the STD method.

By comparison, Zandee (1985) reported that the sea-disposal system used in Calacan Bay in the Philippines costs less than half as much as the operation of the tailings-pond system. Kitsault Mine of Canada which adopted the STD method estimated that STD would save the company as much as $25 million dollars per year in tailing disposal costs relative to the cost of land disposal. The U.S. Department of the Interior concluded that, on average, STD resulted in a 17% reduction in capital costs and a 1.6% increase in operating costs (Coumans, 2002). Thus, to obtain the NPV_E, this study increased the capital costs by 17% and reduced the operation costs by 1.6%. Using this assumption, the NPV_E obtained is US$253.5 million.

Threshold value (TV) analysis

In order to obtain the threshold value of preservation, this study utilizes the data of the annual benefits accrued from conservation in the National Parks of Leuser Sumatra as estimated by Beukering *et al.* (2003) for the value of "r". The average growth rate of annual benefits from preservation in the national park is 2.7%. Using this rate, the TV obtained is US$12.32

[24]Kompas, Kabupaten Halmahera Tengah, 10th July 2003.

million.[25] This represents the TV of the environmental benefits of the preservation of Ake Kobe protected forest in the initial year, growing at 2.7%. The present value of the forest is assumed to be at least equal to the highest NPV of the alternative to preservation, which is the mining activity.

Discussion

The TV represents the minimum value of preservation that would make mining unviable. Therefore, the preservation would be the more desirable alternative if the result of the TV was insignificant. Thus, in this section, an attempt is made to determine whether the value of the protected forest is expected to be greater than US$12.32 million. Several related studies are referred to as benchmarks in appraising the TV obtained.

As stated in the Law 41/1999, every protected forest of the country is listed by the government in a register of designated sites intended "to provide critical ecological services for human life".[26] The Ake Kobe protected forest is in this list of designated sites. The key ecological services, which are provided by protected forests in Halmahera island, are listed in a report published by "BirdLife" entitled "Evaluation of the Proposed Lalobata and Ake Tajawe National Park" in 1999. These services include (Poulsen *et al.*, 1999):

- Watershed area for irrigation purposes in Wasile, Weda, and Maba districts.
- Supporting the Tugutil people who are dependent on the forests for their livelihood.
- Conserving populations of pollinators and seed dispersal agents that are important for regeneration of production and limited production forests surrounding the protected forest area.
- Providing an important refuge and genetic reservoir for economically valuable forest species such as sago Metroxylon spp, indigenous fruits, butterflies, beetles and parrots.
- Conserving the population of the majority of species of birds that occur in North Maluku including four endemic genera of North Maluku and four endemic mammal species, and two rare kinds of fruit bats.

[25] The threshold value is calculated using the following formula $\mathrm{TV} = \frac{\mathrm{NPV_E}}{\mathrm{PV\ of\ \$1 growing\ at\ R\%}}$.

[26] Law 41/1999 Republic of Indonesia.

Table 16.4. Classifications of the total economic value (TEV) of Ake Kobe protected forest.

Use values		Non use value	
Direct use value	Indirect use value	Option value	Existence value
• Recreation • Education/ researches • Supporting the Tugutils	• Ecosystem service i.e. pollination, habitat, nutrient • Flood Control • Ground water recharge and watershed protection • Carbon sequestration • Natural disaster prevention	• Future information • Future use	• Biodiversity

The above ecological services can be valued using the concept of total economic value (TEV) which is a well-established framework for identifying the valuee associated with protected areas as pointed out by IUCN (1998). The TEV of a protected forest consists of its use value and non-use value, indirect use value and option value. Table 16.4 outlines the expected TEV of Ake Kobe protected forest:

The study had insufficient data to fully value each service provided by the Ake Kobe protected forest. In order to explain further the value of each service provided, the following sections attempt to quote the economic value of the services by referring to other studies on estimates of the value of environmental services provided by protected forests in other islands in Indonesia with similar characteristic to those of Ake Kobe protected forest.

Water resources

One of the most important ecological functions performed by protected forest is slowing the rate of surface water runoff. The slow rate of runoff enables vegetation to take up water and thereby delays the time to soil saturation and finally to the runoff flow to the nearest watercourse.[27] Forest soils have a higher water-storage capacity than non-forest soils; compared

[27]http://guide.conservationfinance.org/chapter/index.cfm?Page=5

to a deforested watershed, the more complex structure of the forest ground-surface and underlying soil allow more efficient soil infiltration. Therefore, the forests may increase minimum stream flows during the dry season and prevent flooding during the wet season.[28]

Ake Kobe protected forest protects the watershed area that is important to controlling water supply for the coastal sub-districts such as Wasile, Weda and Maba. These are the focal areas for expansion of wet-cultivation and plantation crops in Maluku province where 80% of the people depend on agriculture.[29] In order to enhance the agricultural productivity in the area, the government has earmarked Rp. 7,655 million to protect the watersheds for irrigation in Wasile, Maba and Weda. Investments in the installation of irrigation systems in these areas would be ineffective should the forest be removed (Poulsen *et al.*, 1999). Moreover, these sub-districts are core areas for the government's "transmigration" (migrant new town) programs with a population of 143,522 people (UNDP, 2001). The water resources flowing from Ake Kobe protected forest is also important to sustain the household and small industry activities in these areas.

A study conducted in Lore Lindu National Park in Central Sulawesi, the island next to Halmahera, could be used as comparison to value the water resource services provided by Ake Kobe forest. The study, which was conducted by Nature Conservancy Indonesia, a well-known non-governmental environmental organization, estimated that 304,607 people from 67,160 households are dependent on water originating from the park. Water from the park also irrigates some 22,338 hectares of agricultural land that on an annual basis produces an estimated Rp. 59.4 billion in crops and plantation products. The water also services livestock and supports inland fishing and fisheries. On a yearly basis, the value of consumption of products from these fishing and fisheries industries is estimated at Rp. 16.4 billion (Nature Conservancy, 2001).

In total, the Nature Conservancy calculated that the household water consumption in the study area was 8.7 million cubic meters per year, with a value of Rp. 5.2 billion. Similarly, about 3.8 million cubic meters, worth an estimated Rp. 1.2 billion are consumed annually by industries in the area. Thus, water arising from this national park has an estimated total

[28] http://guide.conservationfinance.org/chapter/index.cfm?Page=5
[29] http://www.maluku-utara.go.id/default.php?rowIndex=19

current annual value of approximately Rp. 89.9 billion or approximately US$9 million (Nature Conservancy, 2001).

Natural disaster prevention (landslide and flood control)

Another important ecological service provided by the forest is the reduction of soil erosion and sedimentation of waterways. Interception of rain by forest canopies results in less water falling on the ground compared with a deforested watershed. Understorey forest vegetation and leaf litter protect the soil from the impact of rain that falls through the canopy. Extensive root systems hold the soil more firmly in place and resist landslides better than clear cut or heavily disturbed watersheds.[30] Moreover, the function of the forest as watershed protection also prevents the lower areas from flooding during the wet season as the forests help to slow the rate of runoff.

The proposed area for the open pit by PT Weda Bay is located near Mount Mirarong and Mount Tabobo. To uncover the mineral deposits, the company would strip the mountains, remove the surface material and transport it to a disposal area. Both stripping and mining are conducted from one or a sequence of underground benches or seams. A thick deposit, typical of metallic ores, requires many benches and resembles a circular pyramid inverted in the earth (Hartman, 1987). Serious problems such as a flood calamity in the coastal districts could occur due to this activity considering the topographic condition of the area.

The following are the details of the topographic area where the open pit will be installed:[31]

Very steep (slope deg >40%): 4.660 hectares (protected forest)

Steep (slope deg 25–40%): 2,410 hectares (production forest)

From the experience on neighboring islands, the damage due to flood following the massive deforestation in the area can be significant. In the past five years, Indonesia has experienced severe natural disasters which have led to significant losses.

The causes of flooding can be divided into two main types, namely, natural and anthropogenic. El-Nino and La-Nina are both natural phenomena that can cause major natural disasters. However, human factors also play a significant role. Deforestation due to illegal logging has seen the

---

[30] http://guide.conservationfinance.org/chapter/index.cfm?Page=5

[31] Forest Function Scoring by the DPR-RI on Mount. Tabobo and Mirarong, data obtained from JATAM (Jaringan Advokasi Tambang).

destruction of the natural vegetation and halted the soil from absorbing water during the rainy season; therefore, the runoff flows to the cities and residential areas and causes severe flooding (Walhi, 2004). If the forests are left intact, major flood events could be avoided and society would avoid the burden of the costs due to flooding. Therefore, it could be argued that the economic value of services provided by the forests in controlling the flood would be comparative to the costs of avoidance should the disaster have been prevented. Supporting this notion, the study conducted by Beukering *et al.* (2003) estimated the service provided by the park as a flood control agent as being US$70.6 million in 2003.

Carbon sequestration

Forests play an important role in controlling global climate as most of terrestrial carbon is stored in the vegetation and soils of forests (Butcher *et al.*, 2001). The excess carbon which is mainly emitted from burning of fossil fuel can lead to global warming. It is estimated that preserved tropical rainforests could sequester 200–300 tons of carbon per hectare of forest (Butcher *et al.*, 2001). The amount of carbon sequestered by the forest will fall following the conversion of the pristine forest to silviculture and agro-forestry. According to the hypothetical analysis of a Clean Development Mechanism (CDM) project under the Kyoto Protocol (which has the twin objectives of helping developed countries to achieve their emission reduction targets and helping developing countries to promote sustainable development in their own economies), the cost of carbon abatement is approximately US$5–50/ton carbon (Michaelowa and Dutschke, 2000).

Other services

It was not possible to find sufficient data on the value of other services provided by the Ake Kobe protected forest areas, such as ecosystem services, biodiversity richness, and the function of the forest in supporting the Tugutil people.

The protected forest sustains the life of the indigenous Tugutil people who mainly live in the forested areas of Wasile, Maba with a few of them in Kao and Tobelo sub-districts (Poulsen *et al.*, 1999). Tugutil are semi-nomadic forest dwellers who live by harvesting many the forests' products including dammar and gaharu as well as derivatives of sago and palms (Poulsen *et al.*, 1999). Should the GoI choose to approve mining activity over preservation of the forest, it also needs to consider that the loss of livelihood of these people will be significant.

Ecotourism

Ecotourism is an activity that could co-exist with the preservation of the forest areas. The potential to develop ecotourism activities in Ake Kobe forest is high, considering its biodiversity richness. Some companies have been arranging tours for visitors to Halmahera Island to engage in bird watching as the island harbors significant numbers of endemic bird species, which are not be found anywhere else. Considering this real opportunity, the government could start promoting ecotourism within the Ake Kobe protected area.

The tourism industry in Indonesia consistently counts as the country's third largest foreign exchange earner. For example, it generated US$5,172 million in 1995. Visitor arrivals to Indonesia increased by more than 400% in the decade 1985–1994, and had exceeded 5 million per year by the end of 1996. Although ecotourism is not yet a major component of the tourism industry, the earning potential of national parks and other protected areas has been recognized by the government (Sumardja, 1995).

There are several successful examples of Indonesian nature tourism which were established over the past decade without formal direction from the government. They include the orang-utan rehabilitation centre at Bohorok, in North Sumatra; the Komodo "dragons" in Komodo National Park; and Mount Bromo in central Java. Tourism in all of these places began in the 1970s with a few adventure travelers, and has gradually grown, attracting greater numbers of visitors every year (Cochrane, 1997).

The income generated from ecotourism activities in the protected forests might not be significant if the GoI only considers the economic value from only the income generated directly from entrance fees. For instance, in the Bromo Tengger Semeru national park, in 1997/98, only Rp. 348 million was generated from entry fees with a standard entry fee of Rp. 2000 (US$0.2) per person. However, ecotourism is usually followed by a boom in household businesses in the surrounding communities. Surveys conducted in the communities who live near the national park showed that every single family interviewed was involved in an activity relating to tourism. For instance, by taking visitors to the crater of the mountain on horseback, the local were paid Rp. 10,000 to Rp. 20,000 per trip. Moreover, many people also rented rooms to tourists, but generally only during public holidays. For this, they received substantial amounts of Rp. 50,000 to Rp. 70,000 for a small, dark room with primitive sanitation facilities (Cochrane, 1997).

Debt-for-nature swaps

The richness of biodiversity of Halmahera Island has attracted many international agencies which have shown interest in supporting conservation in this area. For example, the British Bird Watching Fair (BBWF) has invested effort and resources in building support among provincial and local government, local communities, and private-sector interests. It was also reported by "BirdLife" that they have obtained World Bank support to focus on this area.[32] The World Bank and Global Environment Facility (GEF) have respectively pledged US$1.1 million and US$1.2 million to support the establishment of the proposed Lalobata-Aketajawe National Park (World Bank, 1998). The project aimed to establish a national park with a multi-sectoral agreement of boundaries and development of effective management operations. Once the park has been legally established the project would assist in financing investment in zoning, basic park infrastructure, strengthening of management and enforcement capacity, and establishment of monitoring programs. Unfortunately, all this work came to a halt because of civil unrest in 1998–1999 which resulted in the deaths of tens of thousands of people on Halmahera.[33]

Given the donor agencies' interest in preserving the forest on the island, it may be possible for the GoI to propose 'Debt-for-Nature Swap' schemes to be applied in the area. These schemes are usually applicable to parks with specific activities or features of interest to foreign donors or conservation investors mainly due to the biodiversity sustained. A debt-for-nature swap (or conversion) is defined as "the cancellation of debt in exchange for a commitment to mobilize domestic resources for the environment" (Haeruman, 2001).

The external debt of the GoI to public and private creditors in 2001 following the Asian economic crisis was estimated at approximately US$130 billion. As the country struggles to even meet the interest payments on this debt, there is greater pressure to exploit the country's natural wealth

[32] http://www.birdlife.net/news/features/2003/08/halmahera.html
[33] http://www.birdlife.net/news/features/2003/08/halmahera.html

(Haeruman, 2001). Debt swaps are an innovative response to this crisis. They enable developing countries to convert external debt obligations into increased support for the conservation of their natural resources.

Conclusion

The following results were obtained from the analyses: (1) the NPV_F which refers to the best estimation of net present value of mining activity without considering the ownership and environmental costs is estimated to be US$369 million. This analysis assumed US$600 of capital cost and US$1.50 operating cost including administration, mining and processing costs. Metal prices in the analysis are US$3 and US$8 per pound of nickel and cobalt respectively; (2) the real benefit accruing to the GoI from the mining activity (NPV_I) is US$323.3 million, as the government is only directly benefits from taxes and royalties; and (3) the study assessed the net social benefits of the mining activity by analyzing the environmental costs that have to be borne by the society in order to obtain the benefits of the mining activity. The NPV_E is estimated at US$253.2 million. The study estimates that NPV_E is the net social benefits of mining activity; and (4) the minimum value of preservation which grows at the average rate of annual benefit from preservation of 2.7% is estimated at US$12.32 million.

The protected forest performs various important ecological and environmental services, each of which provides significant economic value. It could be argued that the total economic value of the Ake Kobe protected forest is expected to exceed the obtained TV, which is estimated at US$ 12.32 million. The study therefore concludes that the preservation of the forest is more desirable than mining activity and recommends the decision makers to prefer the former over the latter. Moreover, if the forest is preserved in its natural condition, it will last for a potentially infinite period of time and remain as a life support system for current as well as future generations. On the other hand, the benefits from mining activities will last for only 20 years, with very limited additional benefits from post-mining operations.

Various activities could also be undertaken in the forest which would not adversely affect their quality. These include ecotourism and debt-swap-for-nature which could further reduce the TV obtained. These activities would provide more benefits while the ecological services can still be enjoyed by society.

References

Beukering, P.J.H, Herman, S.J.C, Marco, A.J, Economic Valuation of Leuser National Park on Sumatra, Indonesia, Ecological Economics 44 (2003) p. 43–62.

Butcher, P.N, James, M.H, James, S.R, Brice, X.S, Mark, A.V, Evaluating the Carbon Sequestration Potential of Tropical Forest, Donald Bren School of Environmental Science and Management, 2001.

Cochrane, J, Factors Influencing Ecotourism benefits to Small, Forest-Reliant Communities: A Case Study of Bromo-Tengger-Semeru National Park, East java, Department of Southeast Asia Studies, University of Hull, UK, 1997.

Coumans, C, STD Toolkit: Introduction, Mining Watch Canada, A Joint Publication of Project Underground and Mining Watch Canada, 2002.

Fisher, A.C, John, V.K and Charles, J.C, The Economics of Environmental Preservation: A Theoretical and Empirical Analysis, American Economic Review, 62 (4) (September 1972): 605–19.

Haeruman, H.J, Financing Integrated Sustainable Forest and Protected Areas Management in Indonesia: Alternative Mechanisms to Finance Participatory Forest and Protected Areas Management, International workshop of experts on financing sustainable forest management, Oslo, 2001.

Hanley, N. and Clive, L.S. Cost-Benefit Analysis and the Environment, Aldershot, Hants, England: Brookfield, Vt.: E. Elgar, 1993.

Hartman, H.L, Introductory Mining Engineering, New York, Wiley, 1987.

IUCN , Threat to Forest Protected Area, Summary of A Survey of 10 Countries Carried Out in Association with The World Commissions on Protected Area, Report, 1999.

IUCN, Economic Value of Protected Areas, Report, (1998).

Michaelowa, A and M. Dutschke, Climate policy and development: flexible instruments and developing countries, Edward Elgar, 2000.

Miller, G.T. Environmental Science: Working with the Earth, 9th edition, Thomson Learning, 2002, p. 467.

Poulsen, M.K, Frank, R.L and Yusup, C, Evaluation of the Proposed Lalobata and Ake Tajawe National Park, BirdLife, 1999.

Sinden J.A. and D.J. Thampapillai, Introduction to Benefit-Cost Analysis, Longmans, Australia, 1995.

Sumardja, E.A., National Parks and Foreign Aid, Conservation Indonesia Vol. 10, No. 4. Jakarta: Worldwide Fund for Nature, 1995.

Thampapillai, D.J. Environmental Economics: Concepts, Methods, and Policies, Oxford University Press, Australia, 2002.

Nature Conservancy Indonesia Program, The Value of Water Resources in Lore Lindu National Park, Central Sulawesi, Indonesia, report, 2001.

UNDP, North Maluku and Maluku Recovery Program, 1st quarterly report, 2001.

Walhi, Tinjauan Lingkungan Hidup WALHI 2004, Bencana Nasional Berkelanjutan: Kegagalan Rejim Menjamin Keselamatan Rakyat, report, 2004, http://www.walhi.or.is/Indonesia/outlook/Final%20outlook% 202004.pdf

World Bank Project Brief on Indonesia: Maluku Conservation and Natural Resources Management (MACONAR), 1998.
World Bank, Indonesia — Environmental and Natural Resource Management in A Time of Transition, Report, 2003.
Zandee, D. Tailing Disposal at Marcopper Mining Cooperation. Asia Mining. p. 35–45, 1985.

17

TIMBER SHORTAGE AND THE SUSTAINABILITY OF THE MALAYSIAN FURNITURE INDUSTRY

TIMOTHY HROMATKA and VICTOR R SAVAGE

Since 1984 the Malaysian furniture industry has grown to RM 7.6 billion worth of annual exports. The industry has been built almost solely on sustainably grown raw material that does not deplete the earth's natural capital: rubberwood. Malaysia was the first country to successfully commercialize rubberwood, a by-product of the latex industry, which was previously considered as waste; however, due to structural changes in the forest based industries and international commodity prices; there is now a market shortage and price increase of rubberwood.

This study looked at the historical development of the industry and the current situation; examining alternative timber species and alternative materials. These two trends are highlighted in case studies looking at acacia and composite panels. To ensure long term wood security for the industry, commercial plantation forestry is necessary. This study traces the beginnings of plantation forestry, analyses the current state of the industry, and make recommendations to how it could be made more commercially viable.

I. Introduction

While industrial systems have reached pinnacles of success, able to muster and accumulate human made capital on vast levels, natural capital, on which civilization depends to create economic prosperity, is rapidly declining, and the rate of loss is increasingly proportionate to gains in material well-being. There is a "critical interdependence between the production and use of human-made capital and the maintenance and supply of natural capital" (Hawken *et al.* 1999). This interdependence is obvious in a natural resource based industry such as the timber and wood products industry. As long as there is a forest, there can be an industry based on the use of that forest, but once the forest is gone, the industry, and all of the human made capital that it generates, will be gone as well.

Will the furniture industry go the way of Malaysia's forest based industries and decline or can the industry recover form the extra-industry disturbances and regain its sustainable foundation? The research attempts to answer this question.

The sustainability of the Malaysian rubberwood-based furniture industry is examined in light of a decreasing supply of the plantation grown timber which the industry is built on. Industry development and growth, current market conditions, and future trends are studied. Causes of the material shortage are examined as well as possible solutions including the long term supply of a sustainably grown timber.

The topic has been approached by exploring the natural (ecological) and industrial supply and demand model from a historical perspective with the intent of seeing the direction of the future of the furniture industry. This is done by examining the forest based industries and their subsequent decline, the growth of the plantation timber-based furniture industry, disruptions to the industry due to extra-industry disturbances, and measures the industry is taking to ensure a sustainable supply of raw materials and the help it needs to do so.

Primary research is conducted by means of industry interviews and a fax survey. Other research that is conducted includes two case studies. The first case study is to look at the largest furniture retailer in the world, IKEA, and how their buying policies have influenced an industry. The second case study involves examining a product currently in production and seeing where material substitution may lead to a reduction in the consumption of raw material; and therefore less strain on the plantation forest resources, and may also have other benefits such as cost savings.

In Malaysia, there are now more than 20 rubber tree clones planted on an area of 1.315 million ha, the majority of which are on the Peninsula. In 1999, more than 80 percent of the 7.2 million hectares of plantations established worldwide for latex production were in Southeast Asia; 70 percent of the total (or 5.2 million ha) being in Indonesia, Malaysia and Thailand (FAO, 1999). For decades, Malaysia had the largest area, followed by Indonesia and Thailand but with increasing wages in Malaysia and decreasing rubber prices, labour-intensive production of natural rubber (NR) in larger estates is slowly shifting to lower-wage countries in the region, while in Malaysia plantations are increasingly replanted with oil-palms (*Elaeis guineensis*). Indonesia is now the world's largest producer of natural rubber from *Hevea brasiliensis* followed by Thailand.

After 30 years of tapping rubber trees, a decline in latex production makes further tapping of the trees uneconomical. The trees are then removed and replaced with new seedlings. Previous to the utilization of rubberwood for timber and timber products the felled trees were used as fuel wood for the drying and smoking of sheet rubber, brick making, and locomotive fuel, or simply burnt away. It was in the late 1970's that intensive commercial utilization of rubberwood started in Malaysia.

II. Methods

In May of 2004 a market survey of furniture manufacturers was conducted with the intention of getting a better "industry feeling" of past use of rubberwood in their products and of the direction that manufacturers intend to take in the future in regards to material substitution and species substitution and utilization as a response to the increase in rubberwood prices. With data obtained from the survey and that gleaned from secondary sources, it was the intent to construct a simple input-output model. Norini (1999) had discussed that in a previous survey of the forest based industries (FBIs) of Malaysia conducted in 1997–1998, that although response was high, there was a reluctance of the participants to disclose information deemed confidential. This included information about the company's turnover, margins, and profitability. With this in mind, the researcher's survey was constructed in such a manner that the potential respondent would be encouraged to reply and questions were constructed to be simple multiple choice without the need for the respondent to reveal detailed facts.

The researcher conducted numerous interviews in the industry in the attempt to gain a better understanding of the dynamics. The persons interviewed were top level managers in their firms and the questions asked were to evaluate where the manufacturer was and where they plan to be in regards to the areas of material and production choices, product design innovation, and export market targeted. The companies were located in Muar, Malacca, Sungai Buloh, and Klang and ranged in size from small (under 50 workers), medium (50 to 150 workers), and large (over 150 workers). Some of the overall themes were those of specialization, higher value added, and material substitution.

A questionnaire was used to collect data from a fax survey. The questionnaire was designed to obtain information on the following:

- Company location
- Company Size
- Products Produced

- Export Markets
- Current and Planned Use of Rubberwood in Production
- Composite Material Substitution
- Timber Species Substitution

III. Survey Results

The survey was sent out by fax to 57 wooden furniture manufacturers throughout Malaysia with an accompanying cover letter. The response to the survey was disappointing with only a 4% response rate. Therefore the amount of data gleaned was not considered to be of any statistical significance to use in any model. The survey may not have been a success because it was not followed up with phone calls or site visits. The researcher was sensitive to the fact that the furniture industry is a competitive business and was reluctant to approach manufacturers soliciting information that may have been viewed as proprietary. This difficulty in obtaining responses was also experienced by Norini (1999). The survey was also conducted in English which may have been difficult for small operations.

In spite of the poor response, some interesting information was gleaned form those that did respond. There was a large manufacturer from Malacca with over 200 employees as well as a small manufacturer in Muar, Johor. Both manufactures are export oriented with a diverse export market and both use over 80% rubberwood as their raw material. The small manufacturer expected the price of rubberwood to continue to rise while the large manufacturer felt the long term price to be stable. Also, both manufacturers use veneered MDF and particle board in their production and both plan to use more of these engineered products in place of solid wood. Both manufacturers have substituted rubberwood with other tropical timber species including: Meranti (*Shorea spp*), Kembang Semangkok (KSK) *(Scaphium spp.)*, bintangor (*Callophylum spp.*), nyatoh (*Palaquium spp*), and acacia (*Acacia mangium*). Therefore, we see a trend away from relying solely on rubberwood as a raw material and substitution with wood composite materials and other tropical timber species. This is a similar trend gleaned from interviews.

i) *Original design manufacturers (ODM)*

A number of dining suite manufacturer's managers were interviewed. These companies are highly specialized in making tables and chairs or simply chairs. Imei Furniture Industries Sdn. Bhd. and Airegold Sdn.

Bhd., manufacture furniture to their own design, (ODM) or use common, classic styles. These two companies are small, located in the Muar region, and use rubberwood as their raw material. Their main competitive advantages are their price, quality, and ability to deliver. While Airegold was maintaining a design direction of heavy hand carving on rubberwood, it was facing competitive pressure from Vietnam and its lower labor wages. Imei on the other hand was starting to produce cleaner designs that made for easier machining and required less labor and was experimenting with using imported ash and oak from the US as a replacement for rubberwood.

Another dining manufacturer interviewed was Tawei Sdn. Bhd. in Klang. They are a medium sized operation and follow the ODM marketing strategy. They have designed many of their products to use composite materials such as MDF and chip board with a layer of veneer over it. Their designs are simple and easy to produce and by minimizing the amount of rubberwood used, they have been able to remain competitive. Tawei explains "with veneered products and those made with veneered/MDF and rubberwood we will continue to make these in Malaysia, but we will make chairs in China using imported oak and local beech".

ii) *Original equipment manufacturing (OEM)*

A component and case goods manufacturer was interviewed employing the OEM strategy. The company is Promptcon Enterprises Sdn. Bhd. and is currently upgrading and expanding their facilities. Their strategy is in species substitution and they are using tropical species such as bintangor (*Callophylum spp.*), KSK (*Scaphium spp.*), and nyatoh (*Palaquium spp.*). For other furniture manufacturers Promptcon is a supplier of veneered MDF and chip board. Their general manager has stated that their business has seen a large increase over previous years suggesting that many of their customers are using more composite materials and less solid wood.

iii) *Material substitution*

Hup Chong Furniture Sdn. Bhd. is a larger manufacturer located in Klang specializing in bedroom furniture. Their executive director has provided some interesting input into the pressures affecting the Malaysian furniture industry and the resulting trends. In 2001 Hup Chong was using all rubberwood in their production except for some tropical hardwood bed frame components which require extra strength. The timber used is kempas

(*Koompassia malaccencis*). In 2002, they started to substitute tropical wood in areas that are not visible. The cost was twice that of rubberwood but required less processing and thus resulted in cost savings. From 2003 to the present, rubberwood is being substituted by tropical timber in any part that the buyer agrees to. Species substituted include Meranti (*Callophylum spp.*) and mindi (*Melia azederach). Callophylum* spp. is now being substituted in parts of products that have a darker finish. Other parts that have been substituted include the sides of case goods and bed headboards that now are being made from veneered MDF. However, due to a surge in demand for veneered MDF, the suppliers cannot keep up and thus a long lead time of up to 45 days has been reported. Another interesting development is that the larger manufacturers, and thus consumers of rubberwood, have been in correspondence with each other as to how to reduce their consumption of the raw material as a move to stabilize the price and supply. CC Chua explained "furniture manufacturers do not cooperate very much for the good of the industry but a while back some local companies and a Taiwanese firm met and agreed that together we should all try to use rubberwood only for parts that we have to. The idea is that we reduce the local demand and stop bidding against each other and driving the price up."

What trends we have seen is that there is a general move to replacing rubberwood with tropical timbers. Although tropical timber is usually slightly more expensive than rubberwood, it is currently more available. The availability of tropical timber is due to smuggling of Indonesian timber into Malaysia with neither Malaysia nor Indonesia having been able to properly address the situation (EIA 2004). This trend does not augur well for the long term sustainability of the Malaysian furniture industry as it is dependent on forests for the raw material and the forest cover in Indonesia continues to decline. Substitution of solid rubberwood with composite materials, however, is a positive move for the sustainability of the industry because composite materials utilize low grade wood or other fibrous material as the raw material.

Twenty years ago, many observers questioned the efforts of the Forest Research Institute Malaysia (FRIM) to develop uses for the large numbers of unproductive rubber trees that were being cut and replaced each year. Some people wondered why Malaysia would bother to use such crooked, small-diameter trees, whose wood was susceptible to blue stain and borer attack and for which there was virtually no market recognition. Today no one questions the logic of FRIMs efforts. In barely two decades, rubberwood has become the preferred raw material for Malaysia's RM[1] 7.6 billion furniture industry and the mainstay of the country's particleboard and fiberboard sectors. "Rubberwood furniture, handcrafts, toys, and kitchen utensils are ubiquitous in shops and stores around the world" (Enters & Durst 2004).

With the dwindling supply of timber from natural forests especially in countries like Malaysia and Thailand there is no doubt that rubberwood is a timely substitute to support the timber-based industries in these countries. The concern for the environment and the destruction of tropical forests through logging have placed further obstacles in the marketing of tropical forest timbers owing to the pressure from anti-tropical hardwood NGOs. Rubberwood, being a timber obtained from sustainably managed and renewable plantations, is acceptable by such NGO's in the traditional timber importing countries. There is little doubt that rubberwood will be an important timber resource supporting the timber and wood-based industry in Southeast Asia even when the timber supply from the natural forests has been depleted.

In Malaysia a review of policies relevant to rubberwood use has been suggested because rubberwood is in a peculiar situation in which the timber falls under the purview of forestry while the tree is an agricultural crop (Salleh 1987). Because of the increasing importance of rubberwood there is an urgent need to review policies and guidelines governing the production of this resource to ensure a consistent and continued supply. This is relevant as it has been pointed out that although Indonesia has large areas of rubber plantations there are problems in harvesting the timber. The reasons are inaccessibility and other infrastructural constraints (Ser 1990). There are indications (in Malaysia) to grow rubber trees for forest plantations to increase the supply.

In 1990 the total annual available volume of rubberwood in the area of the Association of Southeast Asian Nations (ASEAN) was estimated to be about 17 million m^3 (Ser, 1990). From 1982 to 1992 rubberwood production figures for Malaysia increased from 30,000 to 1,872,000 m^3 (Malaysia Ministry of Primary Industries, 1993). Raymond (1993) assessed that, based on a sawn timber recovery of 25 to 45 percent; only 5 percent of rubber tree wood volume is converted into wood product, the remainder being left behind in plantations or burnt. There are various other reasons

[1]One Malaysian Ringgit (RM) is equal to US$0.30 (October 2007)

for the difference between the statistically assessed available timber and the real outputs. Available log volume for diameters above 15 cm ranges from 52 to 162 m^3 per hectare (assessed for nine cultivars by Gan, Ho, and Chew, 1985). The usable wood volume per hectare depends on numerous factors such as clone, site and management. Most of the planted area is owned by smallholders (with less than 20 ha). These plantings are geographically dispersed and accessibility is often poor. Smallholders are said to provide logs of lower quality as they devote less attention to management and proper tapping practices. A study undertaken in Johor, Malaysia, found that only 18 percent of rubberwood logs harvested in smallholder areas were suitable for sawn timber (Zana *et al.*, 1994).

In the 1990s concern regarding the future supply of rubberwood within Malaysia triggered a research programme on rubberwood breeding with the objectives of producing high latex yields and timber. Throughout the 1990s, a general slump in the world latex market, together with rising labour costs, pushed many plantation owners to shift from rubber to oil palm. The resulting glut of rubberwood logs from plantation conversion depressed prices below RM 75 per m^3 for much of the decade. Compared with prices of around RM 380 per m^3 for *Callophylum* spp. logs commonly harvested from the natural forests of Southeast Asia (although the quality of such logs is much higher and the cost of harvesting logs from natural forests is generally much greater). Rubberwood was already priced much lower than traditional timber species, since both plantation owners and wood processors viewed rubber logs largely as a residue of latex production. More recently, however, many processors in Malaysia have been reporting shortages of timber supplies due to high latex prices and a reluctance to harvest the rubber trees. This has pushed the price of rubberwood logs to as high as RM 290 per m^3, compared with the current price of RM 685 per $m^3$for *Callophylum* spp. (Enters & Durst 2004). Demand for rubberwood timber has continued to increase while the total area replanted in rubber has been decreasing, with the predicted result being a shortage of round logs for down stream processors in rubberwood based industries. Since the potential of rubberwood was first made known to timber manufacturers, there have been numerous projections made as to the quantity of timber available for manufacturers.

Based on field research gathered by Norini *et al.* (1990), and Ismariah and Norini (1994), estimates were made as to the total hectarage in rubber plantations, whether it is estate or smallholding, and the yield per hectare for each. Results vary due to the difference in yield estimates from 100 to

180 m^{3}per ha for smallholders and 180 to 220 m^3 per ha for estates. Yields from estates are higher due to better plantation management including less damaging tapping.

Predictions were made on the future availability of rubberwood by examining the replanting and new planting of rubber. With data from the Department of Statistics until 1992, it was calculated the recovery after 25 years based on an average yield estimate is 140 m3 per ha for smallholders and 200 m^3 per ha for estates. Based on the research of Ismariah *et al.* (1994), lumber yield from round logs was then calculated based on a recovery of 20% for estate logs and 15% for smallholder. Norini, Woon, and Sim (2000) estimated an installed capacity for the rubberwood processing industry including furniture and panel products to be 1,590,000 m^3 per year. If this is a close estimation, then there is a deficit for every year, this is being partially supplemented with imports from neighboring Thailand.

The current scarcity has led to price increases in lumber from RM 600–800 per metric ton at the end of 2003 to RM 1300–1600 per metric ton in mid July 2004. The price increase had been due to a very tight supply on the market. Now supply is stabilizing but the price continues to rise. With the increase in rubberwood prices, the market has also seen an increase in the prices for tropical hard and softwoods.

iv) *Conversion of rubber plantations to oil palm*

Strong market demand and high prices for palm oil have stimulated a boom in planting oil palm (*Elaeis guineensis*) in Asia in recent years. Grown in plantations, oil palm produces the palm oil and palm kernel oil used in margarine and other foods and in soaps and cosmetics. Oil palm plantations also produce biomass. In 1999, the area of oil palm plantations in Asia was more than 53 million hectares (Killmann 2001). Malaysia (with 3.3 million ha) and Indonesia (with 1.8 million ha) dominated oil palm planting and production in 1999 and continue to do so today, although new planting is also now occurring in several other Asian countries, including India, Papua New Guinea, the Philippines, and Thailand. Recent data on the extent of oil palm plantations in Asia are uncertain, but it is likely that the total area now exceeds 6.5 million hectares.

Oil palms are grown on a 25 to 30 year rotation before being felled and replanted. At the time of felling, oil palms produce an average of 235 m^3 of stem material per hectare. This means a total of approximately 50 million m^{3}of stem residue will be generated each year in Asia over the coming decades (although the actual volume depends on the rate of plantation replanting, which is often influenced by incentive programs and current market prices for palm oil). In addition, some 100 million tons of palm fronds, 20 million tons of empty fruit bunches, and 5 million tons of palm kernel shells are produced annually as byproducts. In theory, those residues could be available for processing by the forest products industry. In practice, however, nearly all of this production is currently burned or left to decompose. A major problem in converting oil palm stems is their low recovery rate. Palm stems are harvested at a relatively young age and thus build up less medium- and high-density tissue than do, for example, coconut palms. Additionally, fresh oil palm stems have very high moisture, sugar, and starch contents, which accelerates decomposition after felling and generate high transport and seasoning costs (Killmann and Woon 1990). Present technologies do not yet allow for economically viable use of the full range of those residues. However, laboratory research continues, and trials using oil palm fiber in mechanical and chemical pulping processes show promise.

Oil Palm does have some positive uses for some of the forest-based industries including the reconstituted panel industry and the furniture industry. The availability of residues from processing mills in the form of empty fruit bunches, mesocarp fiber, palm kernel shells is great: for every ton of fresh fruit bunches processed, about 230 kg of empty fruit bunches, 190 kg of mesocarp fiber, and 80 kg of kernel shells are produced. In addition, based on the areas of oil palm replanted annually and due to be replanted, the potential supply of oil palm stems is substantial, for example, Norini, Wan Asma and Mohd Azmi (1999) estimate that the number of oil palm stems available annually ranges from 400,000 to almost 5.8 million based on a planting density of 134 palms/ ha. Research and development that has been conducted by a number of institutions has found oil palm stems and mill byproducts to be a suitable raw material substitute in making a range of engineered panels including palm fiber board, particle board, and mineral-bonded particle board.

Palm fiber board is currently being manufactured by Tian Sian Trading Sdn. Bhd. who started out in 2001 with a trial plant at a Sabah oil palm mill producing 200 m^3 per day. They have now finished commissioning another plant in Perak with a much larger daily capacity. In recent conversations with the company, there seems to be a ready market for the product with the price sensitive speaker box manufacturers. Since speaker boxes are coated with a vinyl or foil film, the somewhat rough texture of the palm fibre board is not a problem. The material is also reported to have better resonance compared with particle board, the currently widely used material. As has been pointed out, there is great potential for this material in terms of volume of raw material available. There is some concern as to the suitability of palm fibre board as a substitute to particle board in the furniture industry. One of the problems is that the board contains a large quantity of residual oil that tends to seep thru veneer and causes difficulties to stain or paint the board.

IV. Sustainable Raw Material for the Furniture Industry: Wood Security

Due to declines in domestic tropical timber production and conversion of rubber plantations to oil palm, there is a pressing need for Malaysia to take steps in assuring wood security for its important wood-based industries. There are various options for Malaysia to obtain wood security domestically involving sustainable forest management and commercial forest plantations.

Sustainable Forest Management views the forest as yielding many different products and providing many different ecological services (Pearce *et al.*, 2003) as opposed to sustainable timber management which focuses on long-term, non-declining flows of timber, the difference being in the depth of management and priorities. This research project has grouped the two together as both management systems deal with indigenous tropical forests. For the past three decades, tropical forests have been the vast contributors to Malaysia's total exports from both primary and secondary wood-based industries. However, since most of these forests have been logged before with species composition of second and third generation forests lacking the valuable *dipterocarps*, timber quality and quantity in future harvests will be reduced (Woon 2001). This has been reflected in the reduction of the government's annual coupe from 52,250 ha p.a. for Peninsular Malaysia during the Sixth Malaysia Plan (1991–1995) to 46,040 ha p.a. during the Seventh Malaysia Plan (1996–2000) and to 42,870 ha p.a. during the Eighth Malaysia Plan (2001–2005). This trend will continue resulting in the drastic reduction of the availability of tropical timbers (Woon 2001).

Based on the state of Malaysia's forests and the 30–55 year long cutting cycles required for sustainable forest management; a total sustainable harvest of 2.75 million m^3 is available; roughly one third of the current installed capacity. With a focus solely on timber production, a system of sustainable timber management is very questionable with financial rates of returns only slightly in excess of some "modest" discount rate of 5–10% (Pearce 2003). Therefore, based on current forest cover, reliance on the sustainable forest management of Malaysia's forests will not provide enough wood security to support the primary and secondary wood-based industries of the country. Another alternative is the establishment of commercial forest plantations.

V. Plantation Resources

It is possible not only to sustain but also to increase productivity in successive rotations. This requires clear definition of the end-use objective for forest plantation development and a holistic management view. There is a need to integrate the following strategies into a prudent management plan: tree improvement programs, nursery practices, site and species/provenance matching, appropriate silviculture (site preparation, establishment, weeding, fertilizing, pruning, and thinning), forest protection, and harvesting practices. New Zealand and the southern United States have shown that substantial gains can be made by adopting this holistic approach. In developing countries where resources may be constrained, highly technical solutions may not be essential but it is critical to get the fundamentals correct: careful species and provenance choice, good nursery stock, site preparation, planting techniques, weed control, and, less frequently, fertilizer inputs. Once healthy and uniform plantations have been established, later silvicultural tending may become increasingly important, depending on the end-use objective (FAO 2001).

In Malaysia, plantation forestry is a relatively new field. This is despite the fact that already in the 1950s trial plantations were established using both indigenous and exotic species. The first species experimented with was teak (*Tectona grandis*) followed by pine (*pinus caribaea*) introduced as a pilot project under the assistance of the UN linked FAO in the late 1960s. In 1982, the Compensatory Forest Plantation Project (CFPP) was established to promote the planting of fast growing species such as Acacia (*Acacia mangium*), *Gmelina arborea*, and *Paraserianthes falcataria*. Many countries involved in forest plantation projects have been motivated by foresighted concern over reductions in domestic timber supplies, and for purposes such as environmental protection, export, or reforestation. Plantations established in Malaysia, under the CFPP were aimed to augment the supply of general utility timber and ease pressure on existing forest resources.

Malaysia has limited experience of the establishment of commercial forest plantations as is reflected in a total coverage of about 240,000 ha[2] with 76,000 ha in Peninsular Malaysia and the 164,000 ha in Sabah and Sarawak, as of 2000 (Forest Department) (year – source). In previous years, the estimated volume of timber production from plantations on the Peninsula was less than 50,000 m^3, but with new plantations reaching maturity, this figure rose to 547,000 m^3 by 2000. In Sabah, estimated log production for 1997 was 452,000 m^3 rising to just over 1 million m^3 by 1999 (Sabah Forestry Department, 2000).

In Sarawak, the state government has formed a joint venture with Grand Perfect Sdn. Bhd. to establish 54,000 ha of acacia by 2005 (Anonymous 2004). The Forestry Department of Sarawak has identified the gazetted Planted Forest Zone in the Bintulu Division as the area to be planted. The project is aimed to ensure a continuous supply of wood for pulp and paper mills, however, lumber could also be diverted from these plantations. The project involves the establishment of tree plantations, and in addition, a community development programme, research and development programme, environmental training programme, and training programme.

The hectarage in Malaysia is quite small when compared with the area under tree plantations in neighboring countries: Indonesia, 8.8 m ha; Vietnam, 2.1 m ha; and Thailand, 0.8 m ha (Dan, 1993). Of the total 76,000 ha on the Peninsula, Over 64,000 ha is of Acacia (*Acacia mangium*), followed by *A.excelsa* (*Azadirachta excelsa*), Teak (*Tectona grandis*), and Rubber (*Hevea sp.*) planted for wood production.

A) Acacia as a plantation species

Over 80% of plantation area in Peninsular and East Malaysia has been planted with the Australian exotic, acacia. In the Peninsula, 94% of the hectarage established was through the CFPP. With a projected timber supply deficit in the mid 1990s, the Forestry Department in 1982 launched the CFPP with the aim of producing general utility timber which can be harvested in a 15-year rotation cycle. Being a leguminous species, acacia has shown the ability to grow on most soil types with a low mortality rate. Studies by FRIM found the wood suitable for sawn timber, plywood and veneer, pulping, and reconstituted panels. Acacia also has one of the

[2]In official statistics, rubber tree planted for latex and oil palm plantations are not considered as CFPs but as agricultural crops.

fastest growth rates of 2.0–2.5 cm in diameter per year, higher than all indigenous species, and is able to have a total gross volume increment of 25–30 $m^3\,ha^{-1}\,y^{-1}$. Abdul Razak observes that under the CFPP scheme, the species has not performed as expected because of slow growth and with reports of heart rot on some sites. He states that overall, "its performance for saw log production has remained dismal" (Abdul Razak 1997), but others feel the results have been reasonable (Thai 1995). The CFPP halted planting of acacia in 1992 but other schemes in East Malaysia have planted large tracts of the species for pulp production. Some of the drawbacks of acacia for the furniture industry are its dark core wood, white sap wood, large number of knots, and high core stress causing checking and warping. It is also difficult to dry and to glue, and dulls the edge on saw blades and knives three times faster than rubberwood. But it is easy to sand, and is durable in light outdoor applications, and unlike rubberwood, does not require pressure treatment.

With a larger volume coming on line, the Malaysian furniture and panel board industry is starting to seriously consider acacia. This is in part due to the improvement in timber quality and availability, to the reduction in price per m^3 vis-à-vis rubberwood, and IKEA. The impact of IKEA's buying habits to influence an industry is highlighted in a following case study.

Besides sawn timber, acacia is being successfully used for chippings for the pulp industry (Sabah Softwood Sdn Bhd & Aramijaya Sdn Bhd), chip board, and MDF production (PT Sumalinda Lestari Jaya) and as market acceptance of acacia continues to increase along with technical knowledge on processing and manufacturing with the wood, this species will likely become more prevalent in the Malaysian furniture industry. However, there are limitations of the use of the wood as compared to rubberwood; this primarily being the prevalence of knots and the wood's dark and uneven color which limits the diversity of finish and appearance of the final item of furniture. Another constraint is the yield per cubic meter of timber is much lower than that of rubberwood. When scrutinizing the financial viability of acacia plantations, as with all of the commercial forest species examined, the IRR is low: around 10 to 15%, when compared to other agricultural crops like oil palm (Thai 1995). However, with some more incentives for plantation establishment, we may see an increase in the coverage of acacia as the demand for the timber continues to grow. Current prices for rough cut, 4/4 by random width and random length, kiln dried, is now selling for RM 800–1000 per metric ton. This price has remained fairly stable as the price of rubberwood and tropical light and hardwoods has continued to rise. The current suppliers of acacia lumber mix the sawmilling with chipping of which the majority is sent to Japan for paper production.

Case Study: The Influence of IKEA in the Development of Acacia as a Furniture Wood

The world's largest furniture retailer, IKEA has been instrumental in developing Acacia as a furniture wood. Their environmental responsibility-a refusal to use tropical timber sourced from unsustainable forests, has forced them to look to alternatives to tropical hardwoods such as Balau *(Shorea spp)* and Merbau *(Intsia bijuga)* for their outdoor furniture and accessories lines. Due to the durability of Acacia and increase in wood supplies, Ikea has successfully developed five outdoor living collections and accessories. These products are manufactured in Indonesia, Vietnam, and Malaysia, where now a small Acacia-based furniture industry has been developing to supply almost exclusively to IKEA. This development is due to the huge buying power of IKEA, and to the technical support and guidance that IKEA provides its contract manufacturers. In Malaysia, due to IKEA's lead, we are now seeing a great interest by manufacturers in the use of Acacia, a greater acceptance and confidence in the wood by the buyers, and a larger permeation of the proper technical knowledge for wood processing, drying, and woodworking.[3]

IKEA has been, to a great extent, responsible for the adoption and utilization of Acacia in the furniture industry. This is due to their environmental commitments and corporate responsibility as outlined in The *IKEA Way on Purchasing Home Furnishing Products*, (IWAY) (IKEA, 2002). These are:

i) The supplier shall not utilize wood from intact Natural Forests[4] or nationally/regionally recognized and geographically identified High

[3]Based on interviews by the Author of Ikea's contract manufacturers and Acacia wood processors in Malaysia in 2003 & 2004.

[4]Intact Natural Forests: Forests that are intact, i.e. unfragmented with a minimum block size of 50 000 ha, where no systematic forest management has been carried out within the area and where access to the area is restricted, i.e. no roads, mines, pipelines etc.

Conservation Value Forests[5] unless the forest area is certified according to a standard recognized by IKEA.

ii) The IKEA supplier shall not utilize wood from protected areas (e.g. national parks, nature reserves) unless it can be proven that the wood has been harvested in accordance with management prescriptions for the protected area and/or has been certified according to a standard recognized by IKEA.

iii) The IKEA supplies should not utilize wood from plantations in tropical and sub-tropical region, that have been established after November 1994 by replacing Intact Natural Forests.

iv) The IKEA supplier shall only use high value tropical tree species[6] that are certified according to FSC, the Forest Stewardship Council, or other IKEA recognized level 4 standards.

What these IWAY procedures define is a staircase model with a 4-step approach aimed at improving the suppliers overall performance. To be an IKEA vendor, the supplier must be at minimally, level two.

With IKEA's specifications for the use of acacia as wood for furniture we now see a number of factories having successfully manufactured items. These manufacturers include: Malacca Rubberwood Sdn. Bhd., Flexline Sdn. Bhd., and Cosmopolitan Wood Sdn. Bhd.. Upstream, there are a number of timber suppliers and dryers. Most of these suppliers are concession holders who saw larger diameter round logs for timber and chip the rest for export to the Japanese market in the main. Concessionaires include Aramijaya Sdn. Bhd. in Johor, Asiaprima Sdn. Bhd. in Pahang, and Mangium Plantations Sdn. Bhd. in Sabah. Downstream, a number

[5]High Conservation Value Forests: Forests that possess one or more of the following attributes:

(a) Forest areas containing globally, regionally, or nationally significant concentrations of biodiversity values (e.g. endangered species); and/or large landscape level forests, contained within, or containing the management unit, where viable populations of most if not all naturally occurring species exist in natural patterns of distribution and abundance.

(b) Forest areas that are in or contain rare, threatened, or endangered ecosystems.

(c) Forest areas that provide basic services of nature in critical situations (e.g. watershed protection, erosion control).

(d) Forest areas fundamental to meeting basic needs of local communities (e.g. subsistence, health) and/or critical to local communities' traditional cultural identity (areas of cultural, ecological, economic or religious significance identified in cooperation with such local communities).

[6]High-value tropical tree species: teak, meranti, rosewood, mahogany, etc.

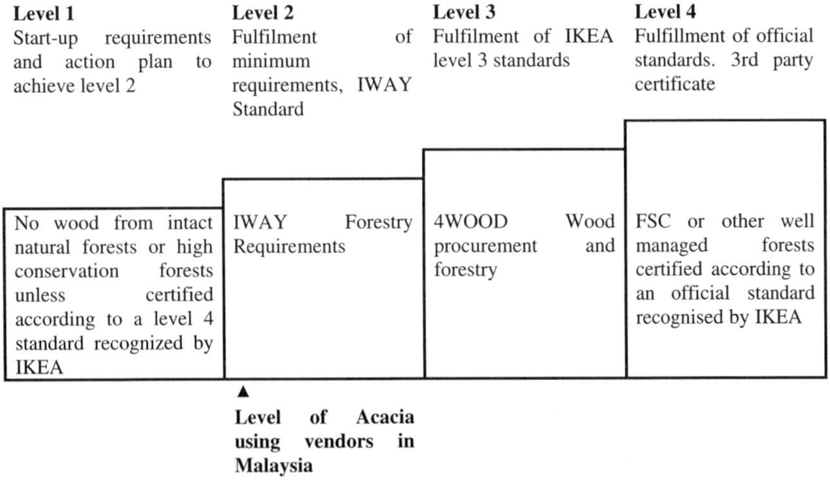

(*The IKEA Way on Purchasing Home Furnishing Products*, IKEA Group, 2002)

Fig. 17.1. IKEA staircase model for wooden merchandise.

of the furniture manufacturers who are currently original equipment manufacturers (OEM) for IKEA are also manufacturing original design manufacturer (ODM) products for other customers. On the furniture market we are starting to see acacia bedroom sets with an antique reproduction styling and distressed finish to "hide" the many knots and sap wood found in acacia timber.

B) Sentang as a plantation species

Sentang (*Azadirachta excelsa*), a native of Sumatra, is a well-known village-tree in the middle and northern regions of Peninsular Malaysia. To the Malays, the tree is important mainly because of its timber which is valued for house building. The young shoots are also consumed as a vegetable (though quite bitter) and the old leaves are used as a traditional medicine. Although not widely planted in Malaysia as a commercial crop, it has been reported to be in mixed plantings with rubber, coconuts, and pineapple in the north of the country and Southern Thailand. In 1998 a preliminary survey was conducted in the Peninsula which showed that at least 1,530 ha had been established with an age range of between 2 and 5 years (Mohd Noor *et al.*, 2000). A 2001 survey revealed 3,676 ha of *A.excelsa* plantation and 970 ha of *A.excelsa* mixed with Teak or Rubber (FRIM 2002) which shows an increase interest in the species. Timber from the species is classified as a light to medium weight hardwood with the hardwood having being pale reddish brown and distinctly demarcated from the yellowish white sap wood. In density and appearance, timber from the species is comparable to that of *Palaquium spp.*. Unlike *Palaquium* spp., *A.excelsa* is not widely used in the wood based industry of Malaysia due to its low availability. However there is interest in growing *A.excelsa* in commercial forest plantations due to its high growth rates with diameter growth of 1.5–2.0 cm per year average. DBH at 15 years is 30–35 cm. *A.excelsa* is also interesting because to date the species is free from major diseases and pest infestations.

C) Teak as a plantation species

Teak (*Tectona*grandis) is one of the oldest commercially grown tree species. However, cultivation in Malaysia is relatively new when compared with the neighboring countries of Myanmar, Thailand, and Indonesia, thus the total hectarage of teak is relatively small. However, teak has been found to grow well in the northern, and recently in the wetter south of the country. Teak has been getting increasing interest as a plantation species due to the outstanding properties of the timber and workability of the wood and is becoming increasingly popular among smallholders due to high expected returns and an established market. Teak wood fetches a premium on the international market of between RM 5,700 and RM 7,600 per metric ton and is relatively fast growing with a 1.5–2.0 cm annual diameter growth and a DBH at 15 years of 25–35 cm. Availability of planting material, proper cultivation techniques, encouragement from the Government, technical support from FRIM, and smallholder adoption schemes are some of the reasons why teak cultivation is on the rise in Malaysia. The cultivation of teak started to gain impetus in the 1980s with the participation of the private sector. On of the reasons for the interest in teak is that it has few natural enemies and is relatively maintenance free, thus a good species to grow in a country where the wages of plantation labourers continues to increase. Due to growing interest in the species there has been work done on cloning techniques with the hope of shorting the gestation period of a teak plantation to a more economically viable period of 15 to 20 years.

D) Rubber as a plantation species

The extensive cultivation of rubber trees (*Hevea brasiliensis*) in Malaysia has been primarily for the latex industry. Only in the 1980s did the timber from the species become of any significant value. As of 2003 there was 1,315,000 ha of rubber under cultivation for latex production and an additional 1,636 ha planted for the main purpose of timber production (Malaysian Ministry of Primary Industries, 1993).

With the continual conversion of rubber plantations to oil palm, and the subsequent fear of rubberwood shortfalls, the Pilot Rubber Forest Plantation Project (PRFPP) was established under the Sixth Malaysia Plan (1991–1995). The main objectives of the plan was to identify potential sub-species and clones suitable for timber production and to develop a supply of planting material, and to identify suitable agronomic inputs and management systems. Research was carried out to determine the most suitable clones and system to maximize yields of a wood-latex plantation, a wood only plantation with no latex production, and a rubberwood agroforestry plantation that intermixed rubber with teak and *A.excelsa*.

Like all other commercial forest plantations examined, without further incentives, the ROI for strictly timber plantations is currently financially unattractive. However, a wood-latex plantation does become more attractive. Since we have already explored the development and utilization of rubberwood from latex production plantations, we will instead focus on the wood-latex plantation since there appears to be no incentive *not* to tap the latex if you are planting rubber trees. As has been researched by the Rubber Research Institute of Malaysia, Forestry Department, and FRIM; "whether the options involve only *Hevea* or combinations of *Hevea* with other wood species, latex is still the major revenue contributor of total revenue" (in Norini, 1999). Under the traditional latex production plantation, revenue from wood harvested after latex production declines, contributes only about 7–8% of the total contribution in terms of net present value (NPV). With a high density latex plantation involving seven years of taping, the total revenue in terms of NPV is around RM 25,700 per ha. With a high density rubber monoculture timber plantation and no latex production, the revenue in terms of NPV is only around RM 9,000 per ha. In 2000, Norini (1999) calculated that based on current economic parameters, a rubber forest plantation would break-even only if the discount rate of 8% was accepted and in Malaysia the norm is 10–15%.

In the teak or *A.excelsa* and rubber agroforestry scenario under a 20 year rotation, latex still contributed the majority revenue of RM 30,000 per ha while the *A.excelsa* and teak contributed RM 12,600 per ha and the rubberwood RM 2,900 per ha. Of the various scenarios explored by the research, the most promising alternatives to the traditional planting of rubber are options involving high density planting over a short cycle of 15 years for latex and wood extraction as well as agroforestry. This involves hedge planting of rubber trees with teak or *A.excelsa*. One of the premises of the 1997 research was of rubberwood saw-log and chip-log prices of 115 and RM45 per m^3 respectively. As log prices rise, the rubberwood timber plantation and agroforestry options would look much better vis-à-vis a traditional latex extraction plantation.

E) Bamboo as a raw material for the furniture industry

Bamboo has a long history of use in Asia. Many new uses are being developed for bamboo that opens numerous opportunities for bamboo growers and processors. Bamboo culms (stems) have traditionally been used as substitutes for timber in construction and scaffolding, and these uses remain important in Asia today. Recent technological developments have cleared the way for bamboo to be used extensively in manufacturing reconstituted panel and board products, such as particleboard, fiberboard, ply-bamboo, laminated boards, bamboo flooring, and pulp and paper (Ruiz-Perez *et al.*, 2001).

Bamboo furniture is rapidly gaining market share, with new and innovative designs contributing to the growth. Bamboo flooring is also becoming popular in western markets. The bamboo trade is growing, particularly between China and the United States. China has more than 4 million ha of dense bamboo areas, of which 1.25 million hectares are managed intensively as plantations (Ruiz-Perez *et al.*, 2001). China is the world's largest producer of commercial bamboo, with production of unprocessed bamboo valued at $1.5 billion in 1999. Processing operations are estimated to have added another USD1.3 billion to this total. The sector provides part- or fulltime employment for more than 5 million people in China. India's bamboo forests extend across 10 million ha, with varying densities (Ganapathy 2000). In India, the annual bamboo harvest approximates 4 million tons, with slightly more than half used in rural construction and scaffolding. Most of the remainder is used for pulp and paper making. Despite some environmental problems in its utilization (mills using bamboo tend to be small and less regulated), bamboo accounts for more than 20 percent of the raw material used in India's pulp and paper industry (Ganapathy 1997). Bamboo is also used extensively in papermaking in Thailand, Bangladesh, and the Philippines.

VI. Incentives for the Establishment of Commercial Forest Plantations

The government of Malaysia has recognized the need to develop commercial forest plantations to ensure a continuous supply of timber to downstream industries. In 2000, the then Primary Industries Minister, Dr Lim Keng Yaik, stated that "Private sector investments in rubber forest plantations must be initiated to compensate for lowering supply from existing plantations, which have gradually been switched to other land use" (Anonymous, 2000).

In the 1994 budget, the government designated commercial forest plantations as a "strategic industry" which enabled plantation owners to qualify for tax incentives under the Promotion of Investment Act of 1986. Although there has been much private sector interest in commercial forest plantations, to this day few have ventured into such projects since the risks currently outweigh the rewards. Investors in these projects face three major types of risks: physical, market, and financial.

The financial risk is perhaps the single most important factor that has prevented the establishment of large scale commercial forest plantations. The high financial risk associated with the long gestation period and uncertainty of the final outcome has discouraged investments in forest plantations. The Malaysian Government has implemented the aforementioned incentives, in the form of tax breaks, to encourage investment in commercial forests but these have not served the intended function. The two main incentives available are Pioneer Status and Investment Tax Allowance. Under the Income Tax Act, Schedule 4A provides for an agricultural allowance which can be used to offset income from other business sources within the company. Commercial forest plantations have been designated as "strategic activities", and as such, qualify for a higher tax exemption. Under pioneer status a company is given a 100% exemption from statutory income for a period of 10 years and the investment tax allowance is a rate of 100% of qualifying capital expenditures incurred within 5 years from the date the first qualifying expenditure is incurred.

Many have expressed the opinion that current government incentives are not enough to encourage the private sector to establishment commercial forest plantations in spite of the enactment of the new initiatives since they fail to address the long gestation period and cash-flow problems associated with such a project (Woon and Lim 2002). One proposal is for the government to provide group relief incentives, therefore, a Malaysian resident company could surrender its adjusted loss, in part or full, in the basis period for the year of assessment to one or more related companies resident in Malaysia. This would encourage larger companies to use their plantations as write-offs and tax shelters once the trees are harvestable. In addition to the group relief suggestion, it is also recommended to extend the PS (what's this) from the current 5 years to 10 years to reflect the long gestation period required for commercial forest plantations.

Upstream moves by the furniture industry

To secure a more steady supply and ensure wood quality, some of the larger furniture manufacturers have been moving further up the supply chain to include finger-jointing and lamination board manufacturing, kiln drying, and saw milling. Examples of companies becoming more involved in the primary processing stages of their raw materials include Latitude Tree Sdn Bhd and Baswell Bhd. In 1999, Latitude Tree set up an operation in the northern Malaysian state of Terengganu to sawmill and kiln dry their own timber. Terengganu was selected due to the state's large area of rubber plantations. In 2001 Baswell opened a similar operation in the neighboring state of Pahang. Larger furniture manufacturers may move further up their supply stream but few to none have ventured into the timber plantation business. In 2000, the then Deputy Prime Minister Abdullah Ahmed Badawi stressed: "To sustain a supply of raw materials, manufacturers may consider the possibilities of setting up joint ventures with owners of rubber plantations" (Anonymous, 2000). But when the owners of furniture factories were asked why they do not endeavor to do so, the primary reason was that most companies in the industry do not have or do not wish to tie up their resources for fifteen years while they wait for trees to grow (author's survey).

In a related industry there is an exemplary example of upstream movement into the plantation business by a manufacturer: pencil maker Faber-Castell. The German company has a large pencil manufacturing operation in Selangor that uses large quantities of softwood from the jelutong trees, a species indigenous to the tropical forests of Southeast Asia. As part of the company's environmental commitment not to contribute to forest degradation, and to ensure a sustainable timber supply, Faber-Castell started a Jelutong plantation in the north of Peninsular Malaysia.

In Malaysia, there is however, some down stream movement of plantation companies. Golden Hope Plantations Bhd, one of the pioneering plantation companies, has subsidiary companies that run sawmills and kiln drying plants. In Nilai, Negeri Sembilan, subsidiaries include Golden Hope Parquet Sdn Bhd, a rubberwood parquet flooring manufacture and Golden Hope MDF Sdn Bhd, a factory which uses the recovery from the parquet company, off cuts from the sawmills, and the branches from the plantations to manufacture a medium density fiberboard. Hume Furniture Industries Sdn Bhd, a large furniture manufacturer and Hume Fiberboard Sdn Bhd, a chipboard manufacturer are both part of the Hong Leong group of companies, another large plantation owner in Malaysia.

In the industry one sees the development of first rubberwood lamination board and now rubberwood based engineered panels. These technological innovations and improvements in manufacturing sophistication have led to a more efficient utilization of the rubberwood resource. The increase in manufacturing efficiency and in the organizational efficiency of companies moving further up the supply stream or down the supply stream have allowed the furniture industry to grow without eroding the sustainability of the resource base. The resource could not be over-exploited because, being a plantation grown timber, the limits in material was already in place as well as clearly defined ownership of the resource.

VII. Case Study: Material Substitution Analysis Using a Simplified Input-Output Model

The intention of this case study is to perform a detailed analysis of a product currently in production in solid rubberwood and also in composite materials. This product highlights a growing trend with Malaysian furniture manufacturers to use more composite materials and less solid rubberwood. This is a trend we have seen in our survey and interviews.

The product is a home bar counter made for and sold by GXG Asia Sdn. Bhd. to the US market. The buyer is reluctant to use tropical timber due to the environmental and legal ramifications of tropical deforestation and illegal logging. Also, the product was first introduced in the market in solid rubberwood, and it is important that the physical appearance of the product changes as little as possible. Other requirements are for the product to be equally durable, functional, and easy to assemble, whether made from rubberwood or composite materials.

i) *Production steps*

Manufacturing of the 82" Cottage bar is very similar if it made from rubberwood FJL board or composite panels. The material of choice for the composite panel bar (MDF bar) is medium density fiber board (MDF) for the front, sides, and bar top as these pieces require routing. The internal parts including the top shelf, shelf, base, footing, and center panel are made from particle board. All of the surfaces are veneered: visible surfaces with

American beech and internal surfaces with merapi, a local species. The only differences with production is that the MDF bar must be edge banded with veneer on the sides of the panels where the internal board is exposed; all other steps are identical with a solid wood product. We have therefore assumed the production costs are the same for both products.

No furniture manufacturer makes MDF or particle board: these are bought on the open market. Some manufacturers will do their own veneering but this is rare with the majority buying cut to size veneered panels from suppliers.

With rubberwood, most manufacturers purchase cut to size FJL boards from suppliers who do the sawmilling, boric acid treatment, moulding, laminating, and sanding. Companies engaged in this type of activity include Limahsoon Bhd. Since they are a newly listed company, they had to open their books to public scrutiny. Their accounts department has reported that they have an 8% share of the Malaysian market and work on a 25% profit margin. Many of the large manufacturers make their FJL board from kiln dried timber thereby gaining extra cost savings.

ii) *Input-output modeling*

The input-output model (I-O model) has been employed by many researchers as a way to quantify both the final demand components and input content of interest in an economy (Leontief 1951). Through such a model one can trace the movement of goods from one sector to another. This study used a simplified I-O model to examine how product substitution can not only result in a lower cost of production but also a higher yield of end product for the same amount of input (in this example measured in RM). This is referred to as input-substituting technology (Goodstein 2002) and is easily demonstrated by organizing our data in an absorption matrix and then employing some simple algebra remembering that the supply will equal demand and thus inputs will equal outputs of some form or another.

The value of creating an I-O absorption matrix is that we can examine a product and compare the effect of material substitution on the cost of the product (in RM). Since this product "cost", the output, is related to the total cost of the inputs, One can thus see how substitutions to our inputs affect our outputs. Our input substitutions are replacing rubberwood finger-joint laminate (FJL) board with MDF and particle board (PB). The output is affected due to the lower cost of the input. The lower input cost is a result of more efficient use of raw material. This is seen by the lower cost, per m3, of MDF and particle board compared to rubberwood FJL board (RW input in the matrix). If the market (the customers) were to accept the substitution of composite panels for the current FJL board, not only would the production costs of the product be reduced, resulting in competitively priced products on the international market, but would also mean that the volume of exports could be increased because the recovery per hectare of rubber plantation would be increased. With our 82" Cottage Bar example, one sees the total hectares required to produce one solid rubberwood bar being 0.0565 ha, while the area required to produce one composite board bar is only 0.011 hectares.

VIII. Conclusion

Unlike unsustainably harvested tropical timber on which Malaysia's forest based industries have been built and now are declining with the exhaustion of that natural capital that feeds them, the rubberwood based furniture industry has experienced until now no crisis.

The material shortage facing the furniture industry is also a result of structural changes to the Malaysian economy, changes in international commodity markets, and a lack of Government intervention. Throughout the 1990s, a general slump in the world latex market, together with rising labour costs, pushed many plantation owners to shift from rubber to oil palm. The resulting glut of rubberwood logs from plantation conversion depressed prices below RM 75 per m^3 for much of the decade. More recently, however, many processors in Malaysia have been reporting shortages of timber supplies due to high latex prices and a reluctance to harvest the rubber trees.

The furniture industry is built on the utilization of rubberwood (*hevea spp.*). Research to determine the potential use of the timber and its applications in products such as fiberboard, wood pulp, and furniture were initiated in the Forest Research Institute Malaysia (FRIM) in 1953. Malaysia was the first to develop the system of wood preservation using boric acid, proper kiln drying procedures, and the production of export quality moldings, wood components, and furniture. It is only since the mid 1980s that rubberwood has been utilized to manufacture furniture and is the reason why the Malaysian furniture industry became an international player.

As demand for rubberwood timber has continued to increase while the total area replanted in rubber has been decreasing, the predicted result being a shortage of round logs for down stream processors in rubberwood based industries. As a result, the price for rubberwood timber has increased some 25–30% from late 2003. In response, my research has shown that manufacturers are looking to material substitution including using more composite materials and veneers and substituting rubberwood with tropical timber.

Although the Malaysian furniture industry is dependent on tropical timber for only 20% of its solid wood inputs; limitations in the primary forest based industries will also limit the development, adaptability, and options of the secondary industries like particle board plants, and the tertiary furniture industry. These limitations will impact the furniture industry as our research has shown a desire for manufacturers to shift production to using more tropical timber, not less. This shift in specie preference is a response to the price increases of rubberwood and the unavailability of its supply. Although there is currently an abundant supply of tropical timber in the Malaysian market, as Environmental Impact Assessment investigations have shown, this has been due to the availability of illegally logged Indonesian timber that is being imported into Malaysia by dubious means. This timber source can therefore not be relied on for the long term.

One way that manufacturers are overcoming the rubberwood shortage with an eye to long-term survival of the industry is to substitute the use of solid wood by veneered composite materials whenever possible. This trend became apparent after analyzing the survey and interviews with company players. We were able to see first hand the price savings and effect of material conservation when a case study on a furniture item was conducted. We analyzed the benefits of, where appropriate, substituting solid rubberwood FJL board with composite panel materials such as veneered MDF and particle board in furniture production. We followed through production a product that is currently being produced using different materials and with some simple analysis were able to calculate the product price of using different materials and the land area required to produce one unit of the similar products.

What the case study has shown is that if the market (the customers) were to accept the substitution of composite panels for the current FJP board, not only would the production costs of the product be reduced, resulting in competitively priced products on the international market, but would also mean that the volume of exports could be increased because the recovery per hectare of rubber plantation would be increased. With our 82"

Cottage Bar example, one sees the total hectares required to produce one solid rubberwood bar being 0.0565 ha, while the area required to produce one composite bar is only 0.011 hectares.

Due to declines in domestic tropical timber production and land use changes, there is a pressing need for Malaysia to take steps towards assuring wood security for its important wood-based industries. There is concern due to the conversion of rubber plantations to oil palm due to the fact that oil palm provides a higher rate of return than rubber and with less labour input. To ensure a long term supply of raw material for the furniture industry, there is need for the establishment of more forest plantations of fast growing exotics like acacia and indigenous re-growth pioneers like bintangor (*Callophylum spp*) and Meranti (*Shorea spp.*). Plantation rubberwood currently forms the foundation of the industry but with decreasing hectares of latex producing rubberwood plantations; it is time for the Government to step in with policy changes and incentives to make commercial timber plantations more appealing to private sector investors. The major barrier to the establishment of plantation forests in acacia or rubberwood-latex is the expected IRR at a low 10 to 15% when compared to other agricultural crops like oil palm. The Government could provide incentives such as group relief incentives; therefore, a Malaysian resident company could surrender its adjusted loss, in part or full, in the basis period for the year of assessment to one or more related companies resident in Malaysia. This would encourage larger companies to use their plantations as write-offs and tax shelters once the trees are harvestable. In addition to the group relief suggestion, it is also recommended to extend the PS from the current 5 years to 10 years to reflect the long gestation period required for commercial forest plantations.

We have seen increasing interest in acacia plantations and furniture production using acacia timber. This trend was highlighted with a case example looking at IKEA's buying preferences and how they have created an, albeit small, sub sector of the furniture industry. Due to IKEA's lead, we are now seeing a greater interest by manufacturers in the use of acacia, a greater acceptance and confidence in the wood by the buyers, and a larger permeation of the proper technical knowledge for wood processing, drying, and woodworking. With more plantations coming on line, it is expected that the usage of acacia will continue to grow as the price of rubberwood has continued to increase and the price of acacia has remained stable.

Although the plantation timber based furniture industry is in a time of flux and change, some would say crisis, there in no need for the Malaysian furniture industry to follow the declining trend of the other FBIs in the country. There are alternatives to using unsustainable tropical timbers which include a utilization of a small amount of timber from sustainably managed tropical forests and continuing the reliance on plantation forests to provide the timber to feed the industry. This may require more Government incentives to encourage the participation of private industry. The other key to sustainability of the industry is to have a better recovery rate and raw material utilization. This is possible with employing more composite panel products in furniture construction.

Further work could be done on the utilization of acacia wood for the furniture industry. This could include proper processing and machining of the specie as there is reluctance in a fairly conservative industry to try new materials or processes. Another area of research that must continue to be explored is the growing and processing of bamboo for furniture manufacturing.

It is possible for the furniture industry to remain strong on a foundation of sustainably grown timber. As an industry colleague and friend of the author stated: "expect Malaysian furniture to be selling strong in the world market for at least the next thirty years", however, these are assertions yet to be substantiated. Required condition for this to remain true is that the industry must maintain a sustainable supply of raw material. It must preserve its natural capital as a precursor for the industry to add human, financial, and manufacturing capital inputs in order to generate continuous wealth. This will be accomplished by creating and maintaining a strong commercial plantation forest industry utilizing rubberwood and other species like acacia. This industry will be able to be nurtured and grow if a conducive environment is maintained by the Government.

References

Abdul Razak Mohd Ali "Prospects for Forest Plantations in Malaysia- Potentials and Challenges" in *Seminar on Commercial Cultivation of Teak, Sentang, Acacia, and Hevea for Timber* Proceedings of the Seminar on 9 January 1997, Kuala Lumpur, Malaysia FRIM 1998.

Anonymous "Malaysia's Industry Matures" in *FDM Asia* April 2000 pp. 41–55.

Anonymous "Sarawak Yearns to Plant 54,000 ha of Acacia mangium by 2005" in *Perkasa* vol. 1/ 2 Jan–Feb 2004.

Bawon Paiman *Performance Rating for Small and Medium Size Furniture Manufacturers Under Integrated Marketing Scheme in Peninsular Malaysia* Master of Science Dissertation Universiti Putra Malaysia, 1999.

Dan, Y.M. "Forestry Statistics, Data Sources and Methods of Data Collection and Verification in Malaysia." Paper presented at the ITTO Workshop on Tropical Forestry and Timber Trade Statistics 1993, Kuala Lumpur, Malaysia. 1993.

Enters and *Durst "What Does It Take? The Role of Incentives in Forest Plantation Development in Asia and the Pacific"* FAO Regional Office for Asia and the Pacific, Bangkok, Thailand 2004.

Environmental Investigation Agency (EIA) *Profiting from Plunder: How Malaysia Smuggles Endangered Wood* 2004.

FAO "Private Sector Forest Plantation Development in Peninsular Malaysia" www.fao.org/DOCREP/005/AC787E/AC787E05.htm 1999.

FAO "Asia-Pacific Forestry Sector Outlook Study" www.fao.org/docrep/003/y0153e/y0153e06.htm 2001.

FRIM "*Status of Forest Plantations in Peninsular Malaysia*" 2001.

FRIM "Area planted with different forest plantation species" from FRIM Plantations Division survey database. (Unpublished) 2002.

Gan, Ho, and Chew *Heveawood: sawntimber production and recovery studies.* Paper presented at the International Rubber Conference, Kuala Lumpur, Malaysia, 1985.

Ganapathy *Sources of non-wood fibre for paper, board and panels production — status, trends and prospects for India.* Asia-Pacific Forestry Sector Outlook Study: Working Paper No. APFSOS/WP/10. FAO, Rome 2000.

Goodstein, Eban S. *Economics and the Environment* 3rd edition, John Wiley & Sons Inc, New York, 2002.

Hawken, Paul, Amory Lovins, L. H. Lovins *Natural Capitalism: Creating the Next Industrial Revolution* Little, Brown and Company, New York 1999.

Ismariah and Norini "Availability of rubberwood resource in Peninsular Malaysia" In Hong and Sim, "Rubberwood processing and utilization," *Malayan Forest Records No. 39* March 1994.

IKEA *The IKEA Way on Purchasing Home Furnishing Products* IKEA Group, 2002.

Killmann *Non-forest tree plantations* FAO Forestry Department: Forest Plantations Thematic Papers: Working Paper FP/6. FAO 2001.

Killmann and Woon *Costs of extraction and transportation of oil palm stems* FRIM Reports No. 54 pp. 13–26. FRIM 1990.

Leontief, W. *The Structure of American Economy, 1919–1939: an Empirical Application of Equilibrium Analysis*, Oxford Univ. Press, New York 1951.

Malaysia Ministry of Primary Industries *Statistics on Commodities 2003* Ministry of Primary Industries, Kuala Lumpur 2003.

Mohd Noor, M., Rasip A.G.,Ahmad Shakri M.S. and Mohd Zali A. "Availability and Utilisation of Plantation Species" in *Proceedings of the National Seminar on Alternatives to Rubberwood* 26 September 2000, FRIM pp. 41–54. 2000.

Norini, H. *Structural Interdependence of Malaysian Industries with Special Emphasis on the Forest-Based Industries* Ph. D Dissertation, University of Wales, Aberystwyth, 1999.

Norini, Ismariah, and Seman. "Availability of rubberwood in Peninsular Malaysia" Proceedings International Rubberwood Seminar. Ed. Hong L.T. *et al.*, FRIM, Kuala Lumpur. pp. 55–69. 1990.

Norini *et al.* "An assessment of planting rubber as forest plantation species", in H.L. Thong and H.C. Sim *Rubberwood — Processing and Utilization.* Malayan Forest Records No. 39, pp. 223–231. 1994.

Norini H., W.C. Woon, and H.C. Sim, "Supply and demand of rubberwood in Peninsular Malaysia." Paper presented at the Seminar on Alternative to Rubberwood. 26 September, FRIM, Kuala Lumpur. 2000.

Pearce, David, Francis E. Putz, Jerome K. Vanclay "Sustainable Forestry in the Tropics: Panacea or Folly" in *Forest Ecology and Mangement* vol. 172, pp. 229–247. 2003.

Raymond, A.G. "Tapping rubberwood's export market potential" Paper presented at the International Forum on Investment Opportunities in the Rubberwood Industry Kuala Lumpur, Malaysia, September 1993.

Ruiz Perez, Fu Maoyi, Yang Xiaosheng, and Belcher, B. "Bamboo forestry in China: toward environmentally friendly expansion" *Journal of Forestry* 99(7): 14–20. 2001.

Sabah Forestry Department. 2000. *Production and Export Statistics of Forest Products.*

Salleh "Rubberwood: The Present and Future Furniture Timber" Paper presented at Woodmac Asia Conference. Singapore, 30 September-1 October 1987.

Ser "Rubberwood Resource in ASEAN and the Potential for its wider Utilization." In Proceedings International Rubberwood Seminar, FRIM, Kuala Lumpur. 1990 pp. 27–39.

Thai, S.K. 1995. *Reforestation: Towards Forest Plantations — Challenges and Prospects, National Conference on Timber and the Wood-Based Industry.* June, 1995, Kuching Sarawak, Malaysia.

Woon, W. C. and Haron Norini "Trends in Malaysian Forest Policy" in *Policy Trend Report* 2002 pp. 12–28.

Woon, W.C. "The impacts of Sustainable Forest Management (SFM) on the wood based industries in Peninsular Malaysia" in *Tropical forestry research in the new millennium: meeting demands and challenges: proceedings of the International Conference on Forestry and Forest Products Research (CFFPR 2001)*1–3 October 2001, Kuala Lumpur, Malaysia pp. 286–292.

Woon, W.C. and C.Y. Lim "Commercial Tree Plantation from a Private Sector's Perspective" in *Challenges facing forestry & forest-based industries in the new millennium: proceedings of the International Conference on Forestry and Forest Products Research (CFFPR 2002)* 17–18 June 2002, Kuala Lumpur, Malaysia pp. 56–64.

Zana, A.U. "Rubberwood from agricultural plantations: a market analysis for Peninsular Malaysia." in *The Planter*, vol. 70, pp. 435–452. 1994.

18

A MODEL FOR AN INTEGRATED ASSESSMENT OF SUSTAINABILITY FOR FOSSIL FUEL-FIRED ELECTRICITY GENERATING UTILITIES

CHRISTIAN RAE PUNZALAN BUSTAMANTE
and GEORGE OFORI

Business plays an important role in the efforts to attain sustainable development. Its production and consumption activities must be able to work with (and not against) ecological systems where it extracts its raw materials and releases its wastes. To be effective in meeting the challenge of sustainable development, adopting a systems approach is recommended wherein all aspects of sustainability are considered. Underlying this systems approach is the need to develop measures to indicate progress towards or away from sustainable development. Many efforts have been made to develop indicators of sustainable development. These efforts adopted different frameworks and used various approaches and measures of sustainability. The lack of consensus in defining indicators may be attributed to the ambiguity of the concept of sustainable development. The limitations in understanding this concept only allow the use of indicators in relative terms. An example is efficiency in the use of resources. Comparisons between similar business units are conducted wherein the less efficient business unit is considered less sustainable due to its less efficient use of resources.

In this study, comparisons among fossil fuel-fired electricity generating utilities in the Philippines were conducted using the composite sustainable development index ($\mathrm{I_{CSD}}$). The utilities were selected due to their resource-intensive operations. The $\mathrm{I_{CSD}}$ is based on a method that aggregates chosen indicators into a single measure of sustainability. Underlying this method is the Analytic Hierarchy Process (AHP) developed to establish priorities among indicators.

Four environmental, six social, and four economic indicators were chosen for the case study. Results of the AHP using an expert survey revealed that the greatest importance was assigned to indicators on air emissions ($m_{\mathrm{SO_2 equiv}}$), occupational health and safety (LWDII), and power plant reliability (EFOR). Aggregation to the $\mathrm{I_{CSD}}$ did not produce realistic results as 20 percent of the entire set of indicators for analysis were missing. Moreover, a set of indicators (LWDII) without minimum and maximum values resulted in errors in the computation. Despite these drawbacks, given suitable information, the $\mathrm{I_{CSD}}$ can serve as a useful tool for decision makers to monitor trends over time and compare performance between similar business units. Other weaknesses of the $\mathrm{I_{CSD}}$ were the subjective selection of indicators and the expert survey. To gain acceptance, selection and weighting of indicators may take place through wider systems of consensus building such as the Global Reporting Initiative

(GRI). This study also observed that power plant reliability indicators can result in conflict with other indicators, mainly because profits for the utilities are closely linked to reliability rather than efficiency. As a result, tradeoffs may occur between economic performance and other aspects of sustainability.

Introduction

The power sector is important in the global economy. The electricity it generates provides one of the most important commodities of the modern world. The sector provides both direct and indirect jobs to people and have fueled the economy of its host countries. The sector also plays an important role in today's society. The use of electricity for lighting, heating, cooling, or powering equipment covers various aspects of human life such as home, office, and outdoor activities. As humanity continues to increase its dependence on modern technologies, there is also a corresponding increase in dependence on electricity. Despite the benefits electricity provides, the power sector is a source of many social and environmental impacts due to the nature of its operations. Each type of power infrastructure, from generation to transmission and then to distribution has unique impacts. Among these types, electricity generation from fossil fuel-based sources has one of the most significant impacts due in particular to its contribution to global warming.

Research objectives

This research develops and demonstrates an integrated assessment model to monitor the performance of power generation companies towards sustainable development goals. It has three main objectives:

- To search for and evaluate approaches that attempt to measure sustainable development performance of a business.
- To identify a suitable method for measuring sustainable development for a power utility.
- To verify and demonstrate the application of this method using fossil fuel-fired electricity generating utilities in the Philippines as a case study.

Research method

Indicators are used to assess parameters related to specific performance characteristics of an organization or company. For power plants, Afgan and Carvalho (2004, 1327) noted the complex nature of the plant systems which require multiple variables for performance assessment. Examples include financial performance, environmental, occupational safety and health, combustion efficiency, monitoring equipment reliability, fire protection systems reliability, and equipment performance. There is a significant number of indicators that need to be reviewed by decision makers of generating utilities. As such, difficulties in making decisions may arise based on a large number of performance measurements (Krajnc and Glavic, 2005b, 552).

To address potential difficulties in decision making, an integrated assessment of sustainability developed by Krajnc and Glavic (2005a, 2005b) is demonstrated in this chapter using fossil fuel-fired electricity generating utilities in the Philippines as a case study. The integrated assessment proposed in this study for generating utilities aims to promote benchmarking activities for companies through performance comparisons between firms. Moreover, the proposed method aims to provide companies with a tool to monitor trends in their performance over time.

Three fossil fuel-fired power plants in the Philippines were invited to participate in this dissertation. All three power plants are located on the island of Luzon. The companies running these plants are foreign investor-owned independent power producers (IPPs) which were commissioned for operation between 1996 and 2000 (DOE, 2006b). Since commissioning, the companies have actively engaged in community partnerships and environment-related projects. Two plants are ISO 14001-certified while the other plant has an existing management system similar to an environmental management system (EMS). For reasons of confidentiality, the companies are identified in this chapter as Company A, Company B, and Company C. The field study conducted with the three companies covered approximately two months from the last week of May 2006 to mid-July 2006.

The integrated assessment method developed by Krajnc and Glavic (2005a, 2005b) aggregates a number of indicators into a composite sustainable development index (I_{CSD}). There are several steps needed to obtain the I_{CSD}. These steps are: selection and grouping of indicators, judging and weighting of the indicators, normalization, calculation of the sub-indices, and finally, aggregation into the I_{CSD}.

In selecting indicators, it is important to be aware of the limitations of efforts to define sustainability. Azapagic and Perdan (2000, 248) stress that people can only aim to measure progress towards or deterioration away from sustainable development rather than define the absolute state of sustainability of industrial systems. Moreover, Azapagic (2003, 309) recommends that indicators should be quantitative whenever possible to facilitate decision making but for some aspects of sustainability, qualitative descriptors may be more appropriate. To be specific, indicators that measure progress regularly such as in one-year intervals are useful in tracking improvements (or deterioration) on a year to year basis (Azapagic and Perdan, 2000, 248). Furthermore, to address the reliability of indicators for researchers, the indicators must be rooted in existing legal reporting requirements or third-party commitments. In this way, the values reported in indicators can be validated or verified.

The next procedural step is determining the grouping of the chosen indicators. An underlying requirement in these regards is choosing a framework to define sustainable development. Welford (1997, 182–206) provides a summary of six models or frameworks of sustainable development for organizations and businesses. For the purpose of this study, the "triple bottom line" or the environmental, economic, and societal framework was adopted as the model because of its wide acceptance in attempts to define sustainability (Krajnc and Glavic, 2005b, 553). The acceptance of the framework is observed in its increasing adoption by GRI especially for sustainability reporting.

The selected indicators are grouped according to the three aspects of sustainability: economic ($j = 1$), environmental ($j = 2$), and societal ($j = 3$) groups of indicators. The notations are assigned for standard application in the succeeding formulas.

In line with the need to be efficient in aspects of sustainability (Callens and Tyteca, 1999, 43), each group (j) of indicators (i) is identified based on positive and negative performance in relation to sustainable development goals. The indicators of positive performance ($I^{+}_{A,ji}$) mean that increasing value of the indicator results in a positive impact in sustainable development. As such, positive performance may be simplified to mean "more is better". The indicators of negative performance ($I^{-}_{A,ji}$) mean that the increasing value of the indicator results in a negative impact in sustainable development. In simplified terms, the indicator means "less is better." To illustrate these points, an example of an indicator of positive performance can be the increasing net profit of a company per unit of production. Alternatively, an example of an indicator of negative performance can be the increasing amount of effluent discharged per unit of production.

An important prerequisite in aggregating indicators into the I_{CSD} is the assignment of weights to individual indicators included in the aggregation.

The method proposed by Krajnc and Glavic (2005a; 2005b) is the pair wise comparison technique based on the Analytic Hierarchy Process (AHP) pioneered by Saaty (2001).

The AHP enables groups or individuals to make decisions on complex issues (such as sustainable development) by simplifying and expediting the natural decision-making processes of the human mind (Saaty, 2001, 5). The AHP is a leading multi-attribute decision model used widely for problem solving and decision making. It has been adopted by various groups such as the United States Army, U.S. Air Force, and U.S. State Department, Xerox, Boeing, General Electric, IBM, Northrop Grumman, U.S. Steel, and the governments of South Africa, Canada and Indonesia (Palmer, 1999, 154).

To describe the process of establishing weights based on the AHP, the first step is to construct a hierarchy where the topmost node is the overall objective of the decision, while subsequent nodes at the lower levels consist of the criteria used in arriving at this decision (Krajnc and Glavic, 2005a, 195). For this research study, the overall objective is sustainable development and the main criterion for achieving the objective is through the performance of environmental, economic, and societal groups of indicators. The next step is to establish priorities within the indicators of a particular group and priorities between groups contributing to sustainable development. Establishing priorities is done through pair wise comparisons.

Pair wise comparisons are made by asking which of the two indicators i and k is more important with respect to the sustainable development of the company. The intensity of preference is expressed on a factor scale from 1 to 9 (see Table 18.1). The value of 1 indicates equality between the two indicators while a preference of 9 indicates that one indicator is 9 times more important than the one it is being compared with. The scale of nine units was chosen because it reflects the degree to which people can discriminate the intensity of relationships between elements (Saaty, 2001, 73).

After assigning preferences, the pair wise comparisons result in a $(N \times N)$ positive reciprocal matrix A, where the diagonal $a_{ik} = 1$ and reciprocal property $a_{ki} = (1/a_{ik})$, i, $k = 1, \ldots, n$ assuming: if indicator i is 'p-times' the importance of indicator k, then, necessarily, indicator k is '1/p-times' the importance of indicator i. To find the normalized weight of each indicator, one way is to normalize each column in matrix A (dividing an indicator's relative weight by the sum of relative weights in the column), and then averaging the values across the rows; this average column is the

Table 18.1. Fundamental scale for pairwise comparison adopted from Saaty (2001, 73).

Intensity of importance	Definition	Explanation
1	Equal importance	Two activities contribute equally to the objective.
3	Moderate importance	Experience and judgment slightly favor one activity over another.
5	Strong importance	Experience and judgment strongly favor one activity over another.
7	Very strong or demonstrated importance	An activity is favored very strongly over another; its dominance demonstrated in practice.
9	Extreme importance	The evidence favoring one activity over another is of the highest possible order of affirmation.
2,4,6,8	For compromise between the above values	Sometimes one needs to interpolate a compromise judgment numerically because there is no good word to describe it.

normalized weight vector W containing weights (W_{ji}) of indicators selected (Krajnc and Glavic, 2005b, 554).

The AHP provides a check for consistency to determine if the pair wise comparisons are consistent. An acceptable level of consistency must maintain a value of 10 percent for n indicators, 9 percent for four indicators, and 5 percent for three indicators. The C_R ensures that the results are within acceptable levels of consistency (Saaty, 2001, 83). Saaty (2001, 81) stressed that should the level of consistency exceed the set limit of C_R, there is a need to reevaluate the weights to arrive at the acceptable level of consistency.

It is not possible to aggregate indicators if they are expressed in different units. Therefore, a normalization procedure is required. In this dissertation, normalization procedure based on Eqs. (18.1) and (18.2) can be used. Equation (18.1) is applied to indicators of positive performance while Equation (18.2) is applied to indicators of negative performance. An advantage of normalized data is that it allows compatibility of different indicators.

Equation (18.1). Normalization procedure for indicators of positive performance.

$$I^{+}_{\mathrm{N,ijt}} = \frac{I^{+}_{\mathrm{A},ijt} - I^{+}_{\mathrm{min},jt}}{I^{+}_{\mathrm{max},jt} - I^{+}_{\mathrm{min},jt}}$$

Equation (18.2). Normalization procedure for indicators of negative performance.

$$I^{-}_{\mathrm{N,ijt}} = 1 - \frac{I^{-}_{\mathrm{A},ijt} - I^{-}_{\mathrm{min},jt}}{I^{-}_{\mathrm{max},jt} - I^{-}_{\mathrm{min},jt}}$$

The normalization procedure does not have to be computed based on the equations presented. Goals may be set by companies based on a target performance. This must be selected based on the realistically achievable option (Krajnc and Glavic, 2005b, 554).

After calculating priorities and normalizing data, the next procedural step is calculating the sub-indices and the I_{CSD}. To do this, the formulas below are needed. For Eq. (18.3), $I_{S,jt}$ is the sub-index for a group of indicators (environmental, economic, or societal) while W_{ji} is the estimated weight of indicator (i) for the group of indicators (j) based on the pair wise comparison technique. For Eq. (18.4), I_{CSD} represents the composite sustainable development index and W_j represents a weight given to the group (j) of indicators.

Equation (18.3). Formula to calculate the sub-indices

$$I_{S,jt} = \sum_{jit}^{n} W_{ji} \cdot I^{+}_{N,jit} + \sum_{jit}^{n} W_{ji} \cdot I^{-}_{N,jit}$$

$$\sum_{ji}^{n} W_{ji} = 1, \quad W_{ji} \geq 0$$

Equation (18.4). Formula to calculate the composite sustainable development index.

$$I_{\mathrm{CSD},t} = \sum_{jt}^{n} W_j \cdot I_{S,jt}$$

Selection of indicators was based on previous research and on the first author's knowledge on specific indicators in the power generation sector. The first approach in selecting indicators was to conduct research on sustainability indicators for the power sector. The WBCSD (2002) considered the role of the electricity utility industry in working towards sustainable development. In particular, it discussed the practices of the industry to attain sustainable development but did not elaborate on quantitative measures on the environmental, social, and economic aspects. As there are no specific studies on quantitative indicators for the power sector, ISD methods from various sources were consulted (Azapagic and Perdan, 2000; Veleva and Ellenbecker, 2001; Krajnc and Glavic, 2003; IChemE, 2006; GRI, 2006). The grouping of indicators was conducted by the first author. To serve as a guide in grouping, the GRI framework was adopted. The version of the GRI used was the GRI G3 version. Although still a draft, it represented key improvements made to the indicators which were significant to include the latest developments.

Aside from the selection and grouping, there was also a need to define the number of indicators for each grouping. The limitation of the AHP, as mentioned by Saaty (2001, 268) is that as the number of indicators in a specific grouping increase, there is a greater chance of inconsistency in judgments. Moreover, the pair wise comparison technique used by AHP will be increasingly difficult. To address this limitation and, at the same time, adequately represent the specific grouping, five to nine indicators are a sufficient number (Saaty, 2001, 268).

As a result of the selection and grouping, a list of environmental, social, and economic indicators was produced. This was presented to the participating companies to choose from. The participating companies were then requested to select between five and nine indicators per grouping. Three conditions were specified to the companies: (1) the indicators are available, easily retrievable, and can be publicly disclosed; (2) raw data used to calculate the indicators are bound by external reporting requirements; and (3) raw data used to calculate the indicators were monitored over a period of three or more years. The conditions were specified for a purpose. The first condition ensured the accessibility of information for use as indicators. The second condition assured the credibility of data provided for verification purposes. The third condition allowed comparison of data over time and the determination of trends whenever possible. Furthermore, the companies were encouraged to nominate specific indicators which they feel should be included in the list of indicators. To be consistent, these nominated indicators must also fulfill the two conditions specified above.

After selection of indicators by the companies and verification by the author, the following step is to determine positive and negative performance indicators within the groupings. The next step after that was conducting the survey.

An expert survey was conducted using the pair wise comparison technique. The survey was prepared in the form of self-administered questionnaires. An important step in ensuring the success of the survey was pilot testing of the questionnaires. A pilot test using hypothetical indicators was conducted from May 15 to May 31, 2006. Three respondents with a background in industry specializing in human resources and sociology, economics and finance, and environment were requested to provide feedback on the social, economic, and environmental questionnaires respectively.

Nine experts from the fossil fuel-fired independent power producer (IPP) sector in the Philippines were invited to participate in the survey. Of the nine, three were environmental experts, three were economic experts, and the last three were experts in societal relations. All experts were practicing their respective fields in the IPP sector at the time of the survey. Upon completion of the questionnaires, the results were processed to determine the weights of each selected indicator. To check for the consistency of judgments made by the respondents, the consistency ratio (C_R) was used to determine the level of inconsistency.

Simultaneous to the expert survey was the collection of temporal data for indicator calculation. As mentioned earlier, three to four years worth of indicators were requested from the participating companies. As a preparatory step to the aggregation to sub-indices and the I_{CSD}, normalization of indicators was conducted. Normalization is essential to address the difficulty of making comparisons of indicators expressed in different units. The normalization procedure converted the indicators to values between 0 and 1. This enabled the sets of indicators to be compared. For indicators of positive performance, Eq. (18.1) was used. For indicators of negative performance, Eq. (18.2) was used.

After determining the weights and normalizing the indicators of the participating companies, the final step was the calculation of the sub-indices (environmental, economic, and social or societal) and the composite sustainable development index.

Equation (18.3) was used in the calculation of the sub-indices. For the calculation of the I_{CSD}, Eq. (18.4) was used. A prerequisite for the calculation of the I_{CSD} was the weighting of the environmental, economic, and societal factors to each other. Weighting of these factors through a survey of experts or other methods can be used to determine the importance of one aspect over the other. For this study, the three factors were assumed to have equal weights. This assumption was based on the definition of sustainable development which implied equality of the three factors in achieving sustainability (Krajnc and Glavic, 2005a, 204).

Results

Indicator selection

Four environmental, six societal, and four economic indicators were chosen for the research. The environmental indicators include mass emission indicators (m_{SO_2equiv} and m_{TSP}), a hazardous waste generation indicator ($m_{haz\ waste}$), and an indicator on fuel consumption (E_{coal}). The societal indicators are employee related indicators ($N_{labor\ service}$, $N_{spec\ employ}$, and $X_{turnover}$) and occupational health and safety indicators (LWDII, RI, and $X_{first\ aid}$). The economic indicators include power plant reliability indicators (EFOR and EAF) and financial performance indicators (C_{NIAT} and f_{GDP}). The chosen indicators fulfilled the three conditions stated previously.

Judging of indicators

A total of four (4) indicators were judged as indicators of positive performance. These are: $N_{labor\ service}$, C_{NIAT}, f_{GDP}, and EAF. On the other hand, a total of ten (10) indicators were judged as indicators of negative performance. These are: E_{coal}, m_{SO_2equiv}, m_{TSP}, $m_{haz\ waste}$, $N_{spec\ employ}$, $X_{turnover}$, LWDII, RI, $X_{first\ aid}$, and EFOR.

Establishing weights for indicators

For the environmental group of indicators, survey results revealed that the greatest weight or importance was attributable to the mass air emissions of sulfur dioxide (SO_2) and nitrogen dioxide (NO_2) per unit of electricity produced (m_{SO_2equiv}). For the societal group of indicators, the Lost Workday Injury and Illness (LWDII) case rate was given the highest importance. For the economic group of indicators, the highest weights were given to indicators of power plant reliability which are the Equivalent Forced Outage Rate (EFOR) and the Equivalent Availability Factor (EAF) with EFOR being the higher between the two.

As for the consistency of judgments made by experts, results of the consistency ratio (C_R), provided by the AHP, showed that the judgments were within acceptable levels of consistency.

Collection of indicators

The participating companies agreed to provide indicators for the period of 2002 to 2005. Due to specific circumstances experienced during data collection, 34 out of 168 values for the indicators were missing in the analysis. The missing values represent about 20per cent of the entire set of values of indicators for 2002 to 2005.

To determine the effect of missing data on the calculations of the I_{CSD} and the sub-indices (I_{ECN}, I_{SOC}, and I_{ENV}), two options were investigated. The first option assigned random numbers to the missing values. The second option assigned zeros for the missing values.

Aside from missing values, problems in normalizing the indicators were encountered when the values for LWDII did not have a maximum and minimum value. The three participating companies reported zeros in their LWDII values for all years. Although zero values for LWDII for all years reflect positive performance of the three companies, the zero values of LWDII indicators returned errors in the normalization procedure. To address this, random values were assigned. The randomly assigned values were taken within the range of 0.0001 and 0.001 to retain the reported zero values of the companies and at the same time address the need for maximum and minimum values in the set of LWDII indicators. As a consequence of the assigning of random numbers, actual performance of the companies in terms of LWDII indicators may not be realistically represented in the sustainability assessment of the three participating companies.

Results of the I_{CSD} *and the sub-indices (I_{ECN}, I_{SOC}, and I_{ENV})*

Missing values do have an effect on the results of the I_{CSD} and the sub-indices (I_{ECN}, I_{SOC}, and I_{ENV}). The effect involved an increase or decrease in values of the I_{CSD} and the sub-indices. As to the degree of change in values, there was no determined pattern. Moreover, results of the I_{CSD} and the sub-indices using zeros or random values for the missing data prevented results from reflecting actual performance of the companies within the time period (2002–2005).

Given a complete set of indicators with corresponding maximum and minimum values, the I_{CSD} and the sub-indices (I_{ECN}, I_{SOC}, and I_{ENV}) can be a useful tool for decision makers to determine their performance across time and compared to similar business units.

Discussion

In selecting a list of environmental, societal, and economic indicators, the author's background and current knowledge of the Philippine power sector may have influenced the decisions made as to which indicators to include and which to exclude. No particular method was employed in the selection. For this reason, a certain degree of subjectivity occurred in indicator selection. Nevertheless, the subjective approach in nominating sustainability indicators for power generating utilities may have been unavoidable considering that there was no currently available research on indicators specific to the power sector.

On the selection of suitable indicators by the participating companies, the selection process was limited to indicators that companies can disclose to the public. This observation is consistent with the experience of other researchers (Azapagic and Perdan, 2000; Veleva and Ellenbecker, 2001; Krajnc and Glavic, 2003) who were limited by the availability of suitable information.

It is worth noting that indicators which describe global warming impacts such as carbon emissions per unit of production were not part of the selected list of indicators. Generating utilities are not required under Philippine environmental laws to report on carbon emissions. For this reason, it was not possible to verify the credibility of the carbon emissions data provided by the companies. As a result, carbon emissions data per unit of production were excluded from the list of indicators even if the companies maintained records. For carbon emissions to be used as an indicator, this research observed that the selection of indicators was dependent on some level to legal requirements of the host country or community of the power plants as these requirements allow indicators to be verified.

This research also affirms the lack of appropriate indicators to describe social issues adequately. Compared to economic and environmental indicators, social or societal indicators were not as developed. As an example, the specific employee number was used to indicate the efficiency (or non-efficiency) of the labor force in producing electricity for consumers. Although this indicator may be taken to mean the efficiency of the labor force in producing goods and services, it does not reveal issues such as potential stress to employees due to pressures to increase productivity. Adding difficulties to the use of social indicators is the fact that some indicators contain sensitive information that companies are not willing to disclose publicly. An example of this is the indicator on social equity. To quantitatively describe social equity, an appropriate indicator would be comparing the difference of salaries between the top officials of the company with the lowest ranking employees of the company. The respondents were not allowed by their companies to disclose the data for such an indicator for reasons of sensitivity.

The expert survey was a useful method to establish weights of indicators. As the survey relied on the opinion of respondents, there was some level of subjectivity in the process. Moreover, it is likely that inputs from a different set of respondents may have a different outcome on the weights. To accommodate a different set of weights, it is not necessary to change the specific steps in the sustainability assessment model. Rather, only the weights need to be changed to arrive at the I_{CSD} and the sub-indices.

For the sustainability assessment model to be applicable to an intended audience such as the Philippine public sector, a consensus-building approach may be an effective way to establish weights. This was observed by Parris and Kates (2003, 576) when they saw that salience, credibility, and legitimacy of indicators increased when a consensus-building approach was used. Should the intended audience of the model be wider than the Philippine public sector, the Global Reporting Initiative (GRI) would be a suitable method to use to establish weights given its strong participatory approach to sustainability reporting. This is consistent with the recommendation of Krajnc and Glavic (2005a) who also called for the application of the GRI to establish weights.

Lastly, it is important to note that inconsistencies in the pair wise technique used in AHP increase when more indicators are selected for sustainability assessment (Saaty, 2001, 268). Given this property of the pair wise technique, future researchers may find the technique too rigid, and not appropriate for application to establish priorities.

It was found in this research that the I_{CSD} was limited by issues on completeness of information. Moreover, it was observed that results of I_{CSD} were affected by missing data. Thus, the results were not reflective of real situations experienced by the companies. Furthermore, assigning random values contributed to inconclusive results. This implies that for the I_{CSD} to represent actual performance of companies, there is a need to ensure completeness of information on the sustainability indicators.

A problem was seen when all companies reported zero results for LWDII. Although this indicates positive performance for the companies, the normalization procedure produced errors because denominator results returned zero values. This may be explained by the fact that there are no maximum and minimum values in the entire set of LWDII indicators for the participating companies. Although random values were technically considered zero when rounded off, the use of random numbers deviates from the actual performance of companies. As such, it is observed in this study that for the application of the I_{CSD} and the sub-indices to companies, indicators used in the assessment must contain maximum and minimum values.

The I_{CSD} was developed to specifically target decision makers. This is the strength of the I_{CSD}. The aggregated measures target specific needs of decision makers to obtain condensed information (Braat, 1991).

Despite problems with missing data and inappropriateness of some information such as the LWDII, this research showed that it was possible to conduct comparisons across time and between similar business units using the sustainability assessment model.

There are issues directed to aggregate measures which say that they fail to demonstrate sustainability using individual indicators. Krajnc and Glavic (2005b) addressed this by using the AMOEBA technique proposed by Brink *et al.* (1991). Further research is needed to compare aggregated indices with measures such as those in AMOEBA technique and other similar approaches which do not use aggregation as part of their methods.

Moreover, one common trait found in research on indicators of sustainable development is that their approach can be characterized by a top down method where indicators applicable across industry are proposed. An example of this is research on indicators for production systems. One limitation of this approach as affirmed by Azapagic and Perdan (2000, 243) is that even if the indicators are applicable across industry, there may be a need to determine specific indicators for different sectors. To advance research on indicators of sustainable development, another approach to complement this is to investigate the specific context of various sectors of business and from this starting point, develop sustainability indicators. By doing this, specific sustainability indicators may be obtained that can potentially reflect sustainability in the specific sector.

The framework used in this research was based on the "triple bottom line". In future research, other models of sustainable development could be used to determine their applicability to the composite sustainable development index. The flexibility of the model can also be tested when different frameworks are used.

It was found in this research that the indicators of power plant reliability were in conflict with indicators of efficiency. Because the profits of the firms are intimately linked with the reliability of the power plant, if these companies are called on to supply electricity to meet the consumer demand at a specific time or period, power firms must meet the demand even if there are problems with operations during that time. These problems may include malfunctioning pollution control technologies. As a result, even if decision makers of the firms express support for attainment of lower emissions per unit of production, these firms need to supply electricity to the power grid to meet demand even though they have specific problems related to pollution removal efficiency.

Another example may involve machinery in power plants. In a specific scenario, machinery may require additional lubrication due to problems arising in use. To solve this problem, the equipment might need to be shut down for maintenance. However, the machinery is needed to remain operational to meet the current electricity demand, although use of additional lubrication is necessary. When an opportunity arises where the electricity demand is not so great, that is the time when the equipment problem can be solved.

It might be beneficial, in advancing research on sustainability indicators, to investigate if currently available indicators fulfill the necessary (but not sufficient) conditions of sustainability as stated by Callens and Tyteca (1999). Conflicts between efficiency and other priorities specific to a sector or companies may need to be further investigated to determine trade offs and driving forces affecting the indicators.

Conclusion

Although missing values of some indicators did not permit the demonstration of actual corporate performance, the I_{CSD} and the sub-indices were able to show its ability to compare companies with each other and show performance trends of individual companies over time using aggregate measures. These aggregate measures were designed to address potential difficulties that decision makers may experience when reviewing a large number of performance measures. Two important findings in this study highlight the need to fulfill prerequisites to properly use the I_{CSD} for its intended purposes. The first is that in applying the I_{CSD}, complete information must be secured so as not to affect the results of the aggregation. The second is that the chosen indicators for the I_{CSD} must have variations in value over time, across similar business units, or both. These two observations affirm the dependence of indicators on the availability of suitable information. On the other hand, it was also observed that power plant reliability indicators (EFOR and EAF) were in conflict with indicators demonstrating efficiency such as less air emissions per unit of production and less fuel consumption per unit of production.

This is because profits of the participating companies were closely linked to reliability in meeting the demand for electricity rather than efficiency in resource consumption. As to the implications of these findings for further research, the context and driving forces behind specific sectors such as the electricity generation industry could be examined. In this way, suitable indicators will be determined to demonstrate necessary conditions of sustainability.

References

Afgan, Naim H., and Maria G. Carvalho. 2004. Sustainability assessment of hydrogen energy systems. International Journal of Hydrogen Energy 29: 1327–1342.

Azapagic, A. 2003. Systems approach to corporate sustainability: A general management framework. Trans. IChemE. 81, Part B (September): 303–316.

Azapagic, A, and S. Perdan. July 2000. Indicators of sustainable development for industry: a general framework. Trans. IChemE78-B: 243–261.

Braat, Leon. 1991. The predictive meaning of sustainability indicators. In In search of indicators of sustainable development, Kuik, Onno; Verbruggen, Harmen. 57–70: Kluwer Academic Publishers.

Callens, Isabelle, and Daniel Tyteca. 1999. Towards indicators of sustainable development for firms: A productive efficiency perspective. Ecological Economics 28: 41–53.

Department of Energy (DOE). 2006. List of existing plants as of December 2005. Power Statistics 2005. http://www.doe.gov.ph/power/Power%20Stat%202005%20update042406.htm: Department of Energy, Philippines, July 1.

Global Reporting Initiative (GRI). 2006. Sustainability Reporting Guidelines (Draft). G3 Version for public comment (2 January 2006-31 March 2006). Global Reporting Initiative, Sources: http://www.grig3.org and http://www.globalreporting.org.

Institution of Chemical Engineers (IChemE). 26 April 2006. The sustainability metrics: Sustainable development progress metrics (recommended for use in the process industries). English. http://www.icheme.org/sustainability/metrics.pdf: IChemE.

Krajnc, Damjan, and Peter Glavic. 2005a. A model for integrated assessment of sustainable development. Resources Conservation and Recycling 43: 189–208.

Krajnc, Damjan, and Peter Glavic. 2005b. How to compare companies on relevant dimensions of sustainability. Ecological Economics 55: 551–563.

Krajnc, Damjan, and Peter Glavic. 2003. Indicators of sustainable production. Clean Techn. Environ. Policy 5: 279–288.

Palmer, Brian. 1999. Click here for decisions. Fortune (New York) 139, 9 (10 May): 153–155.

Parris, Thomas M., and Robert W. Kates. 2003. Characterizing and measuring sustainable development. Annu. Rev. Environ. Resourc. 28: 559–586.

Saaty, Thomas L. 2001. Decision Making for Leaders: The Analytic Hierarchy Process for Decisions in a Complex World. New Edition. The Analytical Hierarchy Process Series. Pittsburgh, PA: RWS Publications.

Ten Brink, B. J. E., S. H. Hosper, and F. Colijn. 1991. A quantitative method for description and assessment of ecosystems: the AMOEBA approach. Marine Pollution Bulletin (Great Britain) 23: 265–270.

Veleva, Vesela, and Michael Ellenbecker. 2001. Indicators of sustainable production: Framework and methodology. Journal of Cleaner Production 9: 519–549.

Welford, Richard. 1997. Models of sustainable development for business. Chap. in Hijacking environmentalism. 179–210. London, UK: Earthscan Publications Ltd.

World Business Council for Sustainable Development (WBCSD). 2002. Sustainability in the electricity utility sector. English. Http://www.wbcsd.org/DocRoot/RBiSuIvtwIJFFpeANshg/20020930_electricity.pdf: World Business Council for Sustainable Development, July.

19

A SCHEMATIC PLAN FOR A GREEN ENERGY POLICY IN SINGAPORE

AMANDA ONG HWEE FANG and YOUNGHO CHANG

This chapter investigates the adequacy of the New Electricity Market of Singapore (NEMS) in internalizing the environmental costs of electricity generation, and proposes a green energy policy to achieve the least-cost technological development and market penetration of renewable energy sources in the NEMS. The findings of the chapter reveal that NEMS is inadequate in internalizing the environmental costs of electricity generation and the least-cost design of the liberalized electricity market may impede the promotion of Renewable Energy Sourced-Electricity (RES-E). However, with the implementation of appropriate policy instruments such as the RES-E support schemes, the liberalization of the electricity market can produce a positive effect of motivating RES-E power producers to be constantly in search of the most cost-effective RES-E technologies, and thereby stimulating technological advancements in RES-E. Electricity market liberalization also provides an opportunity for RES-E power producers to compete for a share of the electricity market, which was previously controlled by government-owned utilities. Following the findings, a Green Energy Policy is suggested to promote the least-cost technological development of renewables through the use of RES-E support schemes and eventually, to develop a national Tradable Green Certificates (TGC) System as part of the international TGC System. Besides environmental benefits, renewables can provide energy security to a resource-scarce Singapore. The RES-E support schemes would also attract investors to invest in Singapore, which would boost the economic development of Singapore and provide employment opportunities. Recommendations for future areas of research and implementation include the application of the RES-E schemes in other areas such as waste management and transport.

1. Introduction

Fossil fuels have been the dominant fuels used to meet the growing demands in transport, electricity generation and manufacturing. Global warming is one of the environmental effects of using fossil fuels posing the greatest threat to present and future generations. Being clean and inexhaustible, the renewable energy sources provide a long-term sustainable solution to the rising world energy demand, the deteriorating global climate and the unsustainable energy consumption patterns. However, the high costs of Renewable Energy Sourced-Electricity (RES-E) technologies is a major obstacle to the development of renewable energy in the electricity market, and the costly RES-E technologies are unable to compete with the conventional power plants using fossil fuels. This problem may be further accentuated by the liberalization and privatization of electricity markets in many countries. The least-cost design of liberalized electricity markets tends to favor conventional power plants using fossil fuels, rather than plants using the costly RES-E technologies. To promote RES-E in liberalized electricity markets, support schemes have been implemented in various countries since the 1990s. Three main types of RES-E support schemes are the Feed-in Model (FIM), the Tender System and the Tradable Green Certificate (TGC) System.

Like other countries, Singapore is in the process of liberalizing her electricity market. To promote the generation of RES-E, there is a need for a change in energy policy and environmental policy, and increased policy integration between the two. As there are few cases of electricity deregulation, this justifies why Singapore is used as a case. This study investigates the adequacy of the New Electricity Market of Singapore (NEMS) in internalizing the environmental costs of electricity generation, and focuses on the development and implementation of a green energy policy based on RES-E support schemes such as the Feed-in Model (FIM), the Tender System and the Tradable Green Certificate (TGC) System. The objective of this policy is to promote the development of RES-E technologies at the least cost. Like the FIM and the Tender System, the TGC System is not new and has gained ground in several countries in recent years, notably in the European Union (EU), the United States (US) and Australia. The current state of discussion in the EU is to harmonise the domestic green certificate schemes and promote international trade of green certificates. This is an important re-affirmation of the international community's commitment towards sustainable development. The recent establishment of the first EU constitution is one positive step towards the attainment of such an ambitious goal in sustainable development.

The rest of this chapter is organized as follows. The establishment of the New Electricity Market of Singapore (NEMS) is first discussed and the adequacy of the NEMS to incorporate RES-E is examined. The presents principles of policy design to promote RES-E in Singapore are next considered. The final section presents a summary of the chapter and some concluding remarks.

2. New Electricity Market of Singapore

2.1 *Establishment of the new electricity market of Singapore (NEMS)*

The electricity and piped gas industries in Singapore had been vertically integrated and government-owned. In May 1963, the Public Utilities Board (PUB) was formed to carry out the functions of supplying water, electricity and piped gas to the whole of Singapore.[1] To promote the efficient supply of competitively-priced electricity, the Government undertook several initiatives in restructuring the electricity and piped gas market to facilitate competition in electricity generation and retailing (EMA, 2004).

Milestones in Market Restructuring

On 1 October 1995, the electricity and piped gas utilities were corporatized to gradually introduce a competitive electricity market. Three generation companies (Tuas Power Ltd, PowerSenoko Ltd and PowerSeraya Ltd), one transmission and distribution company (PowerGrid Ltd), one electricity retail company (SP Services Ltd), one gas supply company (PowerGas Ltd) and one holding company (Singapore Power Ltd) were formed. On 1 April 1998, the Singapore Electricity Pool was established to introduce competition in the wholesale electricity market and was operated by PowerGrid Ltd. Further liberalization of the electricity market was announced by the government in March 2000 to achieve the full benefits of competition (i.e. competitively priced electricity and consumer choice). The initiatives included the separation of the contestable and non-contestable parts of the electricity industry, the formation of an independent system operator and the liberalization of the retail market.

The Energy Market Authority (EMA) and the Energy Market Company (EMC) were established on 1 April 2001 to assume the role of the regulator for the electricity and gas industries and the role of the market operator respectively. Meanwhile, Singapore Power Ltd undertook full divestment of its generation companies (PowerSenoko Ltd

[1]For more details on the historical background of Singapore electricity industry, refer to Overview of Electricity Industry by EMA available at http://www.ema.gov.sg/doc/rs_energy_electricity.doc

and PowerSeraya Ltd) to Temasek Holdings to enhance the market competition by separating the ownership of generation companies from the ownership of the transmission and distribution company.[2]

Market Liberalization

Liberalization of the retail market was implemented in phases. The contestability of the consumers was dependent on their electricity consumption level. From 1 July 2002 onwards, very large consumers with a maximum contracted capacity of 2 MW and above became contestable and had the opportunity to choose their energy retailers. This was followed by the commencement of the Singapore New Electricity Market on 1 January 2003 as announced by the EMA. Further liberalization of the retail market was seen in June 2003 where consumers with monthly consumption of 20,000 kWh and above were notified by the EMA of their contestability status (Phase 1). Subsequently, in the last quarter of 2003, smaller retail consumers with monthly consumption of 10,000 kWh and above also became contestable (Phase 2). The final phase, which is Full Retail Competition (FRC), was expected to be completed in 2004 (EMA, 2004).

Key Players of the NEMS

The key players of the NEMS are as follows:

(a) The industry regulator — EMA;
(b) The market operator — EMC;
(c) The power system operator (PSO) — a division of EMA;
(d) The transmission licensee — PowerGrid (monopoly);[3]
(e) Generation licensees — dispatchable generators above 1 MW;
(f) Market Support Services Licensee (MSSL) — provision of market support services;[4]
(g) Retail electricity licensees — retailers who sell electricity to contestable consumers; and
(h) Consumers — categorized as contestable and non-contestable, depending on their electricity usage.

---

[2]Source from EMA website available at http://www.ema.gov.sg/Electricity/milestones.php

[3]In 2003, PowerGrid Ltd was split into two entities, SP PowerAsset Ltd and SP PowerGrid Ltd.

[4]Currently, SP Services Ltd will be the only MSSL. Other MSSLs may enter the market later.

Design Principles of the NEMS

The wholesale electricity market of the NEMS is designed for all parties involved to operate with confidence, based on principles that would ensure effective operation and reduce risk to market participants and the public. The design principles are (1) robustness, (2) transparency, (3) equity and fairness, and (4) minimization of transaction costs. To accomplish the design principles, key features are put in place in the NEMS, for instance, the real-time dispatch of electricity at half-hourly intervals to achieve a dispatch schedule that is at a minimum cost to the market and the least-cost energy market solution to the dispatch.[5]

2.2 *Inadequacy of the NEMS and the need for market correction*

The following factors explain the inadequacy of the NEMS in internalizing the full costs of electricity, and hence form the basis for the introduction of renewable energy certificates to account for full costing of electricity, which includes the internalization of the environmental costs of electricity generation.

2.2.1 *Design of wholesale market operation*[6]

The energy prices of the wholesale market are determined half-hourly by a computer model called the Market Clearing Engine (MCE) for the upcoming dispatch period.

The objective of the MCE is to find a set of dispatch instructions that minimizes the cost of supplying the loads at all nodes and meets the reserve and regulation requirements. This is the minimum cost market dispatch. Several constraints are taken into account by the MCE in the determination of the electricity prices, and they are:

(a) The offers made by dispatchable generators;
(b) Estimated demand at each node on the network;
(c) Reserve and regulation requirements;

[5]See (Energy Market Authority, 2004), *Introduction to Singapore New Electricity Market,* Chapter 4 (Page 4-1) for the details on the design principles of the Wholesale Electricity Market. http://www.ema.gov.sg/Electricity/introduction_to_sg.php

[6]See (Energy Market Authority, 2004), *Introduction to Singapore New Electricity Market,* Chapter 9 for the details on Wholesale Market Operation. http://www.ema.gov.sg/Electricity/introduction_to_sg.php

(d) State of the power system;
(e) Available capacity of plant;
(f) Rate at which plant can change their output;
(g) Relationship between energy production, reserve capacity and regulation capacity for each plant;
(h) Physical limitations on the flows that can occur on the transmission system;
(i) Losses in relation to power system; and
(j) Constraints in relation to system security.

From the above list of constraints, the least-cost design for market clearing does not consider the fuel type or the generation method, and the emissions resulting from the electricity generation. The emissions from electricity generated from nonrenewable energy sources such as the fossil fuels, can have an adverse impact on the environment. To compete in the wholesale market, generation companies have to be cost-efficient. In one instance, PowerSeraya announced the use of Orimulsion in 2002, a fuel reputed to be the dirtiest oil in the world, in a bid to offer cheap electricity.[7] Such economic gains made at the expense of the environment and future generations are not sustainable in the long term.

2.2.2 *Common property characteristics of environmental goods*

Environmental goods may not fall within the scope of the market mechanism, as their market supply and demand are not always well-defined. All marketable goods have two common characteristics: they can be measured in terms of physical quantity, and they can be valued in monetary terms (Thampapillai, 2002).

Environmental goods such as clean air do not have well-defined property rights, and operating in the market requires those property rights to be assignable. Mechanisms addressing market and policy failures by Government include:

- Command and control regulations that affect quantities, for example, standards, bans, permits, quotas;[8]

[7]Proposal by The New Brunswick Power Commission, Page 4. http://www.gnb.ca/0009/0377/0002/0007-e.pdf.
[8]Permits and quotas do not strictly belong to command and control mechanisms as they may be market-based mechanisms in some instances such as tradable emission permits (TEPs).

- Market-based instructions that affect prices, such as taxes, user fees, subsidies;
- Creating markets, for example, property rights, tradable permits, international offset systems; and
- Engaging the public, including public participation, information disclosure (World Development Report, 2003).

2.2.3 *Short-term energy contracts*

In the regulated market prior to the liberalization of the electricity market, the planning horizon is generally long-term and energy costs are based on cost-recovery. The NEMS is characterized by short-term energy contracts where the consumers would typically enter into a 12 to 24 month contract. This market characteristic will tend to generate high volumes of short term contracts where generation companies compete to provide good prices in the short term. Environmental effects, however, may only be known in the longer run. The increased competition, brought about by the deregulation of the market, may have implications on the environment and lead to the situation where environmentally friendly and sustainable technologies may be driven out of the market by cheaper and non-sustainable technologies such as fossil fuels.[9] This impedes the development of renewables for introduction into the market.

2.3 *Implications of the inadequacy of the NEMS on the environment*

As discussed in the previous section, the inadequacy of the new market in terms of internalizing the environmental costs of electricity generation, which favors the least cost of production, has implications for the environment. Greenhouse gas emissions from burning fossil fuels are threats to the global climate. Reported effects of a changing global climate include the increasing annual average surface temperature of the earth (also known as 'the greenhouse effect'), the depleting ozone layer, the rising sea levels, the threats to biodiversity and natural catastrophes owing to climate change. Such environmental effects can have an adverse impact on the survival of all forms of life on earth. Radical changes to the current energy

[9]Development and application of an optimizing energy and material flow model for strategic planning of energy utilities. www.ise.nus.edu.sg/proceedings/apors2000/fullpapers/15-01.htm

consumption patterns, for instance the change from the use of fossil fuels to renewables, are necessary to avert these environmental problems and potential disasters.[10]

The examination of the new market structure and market rules has revealed an inadequacy of the NEMS in internalizing the environmental issues of electricity generation. In the following section, a framework for a green energy policy is proposed for the development of RES-E as a main solution to promote a sustainable electricity industry and to avoid further environmental problems.

2.4 *Liberalization of the singapore electricity market and RES-E*

The liberalization of the Singapore electricity market could provide an opportunity for power producers to develop RES-E capacity in the electricity generation industry. In a liberalized electricity market, independent power producers (IPPs) are able to generate electricity, to compete with the incumbent generation companies. Prior to the liberalization of the electricity market, IPPs were not allowed to compete in the vertically integrated, regulated and state-controlled electricity industry.

Empowered with the freedom to choose their electricity retailer, energy consumers may choose to purchase electricity generated from renewable energy sources. Power producers and retailers can differentiate the electricity they produce by using renewable energy sources. Through green power marketing, RES-E producers and retailers can advocate cleaner air and a greener environment for the community. Policies that support RES-E would encourage power producers to increase the production of RES-E, and stimulate a greater level of green power marketing activity from the power producers and voluntary demand from the energy consumers on top of their mandatory renewable obligations (if any).

3. Principles of the Policy Design

3.1 *Vision and policy objectives of Singapore green energy policy*

The vision of the proposed Singapore green energy policy is to achieve a green environment and sustainable energy future.

[10]Climate change is a complex and controversial issue and the question of how human activities are changing the global climate is often debated.

The policy objectives are to promote the least-cost technological development of renewables through the use of RES-E support schemes and eventually, to develop a national TGC market as part of the international TGC System.

3.2 *Criteria for the requirements of Singapore green energy policy*

To achieve the policy objectives stated above, the green energy policy should meet the following requirements:

(a) Cost-effective ways of deploying renewables;
(b) Non-discriminating and transparent market for all participants; and
(c) Guaranteed returns for renewable energy investors.

(a) Cost-Effective Way of Deploying Renewables
The promotion of renewables should be implemented using the most cost-effective method. The RES-E support scheme should constantly motivate the producers to search for more cost-effective technologies to generate electricity from renewables. Another consideration is the cost of implementing the RES-E support scheme, which should be kept to the least possible.

(b) Non-Discriminating and Transparent Market
To establish an efficient TGC market, a non-discriminating and transparent market should be available to all participants.

(c) Guaranteed Returns for Renewable Energy Investors
As the RES-E market is new in Singapore and the necessary investments may be perceived as high risk in nature, with uncertain returns, the RES-E support scheme should offer guaranteed returns for electricity generated from renewables as a way of attracting the investors.

3.3 *A step-wise approach*

The development of efficient markets for green electricity is a complex process, faced with an uncertain commercial market or a missing commercial market, and the conflict between the goals of the renewable energy investor and those of the government. The renewable energy investors' interest in the short term is contradictory to the government's long-term plan for the development of sustainable energy sources. Leveraging on the RES-E experiences of other countries, a step-wise approach to the development of the green energy policy framework is proposed to meet the energy demands, the challenges and the complexity of the market.

The stepwise approach comprises the following phases of the development of the RES-E market, which is measured in terms of the percentage of RES-E capacity installed. The level of RES-E capacity installed:

- Phase 1 (P1): From No RES-E Market.
- Phase 2 (P2): Transition Phase — from No RES-E Market to a National RES-E Market.
- Phase 3 (P3): National RES-E Market.
- Phase 4 (P4): International RES-E Market.

3.4 *Rationale of the step-wise approach*

The stepwise approach is a cautious and gradual approach to encourage the development of renewable energy, and its introduction into Singapore's electricity market, which currently has no electricity generated from renewable energy sources, though there are some co-generation power plants. In some countries that have implemented TGC Systems, cogeneration power plants qualify as eligible RES-E sources. For example, in Wallonia, Belgium, co-generation power plants qualify as eligible RES-E sources under the green certificate system: every 3kWh of electricity generated from these co-generation power plants leads to the issuance of one green certificate (OECD, 2001).

Phase 1: From No RES-E Market (From 0% to P1% RES-E target)

In Phase 1, most RES-E technologies have not yet reached mature development stage, and are rather costly. Hence the RES-E technologies generally do not attract investments and are unable to compete under market conditions with conventional fossil fuel power plants. The FIM is selected as the RES-E support scheme for the initial phase as it provides certainty to investors with the guaranteed electricity tariffs. Evidence of the success of the FIM can be seen from the fast and significant results in RES-E development by the EU. The greatest successes in promoting the use of electricity produced from renewable energy sources have been realised by the application of the feed-in system in Denmark, Germany and Spain (Meyer, 2003). Thus, if the highest priority is given to fulfilling goals for the penetration of renewable energy sources, the natural choice would be to rely on the FIM. The burden on the government budget, which is the main weakness of the FIM (and also the Tender System), as discussed in the preceding sections, for instance the Danish case, can be reduced by sharing the premium tariff among electricity consumers. Competitive features may be introduced by using benchmarking principles through the adjustment of the feed-in tariffs at regular intervals by monitoring the advancements and the costs of the most competitive technologies in the market. This will also introduce an element of competition in the FIM, and overcome the problem that the feed-in system does not fully conform to the principles of market competition (Meyer, 2003).

Investments in renewable energy power plants are costly and require a longer horizon than that in conventional power plants using fossil fuels. As the renewables market and RES-E support schemes are also relatively new in Singapore, the potential investor may not have all the information of this new market for decision-making and the experience in operating in the renewables market. Hence, potential investors of RES-E may face some level of price uncertainty and risks if the horizon for the development of the RES-E support is not known. Also, prices of certificates may fluctuate when the supply of certificates is low or when the demand for certificates increases sharply, particularly during the time of proof or the end of the compliance period. In addition, climatic fluctuations could affect the supply of certificates and result in price fluctuations. Renewables such as wind or hydro may be dependent on climatic changes. In low wind seasons, the supply of green electricity from a wind power plant may be low, hence this could lead to a decrease in the supply of certificates in the market. As a result, the price may rise sharply. Flexible mechanisms such as banking and borrowing can be implemented to stabilize the market. To encourage the potential investor to take the first step, the FIM is proposed for Phase 1 as it could provide price stability to the investor over a certain period of time. The fixed premium feed-in tariffs also help to subsidize the costs of renewable energy generation. Feed-in tariffs differ, and the degree of the difference depends on the type of RES-E. This is to prevent windfall profits on renewable power plants that are already cost-effective, even without the support of the RES-E scheme.

Phase 2: Transition Phase — from No RES-E Market to a National RES-E Market (From P1% to P2% RES-E target)

A TGC System with a Centralised Tender System is proposed for Phase 2 because of the problems related to market competition among technologies at different stages of development. These problems may be reduced by combining the TGC System with a Centralised Tender System. One such problem is the fairness of competition among renewable technologies at different stages of development. If an open market with free competition among different technologies were created, wind power would probably dominate the market. Solar electricity would not stand a chance while biomass and small hydro may be competitive in some cases and capture only limited market share. A market that does not take into account the technological development of different types of renewables is not optimal for the long-term promotion of renewable capacity and production. Hence, a TGC market is employed to promote the more mature renewable technologies while the less mature renewable technologies are supported by the Tender System (Meyer, 2003).

Phase 3: A National RES-E Market (From P2% to P3% RES-E target)

Once the specified RES-E target of Phase 2 is achieved, the Tender System would be abolished and Phase 3 would commence. At Phase 3, the RES-E support scheme is the national TGC system and all eligible RES-E technologies are near maturity and are able to compete in the national RES-E market based on a TGC system.

Phase 4: An International RES-E Market (From P3% to P4% RES-E target)

The implementation of an international TGC system is still under discussion in the EU. The likelihood of a community support framework with harmonized rules may be realized, but not before 2005.

3.5 *Key components of the framework for Singapore green energy policy*

In developing the green energy policy, it is important to identify the following key components: Eligible Renewable Energy Sources; RES-E Targets; Liable Entities; Legal Framework; Regulatory Agencies and Bodies; RES-E Fund; and Market Structure.

3.5.1 *Eligible renewable energy sources*

It is important to determine the renewable energy sources that are eligible for development under the proposed programme. For instance, it is suggested that large hydros (that are already competitive in the electricity market) are not to be eligible renewable energy sources. Otherwise, the payments from the certificates would be purely windfall for the large hydro-electric power generators that would simply free-ride on the system.

Hence, it is recommended that eligible renewable energy should include sources such as biomass, solar, photovoltaics, tidal, geothermal and small wind power plants that are less mature technologies. These renewable technologies are rather costly compared to conventional power plants whose short-term returns make them more attractive to investors.

In the case of Singapore, the renewable energy sources can include solar, fuel cell and wind technologies. In the recent years, there has been growing interest in developing renewable energy capabilities in Singapore.

In a recent press release (The Straits Times, 2004), the Minister of State (National Development), Dr. Vivian Balakrishnan noted that one should not rule out the role of ***solar*** or other renewable energy sources and investors who are able to be commercially competitive using solar energy. Such firms would be welcome to enter the electricity market as power producers. He also believed that greater efficiency or volatility of oil prices might make solar energy more attractive in future. Industry players have also indicated the potential of renewable energy sources to lighten 'peak hour' loads. Solar panels could be integrated into buildings as roofs or facades. An example is a building in the new Biopolis which is integrated with solar panels by Sharp.

There has been a growing range of research and development (R&D) on ***fuel cell*** science and technology, and demonstration activities in Singapore such as the SINERGY program which involves activities such as the testing and demonstration of small fuel cell stationary power generation systems in public housing estates. The Agency for Science Technology and Research (A*STAR), Economic Development Board and Nanyang Technological University of Singapore (NTU) are collaborating in R&D under the Fuel Cell Strategic Research Programme at the NTU (Ho *et al.*, 2004).

There is also a potential of tapping the ***wind*** resources in Singapore and on the off-shore islands of Singapore. In 1999, a team of academic staff members and students from Singapore Polytechnic built and installed the first wind electric turbine to demonstrate the feasibility of the technology in Singapore. The project involved the installation of a 1 kW wind turbine system that is combined with the existing solar photovoltaic cells to supply electricity for the lighting of a five-story administration building of the polytechnic. The team has demonstrated the possibility of installing wind turbines on the rooftops of the high buildings in Singapore and on the off-shore islands to generate energy to satisfy a potential 10% of Singapore's energy needs. The project comprised a wind measurement program that started in 1997 and samples of the wind speed had been taken across Singapore and at the polytechnic (AWEA, 1999).

Singapore has several small-scale ***biomass*** power plants. One such plant is a 530 kW unit operated by a local waste-disposal company, ECO Special Waste Management Pte Ltd. In January 2004, a local company, ecoWise Holdings Ltd announced a collaboration between its wholly-owned subsidiary, Bee Joo Industries Pte Ltd, with Vyncke Energietechniek NV and Jebsen & Jessen Process Engineering (M) Sdn Bhd for the supply, installation and commissioning of a 1 MW biomass fired cogeneration plant (Platts UDI Country Profile, 22 July 2004).

It is suggested that incineration plants operated by the National Environment Agency (NEA) should not be considered as qualifying as eligible energy sources. Due to the emissions from the combustion of waste, incineration plants are generally not considered as environmentally-friendly. However, incineration plants for organic waste can qualify as an eligible renewable energy source due to the negligible emissions.

Gradually, renewable energy resources are increasingly being used for power generation in Singapore. However, the use of renewable energy resources is still limited due to the high costs of RES-E technologies. The RES-E technologies are unable to compete with the conventional power plants in the cost-driven liberalized electricity market. The implementation of RES-E support schemes will provide the financial support to generate additional payments for the promotion of RES-E.

3.5.1.1. Segmentation of RES-E technologies

Each type of RES-E technology has a different price elasticity of demand. The mature RES-E technologies tend to have lower price elasticity than the less mature ones. Three examples of RES-E technologies by the stage of development are mature, marginal and less mature technologies and in the order of increasing price elasticity are wind-power, biomass and photovoltaic power (PV). If a single certificate price is applied, this price must be raised significantly to bring the marginal technology into the market. It creates significant price effect i.e. a price increase (and demand reduction) and transfers large amounts of cash from the consumers to the RES-E generators. Some of this transfer is to cover part of the RES-E costs, but an increasing share is rents received by the sub-marginal RES-E generators. The marginal cost of non-mature technologies is rather high so if a one-price-for-all-technologies is applied, it would give windfall gains to mature technologies. This "free-riding" by the mature technologies would make the

TGC System very expensive with the end-users having to pay all the bills; thus, the system would lose its credibility and this would thwart efforts towards attaining the objective of increasing the penetration of RES-E into the electricity market.

Recommended solution

The TGC market is proposed to be segmented according to the type of technology, for example, wind-power, hydro, biomass, and E-solar in Flanders.[11] For every technology group t, one has to study in detail the TGC-supply curve and to find the (k_j^i, F_j^i) pairs that help to force the target market development of the technology over time (years $j = 1, \ldots, n$). The certificate system is defined by two parameters, k and F where k is defined as the quota and F is the fine or penalty imposed for non-compliance. The certificate system is also influenced by the volume of certificates, V. Figures 19.1 and 19.2 illustrate these relationships (Schageffer *et al.*, 2000).

As suggested by Verbruggen (2004), the diversity of RES-E technologies can be addressed by developing a policy in several ways. One way is to diversify and to fine-tune the more direct support measures for the various renewable technologies. When a uniform TGC is maintained, these measures promote RES-E and the TGC System can be used as a closing mechanism and as a source of funding. A second way is to assign a different number of TGCs per MWh of output along the type of technology (project)

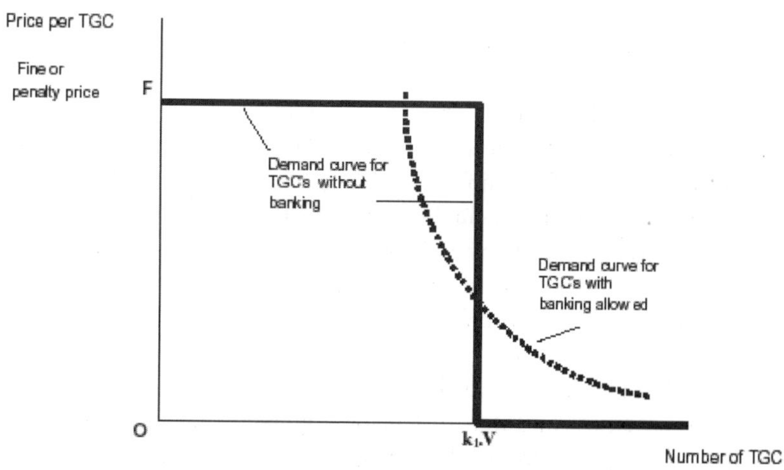

Fig. 19.1. Demand for tradable green certificates (TGCs) by suppliers.

[11]E-solar refers to a mobile solar-powered micro-utility for distributed generation, largely for home and small-office uses.

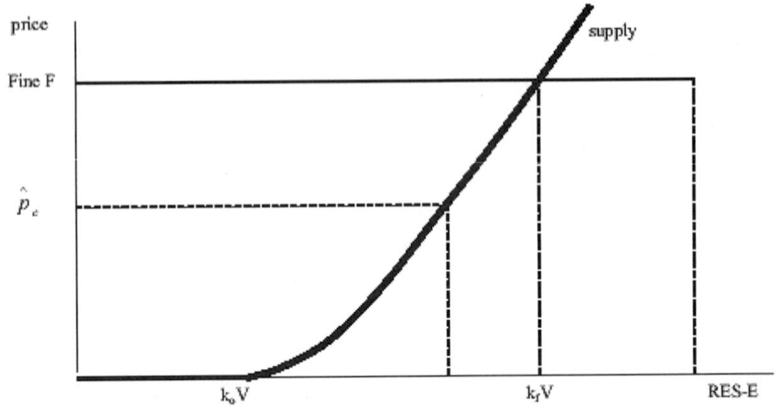

Fig. 19.2. Market equilibrium in the TGC market changes with set k values.

that generates RES-E. When the optimal weight of every individual RES-E technology is found, a single TGC market can be maintained. A third way is to segment the certificate market along RES-E technologies with common cost properties, and to estimate RES-E technologies with common cost properties, and their development over time.

An important point to note is that though clear market segmentation has the advantage that policy makers cannot avoid the effort of fully understanding the sector and the market, it has the disadvantage that the market in every segment becomes thin and so does the possibility of trading. Thus, the success of the TGC system hinges on an in-depth study of the cost structures in the various submarkets (for instance, RES-E segments) and the careful setting of the parameters, k and F.

In a similar way, the feed-in tariffs should be different, depending on the type of RES. This differentiation (or segmentation) of RES-E is important in order to allocate higher subsidies for less mature technologies, and less subsidies for the marginally mature technologies. Different tariffs are applicable to different bands of RES-E technologies. Feed-in tariffs can be established based on estimates of the technological costs and investors' profit margins. For instance, under the German feed-in law of Renewable Energy Act (EEG), feed-in tariffs vary, depending on the type of RES-E technologies as shown in Table 19.1 below (Janssen, 2004).

Similarly, in Singapore, feed-in tariffs can be established for each type of RES-E technology, based on estimates of the technological costs and investors' profit margins. The feed-in tariffs can be gradually reduced over time as the RES-E technologies mature. Such tariffs are also referred to as stepped feed-in tariffs where the feed-in tariffs are decreased over time

Table 19.1. Feed-in tariffs corresponding to different RES-E bands of technologies under the German feed-in law.

Type of RES-E technology	Feed-in tariffs (cents per kWh)
Hydro Power	6.63–7.65
Biogas	6.63–7.65
Biomass	8.5–10
Geothermal Energy	7.14–8.03
Wind Energy	6–8.9
Photovoltaics	45.7

according to the expected learning curve and economies of scale of RES-E technologies (Hass, 2004).

For the Tender System, a minimum price on the basis of technology costs should be set. Technology costs should include an estimate of the investors' costs and profit margins. The minimum price varies for different types of RES-E technologies as the cost of each RES-E technology is different. Hence, a higher minimum price is set for a less mature RES-E technology that requires higher capital costs. This differentiation (or segmentation) of RES-E is important to allocate higher subsidies for less mature technologies, and less subsidies for the marginally mature technologies.

An official method for calculating the minimum price should be established with the necessary information on each RES-E technology and investors' profit margins. For instance, information such as average wind speed and installation costs of wind technology is required for the calculation of minimum price for wind power (Schaeffer *et al.*, 2000). For solar, information should include solar intensity, costs of solar panels and installation costs of solar technology. This calculation is similar to the feed-in tariffs which have been used in countries such as Germany.

3.5.2 *RES-E targets or quota, k*

RES-E targets with specific deadlines should be set and incorporated into the new energy legislation. The targets can be expressed in terms of the percentage of renewable electricity out of total electricity supplied. Reference can be made to the United Nations Framework Convention on Climate Change (UNFCCC) when setting the RES-E targets.

3.5.2.1 Size of quota, k (or RES-E target)

The regulating authorities have to establish the quota k in such a way that an appropriate development of new RES-E capacity will take place

(Morthorst, 2000). If k is set too low, the equilibrium GC price may be too low to secure the development of new RES-E capacity. If k is set too high, the optimal RES-E capacity will be lower than prescribed by the quota. The certificate price plays the important role of an incentive tax when it is stimulating the development of the RES-E market (Verbruggen, 2004).

It is recommended that the quota should be established in such a way that an appropriate development of new RES-E capacity will take place (Morthorst, 2000).

3.5.3 *Liable entities*

To encourage RES-E investments, the energy consumers are the liable entities for all the four phases of the implementation. Placing a mandatory obligation on the generation companies may deter potential RES-E investors from investing in the Singapore electricity market. The other advantages of making the energy consumers the liable entities include market liquidity, market transparency, and public awareness of the environmental benefits of RES-E. The high number of energy consumers would provide the liquidity and transparency for the TGC market. At the same time, public education and participation is achieved by educating the consumers on the benefits of renewable energy and the environmental issues and gaining their support in building a sustainable energy future.

In Phase 1, the FIM is financed by sharing or distributing the premium feed-in tariffs among the energy consumers. The financing mechanism of the FIM is implemented by imposing a tax on energy consumers based on their level of electricity consumption. Such a scheme has been implemented in Spain (Meyer, 2003). Since there is currently no electricity generated from renewable energy sources, the use of the FIM is to develop a certain renewable energy capacity in the generation industry as a first step towards building a TGC market.

In Phase 2, two schemes are implemented where the Tender System is applicable to less mature RES-E technologies and a national TGC system is developed for the more mature RES-E technologies. The Tender System is introduced to provide more financial assistance to less mature RES-E technologies which tend to be more costly as compared to more mature RES-E technologies. The Tender System also provides guaranteed returns to investors by securing a fixed price for the electricity generated based on the less mature RES-E technologies. The Tender System is financed through the same way as the FIM during Phase 1 where a tax is imposed on the energy consumers. The tax collected from the energy consumers will finance the less mature RES-E technologies through the Tender System in Phase 2. At this point, the quantum of tax payable by the energy consumers may be lower since a mandatory renewable obligation would be imposed on the consumers. The tax only applies to the consumption of electricity generated from non-renewables. At the same time, the TGC market is developed to generate additional payments for more mature RES-E technologies and at the same time, promote the cost-effective development of these RES-E technologies through a competitive market.

In Phases 3 and 4, the TGC market is developed at the national level and international level respectively.

3.5.4 *Legal framework*

A legal framework for the green energy policy should be passed. The objective of the energy legislation is to ensure the proper governance, implementation and compliance enforcement of the green energy policy. The new energy act should establish the institutional structure, compliance process, and obligations of the various participants, including RES-E investors, regulatory agencies, energy consumers and energy retailers. The new energy act should also specify the regulations of the reforms in the proposed four phases and indicates the implementing dates and the processes involved.

3.5.5 *Regulatory agencies and bodies*

As the market regulator for the liberalized electricity market, the EMA is the appropriate governmental agency to be the regulator for the various RES-E support schemes and subsequently, the TGC market. As it has direct access to market information and interaction with market participants, the EMC is proposed to be the body issuing the green certificates.

3.5.6 *RES-E fund*

An RES-E Fund is set up and administered by EMA. The objective of the RES-E Fund is to support RES-E development such as the financing of new RES-E installations or technological improvements by RES-E producers. The source of the RES-E Fund is the penalties paid by non-compliant entities for failing to meet their renewable energy obligations. Regulations and guidelines of administering and applying the RES-E Fund would be specified in the energy legislation.

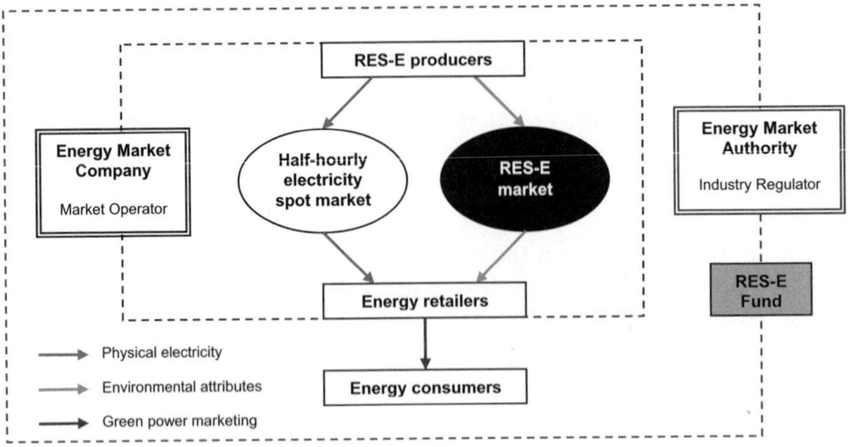

Fig. 19.3. Market structure of the NEMS with Singapore green energy policy.

3.5.7 *Market structure*

Figure 19.3 shows the introduction of an RES-E market into NEMS. The entities comprise the industry regulator (EMA), the market operator (EMC), RES-E producers, energy retailers and energy consumers. As the market operator of the half-hourly electricity spot market, the responsibility of EMC is proposed to include the operation of the TGC market as well. EMA, as the industry regulator, controls and regulates the electricity industry and its participants.

3.5.8 *Other issues*

Besides the exclusion of power plants that are already receiving some form of subsidies from the government, other issues can include the provision of information resources to ensure that the parties involved and the general public have access to information on the policy, legislation and procedures.

The Centralised Tender System that was implemented in Phase 2 may be abolished once the less mature renewable energy sources are cost-effective and are able to compete in the TGC market with all other forms of energy sources. Trading of green certificates should be active and the certificate price should ideally be a single price (with no segmentation of RES-E), indicating fair competition in the market.

3.5.9 *Additional components for the tradable green certificates system*

As the TGC system is a relatively more complicated system as compared to the FIM and the Tender System, there are additional components to establish. These are now considered.

(a) Renewable Obligation

There are various ways of computing the Renewable Obligation. Four possible methods are: (1) a percentage obligation in kWh consumed; (2) a percentage obligation in MW installed capacity; (3) an absolute-amount in kWh; and (4) an absolute-amount in MW installed capacity. Option 1 is proposed for use in Singapore because it is simple to compute, and has been widely applied in the countries that have implemented TGC Systems. To provide security to investors, milestones for the RES-E targets are defined for specific years with a steady growth rate in each period, such as: 3% growth rate in the period of 2005 to 2010; 5% growth rate in 2011 to 2015; and 10% growth rate in the period of 2016 to 2020.

(b) Reference of Renewable Obligation

In Singapore, the reference period should be synchronized with the compliance period in order to simplify the compliance process.

(c) Continuity of Renewable Obligation Over Time

The manner in which the renewable obligations will evolve over the years is important in reducing uncertainty in the market. Eventually, the renewable energy technologies should become competitive with those in the conventional power plants and the targeted percentage of installed capacity in renewables should be reached. The question then would be whether the renewable obligation should be continued or ceased. The continuity of the obligation can be handled in various ways such as (i) fixed sunset date; or (ii) self-sunset date. The second option is proposed in the case of Singapore as it gives more flexibility in policy-making and provides relatively greater security to investors. The flexibility in policy-making also avoids sudden changes in laws and regulations which can hinder the promotion of the TGC market.

(d) Certificates

A green certificate should include the following information: (i) A unique identification number; (ii) Reference to the issuing authority; (iii) Reference to the specific RES-E plant; (iv) Date of issue; (v) Energy source and technology; (vi) Country of RES-E origin; (vii) Indication of public investment or production subsidy (if relevant); (viii) Installed capacity; and (ix) Particulars of the owner holding the certificate.

To address the equity issue that arises from the different stages of development among the RES-E technologies, the number of certificates price will vary for each type of RES-E technology. Specific guidelines on the certificate price of each type of RES-E should be formally established. Hence, for the same amount of energy generated (kWh), the less mature technologies will receive a higher number of certificates than more mature technologies.

There are several ways to compute the green certificate price, for instance the basis of (i) costs; (ii) environmental benefits; or (iii) other consideration such as industrial policy (Schaeffer *et al.*, 2000).

The realized price of green certificates must be below the penalty payment. Hence, the penalty must be set at a level higher than the possible price of certificates to induce the liable entities to meet their renewable obligation by generating renewable electricity than simply paying the penalty.

To prevent price volatility due to over-supply or under-demand situations, it may be better to allow a validity period that extends beyond the year of production. However, it is difficult to conclude, due to limited experience of the certificate system, if certificates without expiry dates would be a better option. Hence, flexible mechanisms such as banking and borrowing should be introduced to prevent price volatility due to over-supply or under-demand situations.

A large certificate size, say 100 MWh, could displace some small producers. It may be better to have a smaller certificate size where smaller producers can participate in the certificate market and producers can have the flexibility to aggregate the certificates. However, small certificates may cause higher administrative costs. The size of certificate is debatable and there is no conclusive evidence of an optimal certificate size. From a comparison of the TGC System conducted by Ong *et al.* (2005), the size of the green certificate varies across the countries that have implemented the TGC System. It ranges from units of 100 kWh to 1 MWh. As the TGC market involves energy consumers from the residential sector, a certificate size of 100 kWh is proposed for Singapore. Generally, a small residential energy consumer, such as a household staying in a Housing Development Board (HDB) flat, would consume below 1,000 kWh of electricity per month.[12] based on data taken from the Department of Statistics, Singapore which shows that in 2000, a HDB household spent between $32 and $60 per month, depending on the size of the HDB flat. Based on a low-tension domestic electricity tariff of 16 cents per kWh, the monthly consumption of an HDB household ranges from approximately 200 kWh to 375 kWh.[13] The denomination of the green certificates in multiples of 100kWh would facilitate the purchase of green certificate by these small residential energy consumers to meet their renewable obligations.

(e) Banking

Banking and borrowing are recommended to prevent price volatility. As the TGC market is relatively new, there is no conclusive evidence whether a limited validity period or no expiry date for banking or borrowing will improve the functioning of the TGC market. Hence, the validity period can be implemented in steps, say 3 years for validity (subject to change) and borrowing could be allowed up to 2 years. The operation of the green market can then be monitored and the validity periods for banking and borrowing can then be adjusted accordingly.

(f) Compliance Process

The compliance period is proposed to be one year. To ensure simplicity and to reduce the administrative costs, the compliance period should coincide with the reference period.

Two options are considered for the time of proof: (i) at the end of the compliance period; and (ii) several times during the compliance period. To reduce administrative costs, the first option is proposed for Singapore. To avoid a situation where market trading is congregated around the time of proof, market mechanisms such as banking and borrowing should be introduced to improve market liquidity and transparency.

Compliance control is required in order to enforce the rules and penalties to ensure that the liable entities conform to all the rules and regulations under the energy act. In designing compliance control, the following key elements are proposed: (i) control bodies; (ii) control mechanism; (iii) penalty; and (iv) use of penalty money.

---

[12]A HDB flat is a type of high-rise public housing provided by Housing Development Board, a government agency responsible for the provision of subsidized, low-cost housing for the public. See www.hdb.gov.sg

[13]The information on expenditure on electricity consumption can be found at the Statistics Singapore website at http://www.singstat.gov.sg/papers/people.html

(i) Control Bodies

The control bodies that are responsible for implementing the compliance process shall be the regulatory bodies established under EMA.

(ii) Control Mechanism

The proposed control mechanism will be implemented in the steps now outlined.

Step 1: The control body, EMA, will calculate the renewable obligation for each liable entity based on the RES-E targets set in the reference/compliance period.

Step 2: EMA will then notify the liable entities of their renewable obligation and the details of what they must comply with, such as the compliance period and the time of proof.

Step 3: At the time of proof, EMA will compare the certificates submitted by each liable entity with the respective renewable obligation assigned for the compliance period in order to check for compliance.

Step 4: In the cases where the liable entities have failed to meet the renewable obligation for the first time, EMA will issue a warning to them to fulfill their obligations within the subsequent three months. If they fail to comply with the renewable obligation after the initial warning, EMA will impose a penalty on the liable entities concerned according to the outstanding number of certificates multiplied by the penalty rate.

In cases where the certificates submitted by the liable entities are in excess of their renewable obligations, the excess certificates will be returned to the liable entities and they can keep them to meet future renewable obligations (banking) or sell them in the TGC market (trading).

(iii) Penalty

There are various options of imposing the penalty such as: (1) fixed price per certificate or per kWh; and (2) x percentage above 1 times the average market price, i.e. $(1 + x)$ times the average market price.

To encourage electricity consumers to purchase the required percentage of electricity supply from renewable energy sources to meet their renewable obligation instead of just paying the penalty, the penalty imposed should be higher than certificate price. Hence, option 2 is recommended for Singapore as the penalty will always be higher than the average market price.

(iv) Use of Penalty Money

The funds generated from the payment of the penalty can be used for various purposes such as contribution towards the renewable energy fund, income for the regulatory bodies to be used in maintaining the renewables market or to be redistributed to the complying liable entities. In the case of Singapore, it is proposed that the money from the penalty will go into the RES-E fund to finance new renewables installations and R&D.

(g) RES-E Fund

The RES-E fund that was established in Phase 1 would continue to be administered by the IB, under EMA. Payments for the certificates by the electricity consumers will be passed to renewable power plants according to the maturity of the technologies when the less mature technologies will receive higher payments.

3.5.10 *Implementation plan*

Table 19.2 shows a summary of the implementation of the four phases to achieve the targeted percentage of RES-E over time.

3.5.11 *Evaluation framework*

Various methods are used in the evaluation of economic instruments, such as the RES-E schemes. In general, evaluation methods are classified into three main categories: (1) theoretical arguments; (2) ex ante assessments; and (3) ex post evidence (OECD, 1997).

Initial advocacy of economic instruments can be in the form of theoretical or conceptual arguments which evaluate the advantages and disadvantages of applying economic instruments to the problem/s concerned. This study is an instance of theoretical arguments. A cost and benefit analysis of economic instruments and conventional command and control regulations may be conducted by building a theoretical model and simulating the behavior of cost-minimizing firms based on a set of assumptions about polluter behavior.

Ex ante assessments of likely outcomes of the proposed economic instruments may be conducted using data that are relevant to the environment (such as polluter behavior or emission levels), electricity prices and consumption levels. Ex post evidence is obtained from evaluations that are carried out when the implementation of the economic instruments is

Table 19.2. Implementation plan of the Singapore green energy policy.

Key components	Phase			
	Phase 1	Phase 2	Phase 3	Phase 4
Development of RES-E technologies	Costly and not mature	Different levels of development (less mature and near mature RES-E technologies)	Mature and competitive in national TGC market	Mature and competitive in international TGC market
RES-E Support Scheme(s)	Feed-in Model	Centralised Tender System (less mature RES-E technologies) National TGC System (near mature RES-E technologies)	National TGC System	International TGC System
Aggregate RES-E targets (%) to be achieved by the end of each phase	P1%	P2%	P3%	P4%
Eligible renewable sources	Less mature RES-E technologies	Less mature and near mature RES-E technologies	All RES-E technologies	All RES-E technologies
Eligible entities	RES-E power producers			
Liable entities	Energy consumers			
Legal framework	Energy act			
Industry regulator and market operator	EMA and EMC			
Market structure	See section 5.9.7			
Evaluation	See section 5.12			
Monitoring and feedback	See section 5.13			

completed in order to assess the effectiveness of the economic instruments in meeting the planned objectives.

3.5.12 *Monitoring and feedback*

EMA should monitor the performance of the RES-E support schemes on a routine basis. A closed-loop feedback channel should be provided for the market participants and liable entities to give feedback to EMA. This is another channel for tracking the progress of the RES-E support schemes to identify and prevent or solve any potential or existing problems.

3.5.13 *International tradable green certificates system*

With the current discussions on the harmonization of domestic TGC systems in the EU countries, it is possible that an international TGC System will be established in future.

The main advantage of harmonizing domestic TGC systems is cost-effectiveness. Under competitive market conditions, TGC Systems will promote the least-cost technological development of RES-E. Hence, a broader geographical basis will allow more efficiency gains, as marginal costs are likely to differ across countries with domestic regimes. For example, Germany has exploited a major portion of its wind potential while the UK's wind potential is relatively less developed (Menanteau *et al.*, 2003). The market mechanism introduces competition which tends to drive electricity prices down. To remain competitive, RES-E producers will select sites with the highest renewable energy potential and require the least costs in order to offer competitively-priced RES-E and maximize their profits. Hence, renewable energy capacity will be developed at the locations with the highest renewable energy potential and the least costs.

However, there are arguments against harmonizing TGC Systems. First, governments that seek to develop domestic RES-E may not favor the harmonizing of systems across boundaries as their own national systems may be used to promote the development of RES-E outside their domestic markets, at the expense of their taxpayers, and to the advantage of foreign power plants. Second, as the imported certificates are proof of reduced emissions elsewhere, the importing country will not benefit from such local benefits such as local improvement in environmental quality. Finally, the renewable energy generation industry is an emerging one which can create employment opportunities and boost national economic development.

The challenge in harmonizing domestic TGC System arises from the differences in the domestic TGC Systems such as eligible sources, penalty levels and subsidies. First, the eligible renewable energy sources are different from country to country. This may lead to the difficulty of importing electricity from sources deemed eligible in the exporting country, but not accepted as eligible sources in the importing one. Second, the different penalty levels across domestic certificate systems may lead to instances of non-compliance in the country with the lowest penalty level. One potential problem is where a liable entity could transfer a green certificate to a country to sell it at a price higher than the penalty. Finally, the level of subsidies differ in each country, hence this will lead to a distortion of competition. The subsidy in an exporting country would benefit the liable entities in the importing countries at the expense of the national budget of the exporting country.[14]

4. Conclusion

In this chapter, the establishment of the NEMS was reviewed and its adequacy in internalizing environmental costs was investigated by analyzing the rules of the operation of the NEMS. The findings show that the least-cost design for market clearing does not consider the fuel type or generation method, and the emissions resulting from the generation of electricity. Since the NEMS is cost-driven, the absence of internalization of the environmental costs of electricity generation can have an adverse impact on the environment if generation companies turn to cheap fossil fuels as their energy sources to produce competitively-priced electricity. A green energy policy is proposed as a solution to the inadequacy of the NEMS in internalizing environmental costs of electricity generation. The liberalization of the electricity market is also seen as an opportunity for renewables to enter the electricity market, which was previously vertically-regulated and was only open to the incumbent state-owned utilities.

The vision and objectives of Singapore's green energy policy are formulated, and its various criteria established. A step-wise approach, consisting of four phases, is proposed for the implementation of the green energy policy and the rationale of the step-wise approach is to adopt a cautious and gradual approach to promote renewable energy development into NEMS, which currently has (almost) no electricity generated from

[14]This is a similar argument to a farm subsidy in the World Trade Organisation (WTO).

renewable energy sources. Key components of the green energy policy are: (a) eligible renewable energy sources; (b) RES-E targets or quota (k); (c) liable entities; (d) legal framework; (e) regulatory agencies and bodies; (f) RES-E fund; and (g) market structure. Design of the RES-E support schemes, which are the FIM, the Centralised Tender System, the National TGC System and the International TGC System are proposed. An evaluation framework is proposed to be applied at regular intervals during the implementation of each phase. The objective of the evaluation is to assess the effectiveness of each RES-E support scheme in meeting the objectives of the green energy policy. Finally, it is recommended that monitoring and feedback should be carried out on a routine basis by EMA to identify and prevent or solve any potential or existing problems.

References

American Wind Energy Association. "*Singapore School Builds Country's 1st Wind Turbine*", *Vol. 18, #837–1 March 1999.* http://www.awea.org/default.htm

British Wind Energy Association. See http://www.bwea.com

Energy Market Authority (EMA), Singapore. See http://www.ema.gov.sg

EMA, 2004. Introduction to Singapore New Electricity Market. See http://www.ema.com.sg

Energy Market Company (EMC), Singapore. See http://www.emcsg.com

Hass, R. *et al.*, 2004. How to promote renewable energy systems successfully and effectively. Energy Policy 32, 833–839.

Ho, H. K., Chan, S.H., Jiang, S.P., 2004. Fuel cell research, development and demonstration activities in Singapore. Fuel Cell Bulletin, Volume 2004, Issue 6, 1–20 (June 2004).

Janssen, R., 2004. Promoting Electricity from Renewable Sources — The German Feed-in Law. 'Partners for Africa' Policy Dialogue, 21–23 June 2004, Durban, South Africa. See http://india.shellfoundation.net/docs/p4a-dbn-3-Session2-Janssen.pdf

Menanteau, P., Finon, D., Lamy, M-L., 2003. Prices versus quantities: choosing policies for promoting the development of renewable energy. Energy Policy 31, 799–812.

Meyer, N.I., 2003. European schemes for promoting renewables in liberalized markets. Energy Policy 31, 665–676.

Morthorst, P.E., 2003. National environmental targets and international emission reduction instruments. Energy Policy 31, 73–83.

OECD, 1997. Evaluating Economic Instruments for Environmental Policy. A publication by Organisation for Economic Cooperation and Development (OECD).

OECD, 2001. Implementing Domestic Tradeable Permits — Recent Developments and Future Challenges. A publication by Organisation for Economic Cooperation and Development (OECD).

Ong Hwee Fang, Amanda. *et al.*, 2005. Promoting Renewable Energy in Liberalized Electricity Markets, National University of Singapore.

Platts UDI Country Profile — Singapore. Updated 22 July 2004. http://www.platts.com

Schaeffer, G.J. *et al.*, 2000. Options for design of tradable green certificate systems, ECN-C—00-32, ECN, April 2000. Available at http://www.ecn.nl/docs/library/report/2000/c00032.pdf

Statistics Singapore (Singapore Department of Statistics). See http://www.stats.gov.sg

Thampapilla, D., 2002. Environmental Economics — Concepts, Methods and Policies. Oxford University Press.

The Straits Times Interactive, "*Renewable energy sources could lighten 'peak hours' load*". July 29, 2004. http://straitstimes.asia1.com.sg

United Nations Conference on Trade and Development, 1992. Combating Global Warming: Study on a global system of tradable carbon emission entitlements.

Verbruggen, A., 2004. Tradable green certificates in Flanders (Belgium). Energy Policy 32, 165–176.

World Development Report 2003. The World Bank Group. See http://www.worldbank.org

20

BIOREMEDIATION OF ARSENIC IN SOME SELECTED AGRICULTURAL CROPS IN BANGLADESH

M. F. AKHTER and M. S. UDDIN

One of the most important global environmental toxicants is arsenic. High concentrations have been reported in groundwater in several countries including Bangladesh. It is the largest known mass poisoning in history, with approximately 35–77 million people being exposed to arsenic-contaminated drinking water throughout Bangladesh. Serious diseases are caused by chronic arsenic poisoning; these include cancers, melanosis, hyperkeratosis, restrictive lung disease, Blackfoot disease, gangrene, diabetes mellitus, hypertension and ischemic heart disease. People are directly exposed to arsenic through underground drinking water and indirectly through the food chain. Vegetables and edible crops grown in arsenic contaminated soil may uptake and accumulate significant amount of arsenic in their tissue which is harmful for growth.

Since scarce evidence was found regarding human risk of arsenic poisoning through the food chain, the current investigation aims at reporting the level of arsenic in some common vegetables grown in Sitakund upazila in Chittagong district, Bangladesh. The vegetables studied were tomato, chili, potato, brinjal, bean, spinach, ladies finger, red amaranth, cucumber and cowpea. Mean arsenic concentration in vegetables was 0.133 ppm.

Like many other elements in soil, arsenic is subject to transformations by microorganisms including arbuscular mycorrhizal fungi (AMF). These are associations between fungi and the roots of vascular plants. Part of this study is devoted to analyze the role of mycorrhiza on the bioremediation of arsenic in agricultural crops. The result showed that arbuscular mycorrhizal fungi have their most significant effect on phosphorous uptake and it could protect plant against high accumulation of toxic elements through their root system. The study also examined the single and interactive effect of arsenic and arbuscular mycorrhizal fungi on plant physical growth, arsenic/phosphorous concentration/uptake and arbuscular mycorrhizal fungi mediated effect on arsenic toxicity. Growing tomato with compatible symbionts (mycorrhiza) minimized arsenic toxicity and increased phosphorus concentration.

Introduction

Bioremediation as defined by the American Academy of Microbiology is "the use of living organisms to reduce or eliminate environmental hazards resulting from accumulations of toxic chemicals or other hazardous wastes"

(Gibson and Sayler, 1992). Bioremediation is not a new technology. It is evident that humankind has practiced composting, sewage treatment and fermentation since the beginning of recorded history. All of these processes utilize microbial processes in a degradation process. Indigenous and enhanced micro-organisms have been shown to degrade industrial solvents, heavy metals, polychlorinated biphenyls, explosives and many different agricultural chemicals, including pesticides (Van Hamme, 2004; Maloney, 2001; Gadd, 2001; Sharma *et al.*, 2007).

Bioremediation can be used to degrade concentrated organic contaminants near their source or as a secondary remediation strategy following specific physical or chemical methods. It can also be applied for the sequestration of metals and radio-nuclides through microbial mediated transformation processes as well as for the remediation of large plumes of diluted contaminants that are broadly dispersed in the environment.

Certain pollutants like metals and radio-nuclides cannot be degraded through biological activity, but can be transformed from one chemical form to another, or move out of the soil by plants to above and below ground plant tissue. Some fungi can convert dissolved arsenic and selenium to gaseous forms through methylation, and in this way remove the compound from the soil. Bacteria have been shown to change the oxidation state of heavy metals such as chromium, selenium and mercury as well as radio-nuclides (uranium) by using these as electron donors or acceptors in cellular metabolism.

More than 90% of the world's vascular plant species, including nearly all crop species, form symbiotic associations with arbuscular mycorrhizal fungi (Smith and Read, 1977). In this symbiosis, plants gain access to soil resources from AMF, which in turn gain carbon compounds from plants. These fungi grow both inside and outside plant roots, and greatly expand the surface area, through which plants can gather soil resources (Gianinazzi *et al.*, 1990), especially, if the plants are growing in nutrient deficient and acidic soil (Howeler *et al.*, 1987). They have an important role in bioremediation by increasing plant uptake of P and other poorly mobile micro-nutrients particularly Zn and Cu (Gnekow and Marscher, 1989; O' Keefe and Sylvia, 1991). These fungi are found worldwide in temperate, tropical and arctic regions, except under waterlogged conditions (Bolan, 1991; Smith and Read, 2000). Of the different types of mycorrhizae, the arbuscular mycorrhizal fungi are by far the most widely occurring, and are very important in the improvement of agricultural and horticultural crops and forest trees in hilly areas (Mridha and Xu, 2001).

In Bangladesh, of the many different heavy metals, arsenic is very important in soil and crop contamination due to use of irrigation water from arsenic affected tube wells (Alam and Sattar, 2000; Alam and Rahman, 2003). Arsenate is often the dominant form of arsenic within the soil solution (Porter and Peterson, 1997). Arsenate acts as a phosphate analogue and is transported across the plasma membrane through phosphate co-transport systems (Ullrich-Eberius *et al.*, 1989). Arsenic tolerant ecotypes of plants have been shown to absorb less arsenic through suppression of the phosphate or arsenate uptake system (Meharg and Macnair, 1990).

The interaction between phosphate and arsenate transport strongly suggests that arbuscular mycorrhizal fungi may influence arsenate uptake. Indeed, plants growing on arsenic contaminated soils tend to be mycorrhizal (Meharg and Cairney, 1999; Sharples *et al.*, 2000). Recently, higher growth in mycorrhiza inoculated plants grown in arsenic contaminated soils was recorded in the mycorrhizal laboratory of Chittagong University, Bangladesh (Akhter *et al.*, 2004).

Bangladesh is an agricultural country. Except for the three hilly districts of Bangladesh (Chittagong, Chittagong Hill Tracts and Khagrachari), the whole country is contaminated with ground water arsenic. The groundwater arsenic contaminates surface soils and plants, and thereby arsenic enters the food chain.

Common agricultural crops grown in arsenic contaminated soil may accumulate significant amounts of arsenic in their tissue. Since scarce evidence was found regarding human risk of arsenic poisoning through the food chain, the current investigation aims at reporting the level of arsenic in some common agricultural crops grown in the homestead garden soil irrigated by arsenic laden tube well water. The status of arbuscular mycorrhizal fungi colonization in agricultural crops and diversity of the fungal communities in arsenic affected soils is needed to find whether there is a relationship with arsenic uptake and mycorrhizal association. Since some evidence was also found regarding effective and economical methods to ameliorate arsenic contamination of crops through management of soils, plants, and their associated microorganisms, there might be scope also to find out probable techniques for bioremediation of those agricultural crops.

Considering the above facts the study was carried out with the following objectives:

(a) To assess the status and diversity of arbuscular mycorrhizal fungi colonization in arsenic contaminated soils,
(b) To assess arsenic uptake in selected agricultural crops,
(c) To assess the effect of arsenic on the physical growth of plants inoculated with arbuscular mycorrhizal fungi, and
(d) To assess the role of arbuscular mycorrhizal fungi in arsenic uptake by plants.

Methods

The experimental work related with the present study was divided into two parts. The first part was conducted in Bangladesh (December 2004) and the second part was conducted in Singapore (from April 2005 to June 2005). A field experiment was conducted in an arsenic contaminated site in Bangladesh to record mycorrhizal colonization as well as arsenic concentration in some selected agricultural crops. In the second part, experiments were set up in the laboratory to compare the ability of mycorrhizal and non-mycorrhizal plants to withstand different level of arsenic concentrations.

Site description

The study site was located in Sitakund upazila (sub district) under Chittagong district in Bangladesh. Sitakund is an industrial area at the entrance of Chittagong city. It is surrounded by hills and riverbanks. Approximately 40,000 people live in Sitakund and surrounding areas. Apart from the indigenous people, most of the people drink tube well water (indigenous people drink water from the open streams in the hills). The groundwater of this area is highly contaminated. As a result, people are not only drinking arsenic contaminated ground water, but also growing crops and vegetables in the contaminated soil. For this reason, Amirabad (a village, near the town) was selected as a study site for detailed analysis of the crops/vegetables produced in the irrigated soils.

The area is very fertile and farmers grow a variety of crops and vegetables every season. Among them, a total of ten crop samples were selected from the study site. Selected crops were *Lycopersicon esculentum* tomato, *Capsicum grossum* (chili), *Solanum tuberosum* (potato), *Solanum melongena* (brinjal), *Vigna sinensis* L Var (bean), *Spinacia oleracea* (spinach), *Abelmoschus esculentus* (ladies finger), *Amaranthus viridis* (red amaranth), *Cucumis sativus* (cucumber) and *Vigna unguiculata* L var. (cowpea). The samples were collected with fine roots and rhizosphere soil. Some rhizosphere soil and fine roots were preserved to assess the intensity of mycorrhizal colonization. The other parts were brought to the laboratory to determine the arsenic uptake.

Preparation of the stock solution of arsenic (As)

Arsenic trioxide (As_2O_3) was placed (1.32 g) in a 1 L volumetric flask and 200 ml of 1 mM Ca^{2+} solutions was added. Then a few drops of 2% diluted hydrochloric acid (HCl) was added and heated in a water bath with stirring for few hours until fully dissolved. Finally, $CaSO_4$ solution was added to make the final volume of 1 L stock solution. Water was mixed with the stock solution to obtain the required concentration of arsenic solution for the experiment.

Soil preparation

Soil with no fertilizer was obtained from a local nursery and sieved through a 3 mm-mesh sieve to remove larger sized non-soil materials. It was then transferred to 500 g pots. A total of thirty pots were required for ten treatments with three replications. Arbuscular mycorrhizal fungi inoculum was collected from the Mycorrhiza laboratory of Chittagong University, Bangladesh and manually added to the selected pots at a rate of 10% (wt basis). Stock solution of required arsenic content was then added to the selected pots. The pots were ready for seed sowing after one day. Growth was left for sixty days.

Experimental design

A Randomized Complete Block Design (RCBD) with three replications was used in the experiment. Five seeds were sown in each pot. The seeds were viable and 100% germination occurred. The treatments used in the experiment were as follows:

T_1: Soil
T_2: Soil + Mycorrhiza
T_3: Soil + 10 ppm (wt by vol.) Arsenic
T_4: Soil + Mycorrhiza + 10 ppm (wt by vol.) Arsenic

T_5: Soil + 30 ppm (wt by vol.) Arsenic
T_6: Soil + Mycorrhiza + 30 ppm (wt by vol.) Arsenic
T_7: Soil + 50 ppm (wt by vol.) Arsenic
T_8: Soil + Mycorrhiza + 50 ppm (wt by vol.) Arsenic
T_9: Soil + 100 ppm (wt by vol.) Arsenic
T_{10}: Soil + Mycorrhiza + 100 ppm (wt by vol.) Arsenic

Harvest

The seedlings were harvested sixty days after germination and physical parameters (total number of leaflets, size of leaflets, collar diameter, shoot height, root length, fresh and oven dry weight) were recorded. Then the treatments that required mycorrhizal assessment were selected and three seedlings from each of them were separated before taking physical parameters. Rhizosphere soil and roots were collected to assess the intensity of mycorrhizal colonization.

Determination of mycorrhizal colonization

Collected roots with adhered soil were submerged in water in a white clean bowl to allow the soil particles to separate away. Water was changed several times for a complete wash. For the determination of mycorrhizal colonization, fine roots were cut into small segments of approximately 1 cm length. Then one hundred root segments were randomly selected for staining.

The root pieces were stained according to the method of Phillip and Hayman (1970) with some modifications followed from Mridha *et al.* (1999). First, the root pieces were heated in 10% potassium hydroxide (KOH) solution for 15–20 min at 90°C. Then they were rinsed in water several times and acidified with 1% HCl solution for 24 h. These segments were then heated in 0.05% aniline blue at 90°C for about 10 min and then destained at room temperature in acidic glycerol.

At first fifteen root segments were mounted on one slide with a cover slip. In total, thirty segments were prepared to identify intensification of mycorrhizal colonization. These segments were observed under a microscope and rated according to their intensity of colonization. These classes gave a rapid estimation of the level of mycorrhizal colonization of each root segment and the abundance of mycelium, vesicle and arbuscule. The values were analyzed to calculate %F, %M, %m, % a and %A, as defined below according to Trouvelot *et al.*, 1986.

Frequency of mycorrhiza in the root system

$$F\% = (\text{no. of fragments myco/total no})^{*}100$$

Intensity of the mycorrhiza colonization in the root system

$$M\% = (95n_5 + 70n_4 + 30n_3 + 5n_2 + n_1)/(\text{no. total})$$

where n_5 = number of fragments rated 5; n_4 = fragments rated 4 etc.
Intensity of the mycorrhiza colonization in the root segments

$$m\% = M^{*}(\text{no. total})/(\text{no. mycorrhizal})$$

Arbuscule abundance in mycorrhizal parts of root segments

$$a\% = (100\,mA_3 + 50\,mA_2 + 10\,mA_1)/100$$

where mA_3, mA_2, mA_1 are the % of m, rated A_3, A_2, A_1, respectively, with $mA_3 = ((95n_5A_3 + 70n_4A_3 + 30n_3A_3 + 5n_2A_3 + n_1A_3)/\text{no mycorrhizal})*100/m$ (mA_2 and mA_1 was calculated as of mA_3)
Arbuscule abundance in the root system

$$A\% = a^{*}(M/100)$$

Assessment of spore population

Spore population in the rhizosphere soil was investigated using wet sieving and decanting method as described by Gerdemann and Nicolson (1963). Rhizosphere soil was collected from a depth of 0–10 cm and mixed thoroughly in an 8 L capacity bucket filled to three quarters with tap water. The bucket was left to settle for about five minutes and then sieved through 400, 240 and 100 micromesh. The supernatant was first sieved through a 100-micromesh sieve into a second bucket to avoid any loss of the materials. This was done to remove any unwanted particles (such as large stones, roots, twigs etc.) from the sample. The same process was repeated after a few minutes through a 240-micromesh sieve into a third bucket. After allowing the suspension to settle down for about 5 min, the supernatant was decanted into the 400-micromesh sieve. The sieved material was transferred into three watch glasses and a small quantity of water was added. Then they were observed under a compound microscope and the number of spores was counted. After that the spores were extracted using fine forceps, and microscopic slides were prepared for identification.

Arsenic analysis

Dry plant sample (5–10 g) was placed in a 100 ml Microkjeldhal flask. 10 ml of concentrated nitric acid (HNO_3) was added and the flask was placed on the digester and gently heated. When the initial brisk reaction was over, the solution was removed and cooled. Then 4 ml of concentrated sulfuric acid (H_2SO_4) was added. After a few minutes, 2 ml of 70% perchloric acid ($HClO_4$) was added. Heating was continued up to the point of emission of dense fumes of sulfur tri-oxide (SO_3). After some time the solution appeared clear. After the solution was concentrated, it was filtered and evaporated to dryness. After this, the residual was dissolved in distilled water and transferred to a 25 ml volumetric flask and made up to the mark with de-ionized water. This was used as the liquid solution of plant sample.

A 20 ml sample was taken into a clean generator bottle. 10 ml of conc. hydrochloric acid (HCl), 2 ml of 15% potassium iodide (KI) solution and 0.5 ml of 40% stannous chloride ($SnCl_2$) solution were then added successively to the generator bottle and left for 15 min. After that, 3 g of granulated zinc was added and the flask was immediately connected to the absorbing unit. The arsine gas generated in the conical flask was passed through cotton (cellulose) soaked with lead acetate and then bubbled into the Ag-DDTC solution in chloroform which turned slowly red. For the completion of the arsine generation, this reaction mixture was warmed for 30 min with frequent stirring. After 30 min of bubbling the red absorbing solution was quantitatively transferred to a 10 ml volumetric flask, and its volume was made up to the mark with chloroform and the absorbance was recorded at 530 nm against a reagent blank prepared using the same procedure without adding the arsenic.

Results and Discussion

Analysis of field samples

Field samples of commonly grown vegetables and crops were collected to assess arsenic uptake in the edible parts. Among all the collected samples, the mean arsenic concentrations in tomato and brinjal were the highest (0.35 ppm and 0.29 ppm respectively). In ladies finger, significant amount of arsenic concentration was recorded (0.2 ppm). Samples of these three species were collected from fields that were usually washed by drainage water from the contaminated tube well. This might be the reason for the high arsenic uptake through their root systems. Arsenic concentrations in chili, bean, spinach, red amaranth and cowpea were insignificant. These species were not exposed directly to arsenic contaminated groundwater as tomato, brinjal and ladies finger were. The degree of arsenic uptake can be attributed to the usage of phosphate fertilizers and pesticides by farmers looking for higher yield. Phosphorus fertilizer can influence arsenic phytoavailability directly through addition of arsenic as a contaminant or indirectly through immediate and long-term effect on soil characteristics (Signes-Pastor *et al.*, 2007; Sisr *et al.*, 2007; Kurosawa *et al.*, 2005; Woolson *et al.*, 1973; Creger and Peryea, 1994).

According to the Upazila health complex (Sitakund, Chittagong), shallow groundwater in the study area was contaminated with arsenic. The study site was a highly contaminated zone marked by a local NGO (Young Power in Social Action) and Upazila health complex. As a result, there was potential for arsenic accumulation in the plant tissues through the localized contaminated groundwater sources. The population in these areas was being affected through drinking of and cooking using contaminated tube well water for a long time. The continued use of contaminated groundwater for drinking and cooking as well as using the fields as crop fields might escalate a buildup of the toxin in the human food chain.

In the study area, farmers were growing vegetables commercially and marketing them throughout the year. Inhabitants found the area fertile (sandy loam, which was very suitable for cultivation of agricultural crops) and has the potential to earn large profits by exporting the crops to city areas. This way, the arsenic contamination in the human food chain may cover a wider area, thus creating a health hazard.

The structural colonization of arbuscular mycorrhizal fungi in crop roots could be in the forms of mycelium, arbuscules and vesicles (Smith and Read, 2000). Mycelium is a network of vegetative hyphae which is largely related to translocation of nutrients and water from the soil to the plant (Hamel, 2007). Among all the species surveyed, the maximum mycelium colonization was observed in cowpea (46.56%), followed by red amaranth. The reason might be due to the soil texture (sandy) of the study area and nutrient availability (poor) in the surveyed site. Arbuscular mycorrhizal fungi prefer nutrient poor sandy soil to develop colonization in the root system and it could support the host plant by supplying nutrients (Auge *et al.*, 1986). They could improve plant water relations in water stress conditions (Barea *et al.*, 1988). Chili, spinach, bean, and potato also showed significant mycelium colonization. The hyphae were very clear in the bean and it spread throughout the root system. Poor mycelium colonization was observed in tomato (2.55%), brinjal (6.99%) and ladies finger, (9.77%). The field was fertile and there was sufficient moisture to support the plant. This might be a possible reason for poor mycelium colonization among those species.

Vesicles are thick-walled mycorrhizal structures of varying shapes and contain abundant lipids and numerous nuclei. It is likely that they are important storage organs and may play a significant role as propagules within root fragments (Smith and Read, 2000). Intensity of vesicular colonization was the highest in cowpea (25.23%). Microscopic slides prepared from root segments of cowpea showed very clear vesicle in the root system. The vesicles were longitudinally attached with the hyphae. Significant amount of colonization was observed in red amaranth (19.2%), spinach (15.12%), chili (12.64%), bean (10.96%), potato (10.33%) and cucumber (6.58%). But the vesicles were not as clear and well arranged as cowpea. Because of soil fertility and moisture content available in the soil, tomato, ladies finger, and brinjal showed poor intensity of colonization. In tomato, the vesicles were large, but the intensity of colonization was lower.

Arbuscules provide a considerable increase in surface area of contact between fungi and plant. It is believed that they are involved in nutrient transfer and are the sites for movement of soil derived nutrients, such as P and Zn to the plant (Smith and Read, 2000).The present study showed moderate colonization of arbuscule (in the root system) in comparison with mycelium and vesicle. The highest intensity of colonization was observed in cowpea at 10.62%. Poor colonization was observed in red amaranth (7.62%), brinjal (6.23%), spinach (5.63%) and chili, (4.5%). The intensity of colonization was negligible in cucumber.

Mycorrhizal spores are the best defined source of inoculum and are the only propagules that can be identified at species level (Smith and Read, 2000). The highest number of spores were observed with cowpea (50/100 g soil), followed by bean (44/100 g soil). Some microscopic slides were prepared to identify the spores. It was observed that all the spores available in cowpea were of *Glomas* sp. In the case of bean, some S*cutillospora* sp. and *Gigaspora* sp. were also observed. Chili and red amaranth showed significantly more or less equal number of spore in the soil. Though arbuscular mycorrhizal fungi mostly form spores in the rhizosphere soil, sometimes spores were found inside the root tissues. Microscopic slides prepared from root samples of chili indicated presence of spore in the root fragments which was very rare in arbuscular mycorrhizal fungi colonization. Poor spore population was observed in cucumber.

Among all the species surveyed, the highest mycelium colonization was observed in cowpea (46.56%) where arsenic uptake was the lowest (0.04 ppm). As the intensity of mycelium colonization in the root system decreased, concentration of arsenic uptake increased. Laboratory analysis showed comparatively low arsenic uptake in red amaranth, spinach and chili, whereas mycelium colonization was significant. The lowest (2.55%) mycelium colonization was observed in tomato, which showed the maximum (0.35 ppm) arsenic uptake. Brinjal and ladies finger also showed significant arsenic uptake from the soil. During the field visit, it was observed that these plants were grown in nutrient rich soil which is not favorable for arbuscular mycorrhizal fungi colonization. As a result they form poor mycelium colonization in their root system. The poor intensity of arbuscular mycorrhizal fungi colonization might be a reason of significant arsenic uptake in these species in the study site.

Arsenic uptake decreased as the rate of vesicle colonization increased. The highest vesicle colonization was observed in cowpea (25.23%) and the lowest was in tomato (2.85%). Brinjal, spinach, chili, also showed higher arsenic uptake with a decrease in the intensity of vesicle colonization.

Cowpea and red amaranth showed comparatively higher arbuscular colonization with an intensity of 10.62% and 7.62% respectively. Chemical analysis of these species showed the lowest (0.04 ppm) arsenic uptake through their root systems. Red amaranth, spinach, brinjal, and chili followed the same trend where arsenic uptake increased as the rate of arbuscular colonization in the root system decreased. Tomato and brinjal showed comparatively lower arbuscules in the root system with the highest arsenic uptake of 0.35 ppm and 0.28 ppm, respectively.

Spore population in the soil of the investigated species was counted. With the exception of potato and cucumber, the other species showed an inverse relationship as with the intensity of mycelium and vesicle colonization. Plant arsenic uptake increased as spore population in the soil decreased. The maximum number of spores was available in cowpea (50 per 100 g of soil) where arsenic uptake was the lowest (0.04 ppm). Microscopic slides of the spores showed abundant *Glomas* sp. in the rhizosphere soils of cowpea, spinach and bean. Spinach and bean showed significant number of spores in the collected rhizosphere soil (42 and 44 respectively) with a comparatively lower arsenic uptake of 0.06 ppm and 0.08 ppm respectively.

Also, tomato and brinjal showed higher arsenic uptake with a significant lower spore population in the soil.

Analysis of laboratory samples

Seedling survivality

It was observed that all the seedlings died in treatment T_3 (soil + 10 ppm arsenic), treatment T_5 (soil + 30 ppm arsenic), treatment T_7 (soil + 50 ppm arsenic) and treatment T_9 (soil + 100 ppm arsenic). The best survivability (90%) was observed in treatment T_2 (soil + mycorrhiza) followed by treatment T_4 (soil + mycorrhiza + 10 ppm arsenic) and treatment T_6 (soil + mycorrhiza + 30 ppm arsenic). Moderate survivability (40%) was observed in treatment T_8 (soil + mycorrhiza + 50 ppm arsenic). Seedling survivability was the lowest in treatment T_{10} (soil + mycorrhiza + 100 ppm arsenic). From the results it could be said that seedling survivability was zero in arsenic contaminated plants (without mycorrhiza). Though arsenic was proven to be toxic to the plants (Mann *et al.*, 1993), it was also thought that high temperature at the roof top, where the experiment was conducted, could have had a significant negative effect on the plant survival. The seedlings were grown during summer when temperature was very high (sometimes exceeding 50°C at noon). As the seedling survivality in treatment T_1 was also very poor, it could be assumed that both high temperature and toxicity were responsible for seedling failure. In contrast, though similar concentrations of arsenic were added to both non-mycorrhizal and mycorrhizal seedlings, only the mycorrhiza inoculated seedlings showed satisfactory performances. Fungi inoculum was added to the selected treatments (treatment T_4, treatment T_6, treatment T_8 and treatment T_{10}) before seed sowing. It is assumed that mycorrhizal fungi increased plant tolerance to extreme temperature which is supported by Pond *et al.* (1984). But it was also observed that seedling survivality decreased as arsenic concentration in the soil increased ($T_4 > T_6 > T_8 > T_{10}$). Plant survivability was the highest in treatment T_2 where no arsenic solution was added and the lowest was in T_{10}, where the maximum concentration of arsenic was added. Considering all of these, it can be said that both arsenic toxicity and high temperature might have negative effect on seedling survivability whereas arbuscular mycorrhizal fungi increased tolerance to stress in unfavorable condition.

Plant growth

Shoot height was recorded twice; the first one after one month of germination and the second measurement was taken after two months. After one month of growth the maximum height (10 cm) was observed with treatment T_2. In the rest of the treatments, it was observed that as the concentration of arsenic in the soil increased, shoot height decreased. Seedlings become thinner and less healthy with high concentrations of arsenic. In treatments T_4, T_6 and T_8, shoot heights were 5.3 cm, 4.5 cm and 4.0 cm, respectively. The minimum (1.5 cm) shoot height was recorded in treatment T_{10}.

Significant growth difference was also observed between treatment T_1 and treatment T_2. After one month of growth, treatment T_2 showed better performance in compared with treatment T_1. Phosphorus might be acting as a limiting factor here as it is an essential nutrient for early growth and development of plants. The soils did not receive fertilizer throughout the study period. In nutrient poor conditions, mycorrhizal hyphae can produce hydrolytic enzymes such as proteases and phosphatase in the root system that might have an important impact on organic matter mineralization and inorganic phosphorus availability (Auge *et al.*, 1986; Allen, 1991). Moreover mycorrhizal hyphae can increase root surface area by nearly 67-fold thus resulting in significant increase in phosphorus uptake from a larger depletion zone (Raghothama, 2005). This might help mycorrhizal seedlings to show comparatively better growth performances than that of non-mycorrhizal seedlings.

Shoot height recorded after the second month of growth also showed similar performance. Seedling height was the best (19 cm) in treatment T_2, followed by treatments T_4 (12 cm), T_6 (14 cm), and T_8 (7.5 cm). Poor shoot height was recorded in the treatments T_1 (6.5 cm) and T_{10} (4.5 cm). Though similar amount of mycorrhizal inoculum was used in all treatments, significant growth differences (in terms of shoot height) were observed among them. It was assumed that different concentrations of arsenic solution added to the soil might have different levels of toxicity on seedling growth. Tang and Miller (1991) reported toxic effects of arsenic on plant growth and showed that plant growth was stunted due to high concentration of arsenic in the soil.

Significant difference in root length was observed between treatment T_2 and control (treatment T_1) at one month growth stage. It was observed that root length was three times higher in treatment T_2 (soil + mycorrhiza) in comparison with control (treatment T_1). After two months, the root length of treatment T_2 increased in an average of 16 cm, which was three times higher than control (treatment T_1) and two times higher than treatment T_4. Other studies also showed that mycorrhizal colonization increased with time and coincides with increased above and below ground biomass and nutrient uptake by the host plant (Barea *et al.*, 1988; Kabir *et al.*, 1998). Moreover, the establishment of the symbiotic interactions also induced changes in root morphology; in particular, the degree of branching increased and the number of lateral roots were greater. Gradual decrease (6 cm > 5 cm > 2.5 cm) in root length was recorded as arsenic concentration in the soil increased in treatments T_6, T_8 and T_{10}.

Plant height (shoots + root height) of the seedlings grown in arbuscular mycorrhizal fungi amended soil was recorded. After one month, the highest growth (18.5 cm) was recorded in treatment T_2 followed by treatments T_4, T_6 and T_8. Plant height was 35 cm in absence of arsenic (treatment T_2) whereas it decreased as 20 cm in the presence of lower amount arsenic (treatment T_4) in the soil. At 30 ppm concentration (treatment T_6), the leaves turned yellow and leaf size decreased. The average height in treatment T_6 was recorded as 16.0 cm. The differences between treatments T_1 and T_8 was insignificant. The poor plant growth in these treatments might be due to nutrient shortage (in the case of treatment T_1) and toxicity (in the case of treatment T_8) in the soil. The lowest (2.5 cm) plant height was recorded in treatment T_{10}, where arsenic concentration was the highest.

Effect of different concentrations of arsenic on collar diameter of the seedlings was recorded twice (after one month and two months respectively). In both the cases, collar diameter was lower in treatment T_1 whereas in treatment T_2, the seedlings showed the best performance in the presence of mycorrhiza. After one month, the average collar diameter of the seedlings was 2.3 mm, which increased to 4.63 mm after two months. It was found that collar diameter of the seedlings decreased as arsenic concentration in the soil increased. One-month growth difference showed that the rate of decrease was much higher after one month in comparison to two months. The lowest collar diameter was recorded in treatment T_1 (0.75 mm and 2 mm after one and two months, respectively). This could be due to lack of nutrients in the soil.

The total number of leaflets was counted during harvest. The average highest number (nine/plant) of leaflets was recorded in treatment T_2 and the lowest number (four/plant) was in treatment T_1. It was found that average shoot height and shoot dry weight were at maximum in treatment T_2. The seedlings in treatment T_1 were grown in a control environment without addition of arbuscular mycorrhizal fungi or fertilizer. As a result, the seedlings were limited by nutrients and water for their growth. This might be the reason for the minimum number of leaflets in treatment T_1. On the other hand, the seedlings in treatment T_2 were supported by arbuscular mycorrhizal fungi that increased nutrient uptake from the soil. As a result, their growth was the best and they contained the highest number of leaflets. Chowdhury and Satter (1993) documented enhanced branching with respect to shoot growth due to mycorrhizal symbiosis in the root, which was supported by Das *et al.* (1997) and Nwoko and Sanginga (1999). It was observed that the concentration of arsenic present in the soil significantly decreased the number of leaflets. Total leaflet number per seedling decreased as arsenic concentration available in the soil increased. It was observed that mycorrhizal seedlings showed better performance with higher number of leaflets in comparison to non-mycorrhizal seedlings. After two months, it was observed that all the non-mycorrhizal seedlings died due to high temperature as well as high arsenic toxicity from the soil. In the case of mycorrhizal seedlings, moderate number of leaflets (seven/plant) was counted in treatment T_4 and treatment T_6, followed by treatment T_8 (six/plant). Mycorrhizal seedlings with the highest concentration of arsenic (100 ppm) produced the poorest number of leaflet per seedlings. The findings tallied with Machlis (1941) who showed that high concentration of arsenic is responsible for leaf wilting as well as other abnormalities including reducing leaflet size. The results are in good agreement with Barea *et al.* (1996) who reported improved growth due to arbuscular mycorrhizal fungi inoculation in arsenic contaminated seedlings.

The improvement in phosphorus uptake and plant growth by arbuscular mycorrhizal fungi has been well documented for many crop plants (Smith and Gianinazzi-Pearson, 1988; Smith and Read, 2000). Such effects have been reported mainly after inoculation of soils with arbuscular mycorrhizal fungi. Howeler *et al.* (1980) and Bolan (1991) reported that arbuscular mycorrhizal fungi improve plant productivity in soils of low fertility largely by enhancing phosphorus uptake in associated plants. In the present experiment, no additional fertilizers were used. Arsenic contaminated plants were allowed to grow on two sets of growing conditions (arbuscular mycorrhizal fungi inoculated and uninoculated) and the role of arbuscular mycorrhizal fungi was monitored up to two months. The complete failure of uninoculated seedlings and satisfactory survival of inoculated seedlings indicates the beneficial role of arbuscular mycorrhizal fungi on seedling growth in unfavorable conditions. Habte and Soedarijo (1996) reported that arbuscular mycorrhizal fungi could help to improve the nutritional condition of the soil and increased plant productivity through increased absorption of mineral nutrients from the soil including phosphorus. This was also supported by Marschner and Dell (1994).

Xavier and Germida (1997) has observed greater root-shoot ratio in arbuscular mycorrhizal fungi inoculated plants in comparison with non-mycorrhizal plants, which was supported by Pattanaik *et al.* (1995), Nurlaeny *et al.* (1996). Arbuscular mycorrhizal fungi can help the plants to expand their root zone and can increase nutrients and water uptake. On the other hand, arsenic slows down plant growth by reducing root zone (Carbonell-Barachina *et al.*, 1998). This was observed in the present experiment where root system was gradually decreased as arsenic concentration in the soil increased.

Many other scientists also supported arsenic toxicity in roots. Sachs and Michael (1971) reported that arsenic is toxic to reticular membranes because it can react with sulphydral groups proteins (Speer, 1973), causing disruption of root functions (Orwick *et al.*, 1976) and cellular death. Carbonell-Barrachina *et al.*, (1995) reported damage to tomato root membrane that caused significant reduction in nutrient uptake and hence reduced seedling growth. Jones (1987) studied the influence of arbuscular mycorrhizal fungi in *Leucaena leucocephala* (white leadtree).They showed that mycorrhiza (*Glomas fasciculatum*) inoculated plants subsequently developed greater root length and root dry weight.

Plant biomass

Shoot fresh weight and root fresh weight were recorded separately after two months of seedling growth. The lowest (0.49 g) shoot fresh weight was observed in treatment T_1 and the highest (4.5 g) was in treatment T_2. Moderate (2.69 g) shoot fresh weight was recorded in treatment T_4 that was contaminated with 10 ppm concentration of arsenic in the soil. Poor shoot fresh weight was recorded from treatments T_6 (mycorrhiza + 30 ppm arsenic) and T_8 (mycorrhiza + 50 ppm arsenic). The findings had greater similarity in case of root fresh weight. The highest (1.22 g) root fresh weight was observed in treatment T_2, whereas it was the lowest (0.1 g) in treatment T_1 (control).

Root dry weight (g) was the highest (1.22 g) when the seedlings were inoculated with arbuscular mycorrhizal fungi and were free from arsenic contamination (treatment T_2). The lowest (0.05 g) root dry weight was recorded both in treatment T_1 and treatment T_{10}. This might be due to lack of nutrient in the case of treatment T_1 and arsenic toxicity in treatment T_{10}. Balasubramaniam and Sekayange (1991) showed that high concentration of arsenic is responsible for damaging the root system in plants that causes poor plant biomass in contaminated sites. Similar finding was observed in the case of shoot dry weight. The highest shoot dry weight (1.8 g) was observed in treatment T_2 followed by treatment T_6 and the lowest (0.15 g) was in treatment T_1. Moderate shoot dry weight was observed in treatment T_4 and treatment T_6, whereas treatment T_{10} showed very poor (0.42 g) shoot dry weight.

The total plant biomass (shoot dry weight + root dry weight) was calculated from shoot and root dry weights. The maximum (2.2 g) biomass was recorded in treatment T_2 (soil with mycorrhiza) and the minimum (0.2 g) was in treatment T_1 (control). The rest of the treatments showed decreased dry biomass with increased soil arsenic concentration. This indicated host's ability to channel energy to shoot production due to an increased efficiency of the roots in association with arbuscular mycorrhizal fungi (Bethlenfalvay *et al.*, 1982). Sengupta and Chowdhury (1995) also showed lower plant biomass in seedlings contaminated with heavy metal.

Mycorrhizal colonization

Analysis of mycelium (hyphae) and vesicular colonization of the treatments showed that as the concentration of arsenic increased in the soil, the percentage of mycelium and vesicular colonization in the root fragments decreased. A similar result observed in the case of arbuscular colonization in the root system. Maximum (11.25%) colonization was recorded in treatment T_2 (control) and the minimum (0.90%) was recorded in treatment T_{10}. In the case of arbuscular colonization in the root fragments, the highest (37.22%) colonization was recorded in treatment T_2 (soil with mycorrhiza) and the lowest (7.65%) was recorded in treatment T_{10} (soil + mycorrhiza + 100 ppm arsenic). From all three types of mycorrhizal structural colonization (mycelium, vesicles and arbuscules) in the roots of the seedlings, it was clear that mycelium was the most common structure and arbuscule was the least common. Another important finding was that as the average concentration of arsenic in the soil increased, mycorrhizal colonization decreased.

Spore population from 100 g of dry soils was sieved and counted. It was observed that spore population decreased as the concentration of arsenic in the soil increased. The maximum (55/100 g soil) number of spore was observed in treatment T_2 (soil with mycorrhiza) and the minimum was observed in treatment T_8 (soil + mycorrhiza +50 ppm arsenic).

Arsenic uptake

There was a positive correlation between soil arsenic level and shoot arsenic concentration in arbuscular mycorrhizal fungi inoculated seedlings. Arsenic concentration in the shoots increased dramatically as arsenic addition levels increased in the soil. The highest arsenic concentration approached 24.15 mg/g in the shoots at higher soil arsenic level (treatment T_{10}) while the lowest was recorded as 2.64 mg/g at the lower soil arsenic level (treatment T_4). It was found that shoot arsenic concentration was nearly ten times higher in treatment T_{10} in comparison with treatment T_4. In contrary to arbuscular mycorrhizal fungi inoculated seedlings, non-mycorrhizal seedlings were failed to survive under experimental conditions. Since the plants were not grown in control environment, this indicates the increased tolerance of host plant in the presence of arbuscular mycorrhizal fungi. It was found that arbuscular mycorrhizal fungi colonization did not appear to be inhibited by arsenic application.

The findings support the previous observations (Al Agely *et al.*, 2005; Trotta *et al.*, 2006; Chen *et al.*, 2006) that plants growing in arsenic contaminated soils are mycorrhizal. One of the important roles for arbuscular mycorrhizal symbiosis in the plants is to obtain phosphorus for their hosts (Smith and Read, 2000; Bucher, 2006, Liu *et al.*, 2005); however it is assumed that this may become a problem when soil has high arsenic concentration. In the present study, arbuscular mycorrhizal fungi inoculated seedlings showed decreased growth at higher level of arsenic addition to the soil. This might be due to lack of phosphorus in the soil resulting in high arsenic accumulation in the shoots. The seedlings were grown in limited soil and root growth was restricted within the pots. Arbuscular mycorrhizal fungi can help plant only when there are scopes to expand hyphal network in a wider depletion zone and can increase phosphorus uptake by absorbing available labile phosphorus from the soil or by transforming plant unavailable phosphorus into available forms (Smith and Read, 2000). Labile and total phosphorus in the soil was not recorded in the present study. So, there are scopes to conduct further experiment in field conditions where the rhizosphere will not be limited to restrict arbuscular mycorrhizal fungi in absorbing nutrients (phosphorus) from a wider depletion zone.

Arsenic and phosphorous belong to the same chemical group and have comparable dissociation constants for their acid and solubility products for their salts, resulting in similar geochemical behavior of arsenic and phosphorous in soil (Adriano, 2001). They have similar carrier system (Al Agely *et al.*, 2005) and compete with each-other for transport sites (Wang, *et al.*, 2002). It is assumed that arbuscular mycorrhizal fungi can help plant uptaking more phosphorus and can suppress arsenic transport to the shoot; however, to our knowledge there is no published work yet supporting this evidence. Sharples *et al.* (2000) reported arsenic suppression in ericoid mycorrhizal fungi (symbiosis between fungi and the roots of plants from the order Ericales) and showed that arsenite efflux enables the fungus to retain its ability to transport inorganic phosphorus from the soil, whereas effluxing absorbed $H_2AsO_4^-$. It is assumed that, like ericoid mycorrhizal fungi, arbuscular mycorrhizal fungi can also act as a filter to maintain low plant arsenic levels, while maintaining an adequate supply of phosphorus to the host; however, this assumption needs confirmation.

Conclusion

Arsenic contamination is a major concern in Bangladesh due to potential health risks from arsenic contaminated groundwater. The study confirms the bioavailability of arsenic in agricultural crops grown in contaminated soils. Arbuscular mycorrhizal fungi, a plant symbiont actively colonized in the roots and rhizosphere of arsenic contaminated crops in both the field and laboratory conditions. The increased growth and arsenic tolerance in arbuscular mycorrhizal fungi colonized seedlings in comparison to non mycorrhizal seedlings leading to the suggestion that the tolerant seedlings are dependent on the fungi for nutrient uptake and resistance to arsenic toxicity. However, the mechanistic basis of arsenic tolerance in mycorrhizal symbiosis with the agricultural crops and potential restriction on the roots remains to be determined.

References

Adriano, D.C. 2001. *Trace Elements in the Terrestrial Environment. Springer*, New York.

Akhter, M.F., Mridha, M.A.U, Bhuiyan, M.K. and Mahmud, M. 2004. Effect of Arsenic on growth of Leucaena leucocephala (Lam) De Wit in Arbuscular Mycorrhiza Fungi Amended Soil. In *Proceedings of 6th Biennial Conference — Bangladesh Phytopathological Society.* 29th July 2004.

Bangladesh Agricultural Research Institute, Gazipur, pp 16. Bangladesh: Bangladesh Phytopathological Society.

Al Agely, A., Sylvia, D.M. and Ma, L.Q. 2005. Mycorrhizae increase arsenic uptake by the hyperaccumulator Chinese brake fern (*Pteris vittata* L.). *J. Environ. Qual.*, 34(6), 2128–2186.

Alam, M.B. and Sattar, M.A. 2000. Assessment of arsenic contamination in soils and waters in some areas of Bangladesh. *Water Sci. Tech.*, 42,185–193.

Alam, M.Z. and Rahman, M.M. 2003. Accumulation of Arsenic in Rice Plant from Arsenic Contaminated Irrigation Water and Effect on Nutrient Content. *Water Sci. Tech.*, 42, 131–135.

Allen, M.F. 1991. *The Ecology of Mycorrhizae.* Cambridge Univ. Press, Cambridge.

Auge, R.M., Schekel, K.A. and Wample, R.L. 1986. Osmotic adjustment in leaves of VA mycorrhizal and non-mycorrhizal rose plants in response to drought stress. *Plant Physiol.*, 82, 765–770.

Balasubramaniam, V. and Sekayange, L. 1991. Effect of the legume hedgerows on soil fertility changes and crop performance in the semi-arid-highlands of Rwanda. Biological Agriculture Horticulture, 8, 17–32.

Barea, J.M., Azcon-Aguilar, C. and Azcon, R. 1988. The role of mycorrhiza in improving the establishment and function of the Rhizobium-legume system under field conditions. In: Beck, D.P. and Materon L.A. (eds.), Nitrogen fixation by legumes in Mediterranean agriculture. ICARDA and Martinus Nijhoff Dordrecht, p. 153–162.

Barea, J.M., Tobar, R.M. and Azcon-Aguilar, C. 1996. Effect of genetically modified *Rhizobium meliloti* inoculants on the development of AM, root morphology, nutrient uptake and biomass accumulation in *Medicago sativa. New Phytol.*, 134, 361–369.

Bethlenfalvay, G.J., Pacovsky, R.S., Bayne, H.G. and Stafford, A.E.1982. Interactions between nitrogen fixation, mycorrhizal colonization and host plant growth in the *Phaseolus-Rhizobium-Glomus* symbiosis. *Plant Physiol.*, 70, 446–450.

Bolan, N.S. 1991. A critical review on the role of mycorrhizal fungi in the uptake of phosphorus by plants. *Plant and Soil*, 134, 189–207.

Bucher, M. 2007. Functional biology of plant phosphate uptake at root and mycorrhiza interfaces. *New Phytologist*, 173(1), 11–26.

Carbonell-Barrachina, A.A., Burlo, F. and Matrix, J. 1995. Arsenic uptake, distribution and accumulation in tomato plants: Effect of arsenite on plant growth and yield. *Journal of Plant Nutrient*, 18, 1237–1250.

Carbonell-Barrachina, A. A., Aarabi, M. A., DeLaune, R. D., Gambrell, R. P. and Patrick, W. H. 1998. The influence of arsenic chemical form and concentration on *Spartina patens* and *Spartina alterniflora* growth and tissue arsenic concentration. *Plant and Soil*, 198, 33–43.

Chen, B.D., Zhu, Y.-G. and Smith, F.A. 2006. Effects of arbuscular mycorrhizal inoculation on uranium and arsenic accumulation by Chinese brake fern (*Pteris vittata* L.) from a uranium mining-impacted soil. *Chemosphere*, 62, 1464–1473.

Chowdhury, M.K. and Satter, M.A. 1993. Homestead and Cropland Agroforestry Practices in the High Ganges River Flood Plain. *In*: Agroforestry Farming Systems Linkage in Bangladesh, BARC-Winrock International, Dhaka.

Creger, T.L. and Peryea, F.J. 1994. Phosphate fertilizer enhances arsenic uptake by apricot linersgrown in lead-arsenate-enriched soil. *Hort. Sci.*, 29, 88–92.

Das, P.K. Sahoo, P.N. and Jena, M.K. 1997. Effect of VA –mycorrhiza and rhizobia inoculation on nutrient uptake, growth attributes and yields of green gram (*Vigna radiata* L.). *Environ. Ecol.*, 15, 830–833.

Gadd, G.M. 2001. Metal transformations. *In*: G.M. Gadd (ed.). Fungi in Bioremediation. Cambridge University Press, UK.

Gerdemann, J.W.and Nicolson, T. J. 1963. Spores of mycorrhizal *Endogone* species extracted from soil by wet sieving and decanting. *Trans. Br. Mycol. Soc.*, 46, 235–244.

Gianinazzi, S., Trouvelot, A. and Gianinazzi-Pearson, V. 1990. Role and use of mycorrhizae in horticultural crop production. Pleanary lecture presented at the XXIII International Horticultural Congress. International Soc. Hort. Sci. Frenze (Italy), Aug. 27–Sept. 1, p. 25–30.

Gibson, D.T. and Sayler, G.S. 1992. Scientific foundation for bioremediation: Current status and future needs. American Academy of Microbiology, Washington, D.C.

Gnekow, M. and Marscher, R. 1989. Use of bioremediation as natural cleansing agent. *Plant Physiol.*, 52, 765–770.

Habte, M. and Soedarijo, M. 1996. Response of *A. mangium* to vesicular arbuscular mycorrhizal inoculation, soil pH and soil P concentration in an oxisol. *Can. J. Bot.*, 74, 155–161.

Hamel, C. and Plenchette, C. 2007. Mycorrhizae in Crop Production. Haworth Food & Agricultural Products Press^{TM}, Binghamton, NY.

Howeler, R.H., Seiverding, E. and Saif, S. 1987. Practical aspects of mycorrhizal technology in some tropical crops and pastures. *Plant and soil*, 100, 249–283.

Howeler, R.H., Weber, E.J., Toro, M.J.C. and Graham, M.1980. The effect of mycorrhizal inoculation on the phosphorus nutrition of Cassava. Cassava Cultural Practices, Proceeding workshop, Salvador, p. 131–137.

Jones, D.L. 1987. Encyclopedia of Ferns. British Museum (Natural History), London.

Kabir, Z., O'Halloran, I.P., Fyles, J.W. and Hamel, C. 1998. Dynamics of the mycorrhizal symbiosis of corn (*Zea mays* L.): Effects of host physiology, tillage practice, and fertilization on spatial distribution of extra-radical mycorrhizal hyphae in the field. *Agricultural Ecosystems and Environment*, 68, 151–163.

Kurosawa, K., Mochizuki, T., Tra Thi Lam Ho, Thanh Huu Nguyen, Haj Nguyen Do, Hirata, M., Egashira, K. 2005. Occurences of arsenic in the surface and groundwater of farm villages near Hanoi, Vietnam and factors relating to it. *Japanese Journal of Tropical Agriculture*, 49(1), 98–106.

Liu, Y., Zhu, Y.G., Chen, B.D., Christie, X. and Li, X.L. 2005. Influence of the arbuscular mycorrhizal fungus *Glomas mosseae* on uptake of arsenate by the As hyperaccumulator fern *Pteris vittata* L. *Mycorrhiza*, 15, 187–192.

Machlis, L. 1941. Accumulation of arsenic in the shoots of Sudan grass and bush beans. Plant Physiology, 16, 521–544.

Maloney, S.E. 2001. Pesticide degradation. *In*: G.M. Gadd (ed.). Fungi in Bioremediation. Cambridge University Press, UK.

Mann, A.R., Masschenlyn, P.H. and Patrick, W.H. 1993. Soil redox-pH stability of arsenic species and its influence on arsenic uptake by rice. *Plant Soil,* 152, 245–253.

Marschner, H. and Dell, B. 1994. Nutrient uptake in mycorrhizal symbiosis. *Plant and Soil*, 159, 89–102.

Meharg, A.A and Macnair, M.R. 1990. An altered phosphate uptake system in arsenate tolerant *Holcus lanatus* L. *New Phytologist*, 116, 29–35.

Meharg, A.A. and Cairney, J.W.G. 1999. Co-evolution of mycorrhizal symbiont and their hosts to metal contaminated environments. *Advances in Ecological Research*, 30, 70–112.

Mridha, A. M. U. and Xu, L. A. 2001. Use of AMF for biocontrol of forest pathogens. *Journal of Microbiology*, 55, 12–15.

Mridha, M.A.U., Sultana, A., Sultana, N., Xu, H.L. and Umemura H. 1999. Biodiversity of VA mycorrhizal fungi of some Vegetable crops in Bangladesh pp. 330–331. In: Takeshi *et al.* (eds.) proceedings International symposium "World food Security and Crop production Technologies for Tomorrow" October 8–9, 1998, Kyoto, Japan.

Nurlaeny, N., Marschner, H. and George, E. 1996. Effects of liming and mycorrhizal colonization on soil phosphate depletion and phosphate uptake by maize and soybean grown in two tropical acid soils. *Plant and Soil*, 181, 275–285.

Nwoko, H. and Sanginga, N. 1999. Department of promiscuous soybean and herbaceous legumes on arbuscular mycorrhizal fungi and their response to Bradyrhizobium inoculation in low P soils. *Appl. Soil. Ecol.*, 13, 251–258.

O'Keefe, D. M. and Sylvia, D.M. 1991. Mechanism of the vescicular arbuscular mycorrhizal plant growth response. *In*: D.K. Arora (ed.). Handbook of Applied Mycology. Marcel Dekker, New York, p. 35–54.

Orwick, P.L., Schrieber, M.M. and Hodges, T.K. 1976. Adsorption and efflux of choloro striazines by Setaria roots. *Weed Research-Oxford*, 16, 139–144.

Pattanaik, R., Sahu, S., Padhi, G.S. and Mishra, A.K. 1995. Effect of inoculation of vesicular arbuscular mycorrhiza on horse gram growth in soil in iron-mine area. *Indian J. Agril. Sci.*, 65, 186–190.

Phillips, J.M. and Hayman, D.S. 1970. Improved procedures for clearing roots and staining parasitic and vesicular-arbuscular mycorrhizal fungi for rapid assessment of infection. *Trans Br Mycol Soc*, 55, 158–161.

Pond, E.C., Menge, J.A. and Jerrel, W.M. 1984. Improve growth of tomato in salinized soil by vesicular arbuscular mycorrhizal fungi collected from saline soils. *Mycologia*, 79, 74–84.

Porter, E.K. and Peterson, P.J. 1997. Arsenic tolerance in grasses growing on mine waste. *Environmental Pollution*, 14, 225–265.

Raghothama, K.G. 2005. Phosphorus and Plant Nutrition: An Overview. *In* Phosphorus: Agriculture and the Environment. Ed. J.T. Sims and A.N. Sharpley. pp. 355–378. ASA-CSSA-SSSA, USA.

Sachs, R.M. and Michael, J.L. 1971. Comparative phytotoxicity among four arsenical herbicides. *Weed Science*, 19, 558–564.

Sengupta, A. and Chowdhury, S.1995.Effect of dual inoculation of Rhizobium and mycorrhiza of growth response of *Sesbania grandiflora* L. in coastal saline and sand dune soil. *Indian J. Forestry*, 18, 33–37.

Sharma, S., Pant, D., Singh, S., Sinha, R.R. and Adholeya, A. 2007. Mycorrhizae in Indian Agriculture. *In*: C. Hamel and C. Plenchette (eds.). Mycorrhizae in Crop Production. pp. 1–25. Haworth Food & Agricultural Products Press^{TM}, NY.

Sharples, J.M., Meharg, A.A., Chambers, S.M. and Cairney, J.W.G. 2000. Mechanisms of arsenate resistance in the ericoid mycorrhizal fungus, *Hymenoscyphus ericae. Plant Physiology*, 124, 1327–1334.

Sharples, J.M., Meharg, A.A., Chambers, S.M. and Cairney, J.W.G. 2000. The symbiotic solution to arsenic contamination. *Nature*, 404, 951–952.

Signes-Pastor, A., Burló, F., Mitra, K. and Carbonell-Barrachina, A.A. 2007. Arsenic biogeochemistry as affected by phosphorus fertilizer addition, redox potential and pH in a west Bengal (India) soil. *Geoderma*, 137, 504–510.

Sisr, L., Mihaljevi, M., Ettler, V., Strnad, L. and Šebek, O. 2007. Effect of application of phosphate and organic manure-based fertilizers on arsenic transformation in soil columns. *Environmental Monitoring and Assessment.* 135, 465–473.

Smith, S.E. and Read, D.J. 1977. Mycorrhizal symbiosis, Academic Press, San Diego, USA.

Smith, S.E. and Read, D.J. 2000. Mycorrhizal Symbiosis, Academic press, London.

Smith. S. E. and Gianinazzi-Pearson, V. 1988. Physiological interactions between symbionts in vesicular — arbuscular mycorrhizal plants. *Annu. Rev. Plant Physiol. Plant. Mol. Bio*l., 39, 221–244.

Speer, H.L. 1973. The effect of arsenate and other inhibitors on early events during the germination of lettuce seeds *(Lactuca sativa* L.). *Plant Physiology*, 52, 129–133.

Tang, T. and Miller, D.M. 1991.Growth and tissue composition of rice grown in soil treated with inorganic copper, nickel and arsenic. *Communication of Soil Science and Plant Analysis*, 22, 2037–2045.

Trotta, A., Falaschi, P., Cornara, L., Minganti, V., Fusconi, A., Drava, G. and Berta, G. 2006. Arbuscular mycorrhizae increase the arsenic translocation factor in the As hyperaccumulating fern *Pteris vitttata* L. *Chemosphere*, 65, 74–81.

Trouvelot, A., Kough, JL. and Gianinazzi-Pearson, V. 1986. Mesure du taux de mycorhization VA d'un systeme radiculaire. Recherche de methods d'estimation ayant une signification fonctionnelle. In: Gianinazzi-Pearson, V. and Gianinazzi, S. (eds.). Physiological and Genetical Aspects of Mycorrhizae, INRA Press, Paris, pp. 217–221.

Ullrich-Eberius, C.I., Sanz, A. and Novacky, A.J. 1989. Evaluation of arsenate and vandate-associated changes of electrical membrane potential and phosphate transport in *Lemma gibba* GI. *Journal of Experimental Botany*, 40, 119–128.

Van Hamme, J.D. 2004. Bioavailability and Biodegradation of Organic Pollutants-A Microbial Perspective. *In*: Singh, A. and Ward, O.P. (eds.). Biodegradation and Bioremediation. Springer, Germany, pp. 37–56.

Wang, J., Zhao, F.J., Meharg, A.A., Raab, A., Feldmann, J. and Mcgrath, S.P. 2002. Mechanisms of arsenic hyperaccumulator in *Pteris vittata.* Uptake kinetics, interactions with phosphate, and arsenic speciation. *Plant Physiology*, 130(3), 1552–1561.

Woolson, E.A., Axley, J.H. and Kearney, P. 1973. The chemistry and phytotoxicity of arsenic in soil:II. Effect of time and phosphorus. *Soil sci. Soc. Am.* Proc. 37, 254–258.

Xavier, L.J.C. and Germida, J.J. 1997. Growth response of lentil and wheat to *Glomus clarum* NT4 over a range of P levels in a Saskatchewan soil containing indigenous AM fungi. *Mycorrhiza*, 7, 3–8.

21

PROSPECTS OF BIOMETHANATION OF SOLID WASTES IN INDIA

RAYAVELLORE SURYAKUMAR ARUN KUMAR
and TING YEN PENG

Solid waste is a significant and growing problem in many urban areas in India. Organic fraction constitutes about 50% of the total urban solid waste. The current systems of waste management are rudimentary and grossly inefficient, ineffective and environmentally damaging. Biomethanation technology has emerged as a viable option for managing the organic solid waste generated in the urban areas. Besides reducing the waste load by about 50%, biomethanation generates biogas, energy derivative, compost derivative, and nutrient recovery. Three different technological approaches have been adopted in three cities of India for managing organic wastes. This research analyses the technical feasibility and operational stability of the three ventures from a long-term sustainability perspective. The research also assesses the incorporation of biomethanation of organic solid-agricultural residues as a technological intervention in rural energy demand and suggests strategies for the utilization of biomethanation technology from a sustainable outlook.

Introduction

Rapid urbanization, increasing urban population and economic growth have resulted in many environmental as well as hygienic evils in India. Solid Waste Management (SWM) is one of the issues which has assumed major importance. The daily per capita waste generation in the Indian cities and towns varies from 0.1 kg to up to 0.5 kg, basically depending upon the extent of urbanization. The economic growth of the population has had a far greater effect on the per capita waste generation than the population itself. In Mumbai, for instance, the population increased from 8.2 million in 1982 to 12.3 million in 1992, an increase of 49%, while the waste generation during the same period increased from 3,200 tones to 5,355 tones per day, an increase of 67% (Control of Urban Pollution Series, Central Pollution Control Board Publications, 2000). This trend is clearly attributed to the changing lifestyles and increase in living standards and newer food habits.

The management of Urban Solid Waste, generally called Municipal Solid Waste (MSW) and the term used in this research, is the responsibility of the Urban Local Body (ULB). The ULB is called a Corporation, Municipality, or Town Panchayat depending on the population of the urban centre it is in-charge of. The ULBs, irrespective of the size of population it is serving, are ill funded with limited revenue sources, and therefore ill equipped to provide adequate collection, transportation, processing and disposal facilities for MSW. The collection efficiency in most of the cities is about 72.5% and more than 70% of the cities lack efficient transport facilities (Pachauri and Sridharan, 1998).

There is no processing system, whatsoever in most of the cities. Few cities like Bangalore, Trivandrum and Mumbai have composting facilities. However, the facilities are incapable of handling the entire waste collected in the city. Open dumping is practiced extensively in almost all the cities, with little, if any, precautions taken against land degradation and contamination of groundwater by leachate. Breeding of rats, cockroaches and other pests are rampant in the dumping areas, resulting in the incidence of vector diseases (World Bank Reports, 2005).

The ULBs are further governed by Municipal Solid Wastes (Management and Handling) Rules 2000, Ministry of Environment and Forests, Government of India, which has extensive guidelines for the ULB to manage waste from their city. The ULBs therefore face a daunting challenge of reinforcing their available minimal infrastructure to ensure hygienic and scientific management and disposal of MSW. The ULBs, due to their limited revenues, must generate enough income for efficient management — either from the waste generators themselves, or by identifying activities that generate resources from the management of waste (Mathur, 2000).

India, in terms of primary energy consumption, ranks 5th in the world, and consumes 3.7% of the world's primary energy (Web resource: BP Statistical Review of World Energy, 2005 — www.bp.com). A GDP growth of 8% is set in the 10th Five Year Plan, and it is expected that the energy demand would grow at 5.2% annually (Energy Policy Committee Report, Strategy paper for Energy forecasting, Planning Commission, Government of India, 2000). Every year, about 590–880 million tones of methane is released to the atmosphere globally, of which about 90% is derived from the decomposition of biomass. India has a huge potential of biomass because it is a land of villages, and agricultural, forestry and agro-industrial residues are plentiful. One study estimated that India has a biomass potential at 12,000 MW of energy (Web resource: Ministry of Non-conventional Energy Sources, Government of India, 2000 — www.mnes.nic.in), whereas another study puts this value at 17,000 MW (Baktavatsalam, 1999).

In this context, Waste to Energy (WtE) projects were initiated by the Ministry of Non-Conventional Energy Sources (MNES) for the management of the organic fraction of MSW, under the auspices of the National Bioenergy Board (NBB), along with the development of technologies for various industrial wastes — pulp and paper, tanneries, distilleries, pharmaceuticals, etc.

Biomethanation is one technological option available which provides ample opportunity for:

1. Pollution prevention;
2. Energy derivative;
3. Compost derivative; and
4. Nutrient recovery.

Biomethanation, in contrast with composting, is a net energy producing process. Composting, an energy consuming process, requires 50–75 kWh of electricity for every ton of waste processed. Biomethanation, on the other hand, produces 75–150 kWh of electricity for the same quantity of waste (Web resource: Biogas and more, 2001 (http://websrv5.sdu.dk/bio/pdf/biogas.pdf)).

In this research, ventures with three different technological approaches and three different scales of operation are examined with the basic principle of these technologies being biomethanation or anaerobic fermentation.

This research is an attempt to evaluate the prospect of biomethanation as a viable technological opportunity for the ULBs to answer the needs for an environmentally safe and hygienic processing of MSW, and at the same time improving the waste management aspects of the entire city. The research aims at providing a guiding documentation process for the ULBs to appropriately select the technology most suitable for their city or town, which would best serve the purpose and best fits the present waste management practice. Further, the research also aims at answering the need for the technology in planning for the rural energy demand.

The objectives of the research are:

1. To analyze the technical effectiveness of the existing technologies in India for solid waste biomethanation and its viability in managing the organic fraction of municipal solid waste in cities and towns as well;

2. To examine the performance of these technologies for urban solid waste management;
3. To examine the technical feasibility of the potential of the technologies as a sustainable energy source as part of the rural energy project; and
4. To recommend the appropriate environment for the successful adaptation of biomethanation of solid waste.

Biomethanation, or anaerobic digestion as it is also known, is one of the oldest technologies as various historic evidences suggest. By the middle of the 19th Century, it was discovered that anaerobic bacteria are involved in the decomposition of organic matters. However, it has only been a century since anaerobic digestion was reported to be a useful method for the treatment of sewage and offensive material. In the rural areas of developing countries, especially in China and India, biomethanation was employed as a sanitary and energy source for centuries. Biomethanation was the most economic measure available to manage waste and generate fuel for cooking and heating. Even today, most of these facilities are prominent sources of energy and waste management systems in the two countries.

The biomethanation technology has not yet fully realized its potential as a possible energy production option. In most industrialized countries, biogas programs, especially for organic solid wastes, are often hindered by operational and maintenance problems, high capital investment and as yet low energy prices (Web resource: FAO, 2005 — www.fao.org/es/esc/common/ecg/330/en/Addendum_B3_Biogas.pdf). In most developing countries, the growth of biogas programs has been slow because of a lack of importance and the need for initial financial, as well as because of readily available and inexpensive non-commercial fuels such as firewood. The decrease in cattle farming practice has also resulted in a decline in the conventional biogas plants.

The degradation of organic matter to produce methane relies on the complex interaction of several different groups of bacteria. Stable digester operation requires that these bacterial groups be in dynamic equilibrium. Changes in environmental conditions can affect this balance, the consequence of which would be a buildup of intermediates which may hinder the overall process. It is of utmost importance to understand the basic microbiological and biochemical pathways.

A wide spectrum of organic wastes can be processed by biomethanation. Earlier, the technology was used for the treatment of sewage sludge and agricultural manures. For sewage sludge digestion, the technology is well researched and established. Farm scale digestion plants treating chiefly animal wastes have seen extensive use all over the world. These farm scale plants are simple stirred tank designs that use long residence times to provide the required treatment. The process of biomethanation is often also called anaerobic digestion, fermentation or gasification. Biomethanation occurs when wet organic matter or biomass accumulate and degrade in an oxygen-depleted environment.

The complexity of biomethanation reveals the biological part of an ecosystem that is also strongly influenced by chemical and physical parameters that is often referred to as environmental factors. The environmental factors that are relevant to the technology of biomethanation include the following (Angelidaki, *et al.*, 2003; Björnsson, 2000).

- Substrate temperature
- Retention time
- pH
- Nutrients
- Carbon/Nitrogen ratio
- Inhibitory factors

The end products of biomethanation process are biogas and digestate, which is a wet solid. The components of the biogas depend on the process of digestion, but are largely methane and carbon dioxide. The solid is a humus-like, stable, organic material, the quality and subsequent use of which is determined by the characteristics of the feedstock to the anaerobic process.

The production of biogas, which contains between 50 and 70% methane, is the most valuable aspect of the biomethanation process. Biogas is composed mostly of methane and carbon dioxide, with trace amount of other gases — hydrogen, nitrogen, carbon monoxide, and saturated or halogenated hydrocarbons. Usually, biogas is saturated with water vapour and may contain dust particles and siloxanes.[1] Water vapour and hydrogen sulphide, which are present in the biogas, are removed to protect machinery against corrosion before the gas is used for electricity generation. Ammonia is often removed as well.

The other alternative to producing electricity from biogas is the removal of carbon dioxide and selling it as natural gas. The market for this option

[1]These compounds are widely used in cosmetics and pharmaceutical products, and as anti-foaming agents in detergents. They react with oxygen during biogas combustion in internal combustion engines forming silicon dioxide which is a white powder.

is established in India, as well as other countries, because of the natural gas infrastructure. Biogas can be used for all applications designed for natural gas, subject to some upgrading, since not all gas appliances require gas with the same quality standards. The use of biogas for cooking is well established throughout the world. Biogas can be used for heating in boilers. The heat thus generated has various applications such as use in industrial processes and for producing steam. Also, boilers do not have a high gas quality requirement.

Biogas can also be used in Combined Heat and Power (CHP) units. CHP units are a good way to produce both electricity and heat efficiently. The utilization of biogas as a fuel for vehicles requires the same type of engine as those used for natural gas. However, the biogas needs to be upgraded because the gas quality demands are higher.

Methane contained in the biogas can also be used as a fuel for fuel cells. The fuel cells produce electricity in two steps; the fuel is converted to hydrogen and then to direct electric current. The by-products of the reaction are water and carbon dioxide.

Biomethanation uses carbon, hydrogen and oxygen from the feedstock, and the essential plant nutrients like nitrogen, phosphorus and potassium largely remain in the digestate. Also, the availability of nutrients is higher in digestate than in untreated organic waste. Nutrients are mineralized to allow improved plant intake. The use of digestate also benefits the humus balance in the soil and depends not only upon its quality but also local conditions and regulations. In order to obtain a high quality product, the digestate can be further processed into compost. It would ensure a complete breakdown of the organic components as well as fixing the minerals, especially nitrogen to humus, which would reduce nitrogen loss.

Biomethanation in India — the need

Reduction in the volume and mass of solid waste is critical especially in the light of limited availability of final disposal sites. Diversion of land for waste disposal would be physically impossible since areas with the largest concentration of solid waste would also be the areas with serious scarcity of vacant land. Although numerous waste and by-product recovery processes have been introduced, anaerobic digestion has unique and integrative potential, simultaneously acting as a waste treatment and recovery process.

A study conducted by the National Environmental Engineering Research Institute, India (Strategy Paper on solid waste management, 2000) estimated the Energy Recovery Potential of different industrial and urban sectors at 1,700 MW. Biomethanation for MSW, with a potential of over 900 MW, is beneficial from the view point of reducing the energy demand, and also in providing an environmentally friendly pathway for processing solid waste containing high proportion of degradable organic waste. Biomethanation is also potentially a compact, cleaner operation and technology with better product range.

Methodology

The research is primarily aimed at understanding the technological feasibility of the biomethanation technologies in the social context of the cities and towns through qualitative research approach.

The approach adopted in sampling is specifically availability sampling, where the subject is selected on the basis of availability. This approach was considered basically to incorporate the subject of variability. Semi-structured interviews were designed to gather detailed, qualitative descriptions of how programs operate and how stakeholders perceive them. Interviews were conducted one-on-one with the stakeholders involved — the ULBs, the local community, the technology provider and the plant operators. Questionnaires were prepared in a semi-structured manner to gather specific information from participants and also to allow new questions to emerge in the discussion. Participatory Rural Appraisal technique was adopted for the information gathering for the rural community. Pictorial representation was made to the rural community engaging them in a constructive discussion about their needs and requirements, relative to their local set up.

Results

Study Area 1: Lucknow

The municipal area of the city is 310 sq.km. The population served by Municipal Council or Lucknow Nagar Nigam (LNN), as it is locally called, is approximately 2.5 million. The quantity of the Municipal Solid Waste (MSW) generated is approximately 1,500 MT per day. The daily collection of solid waste is approximately 1,200 MT which includes demolition waste, with the help of about 4,000 sweepers.

As per norms set aside in Uttar Pradesh Health Manual, the required staff for the work should be 28 per 10,000 populations. The present staff should be around 7,000, which shows that LNN is lacking in staff strength. Out of 110 wards, 28 wards are un-staffed in sanitation for which ULB has engaged 10 persons for each of these wards on contract basis, which is still not enough to meet the requirement.

Two landfill sites namely Purnia and Moti Jheel on the Hardoi Road have been identified and waste is being dumped in an unsystematic manner. Because of rapid urbanization and industrialization of the city, severe strain is put on the natural resources and has ultimately affected the basic civic amenities and quality of life of the people. The MSW is disposed of haphazardly in open dumps. With growing pressure on land due to increasing population it is increasingly difficult for LNN to locate new disposal sites.

In order to overcome this difficulty the LNN has entered into a contract to process MSW generated in the city and to generate power and organic manure from its organic fraction in March 1999. M/s Enkem Engineers India Ltd, has built the power generation-cum-bio-fertilizer plant on a Build-Own-Operate (BOO) basis. The project would generate 5.1 megawatts (MW) of electric power per day (after captive consumption of 0.5 MW) and about 75 tons per day (TPD) of organic manure by treating 300 TPD of waste. The power generated by the company would be transmitted into the grid of the Uttar Pradesh State Electricity Board (UPSEB).[2]

Enkem Engineers arranged for Entec (an Austrian based firm) to provide the designated technology, the digester to produce methane, and a performance guarantee for it. The Ministry of Non-conventional Energy Sources approved the project as a demonstration project under the "*National Programme on Energy Recovery from Urban, Municipal and Industrial Waste*". LNN has agreed to supply a minimum 300 TPD of MSW. The power output would be sold to UPSEB under a 30-year power purchase agreement, extendable for an additional 30 years. The payments from UPSEB are guaranteed by the State of Uttar Pradesh. The corporation also agreed to provide land for the plant and waste at a concession rate.

Biogas Induced Mixing Arrangement (BIMA) Technology is employed. BIMA Technology is particularly designed as a high rate biomethanation unit for high BOD waste materials. The digester converts the waste materials into biogas and sludge. Developed in the 1970s, BIMA digesters can produce biogas from a variety of substrates like animal dung, kitchen

[2]Inputs from discussions with Commissioner, LNN and Mr. Satyanarayana V, IdeCK on 13th July 2005, and 10th June 2005 respectively.

waste, vegetable processing waste, slaughterhouse waste, etc. The digesters are reliable, robust and have low maintenance cost.[3]

Sathish Chandra Rai, Mayor, Lucknow commissioned the Municipal Solid Waste Treatment Plant (MSWTP) with electricity generation unit on 16 August, 2003. The power generation of about 1 MW was initiated as part of the commissioning phase and it was expected that by the end of December 2004, the power generation will increase to 5.1 MW utilizing the entire waste of the city.

The plant operated for about 8 months, including the pre-inaugural phase, and came to a stand still in December 2004. This was first reported in *The Indian Express*, a National Daily on December 24th, 2004. It was stated that segregated waste was not delivered for the plant operations. LNN informed that the contract did not specify anything about the segregated waste, but only said about the required quantity of waste per day. It said that the infrastructure for collection of segregated waste is still not in place for the entire city. Few wards have the door-to-door collection system in place.

The city is now back to square one with its problem of solid waste disposal. Two of the city's older dumping grounds at Jankipuram and Aishbagh are full, while the ones on Hardoi Road are fast running out of capacity. There has been no other infrastructure development in terms of identification of landfill sites or the design for sanitary landfill, which adds to the woes further.

Study Area 2: Vijayawada

The population of Vijayawada is 0.85 million as per the 2001 census (Web resource: Census of India, Ministry of Home Affairs, Government of India, 2001 — www.censusindia.net). With the present population estimated to be over 1 million, Vijayawada is fast approaching the status of metropolitan city. Several development projects are in progress. The Vijayawada Municipal Corporation (VMC) has been recognized as a technology institution by MNES and is actively involved in the appraisal of various wastes to energy technologies. A technical committee was constituted by MNES which made its recommendations.[4]

[3]Discussion with Mr. Subramani, Managing Director, Enkem Engineers Pvt. Ltd., Chennai, India on 24th June 2005.

[4]Interview with Commissioner, Vijayawada Municipal Corporation, Vijayawada on 25th July 2005.

The city generates about 450 tons of waste of which 400 tones are collected. Of this, about 120 tones are supplied for converting into manure. About 250 tones are supplied for generating power. About 20 tones of waste — 16 tonnes of vegetable market waste and 4 tonnes of slaughterhouse waste and sewage from the sewage treatment plant for dilution, are supplied to the biomethanation plant.[5]

The plant was commissioned in February 2004, and is operational 24 hours a day. There have been issues on the quantity of waste not being adequate but the the quality has never been a problem. Separate transportation trips are made containing the market waste and the slaughterhouse waste to the plant site.

The biogas so produced is being used in a 145 kW biogas generator set for generation of electricity. The electricity generated is used for:

1. Overall energy requirement of the plant operations;
2. Energy supplied to the adjacent wastewater treatment plant for its operations; and
3. Additional energy supplied to the state grid.

Study Area 3: Siruguppa

As per the 2001 census, the population of the town was recorded as 43,600 and the present population is estimated at about 50,000. The population has a mix of people from both the states of Karnataka and Andhra Pradesh, since this is a border town.

Siruguppa town is the taluk headquarters. It is governed by a Town Municipal Council, (TMC) which is responsible for water supply and sanitation, MSW management, roads, street lighting and other infrastructure services. However, scarcity of drinking water, especially during the summer months, takes up the entire time and work force of the TMC. TMC depends on a range of revenue sources like taxes on properties and sites, water charges, grants from central and the state governments and loans from financial institutions.

In terms of SWM, TMC is equipped with three tractors, supported by a staff of 36 waste collectors or *pourakarmikas*,[6] supervised by Sanitary

[5]Discussion with the Engineering Department, Vijayawada Municipal Corporation, Vijayawada on 25th July 2005.

[6]Pourakarmikas are workers employed by the government or on contract, to do work such as cleaning streets, collecting garbage, removing posters and other work to keep the city clean.

Inspector. Waste is collected from the main roads and the 20 wards of the town everyday. The TMC collects about 4–5 tractor loads of waste per day from the Siruguppa town.[7]

Technology Informatics Design Endeavour (TIDE), a Bangalore based Non Government Organization, in technical collaboration with Centre for Sustainable Technologies (formerly Centre for ASTRA), Indian Institute of Science, Bangalore, has been involved in the dissemination of the technology of bioreactors. Accordingly, a proposal for setting up a biomethanation processing unit for organic fraction of Municipal Solid Waste (MSW) at Siruguppa was submitted by TIDE to Infrastructure Development Corporation (Karnataka) Ltd., (IDeCK), Bangalore. Immobilized cell bioreactors were developed and deployed.

The Town Municipal Council (TMC) provides the daily feed for the bioreactor.

The plant was commissioned for operations using the organic fraction of MSW on 1st September 2003, and is operational. The biogas generation has been around 7.5 to $10.0\,m^3$ per day, depending on the type and the quantity of the waste. The waste collection is typically by street sweeping, and therefore lot of silt gets along with the waste. Since the organic waste is wet, the silt 'sticks' to the waste making it unusable. Therefore, the waste feed had been isolated to one module, instead of distributing among the three modules.

TIDE has carried out a series of experiments to evaluate and assess the possible use of biogas generated from the bioreactor, so that an optimum system for gas sales could be evolved. Experiments have been conducted for the utilization of biogas for alternate uses such as:

- Lighting using biogas lamps;
- Generation of power using a 1.5 kW power generator set; and
- For cooking

There have been incidents of waste not being supplied to the site at all (Arun Kumar, 2004). The waste transporting vehicles have often been diverted to alternate works, depending upon priorities. Because of the non-availability of sustained supply of waste to the site and a solid support by the TMC, no tangible gas use option for commercial exploitation has

[7]Discussion with Chief Officer, Town Municipal Council, Siruguppa, and observations on 29th July 2005.

been established. At present, the operators and neighbours use the gas for cooking purposes.

Comparative Analysis

A field visit was made to the three study areas during June and July 2005, and several discussions and meetings were held, both formally and informally, and data was collected.

Discussions were also held with several people, both in the Government, as well as in the Non-Governmental, private and NGO sectors. Since the research approach is qualitative, all information collected, both primary and secondary, has been analyzed and referenced in footnotes.

For ease in interpretation, the following nomenclature is used for each of the study areas:

Lucknow — Study Area 1 (SA1)
Vijayawada — Study Area 2 (SA2)
Siruguppa — Study Area 3 (SA3)

A number of potential benefits emerged during the discussions in the field survey. These potential benefits can be seen as driving forces for the introduction of biogas. The benefits have been divided into the following five classes:

• **Energy related benefits**

Quite evidently, biogas has been used as a viable energy resource. At SA2, electricity produced by the combustion of biogas is being used for three purposes — all the power driven activities within the site, operations of the adjacent wastewater treatment plant, and supply to the grid. At SA3, although no tangible use of biogas exists, options like dual fuel operation of diesel generator for pumping water in the adjacent paddy fields and for lighting purposes replacing kerosene lighting systems for evening hawkers are proposed.

• **Fertilizer related benefits**

There has been no substantial monetary benefit yet from the sale of digestate as a fertilizer for field application. The potential however is strong because of two factors: the waste management plant is located outside the city limits, in the vicinity of fields, thus providing a ready market, and shift of the farmers towards the use of organic fertilizer. However, the quality of the digestate is a matter of concern because of the strong possibility of heavy metal contamination although digestate analyzed at SA3 did not reveal any such contamination. The Indian Agricultural Research Institute is studying the implications of application of digestate on to the farmlands (Web resource: Indian Agriculture Research Institute, 2005 — www.iari.res.in).

- **Health related benefits**

There are no direct health benefits. However, there have been sanitation improvements which can be considered as potential health benefits. The elimination of organic waste from being dumped in the landfill sites has substantially reduced scavengers like dogs, pigs and crows. Unattended waste which attracts flies, rats, and other creatures that in turn spread disease is eliminated to a very large extent. Also, the waste workers and rag pickers, who handle the waste directly, are at a lower risk of contracting infectious and chronic diseases.

- **Development related benefits**

The scale of activity is quite low to assess the development related benefits of biomethanation at the SAs. There has been employment created locally and interest groups from other cities are evaluating the possibility of such an intervention in their cities.[8]

One indirect benefit was observed at SA2. The wastewater treatment plant which never operated at its full potential because of regular power disruption is now operating well, after the biomethanation plant was set up.

- **Economic benefits**

The economy of biomethanation technology is often brought forward as one of its major drawbacks. This is factored due to the high capital investment, operational costs, waste segregation costs, as well as low cost of fossil fuels in terms of unit value. However, a long term sustainability point of view is required to assess the economic benefits rationally. The reduction in the carbon emissions, recycling of nutrients, local energy availability, reduction in oil imports, etc. are few substantial benefits of the technological intervention. Clean Development Mechanism of the Kyoto Protocol could be a direct economic benefit.

A SWOT analysis indicates that the strengths are the environmental and the socio-economic benefits for the city as well as for the society through the technological intervention of biomethanation. Besides the benefits

[8]Discussion with Mr. Rama Kumar C.M., Deputy Advisor, Energy and Environment, Karnataka Urban Infrastructure Development and Finance Corporation (KUIDFC), Bangalore on 15th June 2005.

of energy and fossil fuel replacement, and the production of digestate, these technologies provide the perfect opportunity for entrepreneurial intervention suitably supported by the kind of initiatives adopted by the Government.

However, the institutional set up, both in terms of public involvement and public-private partnerships, is lacking in transparency and thereby failing. The lack of institutional capacity is addressed in two stages — the capacity building of the staff within the ULB, the most important stake-holder in the SWM initiatives, and the environmental structural enhancement within the city, with public involvement.[9] Few cities like Bangalore have shown the involvement of citizens in the waste management initiatives through the set up of Resident Welfare Associations (RWAs) (Mull, 2005). Reference providing public education is absent in all the three SAs. It is evident that the major weaknesses posed by the technological intervention are not technical weaknesses but institutional set up. The high capital investment is a matter of concern, especially for investments in smaller towns and cities. This is addressed by the policy set up by the Government of India under the National Bioenergy Board, through the MNES.

However, in spite of the weaknesses and threats, the opportunities for the success of the biomethanation option are quite high owing to the following main reasons:

- The increase in the energy demand;
- The increase in the quantity of MSW;
- Decreasing land for disposal/dumping;
- Water scarcity and threat of ground water contamination;
- Increased environmental consciousness among citizens; and
- Support from MNES.

Discussion

The primary data from two study areas — SA2 and SA3 — revealed that the biogas generated was of acceptable quality. The gas samples were analyzed by gas chromatography at the Centre for Sustainable Technologies, Indian Institute of Science, Bangalore.

[9]Discussion with Mr. Shivashankar, Executive Engineer, Directorate of Municipal Administration, Government of Karnataka on 16th June 2005.

Discussion with Mr. Satyanarayana[10] revealed that the SA1 was operational for about 8 months, producing about 0.5 MW i.e., 10% of the total design capacity. All the three study areas revealed that the biogas produced is of substantially good quality for combustion and energy production.

One important observation made at all the three SAs is that the SWM infrastructure has not been developed holistically. In all the three SAs, there was no source segregation of the waste. Door to door collection systems were absent in SA3, and present only in a few parts of the city in SA1 and SA2. Even the segregated waste was mixed with other waste during the secondary transportation or at the transfer stations due to the lack of transportation vehicles, and partly also because of the lack of knowledge among the waste handlers. The reason for the closure of one of the plants (SA1) has been quoted to be "*lack of good quality*" segregated organic waste.

There have been incidents where Project Developers have initiated door-to-door collection systems to demonstrate to the ULB that it is feasible to collect and implement such a system. The Project Developers at SA3 carried these out in 2 of the 20 city wards for about 3 months, with a collection of a fee from each house, which became part of the salary of the waste collector.

Another important observation made was that all the plants were operated by the technology developers themselves, and the participation of the local Government was minimal, if not completely absent. Talking with the ULB employees,[11] there was a clear indication of their reluctance to take up the operations of the plant, and their content to let the developers handle the O & M, in-spite of the fact that the ULB is one of the important stake-holder representing the Government at the local level.

Energy is a central element in a development agenda. India has more than 60% of its population in rural areas. Economic development is intrinsically related to energy supply to these people. The implementation of rural electrification in providing social benefits are numerous; a few

[10]Mr. Satyanarayana V is Head, Development, Infrastructure Development Corporation (Karnataka) (IDeCK), Bangalore, and also a member of Technology Advisory Group on Solid Waste Management setup by the Ministry of Urban Development and Poverty Alleviation, Government of India. IDeCK financed the project at SA3, and is a subsidiary of IDFC, which co-financed the project at SA1.

[11]From discussions with Chief Officer, Town Municipal Council, Siruguppa, and Commissioner, Vijayawada City Council, Vijayawada on 29th July 2005 and 25th July 2005.

of these benefits are (Reich *et al.*, 2000 — http://www.martinot.info/Reiche_et_al_WP2000.pdf).

1. Improved living standards — amenities and services;
2. Reduced household drudgery, especially for women;
3. Reduced negative impact of energy use on health and local environment;
4. Opportunities for economic entrepreneurship thereby increased employment both direct & indirect on supply side as well as on demand side like rural industries, productive uses, etc.;
5. Synergy benefits due to bundling of services; and
6. Enhanced and more effective communication with the world beyond the village.

Despite this, in the first decade of the 21st century, rural India remains in darkness. More than half of India's poorest households have no access to electricity. In 2001, 78 million of India's 138 million households did not have access to electricity. Rural electrification is hard to accomplish. Rural households are remote and therefore, costly to serve. They do not use very much electricity, which makes such a service relatively unprofitable. The distribution losses have reached staggering proportions in most of the regions, accounting, for more than half of the total power generated, largely due to theft, non-metering, poor collection and inefficiencies in the distribution network. Five regional grids operate in India, with regional grids connecting state transmission networks within a region. Unfortunately, progress in setting up of an integrated national grid has been slow due to poor co-ordination among regional authorities, and technical and financial problems (Ghosh, 2003).

Among the various renewable energy sources, biomass conversion technologies appear to be one of the best suited for conversion to shaft power/electricity. This is due to the fact that biomass is available all year round. In addition, it is cheap, widely available, easy to transport and store, has no environmental hazards and can be obtained from plantation of land which has no competitive use.

Traditional biomass resources still meet an overwhelming proportion of the rural energy demand. Biomass fuels — fuel wood, crop residues and animal dung provide 85–90% of the domestic energy and 75% of all rural energy. The present level of consumption of fuel wood is around 200 million tones per annum (TERI, 2005). Biogas technology has been a particularly useful system in the Indian rural economy, and has the potential to fulfill several end uses. The gas has successfully been useful as a fuel substitute for firewood, dung, agricultural residues, petrol, diesel, and electricity, depending on the nature of the task, and local supply conditions and constraints thus supplying energy for cooking and lighting. Currently, there are thought to be about 3.67 million household and community biogas plants installed around India (MNES, 2005).

From the perspective of biogas generation from organic solid waste in rural India, an assessment of the biomass or non-plantation agri-residue is made. In India, out of the total geographic area of 328 Mha, the net cropped area accounts for about 43% and has stabilized around 140 Mha since 1970. The residue production varies from crop to crop. The total crop residue production in India during 1996–97 was estimated to be 626Mt of air-dry weight. The dominant residues are those of rice, wheat, sugarcane and cotton, accounting for 66% of the total residue production. Sugarcane and cotton residue production is 110 and 50 Mt, respectively (Ravindra *et al.*, 2005).

There has been a significant decrease in the quantity of dung availability in the farms through declining cattle population per capita which is attributed to problems such as difficulty in collecting fodder, and land requirement for a growing number of households and for the cultivation of crops.

These reasons show that there is a clear requirement as well as role of technology which has the potential to power the kitchens in the villages, not only of the families with cattle but also of those without any (Ravindra *et al.*, 2005).

Case Study

A case study is presented to highlight the perspectives of the rural people towards biogas generated from organic solid residues (Arun *et al.*, 2005). Details provided by Biomass Energy for Rural India (BERI), a project jointly supported by the Government of Karnataka, MNES of the Government of India, the India Canada Environment Facility (ICEF), the Global Environment Facility (GEF) and the United Nations Development Programme (UNDP), aims to create a sustainable, decentralized and community-based power generation and distribution system to provide "*comprehensive and qualitative energy services for the rural population*" (Web resource: UNDP, 2005 — www.undp.org).

BERI, as part of its mission towards bio-energy security, has set up two leaf-litter based biogas plants at two villages in Tumkur district, about

90 km from Bangalore. BERI signed an agreement with TIDE, Bangalore with technical assistance from CST, Indian Institute of Science, Bangalore, for the installation of a community leaf-litter based bioreactor for two of its villages adopted under the bio-security program. The case study describes one of the villages.

A small bioreactor generating about 6-8 m^3/d of gas and supplying cooking gas to 7 households with 36 residents was constructed. The bioreactors were commissioned using pre-prepared start-up inoculums that were fed into the system.

Problems were encountered in the field:

1. Construction defects especially in the inlet and outlet sections of the bioreactor were identified;
2. Social conflicts between the households in the village;
3. Gas pressure was not sufficient for the gas to travel through the gas distribution network and biogas burners; and
4. Gas production at the desired capacity took a longer time, resulting in the fall in the excitement of the villagers.

Over a period of time when local adaptation of technology was complete, the irregular and insufficient availability of leaf litter also became an additional problem. All this has resulted in the "*failure*" of the technological intervention in providing the biogas for the villagers.

Further, the project holders perceived that the technology of the bioreactor was something "*off the shelf*". A process of learning was involved which the project holders failed to appreciate, resulting in the let down of the technology. There was a clear need for the project holders to address the issues in a holistic approach, rather than the stopgap process as permitted by the project.

Of the three technologies for biomethanation examined, and study of one of the technology in the context of a rural setting, it is a qualitative assessment to ascertain their viability as a sustainable source of energy to the rural people. Each has its own merits and demerits for incorporation into the rural energy scenario.

From the rural perspective and as evidently shown in the case study, the requirement as an energy solution is a simple working system, with minimal technical or operational intricacies. There is no lack of manual labour for plant operation. Availability of time is not a factor as long as the time spent on the plant operations and collection of biomass/weeds for the plant feed does not exceed the time spent on the collection of firewood for cooking. On

Table 21.1. Relative merits and demerits of the three biomethanation technologies.

Parameter	SA1	SA2	SA3
Technical competence	Yes	Yes	Needs refinement
Suitable for any type of organic feed	Yes	Yes	Yes
Requirements of pre-treatment of feed	Yes, pulverizing carried out	Yes, pulverizing carried out	Yes, cutting into size of less or equal to 1 ft
Operation time	Quick	Quick	Time consuming
Requirement for skilled labour	Yes	Yes	No
Electricity requirement	Yes	Yes	No
Manual labour intensive	No	Partially	Yes
Infrastructural requirements civil works	Substantial	Substantial	Substantial
Requirements for Mechanical and electrical infrastructure	Yes	Yes	No
Utilization of local material	No	No	Yes
Durability of the plant	Yes	Yes	Significantly yes, Because no mechanical/ electrical components

the other hand, there is a shortfall in the skills of the plant operator. The technology best suited to the rural setting is the one developed at Indian Institute of Science, Bangalore.

The major drawback for rural energy per se for the technology implemented at SA1 and SA2 is that they require skilled labour for plant management, and energy for the plant's operations. Owing to the remote location of the village, the level of skills available, the scale of operations, economic suitability, etc., the other technologies are still not ready for intervention in a rural setting.

There are a few criteria, which have to be followed while designing a biogas system for a rural community, besides the technical and the operations design protocols. A few of them are listed below (Matthew and Wim, 2000).

- Involvement of the local people along with the site-specific policies is an essential prerequisite;
- Detailed study on the resource availability in the vicinity of the village;
- Socio-economic condition of the people in that area;
- Demand forecasting and willingness of rural households to pay for good quality energy services;

- Understanding the end-user/market and design the product that meets the needs and addresses the concerns;
- Identifying the most appropriate and cost-effective design for the product;
- Establishing and enforcing solid design, quality and service criteria that will ensure the reliable and cost-effective operation of installed plants;
- Identifying the key institutional players and assisting in strengthening the capacity of these players to effectively carry out their respective roles;
- Securing the commitment and support of financial institutions to work in close partnership for the dissemination and financing of the product;
- Identifying the financial incentives needed to stimulate the market and attract qualified buyers;
- Providing technical and management support to all key players; and
- Instituting coordinating committees to ensure the co-operation and partnership of stakeholders; and
- Maximizing the use of program resources for product support and market development.

Conclusion

The biogas programme in India has evolved over a long period spanning several decades, including national and international crises. While there have been pockets of poor performance, there are also pockets of good performance.

In order to make biomethanation technology a viable option for energy, both in managing the organic fraction of municipal solid waste, as well as satisfying the energy requirements of the rural communities, the framework approach adopted must be oriented towards minimizing environmental degradation, conserving non-renewable resources and realizing the full potential of renewable energy sources. The technological efficacy of biomethanation for organic solid has been proven. It is the barriers of implementation of the technology as a sustainable option that have to be overcome.

On a macro scale,

- A national policy needs to be developed which would consider the issues of selection of appropriate technology, financial support management, public and private sector participation, etc.;

- Research and Development to assess the right technology and its suitability for Indian conditions needs to be carried out more vigorously. It is necessary to analyze critically the reason for "*failure*" of biogas technologies and its impact on biomethanation;
- Renewable energy systems, including biomethanation, have to compete with subsidized fossil fuel energy systems that do not reflect the true environmental costs. This scenario has to change;
- Continuous monitoring and evaluation of the technological developments and adaptation should be carried out for technical, operational and commercial feasibility;
- The present solid waste management systems must be improved to meet the operational requirements of biomethanation;
- Funding should be oriented towards overall development of factors related to biomethanation like solid waste management infrastructure, public awareness and involvement, etc.;
- Urban Local Bodies must be further strengthened with powers to impose fines against defaulters who do not participate in the solid waste management program;
- Detailed survey has to be carried out to ensure steady market and its potential for both biogas and digestate obtained from the biomethanation process; and
- Public awareness should be carried out to highlight the importance of biomethanation in reducing health hazards and sanitation problems.

On a micro scale, at the grass-root level,

- The problem and the need must be understood;
- Local stake-holders must be engaged actively;
- In-house training for the Urban Local Body staff to make them aware of the procedures and techniques; and
- Provision of incentives to attract private organizations in plant operations.

India is the only country in the world to have created a separate ministry for non-conventional energy. The institutional setting is aptly supported by the national policies and framework. A National Bioenergy policy, rational energy pricing, private sector participation, large-scale demonstration programmes and a participatory approach are some of the initiatives requiring immediate attention to realize the potential of biomethanation technology as a sustainable source of energy in the years to come. If biogas technology can stand up to its absolute potential, it can make a real difference for rural people. It has however to be understood that biomethanation is not an automatic or mechanical process, and will yield favourable results only if managed in an appropriate manner.

References

Angelidakin I, Ellegaard L, Ahring B K, 2003. Applications of the Anaerobic Digestion Process, (1) Environment & Resource, The Technical University of Denmark, 2800 Lyngby, Denmark.

Mull JE, 2005. Approaches Toward Sustainable Urban Solid Waste Management: Sahakaranagar Layout, Thesis, Masters in International Environmental Science, Lund University, Sweden.

Arun Kumar RS, 2004, Pre-closure Report — Bioreactors for the Organic Fraction of Municipal Solid Waste at Siruguppa, TIDE, Bangalore.

Arun Kumar RS, Chanakya HN, Bhogle S, 2005, Field experience with leaf litter-based biogas plants, Energy for Sustainable Development, Vol 9, No.2.

Baktavatsalam V, 1999, The Indian market for sustainable energy, IREDA News.

Biogas and more! — Systems and markets overview of anaerobic digestion, July 2001, AEA Technology Environment, Culham, United Kingdom.

Bjornsson L, 2000. Intensification of the biogas process by improved process monitoring and biomass retention, Department of Biotechnology, Lund University Sweden.

Energy Policy Committee report, 2000, Planning Commission, Government of India.

Ghosh D., 2003. Renewable Energy Strategies for Indian Power Sector, Paper presented at Centre De Sciences Humaines (CSH).

Central Pollution Control Board, 2000 Management of Municipal Solid Waste, Delhi.

Control of Urban Pollution Series, Central Pollution Control Board Publications, Ministry of Environment and Forests, 2000.

Mathur MP, 2000, The Spatial Context — Finances and functioning of urban local bodies — A situation report, Paper prepared for Infrastructure Development Report 2000-01- New Delhi.

Matthew SM and Wim J N. 2000 Biogas in rural household energy supply: The Nepal Biogas Support Program, Policy and Best Practice Document No. 4, Ministry of Foreign Affairs, Netherlands.

Strategy Paper on solid waste management, 2000, National Environmental Engineering Research Institute, India.

Ravindra NH., 2005, Assessment of sustainable non-plantation biomass resources potential for energy in India, Centre for Sustainable Technologies, Indian Institute of Science, Bangalore.

Reiche K., 2000, Expanding Electricity Access to Remote Areas: Off-Grid Rural Electrification in Developing Countries, World Power — http://www.martinot.info/Reiche_et_al_WP2000.pdf.

Pachauri R K and Sridharan P V (eds). 1998, Looking Back to Think Ahead: GREEN India — 2047 (Growth with Resource Enhancement of Environment and Nature), New Delhi: Tata Energy Research Institute.
World Bank Reports, 2005, Solid Waste Management Initiatives in small towns — lessons and implications, Water and Sanitation Program.
Municipal solid wastes (management and handling rules) 2000, Government of India.

Web resources

BP Statistical Review of World Energy 2005 — www.bp.com
Census of India, 2001 — http://www.censusindia.gov.in/
MNES 2000; 2005 — http://mnes.nic.in/
FAO 2005 — www.**fao**.org/es/esc/common/ecg/330/en/Addendum_B3_**Biogas**.pdf
TERI, 2005 — http://www.teriin.org/index.php?option=com_casestudy&task=details&sid=1
UNDP 2005 — http://www.undp.org

22

HAZARDOUS WASTE MANAGEMENT IN INDUSTRIAL ZONES, HO CHI MINH CITY: PRESENT SITUATION AND RECOMMENDATIONS FOR IMPROVEMENT

DANG THUY TRANG and TING YEN PENG

Ho Chi Minh City, Vietnam, with 12 industrial zones and export processing zones and an annual growth rate of 15% in the manufacturing sector, produces large amounts of industrial waste, including hazardous wastes. Due to their nature and characteristics, it is particularly important that hazardous wastes are managed in an environmentally sound and sustainable manner. To investigate the current hazardous waste management system in industrial zones in the city, this study focuses on the practice of hazardous waste handling, including all facets of a management system from generation to final disposal. Several problems in this system have been identified. The poor awareness of the hazardous waste generators has led to lack of segregation between hazardous and non-hazardous wastes in their premises. These wastes are then usually collected and transported by non-licenced collectors to be disposed of in a landfill only suitable for domestic wastes. Furthermore, there is currently an absence of incineration facilities and dedicated landfills for hazardous wastes in Ho Chi Minh City. As these problems have been identified and understood, recommendations are made, including development of planning and administrative, legal, financial, and engineering infrastructures. They are proposed for the purpose of moving the current hazardous waste management practice in Ho Chi Minh City towards an integrated system.

Introduction

Ho Chi Minh City (HCMC) is the largest city in Vietnam with a population of 5.12 million (HCMC PC, 2003), covering an area of 2,053 km^2, of which the urban area accounts for about 600 km^2. It is located on the western bank of the Saigon River, approximately 80 km inland from the Eastern Sea. HCMC has a tropical climate with temperature ranging from 21 to 29°C and annual average and maximum temperature of 29°C and 40°C respectively. The city has about 154 days of rainfall each year with an annual average rainfall of close to 2,000 mm. Average annual evaporation is approximately 1,400 mm per annum (HCMC PC, 2003).

HCMC contributes as much as 30% of the industrial output of Vietnam and has an economic growth rate of 11.2% in 2004, or 10.5% per annum between 1997 and 2004 (HCMC PC, 2004a). It is the commercial centre of the country and has seen large increases in population and urban environmental problems over the last 15 years. The city supports about 800 factories of medium to large capacity and as many as 30,000 small to medium-scale enterprises (SMEs). Many large factories have little or no pollution control facilities and the multitude of "back-yard" enterprises is virtually unregulated (HCMC DOSTE, 2002b). Only a small number of the industries are located in 12 industrial zones (IZs) and export processing zones (EPZs) while the rest are scattered among the residential areas.

HCMC is creating on its development pathway a substantial amount of waste, especially hazardous wastes. This problem needs to be addressed; hazardous wastes must be managed in an environmentally friendly manner if sustainable development is to be achieved. An integrated hazardous waste management system is required to be formulated and implemented.

The scope of the study is confined to the management of hazardous waste generated from the EPZs and IZs in HCMC at present. The report examines every facet of hazardous waste management, from its generation, storage, and pretreatment/treatment (if any) within the EPZs and IZs, to the final disposal.

There are currently three EPZs and nine IZs in HCMC, which have attracted 336 foreign direct investment (FDI) projects and a total capital of US$ 511,584,088 as of August 2003.

Vietnam's industry is still predominated by state-owned enterprises. The efforts of the government to partially privatise state-owned enterprises since 1992 resulted in only 17 companies in 1998 (Nguyen and Kim, 2001). Most state-owned enterprises were over-staffed, suffered from a lack of working capital, and could not retrench staff during the restructuring process in conjunction with equitisation. Most of the existing enterprises are old with obsolete equipment, technology and facilities, thus resulting in inefficient use of raw materials, high leakage and serious pollution. They usually only concentrate on maximising their profits, even when this is achieved at the expense of pollution problems. 30,000 enterprises are using production technologies that are 20 years old (Nguyen and Kim, 2001). In an attempt to reduce costs, industries in HCMC are known to import outdated, second-hand equipment despite a prohibition in both the Law and Environmental Protection (1993) and the updated National Law on Environmental Protection (2005) against importing technology that does not meet environmental standards. One source estimates that up to 50% of all industrial equipment imported into Vietnam is second-hand (Vui, 1998). Only 20% of existing enterprises have undergone technological renovation (Socialist Republic of Vietnam, 2000). This is both because of economic constraints and attention being focused on economic development rather than environmental protection.

Another important characteristic of the industries in Vietnam is the planning of most existing IZs without environmental considerations. As a result, industries are scattered across densely populated urban areas (Nguyen and Kim, 2001). Lack of development control has also led to situations where factories with operations considered hazardous are located in residential areas. HCMC is typical of this injudicious zoning; factories and residences in this city are often separated by no more than a few metres.

Because of the above mentioned problems, the pollution loads and the pollution level have significantly and rapidly increased (Nguyen and Kim, 2001). HCMC's authorities are challenged with the task of managing hazardous wastes in an environmentally sound and sustainable manner.

Apart from the common difficulties faced by other cities in the country, namely the lack of financial resources and capable manpower, and the absence of efficient coordination between the government agencies, HCMC also faces other challenges. Some of these include its large physical size, its number of EPZs and IZs (12), its huge number of SMEs (approximately 30,000), and consequently its quantity of hazardous waste generated from these industries. The hazardous waste management system for HCMC in general and for the EPZs and IZs in particular is insufficient, or even totally lacking. Therefore, the problems that such a system is causing to the living environment are of great significance: air pollution, water and land contamination, and its effect on human health, etc. to name a few. This study attempts

- To review the hazardous waste management in the EPZs and IZs in HCMC;
- To identify problems associated with the current practice; and
- To make some recommendations for an integrated hazardous waste management system for HCMC in the future.

A phenomenon, from years 1996 to 1997, is the appearance of IZs. These are zones with modern infrastructures that were built with the aim of attracting foreign and domestic investors and to concentrate industrial production (Pham *et al.*, 1999). As of August 2003, HCMC has three EPZs and nine IZs (out of 60 in the country) with 336 foreign direct investment projects and a total capital of US$ 511,584,088 (HEPZA, 2003; NGIPS, 2003a, 2003b, 2003c). A total of more than a billion dollars has been invested in the construction of factories. About 45,000 employees found jobs in these IZs (Le, 2002a). Many of these workers come from bordering regions and thus there is a great need for accommodation for these people. As a result, ad hoc residential areas have been established right next to these zones with insufficient regard paid to public health and the environment (Truong *et al.*, 1996).

HCMC is planning to move its heavily pollutive industries into IZs. Furthermore, according to the Socio-economic Master Plan for HCMC for the period 1996- 2010, HCMC will grow at 14–15% per annum over the next fifteen years (HCMC PC, 1995). This growth and the re-allocation of industries into IZs will result in an increasing number of IZs. With the influx of migrants to these zones for job opportunities, the demand for accommodation will thus rise accordingly and the impacts of these zones on the neighbouring residential areas will be highest. Yet, early Vietnamese economic planners and policy-makers did not envision the impacts that growth and development could have on human health and the environment. Consequently, environmental considerations, particularly hazardous waste management, were not taken into account during the planning process of the IZs, leading to many problems that will be further discussed below.

Legislation and Regulation

National Law on Environmental Protection

The main regulatory controls governing waste management and environmental protection are embodied in the National Law on Environmental Protection (NLEP) Number 52/2005/QH11 dated 29 November 2005. Section 2 (Articles 70–76) in the NLEP deals specifically with hazardous wastes, stipulating that:

- The Ministry of Natural Resources and Environment (MONRE) shall provide regulations on eligibility of hazardous waste management, regulations on keeping profiles, registration, permission and code for hazardous waste management.
- Hazardous waste generator must self-organize the classification and collection of hazardous wastes for treatment or contracting them to professional hazardous waste management organizations. Hazardous

wastes must be temporarily stored in places managed by generator or by professional management organizations, or stored in places permitted by MONRE.

— Hazardous waste must not be mixed with other waste.
— Hazardous wastes must be transported with specialized equipment, on separate routes and time specified by competent traffic management agencies.
— Hazardous wastes must be treated to meet environment standards using specific technologies and equipment appropriate to each type and chemical and physical characteristics of hazardous wastes. In case there is no treatment technology and equipment available in the country, hazardous wastes must be stored according to specific regulations on hazardous waste storage and technical guidance provided by MONRE until treatment technology is available.
— Only licenced organizations and individuals shall be permitted to participate in hazardous waste transport and treatment activities.
— Hazardous waste landfill sites must comply with a number of requirements with regards to planning, design, technical specifications, equipment, environment conditions and inspection. Ministry of Construction takes the lead in and coordinates with MONRE to issue technical specification and guidelines for inspection and certification of hazardous waste treatment establishments and landfill sites.
— Ministry of Construction takes the lead in and coordinates with MONRE and Provincial People's Committees to formulate the national master plan for collecting, treating, landfilling hazardous waste to be submitted to the Prime Minister for approval. Provincial People's Committees are responsible for allocating sufficient space for constructing hazardous waste landfill sites according to approved planning.

MONRE's remit also extends to ensuring compliance with the environmental aspects of international conventions and agreements to which Vietnam is a party. More than 20 such conventions have been signed in the past 20 years including several which have reference to the movement and disposal of hazardous wastes, such as:

- The Basel Convention on the Control of Transboundary Movements of Hazardous Wastes and their Disposal; and
- MARPOL 73/78 (International Convention for the Prevention of Pollution from Ships);

Decree 175/CP (National Assembly of the Socialist Republic of Vietnam, 1993) sets out the roles of MONRE, the DONREs, other ministries and the People's Committees (PCs) with regard to environmental management and the relations between them.

The Decree recapitulates some of the provisions of the NLEP regarding hazardous wastes, including the following:

- the Decree stipulates that "discharged substances" containing toxic substances should be segregated and appropriately treated;
- requires all production and business establishments (specifically including hospitals) to treat their wastes before discharge, if necessary, to meet environmental standards and to seek approval for the treatment technology used from the competent environmental authority; and
- prohibits the import and export of potentially harmful "discharged substances".

Decision 155 "Hazardous Waste Management"

Decision 155 (Vietnam's Government, 1999b) was issued by the Prime Minister on 16 July 1999. Although it does not deal with hazardous wastes from IZs specifically, it is the most relevant and applicable regulation in Vietnam. The regulation requires hazardous waste producers to register with the local DOSTE and report details of their waste according to a prescribed format. Several of the Annexes of Decision 155 relate to reporting on hazardous waste management. Since this is the major and most important legislative document for hazardous waste generators, collectors, transporters, storers, treaters, and disposers to follow, more details of this Decision and its problems will be reviewed and discussed below.

Strategies

National Strategies for Environmental Protection (NSEP)

An NSEP for the period 2001–2010 has been prepared by Ministry of Science Technology and Environment (MOSTE) and Vietnam National Environment Agency (VN NEA) and submitted to the Government. The strategy identifies three over-arching objectives: (i) continue to prevent and control pollution; (ii) protect, conserve and sustainably use natural and biodiversity resources; and (iii) start improving environmental quality in urban, industrial and rural areas (Le, 2002b).

The strategy lists action programmes, one of which is related to hazardous wastes:

- Improve solid waste management capacity, especially hazardous waste management in densely populated urban areas and IZs.
- Prepare a strategy by 2002 and further develop the current Regulation on Hazardous Waste Management.

National Strategy for Hazardous Waste Management 2001–2010

This strategy includes the planning of engineered landfills as replacement of traditional landfills, and guideline for hazardous waste management. It also identifies the institutional arrangement and managerial and financial requirements to resolve the problems of hazardous wastes.

Strategy on solid waste management in urban areas and industrial zones

This solid waste management strategy in urban areas and IZs of Vietnam towards the year 2020 sets the targets up to 2005 and 2020 regarding hazardous wastes (Vietnam's Government, 1999a).

The targets up to 2005 are (i) to classify and segregate hazardous wastes at source; (ii) to dispose of hospital hazardous wastes by incineration; and (iii) to dispose of industrial hazardous wastes by other appropriate technologies.

The targets up to 2020 are (i) to collect and dispose of hospital hazardous wastes by modern technology; and (ii) to develop two industrial hazardous wastes treatment plans in the IZs in the Northern and Southern Vietnam.

Decision 152 (Vietnam's Government, 1999a) approves the strategy so that it can be implemented. It contains targets for solid waste management including both short term (towards Year 2005) and long term (Year 2020) targets, and measures to achieve these targets.

HCMC programme on industrial solid waste management

The City has conducted a program for relocating heavily polluting industries into centralised IZs. A survey carried out by HCMC Department of Construction (DOC) showed that there are 3,000 polluting companies in the city, over 1,200 of which must be relocated. These industries include chemicals, tannery, bleaching and dyeing, products made from rubber, wood processing and paper, and food processing (Quach, 2003). Industrial air pollution is currently not regarded an urgent environmental problem of the city. Focus is therefore set on levels of impacts on the environment and public health from industrial hazardous and solid wastes in the next ten years. Four objectives have been set up, one of which is of particular interest to IZs: By the end of 2005, the establishment of industrial waste management system at each EPZ or IZ is to be completed, consisting of a plan for industrial waste recycling, disposal and collection (Interconsult International & SINTEF, 2002a).

Categories of Hazardous Waste

It is undisputed that much wider publicity needs to be given to the list of hazardous wastes from the Basel Convention, along with detailed instructions on its interpretation in the real world of industrial activity. Given the low awareness of the meaning of hazardous wastes among industrial workers and managers, it is questionable whether this is the most appropriate way of defining hazardous wastes in Vietnam. Under the present state of awareness and understanding, it is proposed that a priority list of hazardous wastes should be prepared, based on the type, nature and/or volume of the wastes and the current capability and the available infrastructure to deal with these wastes. As awareness increases, the list can be widened to take in other wastes in a phased manner over a period of 5–8 years, with the aim of controlling all listed wastes within 10 years. This would allow the industry to become accustomed to managing its wastes more thoroughly and also give HCMC DONRE time to formulate its monitoring strategies.

Registration of Hazardous Waste Generators

This aspect of control suffers from the same problems as the 'definitions', as generators of hazardous wastes will only register if they are aware of the fact that they are generating hazardous wastes. As the system relies on the generator informing DONREs, rather than DONREs inspecting the waste producer, the only information that will be provided is which wastes the producer wishes to declare. As a consequence, the information will be of limited value, unless HCMC DONRE is able to carry out more site visits and inspections than is currently being accomplished.

Permits for Collection, Transport, Storage, Treatment, and Disposal of Hazardous Wastes

All enterprises wishing to become involved in the collection, transport, storage, treatment, and disposal of hazardous wastes must register their intention with their relevant DONRE and apply for a permit. This also suffers from the above-mentioned drawbacks, but has a further complication. Interviews conducted by The Environmental Resources Management (ERM) in their preparation of Hazardous Waste Inventory for the Southern Focus Economic Zone of Vietnam (ERM, 2002a) reveal that several enterprises have registered with DOSTE for some or all of the permits, but have no current intention to carry out the functions for which they are registered. It is thus impossible to assess how many enterprises are involved in the various waste management activities.

In addition to these difficulties, the Regulation does not stipulate what constitutes 'storage'. Unlike the USEPA which differentiates between two types of storage (90-day-or-less storage and more-than-90-day storage), there is no indication of the quantity or time limit on storage that requires a permit to be applied for. Therefore, permanent storage may be resorted to by the industries when treatment and disposal become too costly for them. On the other hand, some industries that have their wastes collected daily or every two days may find it confusing whether a permit is required.

Industrial Zones (IZs) and Export Processing Zones (EPZs)

HCMC DONRE and the Divisions of Planning and Environment within the HCMC Export Processing and Industrial Zones Authorities (HEPZA) are responsible for the environmental management in the IZs. Both HCMC DONRE and HEPZA are linked to the HCMC PC in political, administrative and financial aspects. DONRE is responsible for implementing national environment laws and regulations. Hazardous waste management, in particular, is not within the scope of the responsibilities of the IZ authorities. In principle, HEPZA is mainly responsible as the governmental representative for managing and ensuring compliance of all the companies/ enterprises located in their IZs in relation to all aspects of the industrial operations.

Public Participation: Identification of Stakeholders

A stakeholder analysis conducted by Interconsult International & SINTEF (2002b) reveals that there are four main parties in waste management of IZs in HCMC: the waste generators (industries within the IZs), the hazardous waste management companies (12 public-owned and private companies in HCMC), the authorities (at different levels of city, district and ward and HEPZA) and the general public. The general public has a double role of being both waste generators and victims of other waste generators such as industries.

Apart from the political and administrative management, there are also the mass organisations such as the Youth Union (YU), the Women's Union (WU), the Farmers' Union in agricultural communes, the War Veterans, and the Elderly Union. In particular, YU and WU are actively working grassroot organisations with intimate local knowledge about the issues and opportunities in all matters that concern their ward/commune (Interconsult International & SINTEF, 2002b).

Monitoring and Enforcement

The implications of the failure to implement the relevant sections of the environmental laws and regulations are very significant. Without the implementation and enforcement of the regulatory controls, the HCMC DOSTE is relatively powerless to compel waste producers to manage their wastes in an environmentally sound manner.

HEPZA co-operates with DOSTE to oversee the proper management of hazardous wastes. They are hampered, however, by the lack of treatment and/or disposal facilities for hazardous wastes and the virtually complete absence of facilities to deal with hazardous wastes. They do not have enforcement responsibility either; this is taken care of by the Inspection Department within DOSTE. DOSTE is charged with the responsibility for inspection and monitoring emissions and the performance of pollution control equipment in industrial premises, but at present, there are too few inspections to exert proper control. Figures released by HCMC DOSTE for 2001, for example, indicate that DOSTE investigated 220 complaints at city level and a further 1,022 at district level (HCMC DOSTE, 2001). A total of 1,495 enterprises were visited/monitored during the year out of an estimated 30,000 enterprises (approximately 5% of the total). Note that all of these visits were made because of complaints from the affected residents, rather than routine inspection. Only 51 out of the 220 city-level complaints were made on industrial premises, none of which was situated in an IZ (HCMC DOSTE, 2001). In other words, the monitoring of pollution, let alone hazardous waste management, seems to be virtually non-existent in HCMC's IZs.

One obvious reason for the minimal level of inspections is the lack of manpower and other resources being made available to this aspect of the work of DOSTE (ERM, 2002c). There is a clear need for additional staff to carry out the required number of inspections. On the assumption that an average of 6 inspections per day can be conducted by each inspector over a 250-day working-year, it would require 20 inspectors to visit all industrial premises just once a year. As two visits per year is the minimum recommended frequency of inspection, it can be seen that many more inspectors are required if this level of inspection is to be achieved. Furthermore, waste management treatment and/or disposal sites need to be inspected much more frequently (as often as once per week is the recommended minimum for the larger operators) (ERM, 2002c). However, as the scope of this study is limited to IZs in HCMC, only one inspector is required to do all the inspection for the 336 enterprises located within HCMC IZs four times a year.

Another less obvious reason for the lack of monitoring is the apparent 'turf-war' that is said to operate between various departments and divisions within HCMC DOSTE (Interconsult International & SINTEF, 2002a). This is partly due to the structure of HCMC DOSTE. The Environmental Management Division (EMD) in DOSTE has 20 experienced and qualified staff engaged in monitoring environmental pollution from industrial activities,[1] but has no enforcement role within their activities. The Inspection Department (ID) however, has 4 staff members who have an 'enforcement-role' but lack the level of expertise of the staff of EMD. It is claimed that ID staff must either accompany EMD staff or make a separate follow-up visit when investigating premises that may be in breach of regulations and be subject to legal proceedings. This is clearly duplication of effort and as EMD staff out-number ID staff, it is also a limitation in monitoring and enforcement activities.

Generation of Hazardous Waste from Industrial Zones

Several attempts have been made to estimate the quantity of wastes generated by the industries in HCMC, the earliest by the Centre for Tropical Technique under HCMC Institute of Tropical Technique and Environmental Protection in February 2000. A survey was done to investigate the quantity of industrial wastes generated from Tan Tao IZ and Tan Thuan EPZ. From their surveys, waste factors have been estimated for each industrial sector as tons of wastes per ha or per licence.

[1]Personal communication with a staff member from the Environmental Management Division, HCMC DOSTE.

Table 22.1. Estimation of industrial waste quantity from IZs and EPZs based on the waste coefficient of tons/ha/year (Method 1).

IZ and EPZ	Occupied Area (ha)	Industrial waste quantity (tons/year)
Tan Thuan EPZ	135.15	43,372.34
Linh Trung EPZ	45.39	14,566.56
Linh Trung II EPZ	39.68	12,734.11
Binh Chieu IZ	20	6,418.40
Cat Lai IZ[2]	—	—
Hiep Phuoc IZ	18	5,776.56
Le Minh Xuan IZ	10	3,209.20
Tan Binh IZ	13	4,171.96
Tan Tao IZ	52	16,687.84
Tan Thoi Hiep IZ	9	2,888.28
Tay Bac Cu Chi IZ	11	3,530.12
Vinh Loc IZ	30	9,627.60
Total	383.22	122,982.96

A second study which was conducted by the Environmental Technology Centre under the then HCMC DOSTE (2000) corroborated this estimation of the waste factor.

The total quantity of industrial wastes from all the 12 zones can be calculated in two ways:

Method 1: This method is based on the average waste factor of 320.92 tons of industrial waste per hectare per year . The results are presented in the Table 22.1.

Method 2: This method is based on the average waste factor of 259.20 tons of industrial wastes per licence per year. The results are presented in Table 22.2.

From the amount of industrial wastes derived from these tables. The assumption that 35 to 41% of industrial wastes are hazardous (Nguyen and Kim, 2001),[4] Method 1 provides the amount of hazardous wastes which ranges from 43,044.04 to 50,432.01 tons per year. Method 2 gives a range of 30,481.92 to 35,707.39 tons of hazardous wastes per year.

[2]Under construction.

[3]Under construction.

[4]Biswas (1989) assumed that as a general rule of thumb, some 10 to 15% of wastes produced by industry are likely to be hazardous. However, the estimation proposed by Nguyen and Kim (2001) is applied in this report, taking into consideration of the fact that it is derived from a Vietnamese context.

Table 22.2. Estimation of hazardous waste quantity from IZs and EPZs based on the waste coefficient of tons/licence/year (Method 2).

IP and EPZ	No. of licences	Hazardous waste quantity (tons/year)
Tan Thuan EPZ	139	36,028.80
Linh Trung EPZ	38	9,849.60
Linh Trung II EPZ	28	7,257.60
Binh Chieu IZ	12	3,110.60
Cat Lai IZ[3]	—	—
Hiep Phuoc IZ	2	518.40
Le Minh Xuan IZ	32	8,294.40
Tan Binh IZ	27	6,998.40
Tan Tao IZ	36	9,331.20
Tan Thoi Hiep IZ	10	2,592.00
Tay Bac Cu Chi IZ	15	3,888.00
Vinh Loc IZ	27	6,998.40
Total	336	87,091.20

The difference between these two methods is significant (between 12,562.12 and 14,715.62 tons per year). High variation between the two methods may arise from uncertainty in the process of arriving at these two waste factors.

A hazardous waste inventory conducted by ERM captures the overall picture of hazardous waste generation in the Southern Focus Economic Zone[5] (ERM, 2002a). In this study, a quantity of 110,000 tons of hazardous wastes is estimated from HCMC and three provinces, which include 33 IZs and EPZs and about 60,000 SMEs (Interconsult International & SINTEF, 2002b). No data on the break-down of hazardous waste quantity from IZs and EPZs is available from the report. Even in this inventory, much uncertainty is also acknowledged.

As these three estimates are compared, the second method which leads to an estimate of 30,481.92 to 35,707.39 tons of hazardous wastes per year seems to be more reasonable and reliable.

Another fact-finding survey has been carried out under the Special Programme for Environment of the Asian Productivity Organisation to study the situation of hazardous substances and industrial wastes prevailing in its member countries, including Vietnam. The finding indicates that the hazardous waste factor in Vietnam is 13,000 tons per billion US$ GDP per year (Hwa, 2001). In Vietnam's National Report to the same Organisation,

[5]The Southern Focus Economic Zone includes Ho Chi Minh City, Binh Duong, Dong Nai and Ba Ria-Vung Tau Provinces.

however, Nguyen and Kim (2001) based their estimates on a factor of 4,500 tons/billion US$ GDP. This factor is used for the Organisation for Economic Cooperation and Development (OECD) countries, which is called the rapid assessment method by GDP and proposed by the Global Waste Survey — Final Report (Ross, 1995). Considering the significant difference between the two factors and their uncertainty, plus the fact that it is based on country GDP, the method is not applied in this report.

In any case, as development of hazardous waste inventory is a continuous process that requires time and resources, this estimation should allow initial planning to be made. More detailed measurement and estimates of waste generation can be carried out along with phased development of waste management facilities.

With the assumption that the annual growth rate in manufacturing sector remains at 15% as it has been over the past 8 years (1995–2003)[6] (HCMC PC, 2004b) and the amount of industrial and hazardous wastes increases proportionately, hazardous wastes generated from IZs and EPZs could mount up to a range of 123,316.4 to 144,456.3 tons in 2010 and 498,883.5 to 584,406.4 tons in 2020 (Method 2). With the uncertainty discussed above, the amount may be much higher due to faster growth in industrial sector, or may decrease due to waste reduction and control programmes. Furthermore, from a broader picture, one can see that it would be much more difficult to control the wastes generated from the 30,000 SMEs in the City than the wastes generated from concentrated zones. This difficulty, coupled with the accelerating number of enterprises, indicates the magnitude of hazardous waste problems that HCMC will have to deal with in the foreseeable future.

Segregation and Storage

The Institute of Tropical Technique and Environmental Protection concluded from their survey that most of the wastes generated from enterprises are mixed. Segregation of hazardous wastes in accordance with Decision 155 is far from a common practice. Storage of wastes is managed by the service company of the zone or by IZ authorities. However, IZ authorities do not seem to pay great attention to the segregation of hazardous wastes, given that six out of the seven interviewed by ERM said that they did not ensure hazardous waste separation was carried out at companies in the

[6]Average growth rate in manufacturing sector between 1995 and 2000 was 14.6%; growth rate in 2001 was 16.2%; in 2002 was 14.8% and in 2003 was 15.3% (HCMC PC, 2004).

IZs (ERM, 2002b). Wastes (usually mixed) are concentrated in a storage site within the zone. This facilitates the collection of wastes but poses a higher risk of interaction between incompatible wastes when segregation is not carried out. It should be noted that there are currently no onsite engineered storage facilities in all six surveyed IZs by the Institute of Tropical Technique and Environmental Protection. These are planned to be developed in the future without indication of a specific timeline. While electric power, water supply, and other infrastructures are integrated in IZ planning, there is an obvious lack of hazardous waste management facilities in the planning stage. If future zones are to attract investments and at the same time ensure health protection for the workers and residents living nearby, it is important for their planners to take this into consideration.

In terms of storage facilities, the study by Interconsult International & SINTEF (2002b) has revealed the use of inadequate packaging. Hazardous wastes are presently stored in rusty, corroded drums, many without lids or with ill-fitting lids. In addition, drums and containers are frequently stored so closely together that it would be almost impossible to detect leaks or seepage until it is too late. Together with extensive lack of inspection programs, these conditions are a major cause for concern.

Few of the storage areas employed effective segregation of different categories of hazardous wastes and most have no bunding or secondary containment. A further problem that can lead to environmental pollution is the practice of storing waste containers immediately on top of or adjacent to surface water drains. Very few of the containers have labels describing their contents (or their source) and often, the labels of the original content (not waste) are still intact and clearly visible (Interconsult International & SINTEF, 2002b).

As a generalisation of the practice of on-site segregation and storage of hazardous wastes by industries, the survey conducted by ERM (2002b) shows that:

- In state-owned and private SMEs, most of the managers do not know the definition of hazardous wastes;
- In larger state-owned and private enterprises, although the lack of awareness is similar, some segregation of hazardous wastes is practiced on site and several have a designated storage facility for hazardous wastes, which is separate from production areas. There are also nominated staff to take care of the wastes taken into or out of the place.

- In foreign-invested and multi-national companies, especially those located within the EPZs, attention is paid to the management of hazardous wastes, which are often segregated and stored properly. Although there are special containers for hazardous wastes with labels, these are usually insufficient. Other types of wastes are classified on site and taken away by a waste treatment company for treatment and disposal.

This conclusion appears to be applicable in IZs as well. Apart from the lack of storage facilities and segregation practice, the absence of an integrated hazardous waste management system in HCMC has led to another problem: even if hazardous wastes are properly segregated and stored on site by the industries, these wastes are mixed with non-hazardous wastes again during the collection or disposal processes. Since there is no treatment and disposal facility for hazardous wastes, they are either disposed of together with non-hazardous wastes, or worse, directly discharged to the environment. This is mainly because of shortage in finance for environmental management, a lack of awareness and an absence of clear motivations or initiatives for an enterprise (given the lack of monitoring and enforcement). The companies generally are not concerned with the performance of the contractor after the waste has left their site (ERM, 2002b).

Collection and Transportation

Although the study on waste collection and transportation is not exhaustive (only 5 out of 12 IZs and EPZs were surveyed), the overview of these activities has been captured. It is the responsibility of the hazardous waste generator to contract a licensed hazardous waste management company for collection services. However, the only licensed company that supplies this service to three out of five surveyed IZs is the City Environment Company (CITENCO). The Urban Environment Companies (URENCOs) play an important role in waste collection in IZs. URENCOs are legally responsible for collection, transportation and disposal of domestic wastes but the current practice is that a significant volume of hazardous wastes is also collected and disposed of with domestic wastes due to the low awareness of waste generators and even URENCOs. The mix-up of domestic wastes with hazardous industrial wastes in enterprises is a common practice. A report from HCMC DOSTE on Impacts of Hazardous Waste Management in HCMC in 2002 confirms that public services companies of IZs and EPZs or private companies take charge of collection and treatment (HCMC DOSTE, 2002a). Most of these companies do not have licences for hazardous collection, transportation, storage and treatment issued by DOSTE. It is further acknowledged in this report that the environmental protection authorities fail to assert control over these companies.

Although all 12 collection companies have registered with the DOSTE as hazardous waste management companies (ERM, 2002b), there has been limited implementation of the reporting and other requirements of Decision 155 to date. Similar to the requirements on the hazardous waste generator, there is very limited enforcement on this group. One problem is that Decision 155 has not specifically defined the frequency of submission of the hazardous waste management reports by waste management companies. In fact, HCMC DOSTE reported in 2002 that until then, they had received no such reports as required by Decision 155 (HCMC DOSTE, 2002a).

Another problem with waste collectors is that most of these companies register with HCMC DOSTE for waste management licences and/or permits, they also collect wastes from many neighbouring Provinces. When issuing licences to these companies, the authorities underestimated the potential risks in transportation of wastes and the capacity of the waste management companies.

Hazardous wastes are usually not transported directly to the treatment/disposal site. Instead, they are transported from the generators to a transfer station where domestic wastes are stored (HCMC Institute of Tropical Technique and Environmental Protection, 2000). This mode of transportation is a serious cause for concern, as hazardous wastes are mixed with domestic wastes in these stations, which may cause great danger to human health and the environment.

In principle, Decision 155 provides a general regulation at the national level. To translate it into local implementation, more detailed guidance is required. However, HCMC has not issued any regulation or guidance on the requirements for the registration of waste collector/transporter, or technical specifications for hazardous waste transportation vehicles, transfer facilities, or emergency response.

Treatment

In HCMC, only Tan Thuan EPZ, Linh Trung EPZ and Le Minh Xuan IZ have constructed wastewater treatment plants (HCMC DOSTE, 2002b). However, they have not yet operated their treatment facilities (ERM,

2002b). Tan Binh IZ is currently reviewing the potential total volume of wastewater in the IZ in order to install a wastewater treatment plant with the appropriate capacity (ERM, 2002b).

Incineration is the main method used for hazardous waste treatment. However, there are no incinerators in operation within the IZs. According to ERM (2002b), five out of seven hazardous waste management companies interviewed have incinerators to treat organic hazardous wastes. There is a cause for doubt over the operational effectiveness and technical requirements of these incinerators. The Master Plan for the Southern Focus Economic Zone (Interconsult International & SINTEF (2002b) views the present system for destruction of hazardous organic wastes in HCMC as *'heavily dependent on utilisation of waste in various boilers and energy-supplying devices, like brick kilns, industrial boilers, ceramic industry, sugar refineries etc.'*

The present system of incineration or other treatment is not appropriate for the safety of the operators and neighbouring residents, and according to Decree 155, is not legal. Because of the inadequate operational temperature (ERM, 2002b), it is likely that furans and dioxins are emitted into the environment from these facilities. This should not be allowed to continue as part of the future development of the hazardous waste treatment system.

Final disposal: Landfill

There are two major disposal sites in HCMC: Dong Thanh and Go Cat. Both sites are operated as "cut and cover" site. A pit is excavated to a depth of about 5 metres and the excavated materials are either carted away for sale or sorted for later use as cover materials. The disposal operation is basically open dumping, with some compaction done and bulldozers used to spread the refuse. Chemicals are applied as a means of odour reduction and pest control.

Dong Thanh landfill is not protected with an impervious liner at the base and sides. As a result, leachate cannot be collected and treated, but seeps into the underlying soil where it constitutes a potential source of groundwater pollution (HCMC DOSTE 2002b).

Go Cat is a new 25-ha landfill that started operation in 2001, receiving about 4,000 tons of wastes daily. Although similar to the Dong Thanh landfill, Go Cat has a liner and a collection and treatment system for leachate (ERM, 2002c). However, because of heavy rain which occurred a few days after its operation, one cell of the landfill was reported to generate 7,000 m^3 of leachate, thus affecting about 210 households in the neighbouring areas (Vnexpress, 2001).

No separate place for hazardous wastes is available in both of these sites. Hazardous wastes are still disposed of with domestic wastes (ERM, 2002c). There are currently no sanitary landfills for hazardous waste disposal in HCMC.

Recommendations

There is currently no environmentally friendly and sustainable hazardous waste management system for IZs in HCMC. An integrated management system is therefore needed to minimise, segregate, treat and ensure appropriate disposal of hazardous wastes and to prevent them from damaging the environment. The following recommendations are made to improve the current situation and plan for future development.

Planning and Administrative Infrastructure

HCMC's EPZs and IZs appear to be unplanned with randomly scattered industries. There are limited collective solid/liquid waste treatment facilities in these zones. These show that the environment has not been taken into consideration in the planning of IZs in HCMC.

Better planning could solve many problems caused by different industries in terms of the nature of their wastes. It could also exploit the opportunity of waste exchange or industrial symbiosis, since it has the advantage of close proximity between the generators within the zone. Better designed treatment facilities within concentrated zones should be integrated into the planning stage. Similar industries should be selected and located within the same zone. The possibility of designing more appropriate and efficient treatment facilities increases with the similarity of the wastes generated from the industries. In contrast to the current situation, high technology industrial parks have to be separated from heavily polluting zones.

Not only planning of industries within IZs but planning of the city or region as a whole should also be taken into consideration. For example, more polluting IZs should not be located close to residential or business areas. Appropriate locations of the zones may significantly reduce the risks posed by hazardous wastes during collection and transportation. Similarly, it is important to ensure judicious siting of waste transfer stations and treatment and disposal facilities near existing industrial areas. Good transport/road infrastructure for easy access must be provided.

With regard to administrative infrastructure, the most challenging task is to close the gap between necessary and available staff resources. This can be achieved by:

- More effective use of existing resources within DONRE by adopting a stricter allocation of tasks between the divisions within DONRE.
- More effective use of existing resources of other city departments by introducing a more distinct allocation of responsibilities between the departments, and wider use of inspectors from district PCs.
- More effective use of existing resources of the environmental division within HEPZA.
- Strengthening of knowledge and competence of the staff involved in hazardous waste management. Capacity building is indispensable in improving the effectiveness of the monitoring and inspection activities of the people in charge; and
- Increase in resources. However effective the resources are, there is an urgent need for DONRE to increase their existing resources. Monitoring IZs is only one of their assignments. There are other 30,000 SMEs and about 800 large-scale enterprises apart from the zones to be monitored. If possible, this task should be transferred to the local environmental divisions at the district and ward levels which have better knowledge of the industries in their areas.

Legal Infrastructure

Existing legislation and regulation should be improved by:

- Deleting registration of hazardous waste generators, to reduce the work load on DONRE (Interconsult International & SINTEF, 2002b). At the moment, only generators who want to register will do so, and there is no resources to enforce this practice or to ensure credibility and accuracy of the information provided;
- Improving the hazardous waste definition by providing a simpler list of priority hazardous wastes. Other hazardous wastes can be included in the list in later years as the awareness of hazardous wastes is improved and the management system is strengthened.
- Clarifying the minimum quantity and/or minimum time period that a generator needs to apply for a storage permit.

The City's hazardous waste management system is in need of supplementary legislation and regulation. The following should be provided:

Collection (collection, segregation, packaging, temporary storage)

- Legislation governing registration procedures for hazardous waste collection;
- Technical requirements for hazardous waste collection vehicles; and
- Technical requirements for on-site hazardous waste storage facilities.

Transportation

- Legislation governing registration procedures for hazardous waste transportation;
- Technical requirements for hazardous waste transport vehicles;
- Technical specifications for the operation of hazardous transport means (on routing, parking, loading and unloading, emergency procedures, etc.); and
- Specification of competence requirement of the staff, in particular the driver.

Transit (transboundary transportation)

- Legislation governing registration procedures for hazardous waste transboundary transportation;
- Technical requirements for hazardous waste transboundary transportation vehicles; and
- Guidance for health, safety and emergency responses.

Guideline on hazardous waste storage (temporary storage before the waste is taken to the treatment site)

- Packaging requirements (types of waste, suitable containers, container materials, labelling, etc.);
- Requirements for storage sites; and
- Safety regulations for hazardous waste storage (technical facilities, warning signs, etc.).

Treatment and disposal

- Proper construction and operation of hazardous waste landfill sites
- Construction and operation of hazardous wastes: ensure the capacity for hazardous wastes in HCMC in the foreseeable future; and
- Separate regulations on hazardous waste incineration and landfills.

Legislation without the backing of enforcement is pointless. Unfortunately, enforcement is often the weak link in the control chain and its improvement of the in HCMC is urgent. It is strongly recommended that enforcement priority should be given to the hazardous waste management companies. By strictly regulating these companies, which are limited in numbers, a more effective use of the resources is possible. Setting a standard for the hazardous waste companies will render them ambassadors of quality in their contact with their clients, the hazardous waste generators from IZs.

Financial Infrastructure

With regard to financial infrastructure for hazardous waste management of IZs in HCMC, it is recommended in the Hazardous Waste Master Plan for the Southern Focus Economic Zone (Interconsult International & SINTEF, 2002b):

- That the system of charges relating to air and water pollution is reviewed to determine its effectiveness as an incentive for waste generators to avoid inappropriate hazardous waste disposal. The total fines for 26 enterprises in HCMC, none of which is located in an IZ, is as low as 28 million VND (about US$ 1,750), an average of US$65 per enterprise (HCMC DOSTE, 2001);
- That the costs incurred by government in administering the hazardous waste management regulations should be recovered from an annual licence or registration fee payable by hazardous waste generators. It is proposed that a 'registration fee' should be payable by enterprises in selected hazardous waste generating industrial sectors;
- That a fixed fee payable by all users should be adopted in the first instance, but that the scope for introducing a graduated, banded fee structure should be examined and a revised system introduced as soon as possible;
- That the scope for introducing a variable fee based on individual movements (consignment fee) of hazardous wastes should be examined;
- That cost recovery should be in line with the "polluter pays" principle, and that individual waste generators should pay the full costs incurred in treating and disposing of their wastes (subject to any indirect subsidies via capital grants);
- That a capital grant from a bilateral donor agency is sought to help finance the development of centralised facilities for HCMC; and
- That during the initial implementation period DONRE ensures that specific categories of enterprise manage their hazardous wastes in an appropriate manner and pay the full costs of the services provided.

Social Infrastructure

It is assumed that setting up an agenda for considering social and public issues in a hazardous waste management system will increase the possibilities to improve a secure handling of hazardous wastes in short and long terms. Such an agenda in the form of a Stakeholder Participant Plan (SPP) as proposed by Interconsult International & SINTEF (2002c) will assist the authorities concerned in ensuring that the planning and implementation of hazardous waste treatment are made according to the existing laws and regulations. When stakeholders are informed and aware of their rights and obligations, the risks related to hazardous waste handling will be reduced. The risks implied are both direct in the form of pollution and dangers to health and environment, and indirect in the form of actions or non-actions due to misinformation, mistrust and misunderstanding among the stakeholders.

A well functioning management system of hazardous wastes will also support streamlining the treatment of non-hazardous wastes since the basic structure is common and the demands for inputs of human and financial resources are similar.

Transparency in management by the authorities and participation in decision-making by the stakeholders in the handling of hazardous wastes will also increase the responsibility of each stakeholder and willingness to accept and follow those decisions.

Opportunities and Constraints

Public participation has been poor in HCMC in development projects. Stakeholder participation has not been a normal practice in a central planning economy. The proposed SPP for hazardous waste management in HCMC is recommended to be extended to comprise waste management in HCMC in general.

There are many opportunities for and constraints upon improving the hazardous waste management process through stakeholder participation in Vietnam.

The opportunities include:

- A political administration system with HCMC PC and HCMC People's Council, from central government/parliament level to City, district, and ward/commune levels. This system also includes line ministries (government), departments (City) and bureaus (district) and officials in charge (ward/commune) as well as mass organisations such as WU and YU;

- An educated public with access to information from central and local authorities, mass organisations and mass media;
- Institutions, including schools, with experiences in public information campaigns and project implementation aiming at environmental awareness, protection and prevention against adverse effects of waste management; and
- Information on the character of the wastes provided to HEPZA in their zones. When pollution occurs, it is easy to trace the source within the zone.

Constraints to take into account are:

- Important environmental considerations have to be balanced against ambitions of fast economic growth;
- The market situation affects management where hazardous wastes are recycled or utilised as energy, but causing environmental pollution during the process;
- Awareness among the stakeholders of what hazardous wastes are, why they are environmentally important to identify, how to separate them from other waste and where to deliver them, is limited; and
- Knowledge among the stakeholders about rules and regulations, rights and obligations is limited.

Raising awareness

The baseline study of the Hazardous Waste Master Plan for the Southern Focus Economic Zone (ERM, 2000b) concludes that the level of awareness and knowledge about hazardous wastes, with a few exemptions, are low or non-existent among these stakeholders. A widespread improvement in awareness and knowledge is therefore needed.

The lack of awareness and knowledge among generators and hazardous waste management companies includes the following:

- Regulation of hazardous waste management;
- Potential impacts of hazardous wastes;
- Basic definitions and classification of hazardous wastes;
- Need for appropriate segregation, storage and minimisation; and
- Emergency planning.

Hazardous waste generators also lack knowledge of hazardous waste minimisation and clean technology. Hazardous waste management companies also need increased knowledge of:

- Hazardous waste handling;
- Collection and transport; and
- Emergency response plans, especially in the case of spillage and road accidents (given that current emergency plans of most waste management companies are about fire protection and fighting-ERM, 2002b).

The DONRE employees need competence to:

- Train and raise awareness among hazardous waste generators, hazardous waste management companies and the public; and
- To train their own staff and other local authorities, including HEPZA.

Training is also needed on control and enforcement and other tasks allocated to HCMC DONRE. A training program needs to be provided to help HCMC DONRE implement the regulation and ensure control in a uniform and systematic way.

Engineering Infrastructure

Hazardous waste minimisation, cleaner production, re-use and recycling

Waste minimisation strategy is an essential part of a corporate hazardous waste management system. This focuses extensively on the preventive part of the problem, and should be emphasised just as strongly as the treatment solutions for the existing waste streams.

Some cleaner production programmes have been in operation in HCMC for some years, but not directly aimed at companies generating hazardous wastes (Interconsult International & SINTEF, 2002b). Available information does not document any substantial results. A cleaner production programme with special focus on hazardous wastes is recommended by the same report to be introduced in selected industries. Industrial sectors should be prioritised according to the following criteria for their wastes:

- Large expected increase in amounts in the future;
- Potential for recovery not being currently done;
- Toxic waste entailing high treatment costs; and
- Waste in large amounts with an uncertain hazardous waste potential.

Once the selected industries have been identified, the following steps are to be taken:

- Clarify potential of the waste prevention and reduction through cleaner production and waste recovery in these industrial sectors;
- Use this information as a basis for preparing a strategic plan for implementation of waste prevention and reduction in the selected sectors; and
- Grant licences to the major recycling companies involved in recovery of packaging materials and containers in accordance with Decision 155.

Hazardous waste storage, collection and transportation

Storage facilities in compliance with specifications stipulated of this study should be provided in each IZ. Industries with inadequate storage facilities have to hire the common storage house in the zone for use. Sufficient information should be provided to the manager of the storage facility for proper handling and management.

More specifically, to improve the current storage situation, the following issues must receive special attention:

- Segregation of hazardous wastes;
- Use of appropriate containers to reduce the risk of accidental spillages during storage, handling and transport. The containers should be rigid to withstand the stresses and strains of handling, tightly sealed (never open) and made from materials appropriate to the waste type in question;
- Proper labelling of the containers. This should include both classification and hazard in accordance with regulation on hazardous waste management and a reference to the corresponding transport document; and
- No hazardous waste should be transferred to any party unless it is in a suitable container.

Collection and Transportation

Application of computer programmes for waste collection and transportation system involving routing and scheduling should be explored. The use of computer models and a geographical information system appears to be applicable in HCMC. Waste collection routing must be integrated in planning stage of new IZs.

Technically, the development of a software package similar to the e-Tracking system in Singapore is feasible and could be replicated in HCMC.

The benefits it brings are much more compared to those of the traditional manual system.

Only licensed companies are allowed to transport hazardous wastes from IZs to the disposal point. Again, these companies must fulfil duties and technical specifications as indicated in the regulations for hazardous waste collection and transportation. Unlicensed public service companies that are collecting hazardous wastes have to limit their services to non-hazardous wastes.

In order to improve current practices, it is important to apply a few precautions, including:

- Upgrading of the transport vehicles to comply with requirements for all relevant hazardous waste categories;
- Training and certification of drivers; and
- Preparation of emergency plans (Interconsult International and SINTEF, 2002b).

Treatment

The main principle should be that the liquid waste is treated in-plant or at a centralised treatment plant within the IZ, and that only the solid sludge from the treatment is hauled to an external site. Different physical and chemical treatment processes must follow appropriate technical specifications and guidelines. The installation and operation costs of treatment plants should not be an excuse for discharge of hazardous wastes to the water bodies.

The sludge produced in connection with the physical chemical treatment should be sent to a dedicated landfill for hazardous wastes.

With regard to solutions of waste incineration, apart from state-of-the-art incinerators, there is a possibility of co-processing hazardous wastes in cement kilns or power plants. Experience should be learnt from other countries such as Thailand, Malaysia, Indonesia and the Philippines.

The discussion of the current hazardous waste management practice in HCMC concludes that the current installed incinerators cannot fulfil the requirements in Vietnam's air emission standards. The incinerators are not appropriate for incineration of polychlorinated biphenyls (PCBs) and other chlorinated wastes, and the risk of excessive emission and unacceptable impacts on the neighbourhood is evident. Furthermore, these incinerators are not very flexible and appropriate for burning liquid hazardous waste.

Given that no other alternative to landfill or storage exists, the current system is in general considered "better than nothing". However, if the stricter standard on hospital waste (TCVN 6560–1999) (Vietnam MOSTE, 1999) is enforced and applied to hazardous waste incinerators, this will trigger the introduction of improved incineration technology. The existing waste treatment companies that have already installed incinerators, after a granted transition period, must comply with the air emission standard, or close down. This will enhance the improvement of new technologies, and open up to investors with a more serious focus on the importance of technical quality and environmental performance.

Clear and consistent procedures on measurements should be prepared, as well as air emission standards for incineration of hazardous wastes. HCMC authorities should support any incineration technology or system that represents an improvement compared to the existing system, provided this may be considered to be in concurrence with the overall strategy.

Final disposal/Hazardous waste landfills

A secure landfill for hazardous waste disposal is a vital physical component of a complete hazardous waste management system. A hazardous waste landfill, combined with storage availability, will represent an essential improvement compared to a system with no such landfill.

Currently no hazardous waste landfill is available in HCMC and there is no plan for development of a sanitary landfill in the City in the National Strategy for Hazardous Waste Management 2001–2010. A company, Sonadezi, in Dong Nai province, adjacent to HCMC, has plans for constructing a hazardous waste landfill in Dong Nai Province. This is the only well documented plan for a secure landfill at present in South Vietnam. A feasibility study (Sonadezi, 2001) has been prepared for a landfill complying in principle with international standards.

Provided the construction is carried out in accordance with the existing description, and Sonadezi acquires the necessary competency in operation and in general hazardous waste management, this may be a useful and important contribution to the hazardous waste management system for HCMC.

Based on the stringent criteria for accepting waste to be deposited in a landfill, only selected types of hazardous wastes may be subjected to final disposal solutions. In this context, landfill is only considered relevant in combination with pre-treatment, i.e. solidification or stabilisation of the waste.

Conclusion

An overview of the present management system of hazardous wastes from IZs and EPZs in HCMC has been presented in this study. The system is inadequate to manage the wastes in an appropriate way. As a result, most of the hazardous wastes produced do not undergo the proper processes they should, i.e. reduction and minimisation, reuse and recycling, segregation, collection and transportation, treatment and disposal. In fact, both the central government of Vietnam and the local government of HCMC have recognised that hazardous waste management is a problem that needs to be seriously addressed and improved upon. This is the first and foremost important step towards public health and environmental protection. Although much has to be done before the present situation is improved to a satisfactory level, efforts are being made by different stakeholders. As a contribution to this recognition and commitment, recommendations are made in this report for the planning and administrative, legal, financial, social and engineering infrastructures of an integrated hazardous waste management system for HCMC.

References

Biswas, A.K., (1989), *Environmental Aspects of Hazardous Waste Management for Developing Countries: Problems and Prospects, Hazardous Waste Management*: selected papers from an international expert workshop convened by UNIDO in Vienna, 22–26 June 1987/edited by Sonia P. Maltezou, Asit K. Biswas, and Hans Sutter. London; New York: Tycooly.

ERM (Environmental Resources Management) (2002a), *Initial Estimates of a Hazardous Waste Inventory in Southern Focus Economic Zone of Vietnam, Hazardous Waste Master Plan*, Ho Chi Minh City, May 2002.

ERM (Environmental Resources Management) (2002b), *Assessment of Current Practices in Hazardous Waste Management in the Southern Focus Economic Zone of Vietnam, Hazardous Waste Master Plan*, Ho Chi Minh City, May 2002.

ERM (Environmental Resources Management) (2002c), *Summary Report on the Impacts of Hazardous Waste Management in the Provinces, Hazardous Waste Master Plan*, Ho Chi Minh City, September 2002.

HCMC PC (Ho Chi Minh City People's Committee) (1995), *Ho Chi Minh City's Socio-economic Master Plan, 1996–2010.*

HCMC PC (Ho Chi Minh City People's Committee and People's Council) (2003), *Overview of Ho Chi Minh City Geography, History, and Culture*, http://www.vpub.hochiminhcity.gov.vn/vie/news_detail.asp?period_id=1&cat_id=2&news_id=25 (accessed 21 June, 2004).

HCMC PC (Ho Chi Minh City People's Committee) (2004a), MCMC-Macro Economics http://www.eng.hochiminhcity.gov.vn/eng/news/default_opennew.aspx?cat_id=591&news_id=496 (accessed 30 May, 2010).

HCMC PC (Ho Chi Minh City People's Committee and People's Council) (2004b), *Socio-Economic Situation of Ho Chi Minh City in 2003*, http://www.vpub.hochiminhcity.gov.vn/vie/news_detail.asp?period_id=1&cat_id=9&ne-ws_id=192#content (accessed June 11, 2004).

HEPZA (Ho Chi Minh City Export Processing and Industrial Zones Authorities), (2003), *The Established EPZs and IZs in HCMC*, http://www.dpi.hochiminhcity.gov.vn/invest/html/hepza.htm (accessed 21 June, 2004).

HCMC DOSTE (Ho Chi Minh City Department of Science, Technology and Environment) Environmental Technology Centre (2000), *Appropriate Management Measures for Industrial Solid Wastes and Hazardous Wastes in Ho Chi Minh City*, Ho Chi Minh City.

HCMC DOSTE (Ho Chi Minh City Department of Science, Technology and Environment) (2001), *Inspection Report in 2001*, Ho Chi Minh City, December 12, 2001.

HCMC DOSTE (Ho Chi Minh City Department of Science, Technology and Environment) (2002a), *Impacts of Current Hazardous Waste Management Practice in Ho Chi Minh City*, Ho Chi Minh City. April 2002.

HCMC DOSTE (Ho Chi Minh City Department of Science, Technology and Environment) (2002b), *Initial Assessment of Existing Information on Hazardous Waste Management in Ho Chi Minh City*, Ho Chi Minh City. March 2002.

Hwa, T. J. (2001) "Integrated Report", *in Hazardous Waste Management: Policies and Practices in Asian Countries*, Tokyo. Asian Productivity Organisation.

Interconsult International & SINTEF (2002a), *Policy, Strategies and Institutional Strengthening, Hazardous Waste Master Plan for the Southern Focus Economic Zone*, November 18, 2002.

Interconsult International & SINTEF (2002b), *Master Plan for Hazardous Waste Management in Ho Chi Minh City, Dong Nai, Binh Duong and Ba Ria – Vung Tau, Ho Chi Minh City*, 21 November 2002.

Interconsult International & SINTEF (2002c), *Social and Public Issues, Hazardous Waste Master Plan for the Southern Focus Economic Zone*, November 18, 2002.

Institute of Tropical Technique and Environmental Protection, Centre of Tropical Technique (2000), *Current Generation of Industrial Wastes in Ho Chi Minh City and Neighbouring Areas*, Ho Chi Minh City.

Le, V.T. (2002a), *Population and Urbanisation in Ho Chi Minh City (Vietnam) towards New Policies on Migration and Urban Development*, Poster paper prepared for the IUSSP Regional Population Conference, Siam City Hotel, Bangkok, Thailand, 10–13 June 2002.

Le, B.T. (2002b), *Hazardous Waste Management Policies and Strategies in Vietnam, Hazardous Waste Master Plan for the Southern Focus Economic Zone,* presented in the Third Workshop, September 11, 2002.

National Assembly of the Socialist Republic of Vietnam (2005), *Vietnamese Law on Environmental Protection*, Number 52/2005/QH11 dated 29 November 2005.

NGIPS (National Graduate Institute for Policy Studies) (2003a), *Number of Foreign Direct Investment Projects*, http://www.grips.ac.jp/vietnam/VNIPs/FDIproj.htm (accessed 21 June, 2004).

NGIPS (National Graduate Institute for Policy Studies) (2003b), *Foreign Direct Investment Registered Capital*, http://www.grips.ac.jp/vietnam/VNIPs/FDIcapit.htm (accessed 21 June, 2004).

NGIPS (National Graduate Institute for Policy Studies) (2003c), *Ho Chi Minh City and Vicinity* (Ho Chi Minh, Binh Duong, Dong Nai and Baria-Vungtau), http://www.grips.ac.jp/vietnam/VNIPs/hcmc.htm (accessed 21 June, 2004).

Nguyen, X.H. and T.N. Kim (2001), "Vietnam" *in Hazardous Waste Management: Policies and Practices in Asian Countries.* Tokyo. Asian Productivity Organisation.

Pham, T., Q.V. Ngo, and V.B. Pham (1999), *Ho Chi Minh City, Growing Cities, Growing Food.* Poster presented at the GTZ/DSE International Conference on Growing Cities Growing Food-urban Agriculture on the Policy Agenda.

Quach, T.D. (2003), *Environmental Pollution Situation and Industrial Relocation Programme in Ho Chi Minh City,* Proceedings of Kitakyushu Initiative Seminar on Industrial Relocation 27–28 August 2003, Ho Chi Minh.

Ross, S.A. (1995), *Global Waste Survey — Final Report.* Coastal Management Centre on behalf of International Maritime Organisation.

Socialist Republic of Vietnam (2000), *Report of Vietnam's Government at the Partnerships' Conference on Environment.* Hanoi, Vietnam.

Sonadezi (2001), *Feasibility Study of of Industrial Waste Treatment Park Investment Project.* Sonadezi Bien Hoa, September 2001.

Truong, S.A., P. Gubry, T.H. Vu, and J.W. Huguet (1996), "Migration and Employment in Ho Chi Minh City," *Asia-Pacific Population Journal*, Vol. 11, No. 2, June 1996, pp. 3–22.

Vietnam's Government (1999a), *Strategy on Solid Waste Management in Urban and Industrial Zones in Vietnam until the year 2020.* Decision issued by Prime Minister 152/1999/QD – TTg.

Vietnam's Government (1999b), *Prime Minister's Decision on the Promulgation of Regulations on Hazardous Waste Management* — No. 155/1999/QD - TTg". Hanoi.

Vietnam MOSTE (Vietnam Ministry of Science, Technology and Environment) (1999), *Vietnamese Standard TCVN 6560-1999- Air Quality — Air Emissions from Medical Waste Incinerators — Maximum Allowable Limits.* Hanoi.

Vnexpress (2001), *Go Cat Landfill Causes Pollution after First Days of Operation*, http://vnexpress.net/Vietnam/Xa-hoi/2001/09/3B9B443F/ (accessed 12 June, 2004).

Vui, P. (1998), *Prevention of Environmental Pollution in Industrial Activities.* National Environment Agency; Hanoi, Vietnam.

23

MANAGEMENT OF ELECTRONIC WASTE IN SINGAPORE

CHAN YEN SEN and TING YEN PENG

As a result of a growing dependence on information technology and coupled with rapid product obsolescence, electronic waste (e-waste) is generated at an increasing rate in many developed and developing countries. E-waste contains a variety of hazardous substances which pose both an environmental and human health threat, and therefore requires an environmentally sound waste management system.

In Singapore, the issue of e-waste is a fairly new phenomenon and the management options available are rather limited. The state regulates the collection and handling of toxic industrial waste originating from trade premises and industries but provides no requirements for consumer e-waste. The state, being a party to the Basel Convention, also regulates the transboundary movement of e-waste. At present, e-waste is disposed of together with general waste into incinerators. Recycling of e-waste is uncommon as only few voluntary recycling programmes are available. By and large, there is lack of monitoring on the generation, collection, treatment and disposal of e-waste in the country.

This paper assesses the present management system of e-waste in Singapore in light of the global trend towards regulating e-waste, greater producer responsibility and more consumer participation. It also provides recommendations for overcoming the inadequacies and shortcomings identified in the current management system.

Introduction

In recent years, electronic waste (or e-waste) has become a critical waste problem for many developed and developing countries in the world. It is not only growing rapidly in quantity but is also hazardous and harmful to human health and the environment. E-waste encompasses a broad and expanding range of used and discarded electronic and electrical devices ranging from large household appliances such as computers, televisions and refrigerators, to personal digital assistants (PDAs), and handphones. These devices contain hazardous substances such as lead, mercury, brominated flame retardants (BFRs) and polyvinyl chloride (PVC) which are toxic and could create serious pollution and environmental health problems if they are inappropriately managed.

Due to the growing dependence on information technology and a shift in consumption pattern towards newer and more efficient electronic devices, e-waste has been generated in high volumes over the years. Coupled with the extreme rates of product obsolescence, made possible by a rapidly evolving technology worldwide, the mounting e-waste problem is further aggravated. Currently, e-waste represents between 2 and 5% of the United States municipal solid waste stream and, according to European studies, the volume is expected to rise by 3 to 5% per year (Arensman, 2000).

In the light of the hazardous nature of e-waste, an integrated approach which comprises a proper mix of technologies, design innovation, legislative mandates, planning and ongoing monitoring is essential. This is to prevent the acceleration of e-waste generation as well as to ensure that the end-of-life treatment of e-waste is safe, clean, efficient and environmentally sound.

Although e-waste has increasingly begun to receive attention from governments, NGOs, manufacturing companies and consumers worldwide, the progress towards an environmentally sound management system has been slow. Nevertheless, some countries, for example, the European Union, Japan, Norway and the Netherlands, have taken the lead in addressing the crisis of e-waste.

In Singapore, the management of e-waste is a fairly new phenomenon. Despite the high consumption rate of electronic and IT devices in the country, it appears little has been done in e-waste management, especially with respect to proper recycling, recovery and disposal of e-waste as well as reduction of the hazardous content in electronic products. Thus, an assessment of the management of e-waste is very important as only then can recommendations be made to improve the current management system. It is against this backdrop that this study report has been conducted: to provide an understanding of the problems of e-waste in the context and to highlight this growing waste stream, and to overcome the inadequacies and shortcomings identified in the current e-waste management system.

This study analyzes the current management of e-waste in Singapore and aims to provide a critical review of the legislative instruments and infrastructure available in the country in handling and treating e-waste, including laws and regulations, source minimization and recycling mechanisms, disposal methods and consumer awareness and participation. A set of recommendations based on an overall integrated approach is given for improving the existing management system.

A questionnaire survey was carried out to solicit primary data from stakeholders important to the management of e-waste. Survey targets were identified among the stakeholders based on their function, involvement, contribution and participation in the various aspects of e-waste management in recent years. Each set of questionnaire was carefully crafted to reach out to the respective survey target group — the authority (1), the manufacturers/producers (9), the recyclers (5) and the grassroots organization (1). Responses were received from five organizations: Nokia Pte Ltd, Hewlett-Packard Singapore (Pte) Ltd, Philips Electronics Singapore, North West Community Development Council (NWCDC) and the National Environment Agency (NEA).

Management of e-Waste in Singapore

This paper examines the existing legislations pertinent to the management of e-waste in Singapore. It also attempts to establish the amount of e-waste in the country by evaluating the high ownership rate of consumer electronic goods. Current end-of-life options for unwanted, obsolete electrical and electronic products are also analyzed. These options include storage at homes and offices, secondary market for second-hand and refurbished equipment, voluntary recycling, and final disposal in incineration plant. Finally, the paper examines the various incentives available for promoting an environmentally sound management system for e-waste in the country.

Legislations relating to waste management

The main government body responsible for matters relating to waste management is the Ministry of the Environment and Water Resources (MEWR). Established in 1972, the mission of MEWR is to provide Singapore with a quality living environment and a high standard of public health, protected against the spread of communicable diseases. MEWR is assisted by two statutory boards, the National Environment Agency (NEA) and the Public Utilities Board (PUB). The NEA is responsible for providing the infrastructure for waste management, as well as enforcing and administering the laws that relate to pollution control (in respect of air, water, noise and waste) and public health whilst the PUB is responsible for providing an adequate and reliable supply of potable water at the most economic cost to sustain Singapore's economic growth.

Implications of the legislation on management of e-waste

Except for the Hazardous Waste (Control of Export, Import and Transit) Act (HWA), the Environmental Public Health Act (EPHA), Environmental Pollution Control Act (EPCA) and Sewerage and Drainage Act (SDA), on the whole, have distinctly grouped waste into two classes:

- general waste originating from houses,[1] trade premises and industries which requires no particular pre-treatment processes prior to final disposal; and
- industrial waste originating from trade premises and industries including those that have the potential to cause serious air pollution or that store quantities of hazardous substances, which require pollution control measure, pretreatment and/or any other processing prior to final disposal

There is no statutory provision requiring hazardous waste originating from households, offices, shops and institutions to be recycled or disposed of properly. Therefore, regardless of their environmentally-damaging content, all refuse including batteries, light bulbs, furniture, computer and printer generated from these sources can be categorized as "general waste" and hence, disposed of in one of the four incineration plants in the country.

As for industrial and trade premises, hazardous waste is regulated based on the concentration of toxic compound (e.g. mercury, lead, chloride, asbestos) generated from some trade or industrial processes. Therefore, a discarded computer, previously used in the administration office of a factory, would not constitute as a toxic industrial waste because it was not part of an industrial or trade process. This is despite the computer containing a variety of hazardous materials with their concentration, when broken down, being higher than the statutory requirement. In other words, unless the broken down or obsolete computer has been dismantled and reduced to functional components (e.g. CRT, circuit boards, etc), it can be disposed of as "general waste".

It is noted that the current legislations are very much end-of-the-pipe oriented with little provision for source minimization and recycling. Except for a few cases, there is generally lack of provision on restricting the use of toxic compounds in consumer products, although there are provisions for import, transportation, storage, supply and usage of hazardous substances

[1] "House" is defined under Section 2 of the EPHA as including "dwelling-house, warehouse, office, shop, school and any other building in which persons are employed".

during a process. Section 24 of the EPCA states that "every person storing, using or otherwise dealing with any hazardous substances and every agent, servant or employee of such person shall do so in such a manner as not to threaten the health or safety of any person, or to cause pollution of the environment". The objective of the clause is to control potential hazards that may arise from inappropriate storage, use or handling of hazardous substances during processes. However, neither the EPCA nor other legislations have provisions to control the potential hazards that may arise from disposing a final product containing hazardous substances.

In relation to the import and export of hazardous waste into or out of the country, the onus lies with the applicant to explicitly declare the waste as "hazardous" and furnish the required information (e.g. physical and chemical characteristics of the waste, the route over which the waste is to be transported, the exporter/ importer of the waste, etc.). The applicant could, however, declare the waste as "goods" and break away from the scrutiny of the legislation. Using a discarded computer as an example again, an exporter/ importer could choose to declare it as:

- Option 1: a "hazardous waste" for disposal or treatment;
- Option 2: an "input material" for recycling; or
- Option 3: a "second-hand" electronic product.

Should option 1 be selected, there is a clear instructions for the applicant to adhere to, and the transboundary movement of the hazardous waste can be stringently regulated. However, should option 2 or 3 be selected, there is no clear procedure in handling such "goods".

It is not apparent if the NEA or other relevant government bodies have policies to deal with the import/ export of discarded electronic items in the name of "recycling" and "reuse or donation". Nevertheless, it is important to recognize the shortcomings of the legislation in this aspect of e-waste management.

The total amount of e-waste in Singapore can be estimated using the formula given below:

$$\begin{aligned}\text{Total E-Waste} &= \Sigma(\text{Import} - \text{Export}) \\ &\quad + \Sigma(\text{Storage} + \text{Reuse} + \text{Recycling}) + \Sigma(\text{Disposal})\end{aligned}$$

The first term refers to the net import of e-waste into Singapore for treatment and disposal. The term, however, can be complicated by the import and/or export of waste in the name of "reuse" or "recycling", and therefore cannot be accurately reflected in the custom record.

The second term refers to a category of e-wastes that have passed the first-hand but have yet to be disposed of in an incineration plant or a dumping ground, or exported out. These include obsolete, unused and spoilt electronic items idling at homes and offices. These also include electronic items donated, given away or sold for second-hand use purposes within the country. Some examples include donating a functional but old printer model to a charity organization, selling a spoilt computer to the "karang-guni"[2] men and selling an old handphone to another user. Finally, the second term also comprises electronic items collected for recycling purposes. Some examples include dropping used printer cartridges in recycling bin and selling malfunctioned copiers to electronic recyclers.

The third term refers to e-wastes that are directly disposed of in incineration plants or Pulau Semakau landfill.[3] These include small electronic toys, batteries, telephones and big items such as broken washing machines and refrigerator thrown into bin chute, bin centre or any refuse collection centres or dumped illegally at public places.

At present, the only likely data available in estimating the amount of e-waste in Singapore is the first term: the net import of hazardous e-waste into Singapore. The Pollution Control Department (PCD) of NEA regulates and monitors the amount of hazardous waste imported into and exported out of the country. Singapore exported 14,354 metric tonnes of hazardous waste and imported none in 2001. The statistics, however, did not distinguish e-waste from other hazardous wastes and is therefore not able to provide any helpful deduction. The remaining two terms are neither monitored nor regulated by NEA. Thus far, there is also no study carried out to find out the amount/volume of e-waste that goes into the second-hand or recycling market or disposed of in incineration plants.

In the absence of relevant data and information, it is difficult to ascertain the exact amount of e-waste in Singapore at this point of time. Nevertheless, there is good reason to believe that the quantity of e-waste

[2]The "karang-guni" men are second-hand retailers, either as individuals or in groups, who go house-to-house to collect and buy old items and recyclables such as old newspapers, clothes, drink cans, radios, computers etc.

[3]In Singapore, all incinerable refuse that are not recovered, reused or recycled are incinerated in one of the four incineration plants — Ulu Pandan, Tuas, Senoko and Tuas South. Non-incinerable waste such as concrete slabs, bulky waste materials and incineration ashes are disposed of at the offshore landfill at Pulau Semakau.

in the country is increasing. This is due to the rising standards of living, increasing affluence as well as affordability of electronic equipments over the years in Singapore. As more people own electronic products such as air-conditioners, PCs, handphones and printers, more would be discarded eventually as a result of malfunction, exhaustion or obsolescence. The following section provides a detailed explanation.

Ownership of electrical and electronic equipment

Households have become more affluent over the last decade, as reflected by the higher ownership[4] of various consumer durables. A study in 2003 concluded that televisions, refrigerators and telephone lines are the most common amenities with almost all households in Singapore having at least one such item (Ng, 2003). The ownership of audio/video CD players jumped from 20% in 1992 to 76% in 2002 and more households are enjoying the cool comfort of air-conditioned homes, with 68% households having air-conditioners in 2002, as compared to only 35% in 1992 (*ibid.*). The ownership of electronic products has increased over the years, with almost every household owning one washing machine, audio/video CD player and handphone in year 2003.

In fact, most households in Singapore now enjoy the convenience of handphones, with 89% of households having at least one handphone in 2003 as compared to a mere 6.5% in 1992. The high ownership of handphones is also reflected in the high mobile phone penetration. Between the years 2002 and 2004, the mobile phone penetration increased from 74.8% to 92.2%, implying that almost every resident in Singapore was a handphone user. The convenience of handphones, coupled with competitive pricing and additional value-added services has helped to boost the ownership rate of handphones among household members.

Besides handphones, the ownership of PCs has also been increasing steadily over the years, up from 20% in 1992 to 70% in 2003. A survey on info-comm usage in households and by individuals showed that in year 2004, the number of households in Singapore owning PCs increased by over 13,000 over the past one year. The proportion of households owning computers, however, remained unchanged at 74% due to the increased number of households in the base population over the same period of time.

[4]The ownership or prevalence rate is defined as any one member of the household who owned or had access to a particular product or service.

The increase in ownership of PCs was largely due to more households members becoming knowledgeable in the use of information technology. Studies have shown that 63% of the total population in Singapore are computer users and slightly more than half used internet services (including both dial-up and broadband internet access) in 2004.

In comparison to household and individual consumers, businesses are the major users of computers and internet in Singapore. According to a preliminary data from IDC Corp, a total of 575,428 PCs — including desktops and notebooks — were shipped into Singapore in the year 2004, of which between 70 and 80% were sold to companies (Business Times, 1 Feb. 2005). The new PCs were brought in to replace older models in offices. Analysts in the hardware and systems research agreed that Singapore has a small and relatively mature IT market which is predominantly driven by replacement PC demand cycles (CNET Networks, Inc., 8 Aug 2000 and Business Times, 1 Feb. 2005). Vendors differentiate their products from one another and constantly introduce marking strategies to encourage users to upgrade their PCs regularly. Presently, more and more office and professional home users are buying notebook to replace the chunky desktop computer.

Up to this point, the discussion on the ownership of electronic equipment has been limited to a few types of electronic equipment (i.e. PCs, handphones, air-conditioner, washing machine, etc.). Nonetheless, the discussion has provided a clear observation, i.e. there is a huge consumption of electronic equipment in Singapore and these electronic equipment would, one day, end up as e-waste. To illustrate the significance of the mounting e-waste in Singapore, consider the data provided by the IDC Corp and assume that 70% of the 575,428 PCs shipped into Singapore in 2004 were desktop (weight 31.75 kg). At the end-of-life, these desktops would make up a staggering mass of 12.8 thousand metric tonnes of computer waste which is equivalent to 0.5% of the total refuse disposed of in year 2004.[5] The mass of computer waste also represents a total of 2,932 metric tonnes of plastic, 806 metric tonnes of lead, 1.20 metric tonnes of cadmium, 0.81 metric tonnes of chromium and 0.28 metric tonnes of mercury, upon disposal.[6]

Given the high obsolescence rate of IT products, consumers' affinity for new equipment, escalating sales of electronic equipment with extensive

[5]Total refuse disposed of in the year 2004 was 2482.6 thousand metric tones.

[6]Example: total amount of plastic in 12.8 thousand metric tonnes of computer waste = 22.9907% $\times$ 12.8 thousand metric tonnes $\times 1000$ = 2932 metric tonnes.

marketing strategy, it is probable that e-waste is generated at a more rapid rate and at a much higher volumes in comparison to other non-electronic goods. Exhibiting the signs of an IT- and electronic-dependent country, e-waste may be, or perhaps already is the fastest growing waste stream in Singapore but has gone unnoticed.

Storing of e-waste at homes and in offices

There are several options for end-of-life treatment of e-waste: storage, reuse and refurbish, recycle and recover, incineration, landfill or export. In Singapore, all these options are available. What is unknown, however, is the scale and prevalence of each end-of-life treatment option. Storing obsolete, unused and broken down electronic equipment is a common way of handling e-waste in Singapore. There are many reasons why people store e-waste at homes, offices and institutions instead of giving them away for recycling, reuse or disposal. Many consumers who once paid highly for these products are unwilling to accept the fact that their unused electronic goods are now largely worthless or have very little residual value. Other reasons include:

- consumers awaiting the opportunity to sell in exchange for money or gift;
- consumers do not know where and how to dispose of the unused equipment; and
- consumers cannot write off the equipment as relinquished asset as yet.

Usually, storage is temporary. The waste will eventually leave the homes, offices and institutions when the "right" times come or when storage space becomes scarce. The e-waste will then be reused and refurbished, recycled and recovered or directly disposed of.

Reuse and refurbishment of e-waste in secondary market

In Singapore, the second-hand market for electronic equipment is fairly lucrative. Whether buying a refurbished computer or a second-hand PDA, or even an old radio and desk lamp, there are avenues to find them online, in shops and even on the streets. There are a few categories of player in the market; they are:

Original manufacturers — Original manufacturers such as HP/Compaq, Dell and IBM run their own online shops to sell high-end refurbished electronic products such as digital cameras, laser jet printers and pocket PCs. The refurbished products are usually sold at a lower price with a basic warranty thrown in for the buyer. Some original manufacturers are also directly involved in trade-in. HP and Canon, for example, occasionally offer trade-in deals to customers who get to buy a new product at a discounted price when they return an old product.

Electronic dealers or shop vendors — Besides selling new products, some electronic dealers or shop vendors also sell second-hand and refurbished items, including computer, printer, hi-fi and TV. Some even offer trade-in for PDA, laptop, handphone, game set and other high-end electronics. In fact, many shops rely on second-hand or refurbished products sales to boost business.

Second-hand shop retailers — Unlike electronic dealers or shop vendors, second-hand shops only buy and sell used products. Customer can walk in and exchange their used items for cash. The items are then displayed and re-sold in the shop at a slightly higher price. Over the past few years, second-hand shops buying and selling handphones have mushroomed in Singapore. This is due to the lucrative overseas market for used phones, especially in developing countries in Asia. Used handphones that Singaporeans regularly discard as outdated models are now refurbished, repacked and sold as new, "in-fashion" handsets in Thailand, Indonesia and Vietnam.

Online brokers/traders — They are many online brokers/ traders in the market providing internet trading and auction services to consumers. While some, for instance, *ebay*, *Yahoo! Auctions* and *Amazon.com Auctions* cater to the global market, others such as *hardwarezone*, *InterAuc!* and *iMall@NTU* are more oriented to the Singapore market. Most of these online brokers/ traders also provide additional services such as comparing prices, reviewing the latest IT gadgets, and creating forum for sharing of information among registered members.

Second-hand collectors/traders — The "karang-guni" men are the traditional second-hand retailers who go house-to-house to collect and buy old items and recyclables such as old newpapers, clothes, radio, computer, etc. The items collected are then sold to second-hand vendors or recyclers in local or overseas market. Over the years, some one-man "karang-guni" businesses have grown into second-hand trading companies. Besides the "karang-guni" men, waste collection companies and recyclers such as V8 Environmental Pte Ltd, SemEnviro Tay Paper Pte Ltd and Eco Resource Recovery Centre Pte Ltd are also active collectors for used electronic goods.

Flea markets — There are a few flea markets in Singapore with the one in Sungei Road being, perhaps the oldest and most famous. A variety of second-hand goods ranging from hi-fi stereo set and TV to fan and hand drills can be found in the flea market. The second-hand goods in the flea market are mostly priced very low.

Donating agencies — Unlike all the above palyers in the market, the donating agencies are usually non-profit making organizations which serve to facilitate the flow of used electronic products from donors to recipients. Some of the common donating agencies in Singapore include the Salvation Army and the Singapore Red Cross Society.

Despite the slight differences among the various categories of players in the second-hand market, most operate similarly. Some are a hybrid of two or more categories. While the value of the second-hand market for electronic and IT products cannot be determined, industry players have observed a rising trend in the second-market for high-end electronic gadgets, for example notebook, PDA, handphone, video camera, digital camera etc. Driven by the desire for the latest technology and cheaper price, more and more consumers are now upgrading and purchasing their latest IT gadget and electronic games in the secondary market.

In general, the secondary market for used and refurbished electronic equipment is showing signs of booming, although not every segment of the industry is enjoying equally good business. More efforts therefore are needed to promote the use of second-hand goods, to overcome the perception of "second-hand is inferior" and to aid more transactions between sellers and interested buyers, donors and deserving recipients.

Recycling of E-waste

There are generally two categories of e-waste recycling in Singapore: (i) recycling of electronic scraps from manufacturers and (ii) recycling of consumer electronic goods.

Recycling of electronic scraps

The recycling of electronic scraps from manufacturers is carried out essentially to comply with legislations, i.e. meeting the discharge limit of trade effluent and safe disposal of the toxic content (e.g. mercury, cadmium) in electronic scraps. In addition, more manufacturers and companies are becoming concerned with data security and proprietary technology. Recycling is therefore needed to ensure fully-secured destruction of sensitive data in electronic scraps and prototypes. In 2004, about 11,000 tonnes of electronic scrap were recycled locally.[7]

Recycling of consumer electronic goods

Since there is no statutory provision, the recycling of consumer electronic goods in Singapore is carried out entirely on a voluntarily basis. Despite an extensive National Recycling Programme put in place since April 2001, the recycling of consumer electronic goods has not taken off in Singapore.

National Recycling Programme

The National Recycling Programme was launched in 2001 in response to the increasing quantity of refuse disposed at the incineration plants and the landfill over the years. It has become extremely difficult to find space in land-scarce Singapore for incineration plants and landfills, along with increasing cost of the construction of these waste disposal facilities. Therefore, recycling is needed to reduce waste output as much as possible so as to relief the constraints on waste management, disposal facilities and the environment.

Under the National Recycling Programme, public waste collectors licensed by NEA are required to tie up with recycling companies to implement door-to-door collection of recyclable material in both Housing Development Board (HDB) estates and landed properties within their servicing sectors. The programme targets the recycling of traditional municipal waste such as paper, cardboard, drink cans, plastic and bottles. Schools, apartments and private condominiums as well as industries are also encouraged to take part in the recycling programme.

It is noted that electronic equipment is not among the recyclables promoted under the National Recycling Programme. Unlike paper, drink cans, glass bottles and plastic bottles, NEA also does not provide public recycling bins for used electronic equipment. In general, there is lack of publicity in promoting the recycling of consumer electronic goods in Singapore. Despite the absence of a national recycling programme for e-waste, there have been several recycling programmes carried out by private companies and grassroots organizations in recent years. Some of

[7]Information was obtained from NEA's reply to the questionnaire given.

these are driven by corporate policies and others are conducted on a short-term basis.

Nokia's Mobile Phone Take-Back Programme

In 2001, Nokia launched a Mobile Phone Take-Back Programme in Singapore to take back end-of-life Nokia handphone, handphone batteries and accessories. Collection bins for these items were provided in all Nokia offices and customer service centres. Customers and general public were informed about the Take-Back Programme and the need to dispose used handphones and batteries in a proper manner through various channels, such as websites, labels on Nokia products, instruction manuals, notices/leaflets at point of sale and email/letter notification.[8] Nokia's take-back programme is to some extent limited by only one collection channel, i.e. drop off at collection bins in Nokia offices and customer service centre. In their reply to the questionnaire, Nokia stated that the company has plans to extend the types of take-back channels for Singapore market but did not give details. While there was no data specific for each category of handphone, batteries and accessories, Nokia revealed that a total of 12 metric tonnes of electronic scrap was collected in Singapore in the year 2004.

HP Planet Partners Return and Recycling Programme

Hewlett-Packard (HP) first introduced HP Planet Partners Return and Recycling Programme in 1997 in Singapore. It was part of HP's worldwide recycling programme, aimed to reduce the environmental footprint of products, minimize waste to landfill and to provide customers a convenient and environmentally sound product end-of-life solution. The programme in Singapore began with the collection of used original laserjet print cartridges in 1997 and was extended to used original inkjet print cartridges and computer and printing hardware products in June 2003.

HP deploys a variety of communication channels to convey the message of recycling. Website, instruction manuals, notices/leaflets at point of sale, email/letter notifications, customer tool kits and recycling events are used to inform customers and general public about the need to dispose of used electrical equipment in a proper manner and the presence of HP Planet Partners Return and Recycling Programme in Singapore.[9] HP also partnered with schools and universities and supported community

---

[8]Information was obtained from Nokia's reply to the questionnaire given.
[9]Information was obtained from HP's reply to the questionnaire given.

events in creating awareness and promoting recycling programmes. Examples include the Supplies Recycling Programme in School and the R&R@North West.

As for collection channels, collection bins are provided for used original HP inkjet and laserjet print cartridges for individual customers. These collection bins are placed at all HP offices/customer care centres, SingPost outlets and participating schools.[10] Corporate customers, on the other hand, are encouraged to join the HP Planet Partners Progromme in recycling HP inkjet and laserjet print cartridges as well as computer and printing hardware products. Free pick-up service is provided with some pre-conditions. The hardware recycling service, however, is not offered to individual customers.

In comparison to other countries like the US, Australia and the European countries, the collection channels available in Singapore are somewhat limited. In the US, for example, individual consumers are provided with postage-paid return and recycling envelope upon purchase of HP inkjet print cartridges. Alternatively, consumers may print the envelope online. In Australia, HP not only offers prepaid mailing services but has also joined forces with Planet Ark in a unique environmental program called 'Cartridges 4 Planet Ark' (C4PA) in collecting and recycling cartridges from any brand of printer. Consumers can drop used cartridges in the green C4PA collection box provided in participating Australia Post or Harvey Norman stores.

In a reply to the questionnaire, HP stated that the company is looking into creating more environmental awareness and extending take-back channels in Singapore. The initiatives cited include expansion of hardware collection program to individual consumers, participation in events like the ENnovation 2005 to showcase HP's recycling programme, continuous partnership with schools on the supplies recycling programme and distribution of educational leaflets. Although no information was available on the amount collected specific to Singapore, there was some data on the amount of used print cartridges and hardware collected in the Asia-Pacific region.

Globally, Planet Partners collected and recycled approximately 54,800 metric tonnes of used products in the year 2004. The Asia Pacific region contributed less than 5% of the total collection.

[10]ibid

R&R@North West: return & recycle electronic waste

Besides private companies, local grassroots organizations are also involved in organizing recycling programmes for consumer e-waste, albeit on a short-term basis. The R&R@North West was a 4-month (July — Oct 2004) community programme organized by the North West Community Development Council (NWCDC), in partnership with Citiraya Industries Ltd, HP, Nokia Singapore and SembWaste Pte Ltd, and supported by NEA, Association of Electronics Industries in Singapore (AEIS) and the Waste Management & Recycling Association of Singapore (WMRAS). The objective of the programme was to raise awareness of electronic waste recycling targeted at 500,000 residents in the North West District.

During the programme, a total of 12,000 pieces of e-waste were collected. Almost 80% of the e-waste collected were electronic accessories. They are small items which cost relatively little. As the purchase price of the electronic item increases, fewer pieces were collected. For example, a CD and floppy disc typically costs less than $0.50 each whilst a printer may cost $80–800 (colour inkjet printer) depending on the brand and model.

Electronic recyclers in Singapore

Like elsewhere in India, China and the US, the recovery and recycling business of e-waste in Singapore are spread over many different market players. Many of these collectors/traders employ a network of brokers and material specific recyclers for further processing, recovery and disposal.

Except for a few major recyclers such as Citiraya and SembCorp Environmental Management, many recyclers lack the appropriate infrastructure and technological know-how in processing e-waste and treating the toxic materials. Therefore, shipment of e-waste in bulk to other countries with recycling processes is sometimes necessary and it is regulated under the Hazardous Waste (Control of Export, Import and Transit) Act. Such an arrangement however contravenes the principle of 'self-sufficiency', i.e. to ensure that environmentally sound hazardous waste management options are pursued to the maximum extent possible within the country of origin. It also goes against the objective of the Basel Convention which calls for minimum transboundary movement of hazardous waste. Therefore, there is a need to strengthen the national capacity to handle and manage hazardous e-waste locally.

While most electronic recyclers in Singapore have their businesses focussed on servicing companies which manufacture and assemble electronic components, scrap collectors and certain end-user which require proper dealing of sensitive data in electronics and computer systems, these only account for a mere 3% of the total electronic waste in the market (The EDGE Singapore, 8 July 2002). Consumer electronics waste account for the remaining 97%. Hence, there is great opportunity for electronic recyclers to tap into the municipal electronic waste and create partnerships with producers such as HP and Nokia which offer recycling programme in Singapore.

One of the business incentives for electronic recyclers is the sales of precious metals recovered from e-waste. E-waste contains a variety of high value metals such as gold, platinum, palladium, silver and copper. Therefore, the challenge facing the electronic recyclers is to recover these precious metals as efficiently as possible. Citiraya, for example, recovers almost 200kg of gold monthly and gold is the primary product and revenue for the company (Singapore Green Plan 2012, Chapter 2). However, the recovery of precious metals may become tougher as fewer precious metals are used in electronic product due to changes in design. Furthermore, not all e-waste are economically viable to recycle as many products are not designed for easy recycling and recovery.

Another challenge facing the electronic recyclers is to process and dispose of the various toxic substances in e-waste. Lead, mercury, cadmium, brominated flame retardants (BFRs) and polyvinyl chlorides (PVC) are known to cause adverse health and environmental problems. In the absence of sophisticated equipment and automation technology, most electronic recycling plants are partially reliant on labour for sorting, shredding, dismantling and other recycling processes. During these processes, workers may ingest, inhale or absorb (through skin) poisonous dust and volatile toxic compounds. Controlling the health and environmental hazards by means of isolation, dilution ventilation, good housekeeping and personal protection, are perhaps the best and most common options for electronic recyclers at present. However, with the recent introduction of the Restriction of Hazardous Substances (RoHS) Directives in EU and green corporate policies among major MNCs requiring the phasing out and/or total ban of hazardous input in electronic equipment, the risk of occupational and environmental health hazards would be able to reduce significantly.

Disposal of e-waste in incineration plants and Semakau Landfill

In Singapore, land scarcity necessitates that all incinerable waste that are not recovered, reused or recycled are incinerated in one of the four incineration plants — Ulu Pandan, Tuas, Senoko and Tuas South. Non-incinerable waste such as concrete slabs, bulky waste materials and incineration ashes are disposed of at the offshore Semakau Landfill. The Semakau Landfill is the only dumping ground in the country since the last landfill site at Lorong Halus on the main island was decommissioned in April 1999. The offshore landfill is expected to meet Singapore's need for landfill space until the year 2030.

With the understanding that, except for toxic industrial waste, there is no legislation requiring hazardous domestic wastes, including e-waste, to be collected and disposed of separately from other municipal waste, it can be presumed that all domestic e-waste that is thrown into refuse bin, bin chutes, bin centre or any approved waste collection centre are incinerated. E-waste contains a variety of hazardous substances such as mercury, lead, PCBs and PVC, and upon incineration, these toxic substances can be emitted into the air and surrounding environment. The hazardous substances can also be deposited in incineration ash, which is later disposed of in Semakau Landfill. The hazardous compounds may leach and contaminate seawater and pose environmental hazards to the marine ecosystem.

Green programmes/initiatives for e-waste

There are several incentives available for industries and companies in Singapore to undertake activities relating to green product or sustainable business practices.

Singapore Green Labelling Scheme

The Singapore Green Labelling Scheme (SGLS) or "eco-labelling" was launched in May 1992. It is a scheme which awards environment-friendly products with eco-labels. Labelling the environment-friendly products helps consumers identify and select them from those which are less so when they make their purchases. The scheme applies to most products, except food, drinks and pharmaceuticals. It does not apply to services and processes.

The SGLS scheme is administered by the Singapore Environment Council (SEC) since June 1999. SEC is a nationally orientated, non-governmental organization that seeks to raise the level of environmental awareness and action within industry and the community. Depending on the category of the product, the criteria of granting the right to display Green Label on a product differs.

Energy Label

The Energy Label was launched in April 2002 as a new initiative under the SGLS. It is supported by the National Energy Efficiency Committee (NEEC)[11] and NEA. The Energy Label is a further development of Green Label in that they provide consumers with factual information on power consumption, annual energy consumption and a user-friendly 3-tick rating for consumers to compare different models of appliances before deciding on purchase. The more ticks an appliance has, the more efficient it is.

Thus far, only two categories of electronic appliances are covered by the Energy Label initiative, i.e. refrigerator and air-conditioner. Both are common household equipment and account for between 40 and 65% of the average household electricity bill (IACEE, 1999).

To draw consumers' attention on energy-saving appliances, Green Corners and Energy Efficient Products Information Centre (or EEPIC) were set up to showcase energy efficient appliances and allow consumers to compare different models more conveniently. There are currently eight of Green Corners/EEPIC in Singapore.

Innovation for Environmental Sustainability (IES) Fund

The objective of the Innovation for Environmental Sustainability (IES) Fund is to encourage and assist Singapore-registered companies to undertake innovative environmental projects that could help to meet the government's goal of environmental sustainability. The Fund provides assistance through grants to cover a percentage of the qualifying cost

[11]The National Energy Efficiency Committee (NEEC) was formed on 1 Apr 01 when MEWR took over the then Inter-agency Committee on Energy Efficiency (IACEE) from Ministry of National Development (MND). The key thrusts of the NEEC are as follows: (i) Promotion of energy conservation through the efficient use of energy in the industrial, building, transportation and households sectors; (ii) Promotion of the use of cleaner energy sources such as natural gas and renewable energy sources; and (iii) Promotion of Singapore as a location for the pilot test-bedding of pioneering energy technologies and as the hub for the development and commercialisation of clean energy technologies.

of the project at three levels of funding. The assistance is based on the various levels of support for different components of allowable cost, up to a maximum of $2 million for each project.

The IES Fund is administered by the NEA since it was first launched in October 2001. Proposed project must meet the following criteria:

- Strong elements of 'innovation' and 'early adoption';
- The outcome of the pilot, if implemented, will assist the country to meet the goal of environmental sustainability;
- Project must not have commenced at the time of application; and
- Project should not exceed 3 years in duration.

Over the years, the number of projects supported by IES Fund has been increasing marginally. In 2004, NEA committed $5.4 million, an increase of 23% from the previous year funding, for 28 projects. None of these projects, however, were related to the management of e-waste.

Environmental Test-bedding Initiative (ETI)

The Environmental Test-bedding Initiative (ETI) was jointly launched by the MEWR and the Economic Development Board (EDB)[12] in March 2003. The intent of ETI is to create a living-working laboratory in Singapore. It serves as a one-stop platform for companies to approach the relevant government agencies such as EDB, NEA and PUB for advice and support on infrastructure, manpower, regulatory and financial matters. For instance, through NEA, companies can gain access to public infrastructure to test and commercialize new products and solutions.

There are two programmes put in place to advance this initiative:

- Singapore Initiative in New Energy Technology, or SINERGY

The programme was launched in May 2001 as a collaborative platform between companies and government entities to advance the development of innovative alternative energy technologies.

- Innovation in Environmental Technology or EnnovaTe

[12]The Singapore Economic Development Board (EDB) is the lead agency that plans and executes strategies to sustain Singapore as a compelling global hub for business and investment. It seeks to facilitate or nurture a vibrant, self-sustaining enterprise ecosystem and a conducive total environment for startups and companies of all sizes. The Board also works closely with other agencies to promote innovation and develop human, intellectual, financial and cultural capital in Singapore.

The programme was unveiled in March 2003 and aims to encourage innovation, development and commercialisation of new environmental technologies, products and solutions in other environmental technology areas, such as water treatment, waste management and air pollution control. To-date, 11 testbedding projects have been undertaken (Singapore EDB, 18 Jan 2005).

Similar to IES Fund, the number of projects supported under the ETI has also increased marginally over the years. The number of project increased from 5 in year 2003 to 13 in year 2004. It is unclear if any of these projects are related to the management of e-waste.

In summary, the present legislations in Singapore are inadequate to deal with the increasing amount of e-waste. There is lack of specific provisions which require proper recycling and disposal of e-waste and reduction or prohibition of hazardous substances in electronic equipment. Globally, many countries are steps ahead of Singapore. European countries have had laws on the take back of e-waste since 1992 (Commission of European Communities, 2000). Asian countries like South Korea and China have also progressed rapidly in the last five years. Japan enacted its Home Appliance Recycling Law in 2001 and both South Korea and China are reportedly drafting laws and implementing policies and programmes on mandatory recycling of e-waste (Interfax, 2004; The Japan Times, 2003).

It is apparent that e-waste is growing in Singapore as a result of rising standards of living and increasing consumption of electrical and electronic goods. There is, however, no active measure to curb consumption or to promote reuse and recycling of e-waste. Green procurement policies like the Law on Promoting Green Purchasing in Japan are not widely practiced in Singapore.

In addition, recycling programmes targeting on e-waste are rare in Singapore. They are often carried out on a voluntary basis. While countries like Switzerland, Norway, Japan and other EU states are equipped with extensive, systematic collection and recycling network, the recycling infrastructure in Singapore is noticeably inadequate. Collection centre and electronic recycling centre are lacking and are not readily accessible to the public.

In addition, there are limited incentives and supports from the government for industries, companies and consumers to undertake green technology investment and electronic design and purchase environmental-friendly electronic products. Furthermore, the level of environmental awareness among the public is rather low and many are not aware of the potential health and environmental hazards associated with e-waste.

Recommendations

This section provides some recommendations on overcoming the inadequacies of the legislation, shortcomings of environmental infrastructure and the lack of public awareness. These recommendations, based on an overall integrated approach, are complementary to each other and the success of one proposal depends on the others.

Strengthening the legislation on e-waste

There is a need to strengthen the legislation on e-waste management in Singapore as the present laws do not provide sufficient coverage for e-waste to be managed in an environmentally sound manner. It is recommended that the government begins with developing a legal and regulatory framework which embodies clear regulatory instruments, adequate to control, monitor and enforce the collection, transportation, treatment, storage, recovery and disposal of e-waste within the country as well as the export and import of e-waste into and out of the country. In the framework, it is necessary to define what e-waste is, taking into consideration that e-waste is subject to transboundary movement. It is best to adopt of a common definition, characterization and classification of e-wastes in the Asia-Pacific region, in compatibility to the Basel Convention and other national legislations such as the EU's Waste Electrical and Electronic Equipment (WEEE) Directive, so as to facilitate easy implementation and enforcement on the subject locally and in the region.

In addition, present legislations on waste management need to be reviewed and revised to accommodate the issue of e-waste. The government should consider restricting the use of certain hazardous substances in electronic equipment and apply prohibition for those substances for which better, cleaner and greener substitutes already exist. Like the RoHS Directive, such impositions would not only minimize the problems of having to deal with hazardous substances at the end-of–life of the equipment but also create incentives for manufacturers to design products that are clean and safe. Many manufacturers, such as Nokia,[13] have shown that it is technically possible to comply with the RoHS Directive. Others such as HP and Phillips are working towards fulfilling the legislative requirement by adopting in-house ecodesign programmes and green supply chain. Some companies, however, find such technology and design practices

[13]Information was obtained from Nokia' reply to the questionnaire.

expensive and economically non-viable. This is because they do not take into account of the external costs associated with the disposal of hazardous substances which are currently borne by taxpayers. Therefore, a mandatory requirement may be necessary to encourage companies to adopt environmental-friendly design practices. It is also a rational measure to ensure that high-standard, green products sold elsewhere are also available in Singapore. This prevents Singapore from becoming the recipient of non-compliant products from other countries.

The government should also consider prohibiting the disposal of e-waste into public disposal facilities. In doing so, there will be a significant reduction in the emission of toxic substances such as furans, dioxins and heavy metal particulates into the environment, thereby improving the air quality in the country. To ensure the prohibition is successful and to minimize the occurrence of illegal dumping or export smuggling, the government needs to provide authorized avenues for consumers to return e-wastes. Some examples include providing public collection bins for used electronic equipment, setting up collection network and requiring manufacturers and/or retailers to collect back used equipment at the point of purchase.

Following the examples in Japan, Switzerland, Norway and EU, manufacturers/producers should be assigned responsibility, financially or otherwise, for the take-back and proper treatment of waste equipment sold in Singapore. Take-back requirement is perhaps one of the most useful mechanisms in ensuring "extended producer responsibility" (EPR). The principle of EPR requires continuous accountability on producer over the entire life cycle of their products. The aim of EPR is to encourage producers to prevent pollution and reduce resource and energy use in each stage of the product life cycle through changes in product design and process technology. By making manufacturers/ producers financially responsible for end-of-life waste management, they will have the incentive to design their products with less hazardous and more recyclable materials. Take-back can be carried out individually or collectively, directly or through a third party and each system has its advantages and shortcomings. Whichever the system may be, consumers must be obliged to play their part, i.e. to return the used and unwanted electronic equipment.

In addition, the government should review the policy on the transboundary movement of e-waste under the purview of HWA. There is a need to monitor the export and import of used electronic goods under the name of "recycling" or "second-hand goods". It is necessary to step up inspection on sham reuse or recycling and this requires the co-ordination of various government bodies as well as international and regional co-operations.

Building up national capacity in e-waste management

Currently, there is inadequate infrastructure and lack of technological expertise in handling and treating e-wastes in Singapore. Hence, to ensure the country is capable of treating and recycling e-waste locally without having to export them in the near future, it is necessary to establish industry-based e-waste collection services, recycling centres and research laboratories. There are ample opportunities awaiting electronic recyclers. In consideration that consumer e-waste which account for 97% of the total electronic waste business has been largely unexplored, it makes good business sense for recyclers to work in partnership with government, producers and waste collectors in setting up take-back system, build more recycling facilities, differentiate services and acquire efficient recycling technology. Furthermore, future product designs will be easier to disassemble and recycle as a result of more environmentally-friendly product design.

Besides the recycling industry, other important aspects of national capacity in e-waste management include research, testing and analytical services. These services play a supporting role to the recycling industry as well as electronic manufacturing industry. Their functions include developing green recycling technology and processes, developing new product design methodology, carrying out sampling and testing and expanding eco-labeling and life cycle assessment.

Globally, many institutions have begun taking up the challenge of developing e-waste knowledge base and recycling technology. In the United States, for example, the "International Symposium on Electronics and the Environment" is the flagship event in the research and development of green electronics. It is an academic conference with an open call for papers, peer-reviewing of abstracts and publication of proceedings by the Electrical and Electronics Engineers, Inc. (IEEE) (*www.iseesummit.org*). In Europe, the main research project addressing e-waste is based at the Swiss Federal Laboratories for Materials Testing and Research (EMPA) in Switzerland. "Knowledge Partnerships with Developing and Transition Countries in e-waste Recycling" was launched in 2003 and has analyzed e-waste recycling frameworks and processes in different parts of the world. The project is envisaged to share knowledge and experience gained through Switzerland's e-waste recycling operation with developing and transition countries like China, India and South Africa (*www.ewaste.ch*).

Stopped Here

To encourage the development of recycling infrastructures and research institutions related to e-waste in Singapore, appropriate incentives should be provided. Incentives could take many forms, for example, tax rebate for investment in the state-of-the-art technology, sharing of technical resources as well as collaboration in research and development. Funding could also be provided for implementing pilot projects in relation to e-waste. In Singapore, institutions and companies can tap into the Innovation for Environmental Sustainability (IES) Fund for financial support. However, it has been shown that the take-up rate is low and none of the 28 projects supported by the Fund is related to e-waste since it was launched in 2001. Therefore, more efforts are needed to promote the IES Fund and woo companies to undertake projects in relation to e-waste management and green electronic product design. Besides the government, funding could also come from non-profit organizations such as the Waste Management and Recycling Association of Singapore (WMRAS).[14]

In addition to recycling centres and research institutions, there is also a need to develop a monitoring and surveillance system to gather data and inventory on e-waste generation, including their transboundary movements. Data collection is vital as it helps to evaluate and assess the cost effectiveness of the management system put in place for e-waste. To ensure data is comparable and meaningful, a mandatory or voluntary system should be set up to encourage important stakeholders, such as electronic and electrical manufacturers, electronic recyclers and e-waste collectors to adopt a standard reporting framework. Where necessary, such information may be audited or verified by a third-party or a competent government authority.

Besides developing the "hardware", it is necessary to build up the "software", i.e. the human resources involved in the management of e-waste.

---

[14]The Waste Management and Recycling Association of Singapore (WMRAS) was established on 8th of August 2001 with the goal to develop a leading waste management and recycling industry in Asia. With the aim to promote business networking opportunities and best practices among its members, the Association has been organizing activities such as talks, mission trips, conferences, and exhibitions.

These include workers, technicians, engineers, researchers, managers, retailers and government regulatory officials. Examples of the activities that should be undertaken include establishing training and information centre, encouraging industrial training programmes, developing e-waste prevention and minimization guideline, launching demonstration or pilot project and creating forum or sharing sessions among interdisciplinary and intersectoral groups.

Finally, it is important for Singapore to continue its active participation in capacity building activities at the regional and international level, particularly in planning and implementing pilot projects on e-waste. On 19–22 November, 2002, Singapore attended the Asia-Pacific Regional Scoping Workshop on Environmentally Sound Management of Electronic Waste in Tianjin, China, the first intergoverment meeting on the issue of e-wastes in Asia. The workshop was a scoping workshop aimed at developing a regional programme for managing end-of-life electrical and electronic products in an environmentally sound manner. Singapore has participated in the Ministerial Conference on the 3R Initiative, held in Tokyo, Japan on 28–30 April, 2005. The conference discussed several issues including national policies to implement the 3Rs, reduction of barriers to the international flow of goods and materials, encouragement of cooperation among various stakeholders, promotion of science and technology suitable for the 3Rs and cooperation with developing countries. By establishing a comprehensive network with experienced countries and other interested organizations, Singapore could learn from their success stories which would help speed up the implementation process of an environmentally sound management system for e-waste.

Boosting up secondary market

The secondary market for used and refurbished electronic equipment in Singapore is doing fairly well. However, more measures are needed to boost this segment of the market so as to reduce the aggregate purchase of new electronic equipment, extend the useful lifespan of electronic products and delay the disposal of e-waste. The following are some suggestions:

- cluster second-hand shops in few areas and promote these designated clusters as attractive spots for local people and foreign tourists;
- support the establishment of retail shops selling second-hand goods and refurbished products by simplifying permit application procedure;

- educate consumers to overcome the perception of "second-hand is inferior" by highlighting the advantages of second-hand goods;
- provide publicity on the avenues available to sell and buy second-hand product;
- encourage the use of secondary, refurbished and/or recycled electronic equipment in government agencies, institutions, offices and at homes as part of green procurement policy; and
- provide matching services for donors and recipients.

In Singapore, the government encourages companies to donate computers (including computer software and peripherals) to schools, libraries, research institutions or charity organizations by offering tax deductions. The donation scheme is administered by the Inland Revenue Authority of Singapore (IRAS) with the assistance from the Infocomm Development Authority of Singapore (IDA). The types of hardware and/or software donated have to be approved by IDA and donating companies are required to apply to IDA to assess the worth of the donated equipment. The donation scheme not only offers financial incentive for companies to give away unwanted computers but also provides opportunity for companies to demonstrate responsible corporate citizenship. Such a donating scheme should be expanded to allow the donation of a more variety of electronic equipment, including handphones, printers, scanners, fax machines, washing machines, refrigerators etc.

While it is in the interest of businessmen and consumers to allow market forces and commercial self-interest to run the secondary market, the government may need to keep an eye on the market, in particular online second-hand traders. This is to ascertain that the second-hand products put up for sale are not stolen or smuggled goods. It is also important to ensure that the second-hand goods collected in Singapore are not illegally dumped in or outside Singapore. On the part of the consumers, they should be educated on tips when buying a second-hand or refurbished electronic product and given consumer rights in seeking redress should the product not meet its claims.

Creating market for green electronic products

In general, green consumerism in Singapore is somewhat lacking due to inadequate supply and poor demand for green products. With respect to electronic products, there are only a handful of electrical and electronic equipment that are awarded green label and energy label under the Singapore Green Labelling Scheme (SGLS). To create a substantial market for green electronic products, there is a need to look into ways to generate more supply of green electronic product quickly as well as to induce demand for such products. Besides increasing the supply, efforts must also be taken to raise the demand for green electronic and electrical products.

The public sector is a major consumer of electrical and electronic equipment in the market. By taking the lead in adopting green procurement policy, the government is able to send a powerful message to private-sector organizations, companies and homes and create a spin-off effect in the supply and demand chain of green electronic products. There will be more choices of green products in the market as manufacturers compete to cater to the needs of the public sector as well as other companies implementing green procurement policy. Prices will also be lowered as a result of demand through bulk purchasing. Lower price will in turn attract even more demand.

In Japan, the nationwide Green Purchasing Network (GPN) has been very successful since the initiative was launched in February 1996. The GPN involves a wide range of businesses and public sectors including major producers (e.g. Matsushita Electric, Sony, NEC etc), consumer associations, environment NGOs, government bodies and specialists. As of August 2004, the number of green purchasing network members was 2,817, of which there were 2,195 corporations, 356 governmental bodies and 266 NGOs and NPOs (Nakahara, 03 Sept 2004). The movement towards green purchase in Japan was accelerated by the enforcement of the Law on Promoting Green Purchasing in 2001. Under the Law, all the State ministries and agencies must draw up annual procurement policy, practice the plan and report the result to the Ministry of the Environment. Private enterprises and consumers are strongly encouraged to implement green purchasing voluntarily. As a result of the Law and active participation of government in green purchasing, many manufacturing companies have accelerated the development of green products and brought about a domestic market size of green products of 38 trillion yen (or US$ 320 billion) (ibid.). At present, the Taiwan and Korea governments are also actively adopting green procurement policy.

Enhancing the recycling programmes for e-waste

In Singapore, the participation rates in recycling programmes are relatively low. Under the National Recycling Programme, the government aims to achieve an overall waste recycling rate of 60% by the year 2012. Yet, the recycling rate had been increasing only marginally from 45% (2002) to

48% (2004) (*www.nea.gov.sg*). The biggest obstacle to recycling is perhaps the lack of environmental awareness among the public. Many people do not realize and fully understand the necessity to recycle. Other barriers include inadequate collection channels, insufficient promotion on recycling and absence of mandatory requirement for recycling.

There is no national recycling programme for e-waste and only few companies run voluntary product take-back programmes in Singapore. Therefore, more concerted efforts are needed to encourage the recycling of e-wastes among government institutions, companies and households. The government should adopt a macro-management approach and create more commercial opportunities for the industry to take up the recycling programme for e-waste. This will help to ensure that the recycling programme is commercially viable and sustainable.

Raising the environmental awareness among the public

Unlike countries such as Germany, Denmark and Switzerland, Singapore has yet to nurture a society with "green" ethics that practises recycling effortlessly. There are many factors contributing to the development of green ethics. In Switzerland, for example, there is a long history of environmental legislation; the laws provide incentives for the people to separate and recycle their glass, aluminium, plastics, paper, tin and green waste. Recycling is free whereas throwing away rubbish costs money. Therefore, less rubbish thrown thrown out translates to a lower disposal cost (BBC News, 2005). Germany has similar laws governing the recycling and disposal of waste. In addition, the Germans generally take green issues seriously. According to a survey, 90% of Germans are willing to sort out their rubbish although the practice is not compulsory for private citizens (*ibid.*). In Denmark, the environment policy has been to regard waste as a resource, with nearly 10,000 Danes in the business of collecting waste — more than 0.1% of the entire population (*ibid.*). It is therefore in the interest of the people of Denmark to recycle as much as possible.

While Singapore may have relatively short history of environmental policies and green ethics, it is high time for the government, industries and NGOs to promote greater public awareness of and concern for the natural environment, and to encourage the community to display a greater sense of environmental responsibility.

With respect to e-waste, public education should begin with the understanding that used electronic equipment need to be disposed of in a proper manner. The public should learn that the hazardous substances contained in electrical and electronic products could create adverse environmental and human health effect if the products are disposed of like all other general waste. They should also learn about the 'where' and 'how' to return, recycle and dispose of used electronic equipment.

In addition, the public should be informed about other consumption alternatives, for example, the repair and reuse of a faulty equipment or the purchase of a second-hand or refurbished product instead of buying a new one, and the selecting of electronic products with Green Label and Energy Label, and leasing instead of a purchase, and so on. It is also important for consumers to handle electronic equipment as recommended so as to ensure the equipment is in good working condition throughout its useful lifespan, thus reducing new purchases.

Essentially, Singapore should aim to establish a recycling-based society where members of public take pride in environmental protection. The responsibility of raising the environmental awareness among the public lies not only with the government, but also with companies, manufacturers, retailers, waste collectors, consumer groups, NGOs, community organizations, educational institutions, etc.

Conclusion

The purpose of this study is to understand and assess the present management system of e-waste in Singapore. To achieve the purpose, a research methodology consisting of literature analysis and questionnaire was undertaken to gather information with regards to legislations, ownership of electronic equipment, second-hand market, recycling industry, end-of-life treatment and green initiatives. It was found that there is limited data and information pertaining to e-waste in the local context as the issue has received fairly little attention from the government, industries, companies, households and media. Furthermore, response to the questionnaire was poor. This is perhaps due to the sensitivity of the information and mostly, the lack of data on the part of the respondents.

Despite the limited information available on e-waste in Singapore, this study has provided some useful information on the extent of the problem as well as the effectiveness and shortcomings of the present management system for e-waste in the country. From the study, several conclusions could be made as follows:

- E-waste is growing in Singapore.

- The present legislations in Singapore are inadequate to deal with the complex nature of e-waste.
- Singapore lacks appropriate infrastructure and technological know-how in handling and treating e-wastes locally.
- To date, there have been few activities promoting the reduction, reuse and recycling of e-waste.

References

Arensman, R. (2000), "*Ready for Recycling?*" In: *Electronic Business, The Management Magazine for the Electronics Industry*, November, cited as Endnote 7 in Exporting Harm: The High-Tech Trashing of Asia, Feb. 2002.

BBC News (25 June 2005) *Recycling Around the World* [Internet] Available from: <http://news.bbc.co.uk/2/hi/europe/4620041.stm> [Last accessed 24 July 2005].

Business Times (1 Feb, 2005) *S'pore PC Sales Cross 500,000 Mark in 2004* [Internet] Available from: http://it.asia1.com.sg/newsdaily/news001_20050201.html [Last accessed 24 July 2005].

CNET Networks, Inc (Aug 08, 2000) *Singapore PC Market May Be "Maturing"* [Internet] Available from: http://news.com.com/Singapore+PC+market+may+be+maturing/2100-1001_3-244219.html [Last accessed 24 July 2005].

Commission of European Communities (2000) *Explanatory Memorandum of the Proposal for a Directive of the European Parliament and of the Council on Waste Electrical and Electronic Equipment and on the Restriction of the use of certain Hazardous substances in Electrical and Electronic Equipment*, Brussels, Jun 13].

Computer Times (15 Jan 2003) *Resale Value* [Internet] Available from: http://it.asia1.com.sg/specials/issues20030115_001.html [Last accessed 24 July 2005].

Computer Times (28 May 2003) *There is Gold in the Old* [Internet] Available from: http://it.asia1.com.sg/specials/issues20030528_002.html [Last accessed 24 July 2005].

IACEE Report (1999) [Internet] Available from: http://www.neec.gov.sg/aboutneec/report.shtm [Last accessed 22 July 2005].

Interfax (2004) *China to Establish Scrap Electronics Recycling Systems* January 12 [Internet] Available from: <http://www.recyclingtoday.com/news/news.asp?ID=5174> [Last accessed 24 July 2005].

Nakahara, H. (03 Sept 2004) *Green Procurement in Japan* IN: Workshop on Green Procurement, Kuala Lumpur, Malaysia.

Ng, M.K. (2003) *Consumer Durables in Households, 1992–2002* In: Statistics Singapore Newsletter September 2003, Singapore Department of Statistics, p. 2–4.

Singapore Economic Development Board (EDB) (18 Jan 2005) *Record Year of Investment Commitments in Headquarters, Professional and Engineering &*

Environmental Services [Internet] Available from: http://www.sedb.com/edbcorp/sg/en_uk/index/in_the_news/press_releases/2005/record_year_of_investment.html [Last accessed 24 July 2005].

The Ministry of the Environment and Water Resources (MEWR), *Singapore Green Plan 2012 (2002 edition).*

The EDGE Singapore (8 July 2002) *Turning Waste into Money* The Edge Publishing Pte Ltd.

The Japan Times (1 Oct 2003) *New Law Requires Recycling of PCs at Consumers' Expense* [Internet] Available from: http://www.japantimes.co.jp/cgi-bin/getarticle.pl5?nn20031001b4.htm [Last accessed 24 July 2005].